STEEL HEAT TREATMENT HANDBOOK
Second Edition

STEEL HEAT TREATMENT

EQUIPMENT AND PROCESS DESIGN

STEEL HEAT TREATMENT HANDBOOK

Second Edition

George E. Totten, Ph.D., FASM

STEEL HEAT TREATMENT
METALLURGY AND TECHNOLOGIES

STEEL HEAT TREATMENT
EQUIPMENT AND PROCESS DESIGN

STEEL HEAT TREATMENT HANDBOOK
Second Edition

STEEL HEAT TREATMENT

EQUIPMENT AND PROCESS DESIGN

EDITED BY

George E. Totten, Ph.D., FASM

Portland State University
Portland, Oregon, U.S.A.

Taylor & Francis
Taylor & Francis Group
Boca Raton   London   New York

CRC is an imprint of the Taylor & Francis Group,
an informa business

Cover Figure: The photograph shows a hybrid continuous (electrically heated/gas-fired) mesh belt sintering furnace with controlled cooling to enable desired microstructure and hardness on sintered parts without subsequent heat treatment. (Courtesy of Fluidtherm Technology (P) Ltd.)

CRC Press
Taylor & Francis Group
6000 Broken Sound Parkway NW, Suite 300
Boca Raton, FL 33487-2742

First issued in paperback 2019

© 2007 by Taylor & Francis Group, LLC
CRC Press is an imprint of Taylor & Francis Group, an Informa business

No claim to original U.S. Government works

ISBN-13: 978-0-8493-8454-7 (hbk)
ISBN-13: 978-0-367-39028-0 (pbk)

Library of Congress Cataloging-in-Publication Data

Steel heat treatment : equipment and process design / George E. Totten, editor.
    p. cm.
  Includes bibliographical references and index.
  ISBN-13: 978-0-8493-8454-7 (alk. paper)
  ISBN-10: 0-8493-8454-0 (alk. paper)
  1. Steel--Heat treatment . 2. Mineral industries--Equipment and supplies. I. Totten, George E. II. Title.

TN751.S685 2006
672.3'6--dc22
                                             2006023583

Visit the Taylor & Francis Web site at
http://www.taylorandfrancis.com

and the CRC Press Web site at
http://www.crcpress.com

Preface

The first edition of the *Steel Heat Treatment Handbook* was initially released in 1997. The objective of that book was to provide the reader with well-referenced information on the subjects covered with sufficient depth and breadth to serve as either an advanced undergraduate or graduate level text on heat treatment or as a continuing handbook reference for the designer or practicing engineer. However, since the initial release of the first edition of the *Steel Heat Treatment Handbook*, there have been various advancements in the field that necessitated the preparation of an updated text on the subject. However, with the addition of new chapters and expanded coverage, it was necessary to prepare this updated text in two books. *Steel Heat Treatment: Metallurgy and Technologies* focuses on steel heat treatment metallurgy while this text addresses heat treatment equipment and processes. *Steel Heat Treatment: Equipment and Process Design* contains updated revisions of the earlier *Steel Heat Treatment Handbook* while other chapters are new to this book. The chapters included in this book are:

SECTION I—EQUIPMENT

Chapter 1. Heat Treatment Equipment
Chapter 2. Design of Steel-Intensive Quenching Processes
Chapter 3. Vacuum Heat Processing
Chapter 4. Induction Heat Treatment: Basic Principles, Computation, Coil Construction, and Design Considerations
Chapter 5. Induction Heat Treatment: Modern Power Supplies, Load Matching, Process Control, and Monitoring
Chapter 6. Laser Surface Hardening

SECTION II—TESTING METHODS

Chapter 7. Metallurgical Property Testing
Chapter 8. Mechanical Property Testing Methods

This book is intended to be used in conjunction with *Steel Heat Treatment: Metallurgy and Technologies*. Together, both books provide a thorough and rigorous coverage of steel heat treatment that will provide the reader with an excellent information resource.

George E. Totten, Ph.D., FASM
Portland State University
Portland, Oregon

Editor

George E. Totten, Ph.D. is president of G.E. Totten & Associates, LLC in Seattle, Washington and a visiting professor of materials science at Portland State University. He is coeditor of a number of books including *Steel Heat Treatment Handbook*, *Handbook of Aluminum*, *Handbook of Hydraulic Fluid Technology*, *Mechanical Tribology*, and *Surface Modification and Mechanisms* (all titles of CRC Press), as well as the author or coauthor of over 400 technical papers, patents, and books on lubrication, hydraulics, and thermal processing. He is a Fellow of ASM International, SAE International, and the International Federation for Heat Treatment and Surface Engineering (IFHTSE), and a member of other professional organizations including ACS, ASME, and ASTM. He formerly served as president of IFHTSE. He received the B.S. and M.S. degrees from Fairleigh Dickinson University, Teaneck, New Jersey and the Ph.D. degree from New York University, New York.

Contributors

Micah R. Black
INDUCTOHEAT, Inc.
Madison Heights, Michigan

Jan W. Bouwman
Ipsen International, GmbH
Kleve, Germany

Raymond L. Cook
INDUCTOHEAT, Inc.
Madison Heights, Michigan

Bernd. Edenhofer
Ipsen International, GmbH
Kleve, Germany

N. Gopinath
Fluidtherm Technology Pvt. Ltd.
Chennai, India

Janez Grum
University of Ljubljana
Ljubljana, Slovenia

Daniel H. Herring
The Hering Group, Inc.
Elmhurst, Illinois

Nikolai I. Kobasko
Intensive Technologies Ltd.
Kiev, Ukraine

Don L. Loveless
INDUCTOHEAT, Inc.
Madison Heights, Michigan

D. Scott MacKenzie
Houghton International
Valley Forge, Pennsylvania

Wytal S. Morhuniuk
Intensive Technologies Ltd.
Kiev, Ukraine

David Pye
Pye Metallurgical Consulting, Inc.
Meadville, Pennsylvania

Valery I. Rudnev
INDUCTOHEAT, Inc.
Madison Heights, Michigan

George E. Totten
Union Carbide Corporation
Tarrytown, New York

Boris K. Ushakov
Moscow State Evening
   Metallurgical Institute
Moscow, Russia

Xiwen Xie
Beijing University of Aeronautics and
   Astronautics
Beijing, China

Contents

Section I

Equipment

1 Heat Treatment Equipment*

George E. Totten, N. Gopinath, and David Pye

CONTENTS

*Original edition written by George E. Totten, Gary R. Garsombke, David Pye, and Ray W. Reynoldson.

1.1 INTRODUCTION

Some of the most important decisions that the heat treater will make are related to the selection of furnaces and ancillary equipment. These decisions involve selection of the energy source, gas or electricity, which is vital to the overall profitability of the heat treatment process. Another is the selection of the furnace transfer mode, batch or continuous, and the particular furnace type. If a furnace is rebuilt, the proper choice and installation of the refractory material are vital.

In this chapter, the selection and operation of heat treatment equipment are addressed. The focus of the discussion is on the furnace, furnace atmosphere generation, and ancillary equipment. Discussion subjects include:

Furnace part transfer mechanisms
Furnace heating (heat transfer principles and application to furnace calculations)

Atmosphere generation (atmosphere sensors; thermocouples)
Refractory materials
Fans
Fixtures
Parts washing
Quenching systems
Furnace safety
Salt bath furnaces
Fluidized bed furnaces

1.2   FURNACE TRANSFER MECHANISMS

Heat treatment furnaces are classified as batch, semicontinuous, or continuous. In batch furnaces, which are the most common and most versatile in the heat treatment industry, the work is typically held stationary in the furnace vestibule. The furnace is loaded or unloaded in a single (batch mode) operation.

In continuous furnaces, the load moves through different zones, usually with varying temperatures as shown in Figure 1.1. Parts that are heat treated are moved through the furnace in a continuous process.

In semicontinuous furnaces, parts move through in a continuous but stepwise manner. For example, they may move through the furnace in a tray or a basket. As the tray or basket is charged from the furnace, it is quenched. After the quench cycle is completed, the tray or basket moves on, the next one is discharged from the furnace and quenched, and the process continues in a stepwise manner. A comparison of the features of a few select examples of each furnace is provided in Table 1.1.

For the purposes of this discussion, semicontinuous furnaces are considered to be continuous. Batch-type furnaces discussed include box, pit, integral quench (IQ), and tip-up (both circular and car-bottom). Continuous-type furnaces discussed here include walking beam, rotary hearth, pusher, roller hearth, conveyor, shaker hearth, screw conveyor, and rotary retort. Table 1.2 provides a selection guide for these types of furnaces as a function of

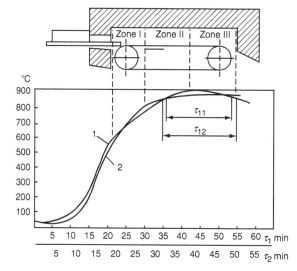

FIGURE 1.1 Temperature variation with distance through a continuous furnace. (From A.N. Kulakov, V. Ya. Lipov, A.P. Potapov, G.K. Rubin, and I.I. Trusova, *Met. Sci. Heat Treat. Met.* 8:551–553, 1981.)

TABLE 1.1
Comparison of Types of Heat-Treating Furnaces for Small Parts

Class of Equipment	Furnace Type	Versatility in Use	Labor Needs	Atmosphere Quality	Quenchability
Batch	Integral quench	E	F	E	G
Batch	Salt pot	F	H	F	LD
Batch	Rotary drum	F	F	L	G
Semicontinuous	Pan conveyor	G	F	F	F
Semicontinuous	Tray pusher	G	F	G	G
Continuous	Belt shaker	F	L	F	E
Continuous	Shaker hearth	F	L	F	E
Continuous	Rotary retort	LD	L	F	E

E, excellent; G, good; F, fair; H, high; L, low; LD, limited.

Source: From T.W. Ruffler, Bulk heat treatment of small components, 2nd Int. Congr. Heat Treat. Mater.: 1st Natl. Conf. Metall. Coatings, Florence, Italy, September 20–24, 1982, pp. 597–608.

the heat treatment process. A comprehensive review of furnace applications and a summary of suppliers are available in Ref. [4].

1.2.1 BATCH FURNACES

1.2.1.1 Box Furnace

The box furnace, such as the one shown in Figure 1.2, is the simplest heat-treating furnace. It is used for tempering, annealing, normalizing, stress relieving, and pack-carburizing. It is capable of operating over a wide range of temperatures, 95–1095°C (200–2000°F).

TABLE 1.2
Furnace Selection Guide

Function	Production Process	Atmosphere	Furnace Types
	Continuous		Conveyor; pusher; rotary hearth; shaker hearth; roller hearth cast-link belt
Annealing steel	Batch		Car type
Normalizing steel	Batch		Semitype look-up
Spheroidizing steel	Batch		Semimuffle oven
		Air	Full-muffle oven
			Vertical muffle oven[a]
		Salt	Round pot
			Rectangular pot
Blueing steel	Batch	Steam or air	Round pot
			Rectangular pot
Bright annealing steel (also copper, brass, etc.)	Continuous	Air	Conveyor atmosphere controlled
		Air	Pusher atmosphere controlled

Continued

TABLE 1.2 (Continued)
Furnace Selection Guide

Function	Production Process	Atmosphere	Furnace Types
	Batch	Air	Full-muffle oven atmosphere controlled
Carburizing steel	Continuous		Pusher
	Batch		Car type
			Semimuffle oven
Cyaniding steel (or liquid carburizing)		Cyanide or salt	Round pot
Forming and forging steels (steel, brass, copper, etc.)	Continuous		
	Slugs, billets		Rotary hearth
	Billets		Pusher type
	Batch		
	End or center heating of slugs		Open slot
	Large billets, heavy forgings, plates structural steel shapes, rods, bars, etc.	Oven type, direct fired	
Hardening steel	Continuous		Conveyor; pusher; rotary hearth; shaker hearth; roller heart cast-link belt
	Batch	Air	Car type; semimuffle oven; full-muffle oven; vertical muffle oven
Stress-relieving steel	Continuous		Conveyor air-recirculating; pusher air-recirculating
	Batch		Car-type air-recirculating
Tempering/drawing	Continuous		Conveyor air-recirculating; conveyor; pusher air-recirculating; Pusher
	Batch		Car-type air-recirculating[b]; basket air-recirculating[c]

[a]The vertical muffle oven furnace is not commonly encountered.
[b]Used for high-temperature stress relieving and subcritical annealing.
[c]Used for large quantities.
Source: From *Heat Treating Furnacers and Ovens*, Brochure, K.H. Hupper & Co. (KHH), South Holland, MI.

1.2.1.2 Integral Quench (Sealed Quench) Furnace

IQ furnaces are among the most commonly used and most flexible furnaces for processing small parts. They can be used for either neutral or atmosphere hardening processes in addition to normalizing and stress relief.

IQ furnaces are similar to box furnaces except that the quench tank is located at the discharge end of the furnace as shown in Figure 1.3. As the parts are removed from the furnace, the baskets used to hold the parts that are heat treated are lowered into the quenchant with an elevator.

The attached quench tank can be a disadvantage because the bath loading and height of the baskets restrict the size of the parts that can be quenched [2]. Another limiting factor is the size of the heating chamber.

(a)

(b)

FIGURE 1.2 (a) Electrically heated box furnace. (Courtesy of AFC-Holcroft, LLC, www.afc-holcraft.com.) (b) Small "high-speed" laboratory box oven used to heat small parts and can also be controlled to model heat-up, soaking and cool-down cycles to develop stress relieving and tempering processes for production-scale continuous stress relieving ovens. (Courtesy of PYROMAÎTRE Inc., www.pyromaitre.com.)

Economics typically dictate that the largest furnace available with respect to the amount of work to be treated be used. Furnace design developments have led to ever-greater automation, reduced cycle times, and greater fuel efficiency of these furnaces [5]. Cycle times at

(a)

(b)

FIGURE 1.3 (a) Sealed integral quench furnace. (Courtesy of AFC-Holcroft, LLC.) (b) Compact batch integral quench furnace. (Courtesy of Fluidtherm Technology.)

temperature are reduced even with newer furnace designs. High-temperature carburizing is not practiced because of the limitations of materials engineering.

1.2.1.3   Pit Furnaces

Pit furnaces are circular furnaces that can be either floor- or pit-mounted. They are used for such processes as annealing, carburizing, tempering, normalizing, and stress relieving. Some of the advantages of the circular furnace are the smaller heat release areas in comparison to

FIGURE 1.4 Pit furnace with retort for steam tempering of PM parts. (Courtesy of Fluidtherm Technology.)

box or rectangular furnaces, more uniform atmosphere, improved temperature distribution, and smaller furnace body weight [6]. A pit furnace is shown in Figure 1.4. Such furnaces may also be used as pit carburizers or nitriders.

1.2.1.4 Car-Bottom Furnaces

Car-bottom furnaces are used for thermal processing of very large parts such as gears [11–12] and forgings [5]. They may be used for carburizing, annealing, hardening, normalizing, stress relieving, and tempering [3]. The bottom of the furnace is a refractory-covered flatbed railcar that moves on rails in the shop. Some furnaces are loaded and unloaded from the same end. Others are loaded at one end and unloaded at another. Still other furnaces may be loaded from the side to permit the use of more than one railcar. Examples of car-bottom furnaces are shown in Figure 1.5.

1.2.1.5 Tip-Up (Lift-Off) Furnaces

Some heat treatment furnaces are designed so that the top can be hydraulically lifted over the load to facilitate removal by a forklift or removal from a car-bottom railcar. These furnaces may be circular [6,7] or a variation of the car-bottom furnace.

1.2.2 CONTINUOUS FURNACES

Continuous furnaces are particularly suited to continuous heat treatment processing of parts that are the same or at least similar. A nomograph is provided in Figure 1.6 that interrelates heating time, furnace load, furnace length, and production rate for a continuous furnace [8]. In this section, we give an overview of the features of different furnace designs. Figure 1.10 illustrates examples of various types of continous furnaces.

1.2.2.1 Walking Beam Furnaces

The movement of parts through a walking beam furnace, illustrated in Figure 1.7, is by repeated lifting, moving, and lowering of the parts in the furnace in a walking action [4]. Walking beam furnaces are particularly suitable for continuous thermal processing of large parts and heavy loads.

FIGURE 1.5 Car bottom furnaces for heat treatment of castings and forgings. (Courtesy of L & L Kiln Manufacturing Inc., www.hotfurnace.com; courtesy of Pyradia Inc., www.pyradia.com; courtesy of Fluidthem Technology; courtesy of HTF, Inc., www.heattreatfurnaces.com.)

1.2.2.2   Roller Hearth Furnace

In a roller hearth furnace, the load is moved through an externally driven heat-resistant alloy rollers. This furnace is best suited for continuous heat treatment of large parts and plates [4] and is illustrated in Figure 1.8 and Figure 1.9.

1.2.2.3   Pusher Furnaces

Pusher furnaces have one of the simplest furnace designs. The load is hydraulically or pneu-matically mechanically pushed on rollers or skid rails inside the furnace. One variant of this

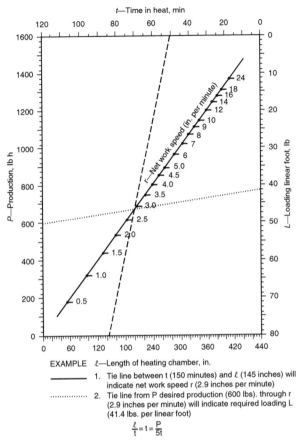

FIGURE 1.6 Production rate nomograph for continuous furnaces. (From *Sintering Systems*, Product Bulletin No. 901, C.I. Hayes, Cranston, RI.)

furnace design is the tray pusher furnace [2], which can be used for both neutral carburizing and carbonitriding in addition to annealing, normalizing, tempering, and stress relieving [9].

1.2.2.4 Mesh Belt Conveyor Furnaces

Belt conveyor furnaces may vary in size from very small to very large. They are similar to roller hearth furnaces except that the load is continuously moved through the furnace on a metal mesh, roller chain, or link chain as shown in Figure 1.11 [10]. The load may be moved continuously through other process lines such as quenching and tempering. Figure 1.12 illustrates some of the designs of mesh belts that are available.

1.2.2.5 Shaker Hearth Furnaces

The shaker hearth furnace (Figure 1.13) is another example of a continuous furnace [2,10] that is used primarily for heat treatment processing of small parts. Parts are moved through the furnace by a mechanically induced vibrating mechanism that has the advantage of not requiring a belt or chain to move back to the beginning.

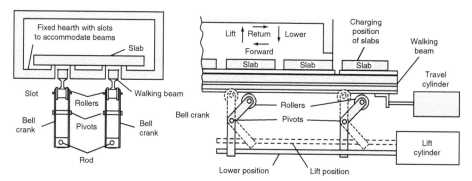

FIGURE 1.7 Illustration of a walking beam furnace mechanism.

1.2.2.6 Screw Conveyor Furnace

In the screw conveyor furnace, parts proceed through the furnace on one or more screw mechanisms [2,10] as shown in Figure 1.14. These furnaces are suitable for hardening, tempering, annealing, and stress relieving long, thin parts, which require careful handling during heating.

1.2.2.7 Rotary Hearth Furnace

The rotary hearth furnace differs from previously discussed furnaces in that the load is transferred through the heating zone in a rotary motion on a moving furnace hearth. This is illustrated in Figure 1.15, where the work to be heat treated is loaded into the furnace near the discharge point. An example of a rotary hearth furnace is provided in Figure 1.16. In the commercial design illustrated in Figure 1.17 several rotary hearth furnaces are connected in sequence to obtain even greater furnace and floor-space flexibility. This type of furnace can be used for a variety of processing methods such as atmosphere hardening or carburizing.

1.3 FURNACE HEATING: ELECTRICITY OR GAS

One of the first steps in furnace design is to select the energy source, typically electricity or gas, which will be used to heat the furnace. The economics of gas and electricity vary with availability and cost. In this section, the calculation of furnace heating economics is discussed, followed by a brief overview of electric and gas furnace heating designs.

1.3.1 FURNACE HEATING ECONOMICS

The first step in determining the economics of heating a furnace is to perform an energy balance [13]. The heat loss factors to be considered relative to total heat input [14,15] are

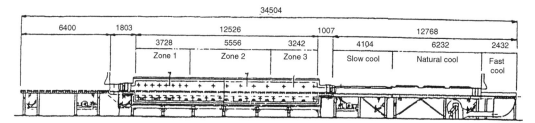

FIGURE 1.8 Heating/cooling zones and transfer mechanism in a roller hearth furnace.

FIGURE 1.9 Roller hearth furnace. (Courtesy of Can-Eng Furnaces, www.can-eng.com.)

1. Wall losses
2. Heat to atmosphere
3. Heat to trays and fixtures
4. Heat to stock

Heat losses to the atmosphere such as fresh air, infiltration, and exhaust are calculated from

$$\text{Btu/h} = \text{scfm} \times 1.08\Delta T$$

where Btu/h is the heat transfer rate, scfm is the standard cubic feet of air per minute, 1.08 is the conversion constant, and ΔT is the change in temperature of air (°F).

Heat losses to materials passing through the furnace such as trays, fixtures, and steel are calculated from [14]

$$\text{Btu/h} = WC_\text{p}\Delta T$$

where W is the weight of materials passing through the furnace in lb/h, C_p is the specific heat of the material [Btu/(lb °F)], and ΔT is the change in temperature between entry and exit of the furnace (°F).

The heat loss through the furnace wall as shown in Figure 1.2 is calculated from [14]

$$\text{Btu/h} = AU\Delta T$$

where A is the surface area of the furnace (ft^2) and U is the overall heat transfer coefficient [Btu/(h ft^2 °F)] and is calculated from

$$U = \frac{1}{1/h_1 + x/k = 1/h_2}$$

where h_1 is the inner film coefficient (Btu/h ft^2 °F), h_2 is the outer film coefficient (Btu/h ft °F), k is the panel thermal conductivity (Btu/h ft^2 °F), and x is the panel thickness (ft).

The heat input is calculated from the percent energy available, both latent and sensible:

$$\text{Gross Btu/h} = \frac{\text{Btu/h}}{\% \text{ energy available}}$$

The total energy available from the fuel gas or oil will decrease with increasing furnace exhaust temperature, although the amount of decrease depends on the combustion system and raw material source for gas. This is illustrated in Figure 1.18.

(a)

(b)

FIGURE 1.10 Examples of continuous furnaces. (a) Walking beam furnace for high temperature sintering. (Courtesy of Fluidtherm Technology P. Ltd www.fluidthem.com.) (b) Larger Walking beam furnace. (Courtesy of AFC-Holcroft, LLC, www.afc.holcroft.com.)

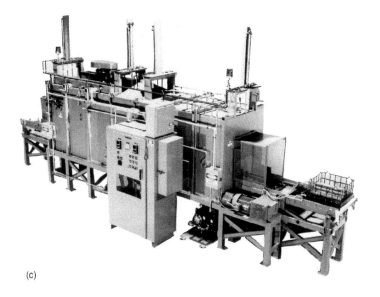

(c)

(d)

(e)

FIGURE 1.10 (Continued) (c) Continuous batch pusher type tempering over. (Courtesy of L & L Kilm Manufacturing Inc., www.hotfurnace.com.) (d) Pusher furnace for large steel billets. (Courtesy of Bricmont Incorporated, www.bricmont.com.) (e) Fully automated pusher furnace. (Courtesy of Schoonover Inc., www.schoonover.com.)

(f)

(g)

FIGURE 1.10 (continued) (f) Screw conveyor furnace and (g) Slat conveyor furnace. (Courtesy of AFC-Molcroft, LLC, www.afc-holcroft.com.)

The first step in the selection of a gas or electric furnace can be made once these calculations are performed for both energy sources.

1.3.2 ELECTRIC ELEMENT FURNACE HEATING

Electrical heat has a number of advantages over gas. These include lower furnace cost and higher furnace efficiency because there are no exhaust losses, advantageous regulations that permit unattended operation, lower maintenance costs, improved temperature uniformity, the relative ease of replacing electric elements, and wider operating temperature range [16,17].

In this section, design recommendations and general properties of the most commonly used heating elements are discussed.

(a)

(b)

(c)

FIGURE 1.11 (a) Continuous wire mesh belt hardening and tempering plants for bulk produced components. (b) Continuous mesh belt austempering plant with molten salt bath quench. (c) Continuous mesh belt sintering furnace with controlled cooling rate module. (Courtesy of Fluidtherm Technology.)

FIGURE 1.11 (continued) (d) Continuous mesh belt frunace for manufacture of small parts. (Courtesy of AFC-Holcroft, LLC, www.afc-holcroft.com.) (e) Electrically heated high-speed (rapid-heating) stress relieving and tempering over.)

Electrical heating is dependent on the ability of electric current to flow through a conductor, which may be either metallic or nonmetallic. The heating rate of a material depends on the current density and the specific resistance of the material. (Resistance is inversely related to conductance.) The basic electricity equations that illustrate these concepts are provided in Table 1.3.

When designing an electrically heated furnace, three variables that affect element performance must be considered [19]:

1. Electrical characteristics of the element material
2. Watt loading of the elements
3. Furnace atmosphere

Currently, there are primarily four materials used for heating element construction: nickel–chromium (80Ni, 20Cr) [17,20], iron–chromium–aluminum [17], silicon carbide [19,20], and molybdenum disilicide ($MoSi_2$) [21].

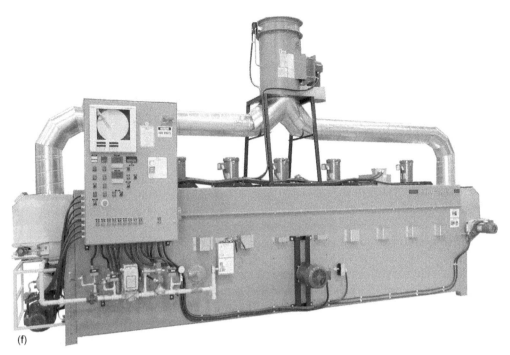

(f)

FIGURE 1.11 (continued) (f) Gas-fired high-speed (rapid-heating) stress relieving and tempering oven. (Courtesy of PYROMAÎTRE Inc., www.pyromaitre.com.)

Except in reducing atmospheres, plain carbon steel elements cannot be used at temperatures above 800°F and therefore are rarely used in heat treatment operations. In reducing atmospheres, they can be used up to 1200°F [20]. Because of these limitations, elements constructed from plain carbon steels are not used in heat treatment furnaces.

Electrical resistance alloys based on nickel–chromium (Ni–Cr) were developed for high-temperature furnace heating [18]. Alloys of this type are suitable for use in furnaces at temperatures up to 2190°F (1200°C) [17].

To obtain even greater lifetimes and higher maximum operating temperatures, alloys based on iron, chromium, and aluminum (Fe–Cr–Al) were developed and are marketed under the trade name Kanthal [17,20]. The aluminum alloying element is used to form a chemically resistant protective layer in the furnaces [20]. Kanthal elements are the most commonly used elements in electric furnaces employing reactive atmospheres. Table 1.4 lists the maximum recommended operating temperature for Kanthal elements in various environments.

Kanthal elements can be used at temperatures up to 2550°F (1400°C) [17]. A comparison of the use of Ni–Cr and Kanthal electric heating elements is provided in Table 1.5. There are positive features to both materials. Kanthal provides greater high-temperature creep strength and longer life at high temperatures and is more chemically resistant.

Watt loading increases with the cross-sectional area of the element. This is illustrated in Figure 1.19 for a Kanthal element at 900°C (1650°F) furnace temperature.

The resistivity of a heating element increases with temperature, as illustrated in Figure 1.20 for a Kanthal element. These curves may vary with the material and even within a class of materials and should be obtained from the manufacturer of the specific heating element considered.

Some materials such as graphite [16] and silicon carbide [19] that are used as refractories become electrical conductors at high temperatures [14]. Although they are subject to embrittlement, silicon carbide elements may be used in heat treatment furnaces [20].

Silicon carbide elements are known by the trade name Globar [18] and are constructed from high-density silicon carbide crystals that are relatively chemically resistant. However, Ni–Cr exhibits superior high-temperature mechanical properties [17]. Advantages of silicon carbide elements include slowness of resistance changes with aging, wide use temperature range (up to 3000°F), suitability for use with high wattage density, and good lifetimes (6 months to 2 years) [19].

The watt loading properties of Globar heating elements are summarized in Table 1.6. Figure 1.21 shows that Globar elements exhibit higher watt loading in oxidizing air environments. The Globar element may accumulate carbon when used with some endothermic atmospheres. Excessive carbon can be removed by preventing replenishment of the atmospheric gas and introducing air periodically to burn off-residual carbon [19]. The high-temperature zone should be kept free of excessive moisture and carbonaceous vapors [19].

Electrical resistance properties of Globar elements are summarized in Figure 1.22. Electrical resistance decreases with increasing temperature up to approximately 1200°F (650°C). Above this temperature it increases with temperature. The different negative resistance values shown in Figure 1.22 are due to the effect of trace impurities in silicon carbide [19].

Another, less often used, heating element material is molybdenum disilicide, $MoSi_2$. These elements typically contain 90% of $MoSi_2$ and 10% of metallic and ceramic additions [21]. They are used in furnaces for both high- and low-temperature processes and are suitable for use with atmospheres of pure hydrogen and cracked ammonia (with a very low dew point) [16].

An electrical resistance curve for an $MoSi_2$ element is provided in Figure 1.23. As with the Kanthal and Globar elements shown earlier, electrical resistance generally increases with increasing temperature [21].

Element watt loading with respect to temperature is plotted in Figure 1.24, which shows that relatively high loadings at high temperature are possible [21].

In a furnace atmosphere, $MoSi_2$ reacts with oxygen above 1800°F (980°C) to form a layer of silicon dioxide (SiO_2), which protects these elements against further chemical attack.

The selection of electric elements in furnace design is carried out in 13 steps [28]. The necessary equation for each step is given in Table 1.7. Although optimal furnace design is considerably more complicated than this process and should be reviewed with an appropriate engineering consultant, these calculations do provide a first approximation of the likely design requirements for an electric furnace.

Thomonder [23] reviewed the design and construction recommendations for electric furnace design using Kanthal elements. Some of these recommendations are as follows:

1. Proper refractory material should be used for each part of the furnace. Generally a low iron content brick or low fiber modulus (1 mcf) refractory material is used for an electric furnace. Brick refractory should exhibit an electrical resistance of at least 4×10^4 Ω cm at 1200°C. The voltage drop through the brick section should be less than 25 V/cm, if possible.
2. Wire element size depends on the watt loading, and the proper size should be used to obtain optimal lifetime from the element. A wire thickness of at least 3 mm (0.12 in.) is often used.
3. Wire is generally used for relatively low amperage applications.

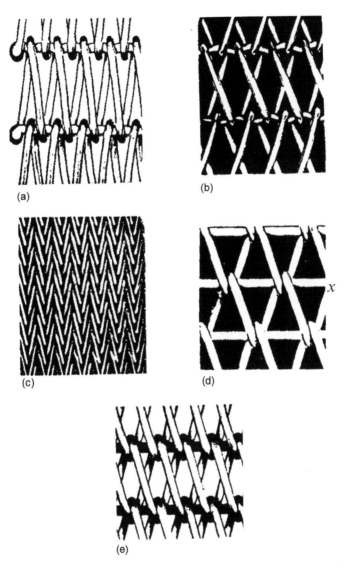

(a)

(b)

(c)

(d)

x

(e)

FIGURE 1.12 Examples of mesh belt types. (a) Balanced weave; (b) double balanced weave; (c) compound balanced weave; (d) rod-reinforced; (e) double rod reinforced. (Courtesy of The Furnace Belt Company Ltd.)

4. Spiral diameter should be four to six times the wire diameter for furnace temperatures greater than 1000°C (1830°F) and four to seven times the wire diameter at furnace temperatures below 1000°C (1830°F).
5. Thickness for strip elements should be at least 1.5–2.5 mm (0.59–0.09 in.).
6. Terminal area should be approximately three times the area of the heating zone to minimize the potential of wires breaking off. The area of the wall should be at least as large as the heating zone.
7. Electric furnace efficiency can generally be assumed to be 50–80%. However, this approximation may be insufficient for calculating power requirements with respect to a small load and large chamber. In such cases, Figure 1.25 may be used to approximate furnace power requirements [23].

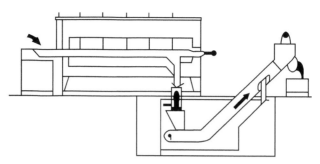

FIGURE 1.13 Schematic of a vibratory retort continuous furnace.

8. Element loading is dependent on furnace wall construction, atmosphere, temperature, and load capacity (throughout per hour). Figure 1.26 may be used to determine element temperature [24]. (This figure should be used only for unrestricted radiation.)

1.3.3 Gas-Fired Furnaces

Although electrically heated furnaces may be much more efficient (>85%), they may also be significantly more expensive to operate than a less efficient (50–70%) gas-fired furnace [16,25]. However, this depends on the cost of gas and on whether the natural gas is spiked.

For example, in a 1981 paper, the cost of operating a gas-fired surface combustion super all case furnace was compared with the cost of operating an electric furnace performing the same heat treatment operations. The results of this study showed that the gas-fired furnace cost approximately 80% as much as the electric furnace to operate. The operational cost of the gas-fired furnace could be further reduced to approximately 30% of that of the electric furnace with a heat recovery process in which the combustion air was preheated with the flue gases. Although this may be practical and cost-effective for large four-row pusher furnaces, that would not be the case for smaller batch, temper, and medium temperature furnaces. An illustration of fuel savings of this process is shown in Figure 1.27. A nomogram for the calculation of fuel savings by the combustion air preheating process is provided in Figure 1.28.

In a more recent study, it was shown that the cost of operating a gas-fired furnace could be reduced to approximately 8% of that of operating an electric furnace [25]. Furthermore, the use of gas had a number of additional advantages. For example, (1) it is possible to use a more useful heat input into the heating process, (2) the use of gas increases the heat treatment process rate, (3) natural gas is reliable and burns cleanly, and (4) furnace conversions are fast and low cost, and gas-to-electric conversions of any size of heat treatment equipment are readily

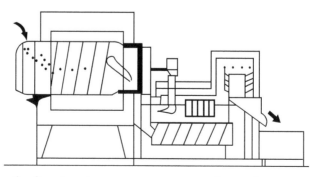

FIGURE 1.14 Schematic of a rotary drum screw conveyor continuous furnace.

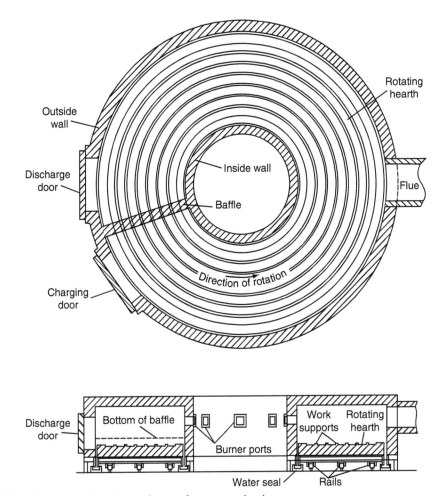

FIGURE 1.15 A rotary hearth continuous furnace mechanism.

performed [25]. Disadvantages of this conversion included: (1) conversion downtime causes production loss, (2) it is necessary to install flue ducts, (3) temperatures are higher around the equipment, and (4) it is necessary to install flame safety controls and train operators [25].

In this section, methods of improving gas-fired furnace efficiency through combustion control and waste heat recovery are discussed.

1.3.3.1 Gas Combustion

Combustion is an oxidative chemical reaction between a hydrocarbon fuel source such as methane (CH_4) and oxygen. If sufficient oxygen is available, there are no impurities in the gas, and the reaction is complete, then the sole reaction products are carbon dioxide (CO_2) and water (H_2O) [25]. Figure 1.29 shows that increasing the combustion temperature increases the degree of completion of the combustion reaction (increasing CO_2) [27].

$$CH_4 + 2O_2 \rightarrow CO_2 + 2H_2O$$

Incomplete combustion will occur and insufficient oxygen is available, and carbon monoxide (CO) will be produced.

FIGURE 1.16 Rotary hearth furnace. (Copyright 2005, O'Brien & Gere, www.obg.com)

$$2CH_4 + O_2 \rightarrow 2CO + 4H_2$$

If additional oxygen is added, the overall reaction can be driven to produce the products of complete combustion [28].

$$2CO + O_2 \rightarrow 2CO_2$$

$$2H_2 + O_2 \rightarrow 2H_2O$$

To achieve the desired degree of combustion and combustion efficiency, excess air is used. The completion of combustion and the percent available gross fuel input depend on the amount of excess air (combustion ratio) as shown in Figure 1.30 [29,30] (see also Ref. [31]).

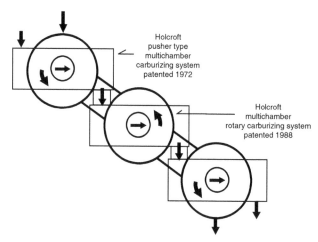

FIGURE 1.17 The Holcroft multichamber rotary carburizing system. (Courtesy of Holcroft—A Division of Thermal Process System Inc.)

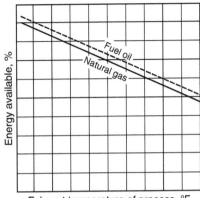

FIGURE 1.18 Dependence of available heat energy on exhaust temperature for fuel oil and natural gas. (From S.N. Piwtorak, Energy conservation in low temperature oven, in *The Directory of Industrial Heat Processing and Combustion Equipment: United States Manufacturers, 1981–1982*, Energy edition, Published for Industrial Heating Equipment Association by Information Clearing House.)

The effect of the combustion ratio on the amount of heat available from the combustion process is determined from Figure 1.31 [29]. For example, if the excess air increases from 5 to 30% with a flue temperature of 1500°F, available heat decreases from 58 to 50% [29]. The percent decrease in available heat is

Available at 5% excess air

$$\frac{\text{Available heat at 30\% excess air}}{\text{Available heat at 30\%}} = \frac{58 - 50}{50} \times 100 = 16\%$$

The decrease in available heat is due to the energy required to heat the additional air. Figure 1.32 provides an estimate of energy savings attainable by minimizing the excess of air used for combustion [30]. It should be noted that any opening in the furnace wall produces a chimney

TABLE 1.3
Electricity Equations

$$W = EI = I^2 R = \frac{E^2}{R}$$

$$E = IR = \sqrt{WR} = \frac{W}{I}$$

$$I = \frac{E}{R} = \frac{\sqrt{W}}{R} = \frac{W}{E}$$

$$R = \frac{E}{I} = \frac{E^2}{W} = \frac{W}{I^2}$$

Btu = kW × 3412

$$\text{kW} = \frac{\text{Btu}}{3412}$$

W, heat flow rate (W); E, electromotive force (V); I, electric current (A), R, electrical resistance (Ω).

Source: From *Hot Tips for Maximum Performance and Service—Globar Silicon Carbide Electric Heating Elements*, Brochure, The Carborundun Company, Niagara Falls, NY.

TABLE 1.4
Maximum Recommended Fe–Cr–Al (Kanthal) Element Temperatures with Different Atmospheres[a]

Atmosphere	Temperature	
	°F	°C
Air	3090	1700
Nitrogen	2910	1600
Argon, helium	2910	1600
Hydrogen[b]	2000–2640	1100–1450
Nitrogen/hydrogen (95/5)[b]	2280–2910	1250–1600
Exogas (10% CO_2, 5% CO, 15% H_2)	2910	1600
Endogas (40% H_2, 20% CO)	2550	1400
Cracked and partially burned NH_3 (8% H_2)	2550	1400

[a]The values shown are for Kanthal Super 1700 and will vary with manufacturer and grade.
[b]The useful element temperature varies with the dew point of the atmosphere.
Source: From *Kanthal Super Electric Heating Elements for Use up to 1900°C*, Brochure, Kanthal Furnace Products, Hallstahammer, Sweden.

effect, which results in heat loss that may significantly perturb the air–fuel balance as shown in Figure 1.33 [30].

The exit gas temperature will increase with the furnace temperature because the theoretical combustion temperature increases with the amount of air (oxygen) as shown in Figure 1.34. Figure 1.35 shows that the combustion efficiency is significantly improved by oxygen addition (enrichment) into the combustion mixture [32].

Common fuel sources are city gas, natural gas (~85% CH_4), propane, and butane. Table 1.8 provides burner combustion data for these gases.

TABLE 1.5
Comparison of Ni–Cr and Fe–Cr–Al Elements[a]

Process Variable	Element	
	Ni–Cr	Fe–Cr–Al[b]
Furnace temperature, °C (°F)	1000 (1830)	1000 (1830)
Element temperature, °C (°F)	1068 (1955)	1106 (2025)
Hot resistance Rw	3.61	3.61
Temperature factor Ct	1.05	1.06
Cold resistance, R_{20}	3.44	3.41
Wire diameter, mm (in.)	5.5 (0.217)	5.5 (0.217)
Surface load, W/cm^2 (W/in.)	3.09 (19.9)	3.98 (25.7)
Wire length, m (ft), for three elements	224.9 (738)	174.6 (573)
Wire weight, kg (lb), for three elements	44.4 (98)	29.6 (65)

[a]120-kW furnace. Three elements at 40 kW each, 380 V.
[b]These are Kanthal AF elements.
Source: From *Sintering Systems*, Product Bulletin No. 901, C.I. Hayes, Cranston, RI.

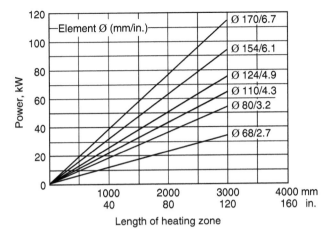

FIGURE 1.19 Possible loading at 900°C (1650°F) furnace temperature. (From *Kanthal Handbook—Resistance Heating Alloys and Elements for Industrial Furnaces*, Brochure, Kanthal Corporation Heating Systems, Bethel, CT.)

Specific gravity [(density gas)/(density air)] is used to calculate flow and combustion products in gas mixing.

Air requirement is used for the calculation of total air required for the combustion of a particular gas.

Flame temperature reflects the energy available in the combustion process. The flame temperature is dependent on the amount of oxygen available during combustion as shown in Figure 1.36 [33].

Flame propagation speed reflects the ability to obtain a stable flame.

Limits of inflammability are used to determine the safety of the use of a particular gas–air mixture.

The combustion air/fuel ratio can be controlled either by precise metering of the fuel and air entering the burner or by flue gas analysis. The preferred method of flue gas analysis is usually to use an oxygen sensor [34].

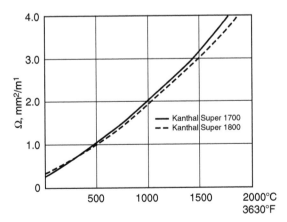

FIGURE 1.20 Resistivity of Kanthal Super 1700 and 1800. (From G.C. Schwartz and R.L. Hexemer, *Ind. Heat.*, March 1995, 69–72.)

TABLE 1.6
Recommended Operating Limits of Globar Elements and Effect of Various Atmospheres

Atmosphere	Recommended Operating Limits		Effect on Elements
	Temperature (°F)	Watt Loading	
Ammonia	2370	25–30	Reduces silicon film; forms met
Argon	Max	Max	No effect
Carbon dioxide	2730	25–25	Attacks silicon carbide
Carbon monoxide	2800	25	Attacks silicon carbide
Endothermic			
18° CO	Max	Max	No effect
20° CO	2500	25	Carbon pick-up
Exothermic	Max	Max	No effect
Halogens	1300	25	Attacks silicon carbide
Helium	Max	Max	No effect
Hydrocarbons	2400	20	Hot spotting from carbon pick-up
Hydrogen	2370	25–30	Reduces silicon film
Methane	2400	20	Hot spotting from carbon pick-up
Nitrogen	2500	20–30	Forms insulating silicon nitrides
Oxygen	2400	25	Oxidizes silicon carbide
Sulfur dioxide	2400	25	Attacks silicon carbide
Vacuum	2200	25	Below 7 millions, vaporizes silicon carbide
Water			Reacts with silicon carbide to from silicon hydrates
Dew point	60	2000	20–30
	50	2200	25–35
	0	2500	30–40
	−50	2800	25–45

Source: From G.C. Schwartz and R.L. Hexemer, Designing for most effective electric element furnace operation, *Ind. Heat.*, March 1995, pp. 69–72.

Although perfect combustion is achieved by mixing the exact quantities of fuel and air to produce only CO_2 and water, this is often not practical. Typically the fuel–air mixture if either rich or lean. A rich mixture contains excess fuel, and since CO and H_2 are produced, it produces a reducing atmosphere. If excess air is used, an oxidizing atmosphere will result. In addition to the proper fuel/air ratio, it is important to provide sufficient time at temperature for complete combustion of the fuel to occur, as shown in Figure 1.37 [35].

As an approximation for furnace calculations, it is often assumed that approximately 1 ft^3 of air is required for each 100 Btu of heating value [36]. For 1 gal of fuel oil, approximately 1500 ft^3 of air is required [36].

1.3.3.2   Burner Selection

The two most common types of burners for heat treatment furnaces are direct-fired, high velocity (Figure 1.38) and indirect-fired, radiant tube (Figure 1.39). With direct-fired burners, fuel combustion occurs in the furnace vestibule. The circulation of the hot gases, which may be oxidizing or reducing, depending on the air/fuel ratio, provides the temperature uniformity within the furnace.

Direct-fired burners are not favored in heat treatment processes such as carburizing where atmosphere control is critical, and also the generation of reducing gases is not a very efficient

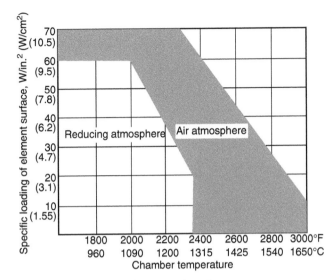

FIGURE 1.21 Recommended loading for Globar heating elements.

furnace heating process. In these cases, more efficient combustion may be attained inside a protective radiant tube.

1.3.3.2.1 High-Velocity Burners

To obtain uniform microstructure and properties and to minimize undesirable residual stresses that arise from thermal gradients, it is important to facilitate a uniform temperature distribution within the furnace. Some methods that have been used to accomplish this are [37,39]:

1. Use of high-temperature fans.
2. Use of baffled walls in both the upper and lower parts of the furnace.

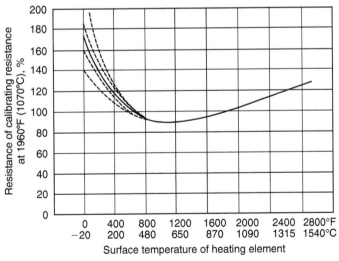

FIGURE 1.22 Electrical resistance vs. surface temperature for Globar heating elements. (From G.C. Schwartz and R.L. Hexemer, *Ind. Heat.*, March 1995, 69–72.)

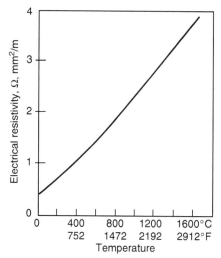

FIGURE 1.23 Electrical resistance curve for molybdenum silicide heating elements.

3. Utilization of fewer heating zones. The use of fewer burners instead of the traditional array of multiple burners may produce excellent thermal uniformity as shown in Figure 1.40 [37,40].
4. Hot-charging and preheating steel with flue gases.
5. Utilization of high-velocity burners.

In high-velocity burners, fuel is burned either completely or partially in ceramic-lined combustion chamber (see Figure 1.41). The exit velocity of the combustion products is 50–200 m/s. The high-speed jet promotes gas mixing and temperature uniformity in the vestibule. Depending on the design and type of gas–air mixing, high-velocity burners are classified as parallel-flow, cross flow, cyclone, or turbulent jet designs. Examples of each are illustrated in Figure 1.38.

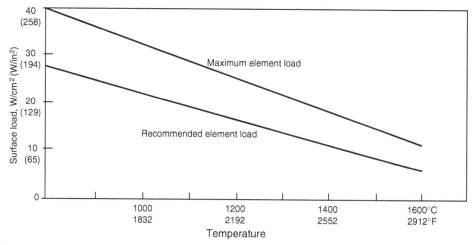

FIGURE 1.24 Recommended loading for molybdenum silicide heating elements.

TABLE 1.7
Electric Element Furnace Design Equations

1 Process heat (kW) $= \dfrac{W \times C_p \times \Delta T}{3412}$	(a)

where W, load weight, ΔT, °F heat-up/h, C_p, specific heat

2 Heat loss (kW) $= \dfrac{(\text{heat loss/ft}^2) \times A}{3412}$	(b)

where A, total furnace area

3 Heat storage (kW) $= \dfrac{(\text{heat storage/ft}^2) \times A}{3412}$ (c)

4 Total power requirement $=$ process heat $+$ heat storage   (d)

5 Watts/element $=$ heating section area (in.^{2}) $\times$ watts/in.^{2} (e)

6 Number of elements $= \dfrac{\text{total power requirement (W)}}{\text{W/element}}$ (f)

7 Volts/element $= [\text{W/element}) \times (\text{resistance/element})]^{1/2}$ (g)

8 Total volts $=$ (volts/element) $\times$ (number of elements in series)   (h)

9 $\dfrac{\text{Maximum amperes/element} = \text{volts/element}}{\text{resistance/element}} \times 1.56$ (i)

   (j)

10 Total amperes $= \dfrac{\text{amperes}}{\text{element}} \times$ number of elements in parallel circuit

11 Delta circuit:

    Three-phase volts $= 1.73 \times$ single-phase volts   k (i)

    Three-phase amperes $=$ single-phase amperes   k (ii)

12 Wye circuit:

    Three-phase volts $= 1.73 \times$ single-phase volts   l (i)

    Three-phase amperes $=$ single-phase amperes   l (ii)

Transformers with taps:

13 Total voltage $=$ lowest tap   m (i)

    Voltage $\times 1.0 =$ normal tap   m (ii)

    Voltage $\times 2.0 =$ high tap   m (iii)

Source: From G.C. Schwartz and R.L. Hexemer, *Ind. Heat.*, March 1995, 69–72.

Combustion efficiency varies with the rated furnace load of the combustion chamber, Q/V, where combustion Q is the total quantity of heat released in the burner chamber and V is the volume of the burner combustion chamber. In a recent study, Keller [38] reported that

1. As the specific load increases, combustion shifts from the burner toward the furnace.
2. Combustion efficiency decreases as the rated furnace load increases (see Figure 1.43a).
3. Specific load of the combustion chamber increases with fuel flow rate (see Figure 1.42b).
4. Burner with the largest rated load and largest inside diameter produces the lowest specific load (see Figure 1.43).
5. Combustion efficiency and the mean load on the combustion chamber vary inversely with the diameter of the combustion chamber (see Figure 1.43).

High-velocity burners have a number of advantages for use in heat treatment furnace applications [42]:

1. Velocity of the exiting combustion products reentrains the existing furnace atmosphere, maximizing the available heat.

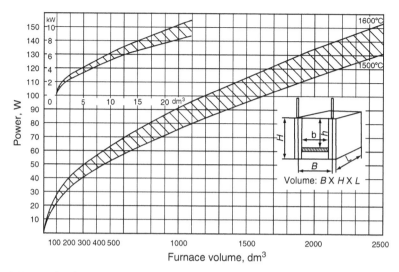

FIGURE 1.25 Determination of approximate furnace power based on furnace volume.

2. Depending on the type of furnace, the relatively short flame of a high-velocity burner reduces the risk of flame impingement on the load.
3. Turbulence created by the burner combustion products facilitates temperature uniformity.
4. Turbulence of the furnace gases facilitates heat transfer from the atmosphere to the load.

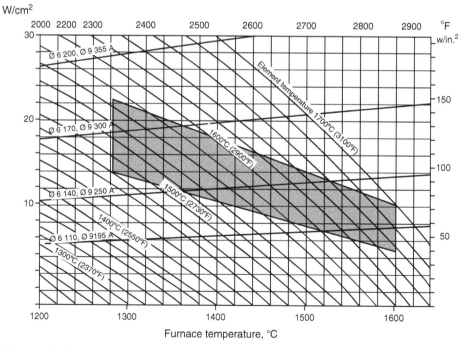

FIGURE 1.26 Surface loading graph for Kanthal heating elements.

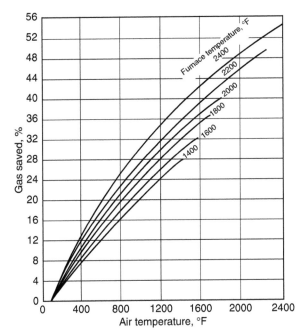

FIGURE 1.27 Potential fuel savings available with exhaust gas recovery. (From R.A. Andrews, *Heat Treating*, February 1993, 2–23.)

Some general rules that have been proposed for a heat treatment combustion system are [43]:

1. There should be a maximum of two burners, one at each end of the furnace, with firing occurring at a high position along the long walls.
2. All flues should be placed under the center of the load.
3. High-velocity discharge should be sufficient to provide temperature uniformity.

Note: The validity of these general rules is dependent on the particular furnace designed. For example, exceptions to these rules include large furnace systems such as continuous carburizing and continuous tempering furnaces.

When lower heating temperatures ($<700°C$ ($<1300°F$)) are required, it is often difficult to obtain the necessary furnace uniformity with a low burner velocity. One method of improving furnace uniformity in such cases is to use excess air [43]. A variable excess air furnace combustion system has been described that achieves superior temperature control and greater combustion efficiency. A 42% energy savings was reported [44].

Another method of increasing the heating performance and fuel efficiency of a high-velocity burner at lower temperature is to use a computer-controlled pulse-firing system [42,45]. Since high-velocity burners are more efficient when they are on, the idea is to turn them on or off with a timed pulse. The burned firing frequency may vary from 3 to 6 s between pulses [35].

1.3.3.2.2 Radiant Tube Burners
When heat treatment processes requiring exclusion of combustion gases from the furnace load and also requiring precise temperature control are carried out; the use of direct-fired burners is unacceptable. These furnace applications generally require isolation of the burner combustion process. This is accomplished by encasing the burner, often pulse-fired, in a radiant tube such as the single-ended radiant (SER) tube design shown in Figure 1.39 [24].

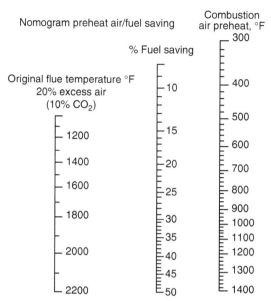

FIGURE 1.28 Nomogram to determine expected fuel savings from preheating combustion air. (From *Single-End Recuperative Radiant Tube Combustion Systems*, Brochure, Pyronics Inc., Cleveland, OH.)

Currently, many manufacturers favor radiant tubes constructed of ceramic materials such as reaction-bonded silicon carbide [38,41].

1.3.4   HEAT RECOVERY

Combustion efficiency can be significantly improved by preheating the incoming cold air with the hot flue gas. The effect of the temperature of the flue gas and percentage of excess combustion air on gas efficiency is illustrated in Figure 1.30 [39]. Alternatively, fuel savings may be calculated using the nomogram provided in Figure 1.28 [26].

There are two principal methods for recovering and reusing heat that is normally lost through flue gas emission: recuperation and regeneration. Recuperation uses a heat exchanger to transfer heat from the hot flue gas to the incoming cold combustion air. Regeneration increases combustion efficiency by using the hot flue gas to both preheat combustion air and further increase the burner flame temperature. Figure 1.44 shows the enhancement in combustion efficiency gained by recuperation and regeneration. An overview of both those processes follow.

1.3.4.1   Recuperation

The use of recuperator systems to provide substantial improvement in both batch and continuous heat treatment processes [13,46] has been reported for both new furnace systems [25] and retrofitted older systems [47,48].

The improvement of combustion efficiency by preheating air for natural gas combustion is illustrated in Figure 1.45. The recuperative process that increases combustion air temperature through heat exchange with the hot exiting flue gas is illustrated in Figure 1.46.

There are three types of heat exchangers for gases: continuous flow, parallel-flow, and cross-flow recuperators. These heat exchanger flue gas and airflow patterns are illustrated in Figure 1.47.

Examples of commercial recuperator systems are provided in Figure 1.48 and Figure 1.49. Convective heat transfer efficiency increases with flow rate through the heat exchanger (Figure

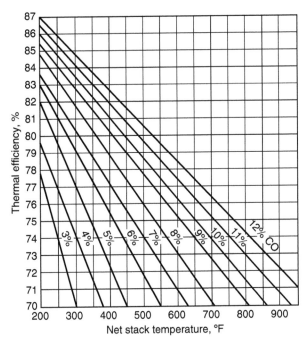

FIGURE 1.29 Thermal efficiency as a function of exhaust temperature. *Note*: This chart applies only to cases where the percentage of CO_2 is less than ultimate because the fuel/air ratio is leaner than that required for perfect combustion.

1.50) and is nearly independent of gas temperature [50]. Larger heat exchanger surface areas are required with increasing air preheat temperature as shown in Figure 1.51 [50].

Recuperators may be constructed either from high-temperature, corrosion-resistant metallic materials or nonmetallic ceramics. However, the use of ceramic materials is much less dependent on the service temperature and contact with corrosive flue gases [51].

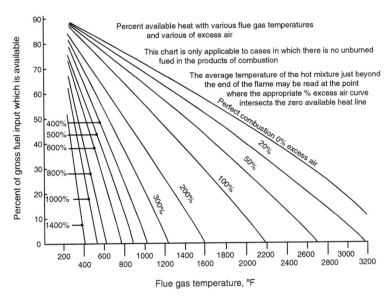

FIGURE 1.30 Available heat vs. excess air for flue temperatures of 200–3200°F.

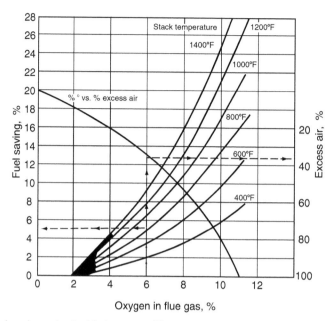

FIGURE 1.31 Fuel savings obtainable by controlling excess air.

1.3.4.2 Regeneration

The performance of recuperator systems is generally limited by the surface area of the heat exchanger [21] and the upper temperature of the preheated air due to potential oxidation of the preheated recuperator surfaces [52]. These and other limitations are circumvented by the more efficient regenerative combustion system.

Numerous publications have discussed the use of regenerators in both batch [52–54] and continuous [40,55,56] systems to achieve substantial energy cost reductions [57]. This section provides an overview of the operation of regenerator burner combustion systems.

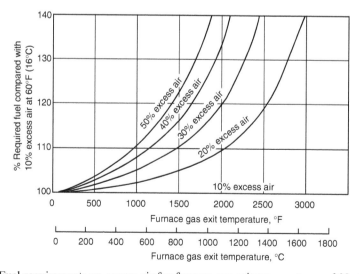

FIGURE 1.32 Fuel requirements vs. excess air for furnace gas exit temperatures of 32–3000°F.

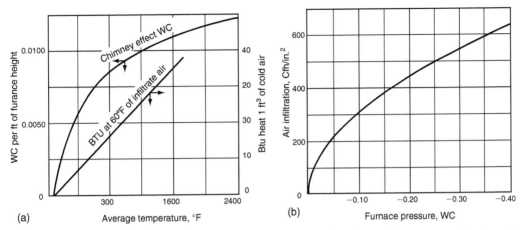

FIGURE 1.33 Calculation of chimney effect: (a) is used to determine the negative pressure of the hearth; (b) is used to calculate the flow rate due to negative pressure.

A regenerative combustion system generally consists of two regenerators, two burners, a flow reversal valve, and the necessary control system [58]. (*Note*: Systems such as these are not often encountered in heat treatment furnaces.) A schematic of a typical system is shown in Figure 1.52. Pulse-firing burners are often selected to minimize the potential for radiant tube burnout [34]. Temperature fluctuations are controlled by the mass of the radiant tube.

The regenerator is a two-chamber system that contains firebrick [50] or a ceramic material for heat storage. These materials may be in the form of balls [58], honeycomb [50,55], or even a granular refractory [58]. The performance of a regenerator depends on [36,47]:

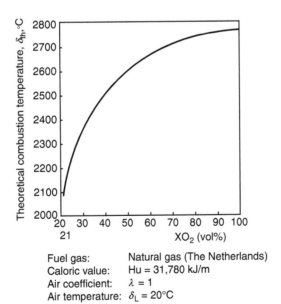

Fuel gas:	Natural gas (The Netherlands)
Caloric value:	$Hu = 31{,}780$ kJ/m
Air coefficient:	$\lambda = 1$
Air temperature:	$\delta_L = 20°C$

FIGURE 1.34 Theoretical combustion temperature as a function of oxygen content in combustion air.

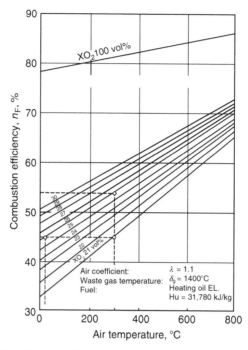

FIGURE 1.35 Influence of combustion air temperature and oxygen content on combustion efficiency.

The size of the furnace and regenerator
Reversal time
Thickness of the firebrick or other refractory material
Conductivity of the refractory
Heat storage ratio
Geometry of the regenerator
Temperature and flow rate of the gases

The basic principle of operation of a regenerator system is that flue gas gives up its heat to the refractory in one of the regenerators. At the same time, combustion air is heated by the hot

TABLE 1.8
Burner Combustion Data for Common Fuel Sources

Gas	Specific Gravity[a]	(Btu/ft³)	Air Required (ft³ Air/ft³ Gas)	Flame Temperature (°F)	Flame Speed (in./s)	Limits of Inflammability, % Gas in Air Low	High
City gas	0.5	500	5	3600	60	5	40
Natural gas	0.6	1000	10	3550	25	4	13
Propane	1.52	2500	25	3650	30	2	10
Butane	1.95	3200	32	3660	30	2	9

[a]Specific gravity = (density of gas)/(density of air).

Source: From R.G. Martinek, *Eclipse Industrial Process Heating Guide*, Brochure, Eclipse Fired Engineering Co., Dan Mills, ON, Canada.

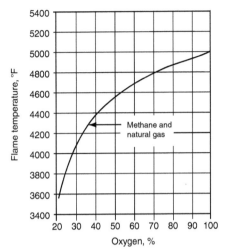

FIGURE 1.36 Effect of oxygen content on the flame temperature of methane and natural gas.

refractory material in the other regenerator. After approximately 20 min, the flows are reversed, and the hot refractory in the regenerator that previously had hot flue gases passing through it is used to heat the combustion air. Conversely, the cooled refractory in the other regenerator chamber is then reheated by hot flue gases. Regenerator microprocessor-controlled cycle times of 20 s have been reported [57].

The burner typically used for regenerator heat treatment furnace applications is of the radiant tube type shown in Figure 1.53.

1.3.4.3 Rapid Heating

Rapid heating is any heating method that accelerates conventional furnace heating. It may be accomplished in gas-fired furnaces by [59]:

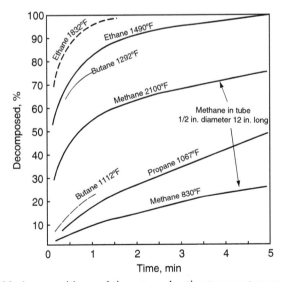

FIGURE 1.37 Effect of fuel composition and time at combustion temperature on combustion efficiency.

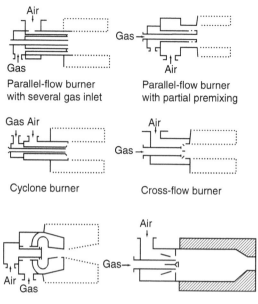

Parallel-flow burner
with several gas inlet

Parallel-flow burner
with partial premixing

Cyclone burner

Cross-flow burner

Turbulent-flow burner

FIGURE 1.38 Several types of direct-fired high-velocity burners. (From K. Keller, *Formage Trait. Met.*, November 1977, 39–44.)

1. Lifting the stock off the furnace hearth
2. Rotating stock to eliminate cold surfaces
3. Separating stock in the furnace
4. Increasing the heat flux on the metal surface by increasing the furnace temperature

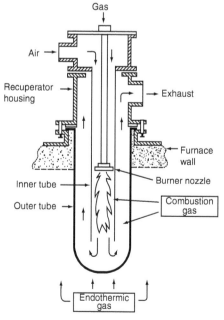

FIGURE 1.39 A single-ended radiant (SER) tube burner. (From T. Darroudi, J.R. Hellmann, R.E. Tressler, and L. Gorski, *J. Am. Ceram. Soc.* 75(12):3445–3451, 1992.)

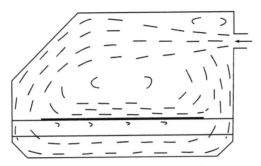

FIGURE 1.40 Furnace gas flow patterns from one-sided burner firing arrangement.

5. Increasing the flow velocity in the furnace heated by high-velocity burners by (a) matching the internal shape of the furnace to the shape of the stock or (b) having the burner jet impinge on the stock

The heat transfer rates attainable by these methods are summarized in Table 1.9.

The nomogram in Figure 1.54 is provided for the estimation of soaking times for steel greater than 75 mm (3 in.) in diameter in the temperature range of 1000–1250°C (1830–2280°F) following rapid heating to the soaking temperature [59]. Although rapid heating is not often used in heat-treating furnaces, but it is used in forge-heating furnaces.

1.4   HEAT TRANSFER

The heat transfer process that occurs when a part is heated in a furnace is depicted in Figure 1.55. Typically the heat transfer rate is rapid initially and decreases as the temperature of the center of the part approaches the surface temperature, which achieves the furnace temperature more rapidly. Ideal furnace design permits thermal equilibrium to be reached as quickly as possible while minimizing thermal gradients within the past during the heating process.

Heat transfer in furnaces occurs by convection, radiation, and conduction. The application of these modes of heat transfer in furnace heating is illustrated in Figure 1.56.

1.4.1   CONVECTIVE HEAT TRANSFER

Heat flux (q) in a furnace is dependent on the change in temperature (dT) with an incremental change in distance (dx):

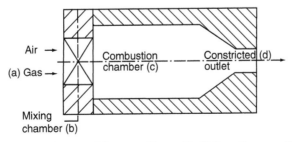

FIGURE 1.41 Schematic of a high-speed burner. (From K. Keller, *Formage Trait. Met.*, November 1977, 39–44.)

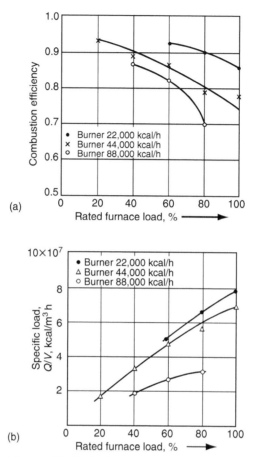

(a)

(b)

FIGURE 1.42 Variation of the specific load of the combustion chamber with fuel flow rate. (From K. Keller, *Formage Trait. Met.*, November 1977, 39–44.)

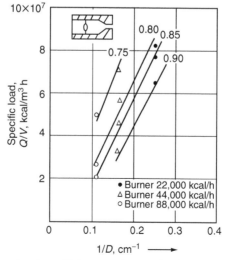

FIGURE 1.43 Variation of combustion efficiency and mean load with chamber diameter.

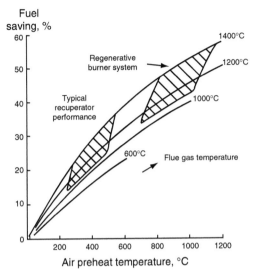

FIGURE 1.44 Relative impacts of recuperators and regenerative burner systems on fuel savings. (From D. Hibberd, *Metallurgia*, February 1968, 52–58.)

$$q = -K \frac{\mathrm{d}T}{\mathrm{d}x}$$

Heat flux q is related to total power Q by

$$Q = Aq$$

where A is the total area of the part. Total power (Q) is often preferred because it accounts for the variation of heat flux with total area. Figure 1.57 illustrates the calculation of total power transferred through members of simple shapes that can be combined to approximate power losses in the actual equipment.

The variation of thermal conductivity with temperature is shown in Figure 1.58. The variation within the area bounded by *abcda* in Figure 1.58 can be approximated as

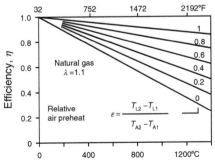

FIGURE 1.45 Effect of temperature of furnace exhaust gas entering recuperator on combustion efficiency. (From J.A. Wünning and J.G. Wünning, *Ind. Heat.*, January 1995, 24–28.)

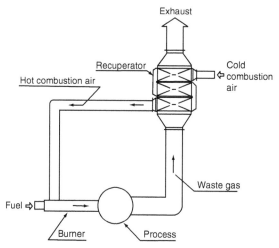

FIGURE 1.46 Schematic of a recuperator process. (From F.M. Heyn, *The Directory of Industrial Heat Processing and Combustion Equipment: United States Manufacturers, 1981–1982*, Energy edition, Published for the Industrial Heating Equipment Association by Information Clearing House.)

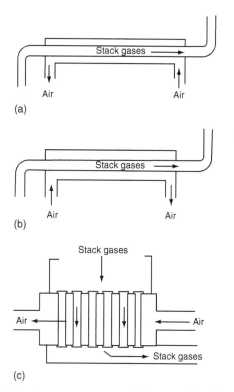

FIGURE 1.47 Recuperator heat exchanger processes. (a) Counterflow; (b) parallel flow; (c) cross flow. (From W. Trinks, *Industrial Furnaces*, 4th ed., Vol. 1, Wiley, New York, 1950, pp. 220–262.)

FIGURE 1.48 Comparison of (left) a recuperator and (right) a typical exhaust stack. (Courtesy of Holcroft—A division of Thermo Process Systems Inc.)

$$\int_{T_c}^{T_h} k(T)\, dT$$

Once the variation of thermal conductivity with temperature is known, the heat transferred through a flat wall can be approximated as shown in Figure 1.59. Heat transfer through a curved wall may be approximated from the flat wall expression as shown in Table 1.10.

The thermal conductivity of various metals and refractory materials with respect to temperature are shown in Figure 1.60 and Figure 1.61, respectively. The mean heat transfer coefficient accounting for fluid flow properties in natural convection can be calculated from [62,63]

$$\bar{h} = \frac{k}{L} C \left[\frac{g\beta(T_s - T_b)L^3}{\nu^2} N_{Pr} \right] m = \frac{k}{L} C (N_{Gr} N_{Pr})^m$$

where k is thermal conductivity, L is the length (for vertical planes and cylinders, L, height of surface; for horizontal cylinder), L is the diameter (for horizontal squares, L, length of a side), g is the gravitational constant, β is the coefficient of volumetric thermal expansion, T_s is the surface temperature, T_b is the temperature of the boundary layer, ν is the kinematic viscosity, N_{Pr} is the Prandtl number (N_{Pr}, ν/a), N_{Gr} is the Grashof number (N_{Gr}, $g\beta(T_s - T_b)x^3/\nu^2$). C and m are constants determined from Table 1.11 and Figure 1.62 for vertical planes and cylinders, horizontal cylinders, and square surfaces [63]. When the mean value of the heat transfer coefficient is calculated, it is assumed that

FIGURE 1.49 Side-mounted recuperators illustrate airflow both parallel and counter to the exhaust. (Courtesy of Holcroft—A Division of Thermo Process Systems Inc.)

$$T_f = \frac{(T_s - T_b)}{2}, \qquad \beta = \frac{1}{T}$$

where T is in kelvins.

For laminar ($N_{Pr} < 2300$) flow through smooth tubes, the equation for the mean heat transfer coefficient is [62]

$$\bar{h} = \frac{k}{d}\left[3.66 + \frac{0.0668(d/L)N_{Re}N_{Pr}^3}{1 + 0.04[(d/L)N_{Re}N_{Pr}]^{2/3}}\right]$$

where L is the tube length and d the inside diameter for circular tubes or $d = 4 \times$ (area of cross section)/(inside circumference) for noncircular tubes.

For turbulent flow through smooth tubes [62],

$$\bar{h} = 0.023\frac{k}{d}N_{Re}^{0.8}N_{Pr}^n$$

where $N_{Re} > 4000$ and $0.6 \leq N_{Pr} \leq 100$; $n = 0.4$ for heating and 0.3 for cooling.

The equation for the mean heat transfer coefficient for gas flow across a cylinder is [64]

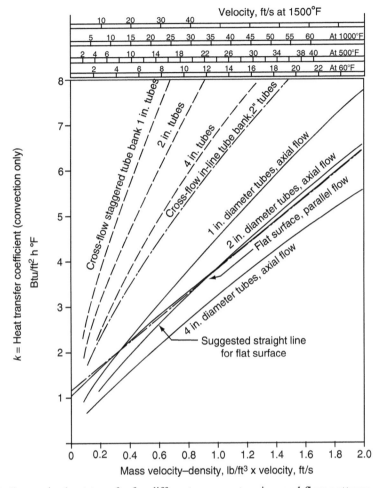

FIGURE 1.50 Convective heat transfer for different recuperator sizes and flow patterns.

$$\bar{h} = C\frac{k}{d}N_{\text{Re}}^{n}C\frac{k}{d}\left(\frac{vd}{v}\right)^{n}$$

where the mean film temperature (T_f) is

$$T_f = \frac{T_s + T_b}{2}$$

The values for C and n are given in Table 1.12 [64].

The constants for calculating the mean heat transfer coefficient for flow across non-cylindrical shapes are provided in Table 1.13 [64].

Heat transfer coefficients for various furnace heating media used for heat treating are provided in Table 1.14.

1.4.2 Radiant Heat Transfer

Radiant heat transfer is dependent on the amount of radiation emitted or absorbed, the wavelength of the radiation, and the temperature and physical condition of the surface. The rate of radiant heat transfer (Q) between two surfaces at T_1 and T_2 is [64]

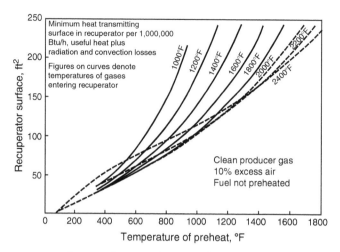

FIGURE 1.51 Recuperator surface requirements for natural gas.

$$Q = A_1 F_{12} \delta (T_1^4 - T_2^4)$$

where A is the area, δ is the Stefan–Boltzmann constant, and F_{12} is a constant that depends on emissivity and the geometry and is usually determined empirically.

The δ radiative heat transfer coefficient (h_r) can be calculated from

$$h_r = F_{12} \delta (T_1^3 + T_1^2 T_2 + T_1 T_2^2 + T_2^3)$$

One of the greatest sources of radiative heat loss is when the furnace door is opened and the rate of heat loss can be calculated from [66]

$$q = 0.173 A e \left[\left(\frac{T_0}{100} \right)^4 - \left(\frac{T_a}{100} \right)^4 \right]$$

where A is the effective area of the door opening, e is the emissivity, and T_0 is the absolute temperature of the air.

Some illustrative values of surface emissivity are provided in Table 1.15.

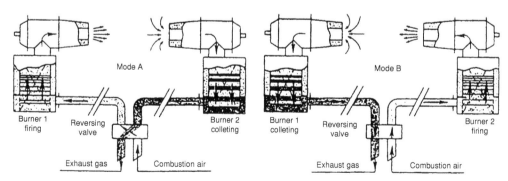

FIGURE 1.52 Schematic of a regenerative burner process. (From T. Martin, *Ind. Heat.*, November 1988, 12–15.)

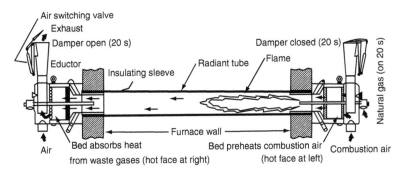

FIGURE 1.53 Schematic of regenerative ceramic radiant tube burner. (From J.D. Bowers, *Adv. Mater. Proc.*, March 1990, 63–64.)

1.4.3 Conductive Heat Transfer

The rate of conductive heat transfer (q) depends on thermal conductivity k, temperature T, and the distance from the heat source X:

$$q = k \frac{dT}{dX}$$

For plane surfaces or thin layers,

$$q = \frac{k}{t}(T_1 - T_2)$$

where t is the thickness of the material.

Figure 1.63 provides an illustrative summary of these heat transfer processes for an electrically heated furnace.

Heat transfer within the part heated may be modeled from the differential equation [61]

$$\rho C \frac{\partial \theta}{\partial t} = k \frac{\partial^2 \theta}{\partial x^2}$$

where ρ is the density, C is the specific heat, θ is the temperature, k is the thermal conductivity, and x is the distance.

It is important to note that these data are temperature-dependent. This dependence should be accounted for when conducting any furnace thermal modeling.

TABLE 1.9
Typical Gas-Fired Furnace Rapid Heating Convective Heat Transfer Coefficients

Heating Method	Gas Velocity		Convective Heat Transfer Coefficient	
	m/s	ft/s	W/(m² K)	Btu/(ft² h °F)
Jet impingement	150	500	250–500	50–100
Furnace stock matching	50	150	100–250	20–50
Conventional furnace	<5	<15	<25	<5

Source: From N. Fricker, K.F. Pomfret, and J.D. Waddington, Commun. 1072, Inst. Gas Eng., 44th Annu. Meeting, London, November 1978.

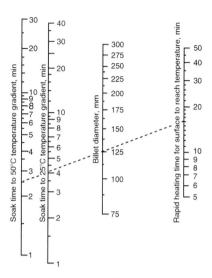

FIGURE 1.54 Nomogram for rapidly heating steel billets to 1000–1250°C.

Furnace designers use computerized computational methods for the solution of these and related equations. A detailed discussion of these methods is beyond the scope of this text. However, the methods discussed here provide an excellent approximation to the solution of many routine heat transfer problems encountered in heat treatment shops.

The temperature rise of a simple shape can be estimated using Heisler charts, which are constructed with the following information [60].

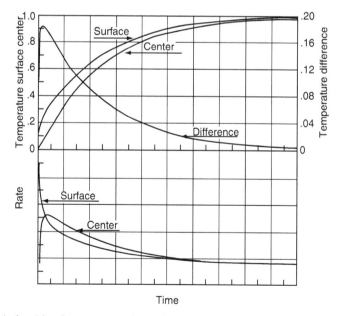

FIGURE 1.55 Relationship of temperature rise and heating rates with respect to time. (From V. Paschkis and J. Persson, *Industrial Electric Furnaces and Appliances*, Interscience, New York, 1960, pp. 14–25.)

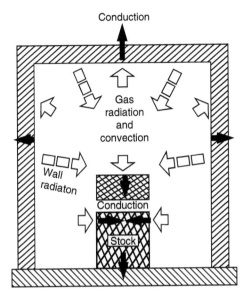

FIGURE 1.56 Furnace heat transfer processes. (From D.F. Hibbard, *Melt. Mater.* 3(1):22–27, 1987.)

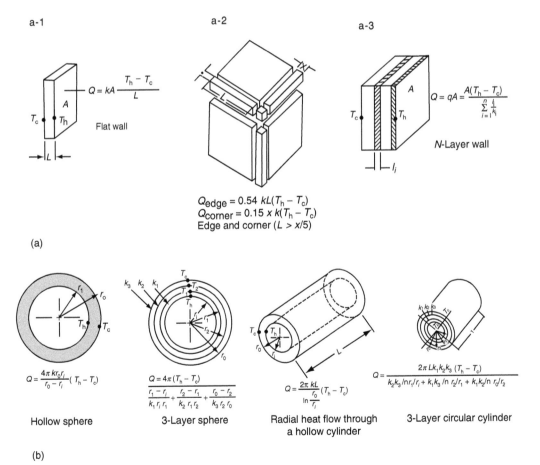

FIGURE 1.57 Heat transfer equations for commonly encountered shapes in heat treatment. (From AFS, *Refractories Manual*, 2nd ed., American Foundrymen's Society, Des Plaines, IL, 1989.)

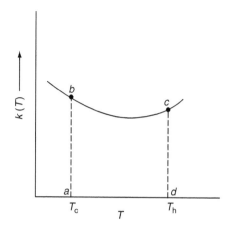

FIGURE 1.58 Variation of thermal conductivity with temperature.

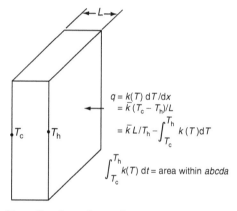

FIGURE 1.59 Calculation of heat flux for a flat wall.

1. Diffusivity (α):

$$\alpha = k/C_\rho$$

where k is the conductivity, C is the specific heat, and ρ is the density of the material.

2. Fourier number (N_{Fo}):

$$N_{\text{Fo}} = \alpha T/L^2$$

TABLE 1.10
Flat Wall Thickness Equivalents for Curved Surfaces

Shape of Wall	Assume Flat Wall Thickness of	Assume Flat Wall Area of
Cylindrical	$r_o - r_i$	$\pi(r_o + r_i)$
Spherical	$r_o - r_i$	$4\pi r_o r_i$

Source: From AFS, *Refractories Manual*, 2nd ed., American Foundrymen's Society, Des Plaines, IL, 1989.

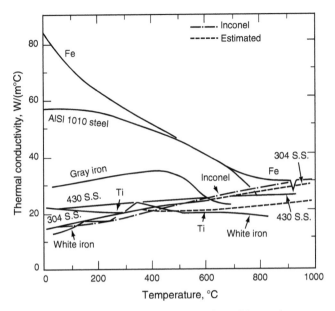

FIGURE 1.60 Thermal conductivity of various ferrous metals and Inconel.

where T is the heating time and L is the critical dimension of the shape. The critical dimension for slabs, cylinders, and spheres (assuming no end effects) is determined as follows:

Slab	Half the thickness
Cylinder	Radius
Sphere	Radius

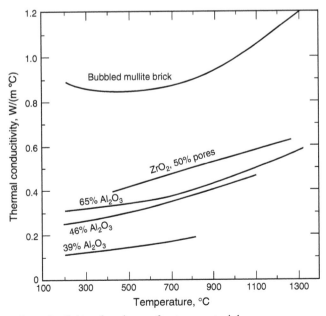

FIGURE 1.61 Thermal conductivity of various refractory materials.

TABLE 1.11
Equation Constants for Natural Convention

Surface	$N_{Gr}N_{Pr}$	C	m	Note
Vertical	10^{-1}–10^4			Use Figure 1.57
	10^{-4}–10^9	0.59	1/4	
	10^9–10^{12}	0.13	1/3	
Horizontal cylinder	0–10^{-5}	0.40	0	
	10^{-5}–10^4			Use Figure 1.57
	10^4–10^9	0.53	1/4	
	10^9–10^{12}	0.13	1/3	
Horizontal square surface	10^5 to 2×10^7	0.54	1/4	Upper surface if heated; lower surface if cooled
	2×10^7 to -3×10^{10}	0.14	1/3	
	3×10^5 to -3×10^{10}	0.27	1/4	Lower surface if heated plate; upper surface if cooled plate

Source: From W. Trinks, *Industrial Furnaces*, 4th ed., Vol. 1, Wiley, New York, 1950, pp. 220–262.

3. Relative boundary resistance (m):

$$m = k/hL$$

where h is the boundary conductance.

4. Temperature function (TF):

$$TF = (t_f - t)/(t_f - t_i)$$

where t_i is the initial temperature of the part, t_f is the constant furnace temperature, and t is the temperature at any point in the part.

The Heisler charts for spheres, cylinders, and slabs are shown in Figure 1.64 for short times where $N_{Fo} = 0$–0.2 and $m < 100$ and in Figure 1.65 for long times, where $N_{Fo} > 0.2$ and $m < 100$. When $m > 100$, Figure 1.66 should be used. The relationship between the ratio of TF

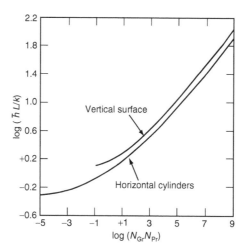

FIGURE 1.62 Correlation of natural convection for vertical and horizontal surfaces.

TABLE 1.12
Constants for Gaseous Cylindrical Cross Flow

N_{Re}	C	n
0.4–4	0.891	0.330
4–40	0.821	0.385
40–4,000	0.615	0.466
4,000–40,000	0.174	0.618
40,000–400,000	0.0239	0.805

TABLE 1.13
Constants for Calculation of Gaseous Cross Flow of Noncylindrical Tubes

Tube Shapes	N_{Re}	C	n
u_∞ ◇ d	5×10^3–1×10^5	0.222	0.588
u_∞ □ d	5×10^5–1×10^5	0.092	0.675
u_∞ ⬡ d	5×10^3–1.95×10^4 1.95×10^4–1×10^5	0.144 0.0347	0.638 0.782
u_∞ ⬡ d	5×10^3–1×10^5	0.138	0.638
u_∞ | d	4×10^3–1.5×10^4	0.205	0.731

on the surface (TF$_s$) and in the center (TF$_c$) of the heated object and m is shown in Figure 1.67 [61].

Furnace temperature and heating time can be interrelated using the uniformity factor U, which is defined as [61]

$$U = \frac{t_s - t_c}{t_s - t_i}$$

TABLE 1.14
Heat Transfer Coefficients for Various Furnace Heating Media

Medium	Heat Transfer Coefficient (Btu/(ft^2 h °F))
Air circulation furnace	2–8
Jet heating/cooling	20–50
Batch and pusher furnaces[a]	15–80
Gaseous fluidized bed	50–110
Stirred salt bath	200–600
Liquid fluidized bed	1300
Lead bath	1000–6000

[a]Convection/radiation.
Source: From J.P. Holman, *Heat Transfer*, 2nd ed., McGraw-Hill, New York, 1968.

TABLE 1.15
Total Emissivities of Selected Surfaces

Material	Emissivity
Refractory	0.8
Carbon	0.9
Steel plate	0.95
Oxidized aluminum	0.15

Source: From R.N. Britz, *Ind. Heat.*, January 1975, 39–47.

where t_s is the temperature at the surface of the part, t_c is the temperature at its center, and t_i is its initial temperature. Uniformity factors for spheres, cylinders, and slabs are shown in Figure 1.68.

1.4.4 FURNACE TEMPERATURE UNIFORMITY

The characteristic thermal flow patterns in a furnace may be significant and will cause nonuniform heating of parts. This nonuniformity may be due to various factors, including [62]:

1. Interaction between burners
2. Unstable flow
3. Variation of mass circulation rates and thermal distribution within the load
4. Stagnant regions of high or low temperature
5. Combustion patterns

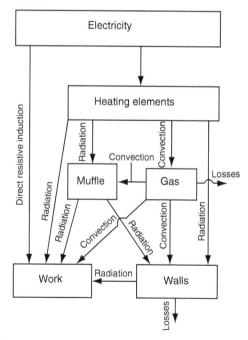

FIGURE 1.63 Summary of heat transfer processes for various furnace components. (From D. Nicholson, S. Ruhemann, and R.J. Wingrove, in *Heat Treatment of Metals,* Spec. Rep. 95, Iron and Steel Inst., 1966, pp. 173–182.)

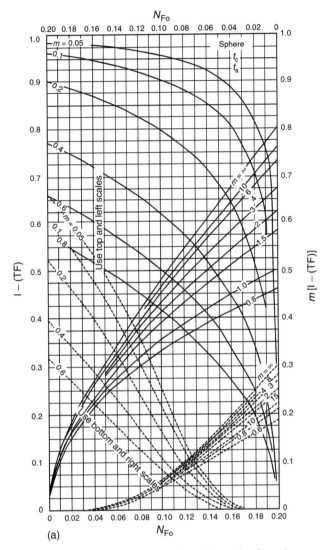

(a)

FIGURE 1.64 (a) Short-time temperature function (TF) relationships for spheres.

Continued

Two important criteria in considering appropriate furnace design are the variation of size and type of the materials heated (see Figure 1.69) [65,66], if more than one, and the required production rate [67]. Another factor that must be considered is the necessity for various heat treatment processes such as austenitizing, normalizing, and stress relief in a single furnace.

The first step in determining the best furnace design is to conduct an energy balance to determine the relative efficiencies of different furnace designs considered. One method of conducting this assessment is to model the various heat transfer processes in the furnace and conduct an energy balance [68]. The energy balance may be illustrated using a Sankey diagram such as the chart depicted in Figure 1.70, which shows that furnace efficiency [(useful output)/(fuel input) × 100] is dependent on

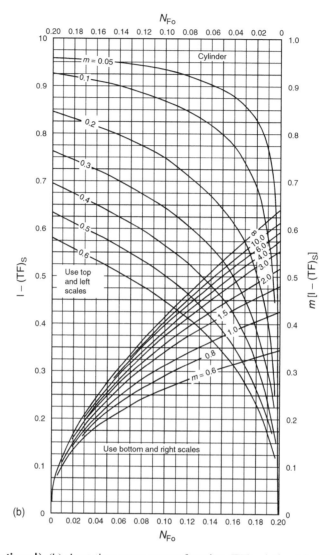

FIGURE 1.64 (Continued) (b) short-time temperature function (TF) relationships for cylinders.

Continued

1. Thermal energy in the material
2. Structural losses
3. Waste gas losses
4. Heat recovery
5. Unaccounted losses

One method of improving furnace temperature uniformity is to use forced circulation of the heated gaseous atmosphere [65,66]. The effect of increased flow velocity on furnace temperature uniformity is shown in Figure 1.71. The required airflow to maintain a given temperature tolerance, typically 5–15°F, may be calculated from [65,66]

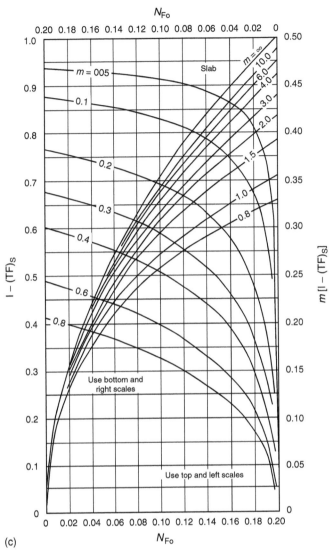

FIGURE 1.64 (Continued) (c) short-time temperature function (TF) relationships for slabs.

$$\text{Airflow (cfm)} = \frac{HA}{625.7U} \times T_A$$

where H is the heat loss in Btu through $1\,\text{ft}^2$ of furnace wall per hour, A is the furnace wall area (ft^2), T_A is the absolute furnace temperature ($460° + °\text{F}$), and U is the maximum allowable variation in furnace temperature ($°\text{F}$). This calculation assumes that the heated air has the necessary Btu content to heat the load for the furnace cycles calculated and that the heat losses through the furnace walls are included [65].

1.4.5 SOAKING TIME

Soaking time is dependent on (1) gas metrical factors relating to the particular furnace and load, (2) type of load, (3) type of steel, (4) thermal properties of the load, (5) load and furnace

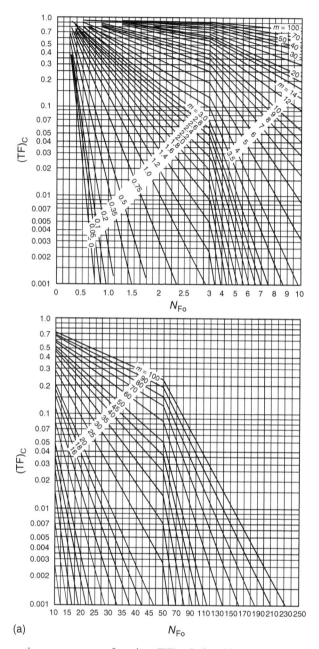

(a)

FIGURE 1.65 (a) Long-time temperature function (TF) relationships for spheres.

Continued

emissivities, (6) initial furnace and load temperatures, (7) characteristic fan curves, and (8) the chemical composition of the atmosphere [69,70].

Aronov et al. [69,70] modeled soaking times as a function of these parameters and developed menu-driven software for predicting furnace soaking times based on load characterization. Load characterization diagrams are provided in Figure 1.72. These models are based on the generalized equation for soaking time (T_s),

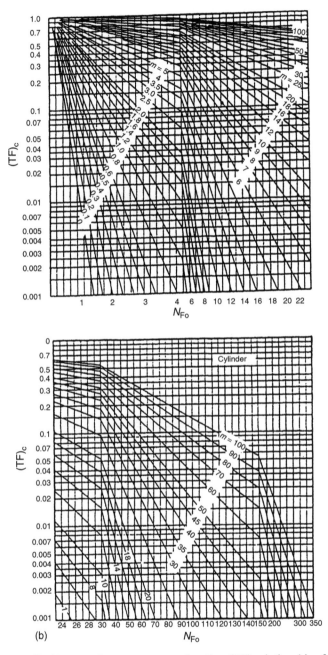

FIGURE 1.65 (Continued) (b) Long-time temperature function (TF) relationships for cylinders.

Continued

$$T_s = T_{sb}K$$

where T_s is the calculated soaking time (min), T_{sb} is the baseline soak temperature condition selected from graphs such as those in Figure 1.73a through Figure 1.73d [69,70], K is the correction factor for the type of steel ($K = 1$ for low-alloy steel and 0.85 for high-alloy steel).

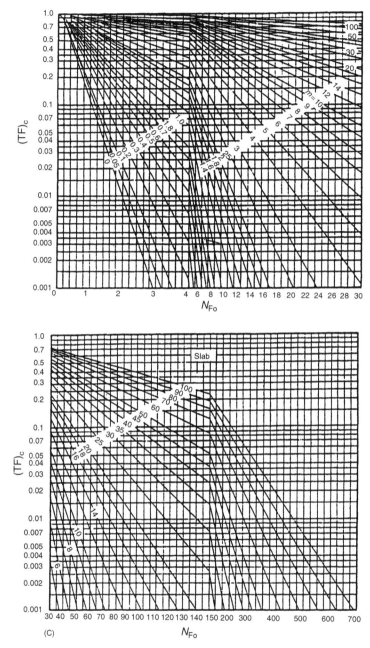

FIGURE 1.65 (Continued) (c) Long-time temperature functin (TF) relationships for slabs.

1.5 THERMOCOUPLES

Of the various methods of temperature measurement in the heat treatment shop, the use of thermocouples is one of the most common. The thermocouple is based on the thermoelectric effect that exists when two conductive wires (A and B) at different temperatures (t_1 and t_2) are connected to form a closed circuit. An electromotive force (emf) is developed whose magnitude and direction depend on the contacting materials and the temperature difference

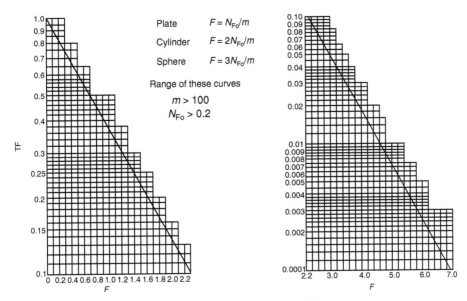

FIGURE 1.66 Temperature function (TF) when m is large ($m > 100$).

between the two points [71]. This is illustrated in Figure 1.74. When the wires depicted by A and B are different, current will flow as long as the temperatures t_1 and t_2 are different. This is called the Seebeck effect [73]. The Seebeck voltage (Δe_{ab}) is defined [74] as

$$\Delta e_{ab} = \alpha \Delta T$$

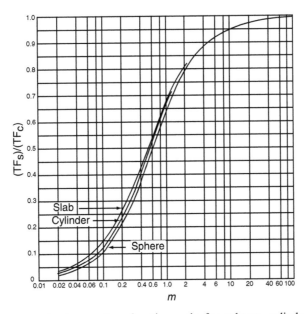

FIGURE 1.67 Surface to center temperature function ratio for spheres, cylinders, and slabs. (From V. Paschkis and J. Persson, *Industrial Electric Furnaces and Appliances*, Interscience, New York, 1960, pp. 14–25.)

where α is the Seebeck coefficient (see Table 1.16) and ΔT is the difference in temperature $(t_2 - t_1)$. Table 1.17 shows that in view of the small voltages involved, very sensitive measurement instruments are required [74].

It is not possible to measure the voltage (Δe_{AB}) by simply connecting the two wires to a voltmeter because the voltmeter itself introduces a significant junction potential. This problem is solved by adding the junction potential to a reference potential, which is typically taken as the freezing point of water ($T_{ref} = 0°C$). The measured junction potential now becomes [74]

$$V = \alpha(T_1 - T_{ref})$$

Most thermocouples use an external reference junction compensation with either hardware or software compensation instead of an ice-water bath [74].

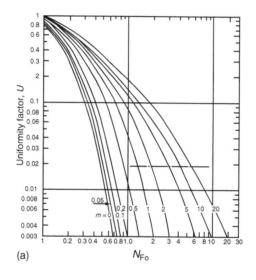

(a)

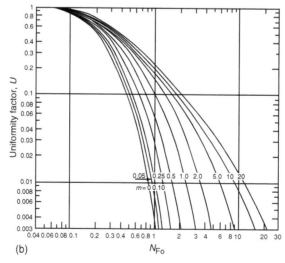

(b)

FIGURE 1.68 Uniformity factors for (a) spheres; (b) cylinders.

Continued

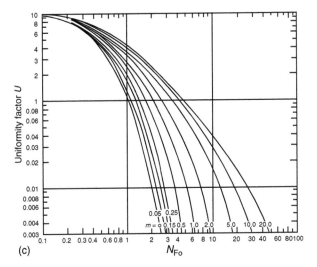

FIGURE 1.68 (Continued) Uniformity factors for (c) slabs.

Table 1.18 [76] summarizes the composition and maximum use temperature for various standard thermocouples [75]. However, the maximum recommended use temperature is dependent on the size of the thermocouple wire as shown in Figure 1.75 [75].

The voltage for a thermocouple may be read directly from voltmeter using either hardware or software compensation or calculated from [74]

$$T = a_0 + a_1 x + a_2 x^2 + a_3 x^3 + \cdots + a_n x^n$$

where T is temperature, x is the thermocouple voltage, the a's are thermocouple-dependent polynomial coefficients (Table 1.19), and n is the maximum order of the polynomial; as n increases, accuracy increases, e.g., when $n = a$, the accuracy is $\pm 1^\circ C$.

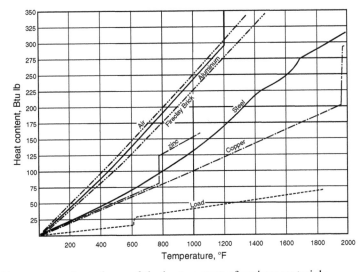

FIGURE 1.69 Temperature dependence of the heat content of various materials.

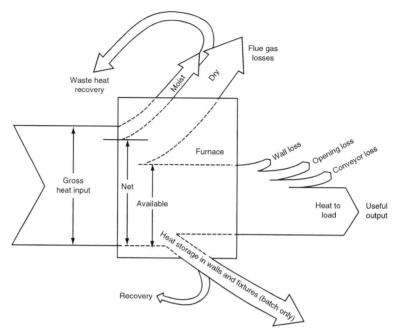

FIGURE 1.70 Illustration of the use of a Sankey diagram to track furnace heat losses.

The thermocouple response times are also dependent on the size of the wire as shown in Figure 1.75 [77] and the heat transfer medium as shown in Table 1.20.

There are three conventional styles of thermocouples as shown in Figure 1.76 [77].

1. *Exposed.* These thermocouples are used when very fast response times are necessary. They are characterized by their exposed thermocouple junctions.
2. *Grounded.* These thermocouples are characterized by grounding to the thermocouple sheath, which provides both excellent response time and protection of the thermocouple junction.

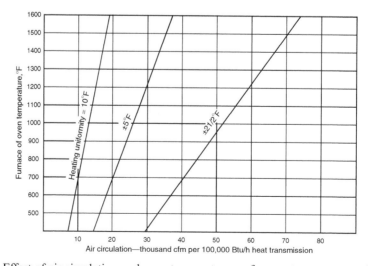

FIGURE 1.71 Effect of air circulation and oven temperature on furnace temperature uniformity.

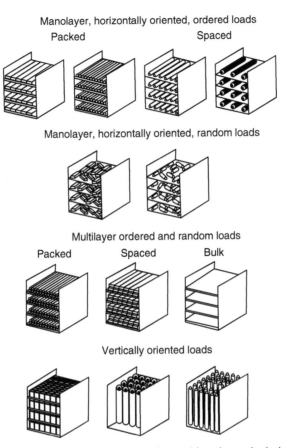

FIGURE 1.72 Aronov load characterization diagram for soaking time calculations.

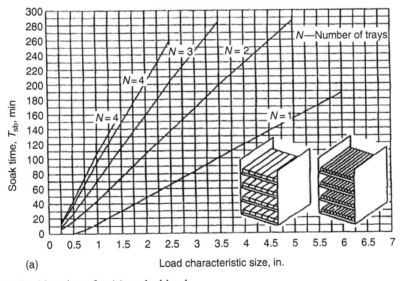

FIGURE 1.73 Soaking times for (a) packed load.

Continued

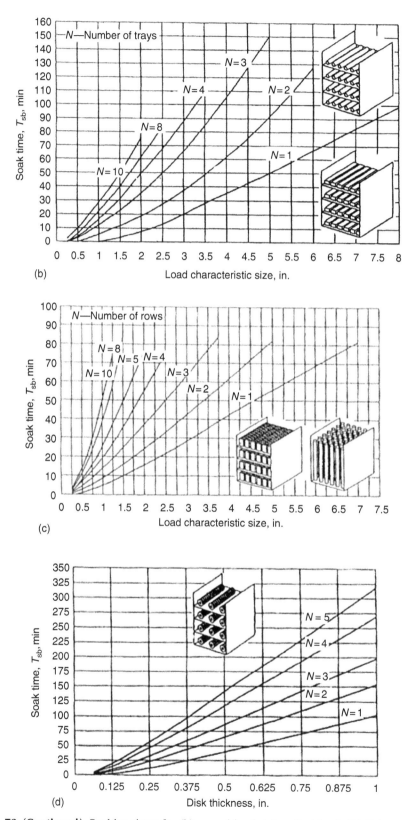

FIGURE 1.73 (Continued) Soaking times for (b) spaced load; (c) vertical load; (d) disks.

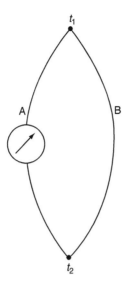

FIGURE 1.74 A thermocouple circuit. (From H.D. Baker, E.A. Ryder, and N.H. Baker, *Temperature Measurement in Engineering*, Vol. 1, Omega Press, Stamford, CT, 1975.)

3. *Ungrounded* (insulated). The thermocouple junction is electrically isolated from the protective sheath (which is usually stainless steel or Inconel) and is used when electrical noise hinders measurement of the thermocouple voltage. These thermocouples typically have somewhat slower response times than grounded thermocouples.

Thermocouple assemblies can be constructed in various ways such as those depicted in Figure 1.77. The most common types of insulation for high-temperature applications are fiberglass, fibrous silica, and asbestos. Asbestos, because of its toxicity, is no longer commonly used.

It is important that thermocouple measurement be both accurate and precise. It is possible, as illustrated in Figure 1.78, for thermocouple readings to be reproducible and precise but still wrong. Therefore, the temperature reading of a thermocouple should be traceable to an NIST standard [75]. The uncertainty of the thermocouple calibration is calculated from [75]

$$U_{TE} = [(U_{NIST})^2 - (R_{TE})^2]^{1/2}$$

where U_{TE} is the uncertainty of calibration in the user's laboratory, U_{NIST} is the uncertainty of calibration at NIST, and R is the reproducibility in the user's laboratory.

Ideally, thermocouples will be calibrated at temperatures between the use temperature and room temperature. Thermocouples of types J, K, E, and N may be calibrated at temperatures between ice water (32°F; 0°C), boiling water (212°F; 100°C), the melting point of tin (449°F; 232°C), and the melting point of zinc (787°F; 420°C) [73].

1.6 ATMOSPHERES

Furnace atmosphere selection, generation, and control are among the most important steps in the heat treatment operation. For example, control of oxide formation, facilitation of the formation of the desired steel surface chemistry, and prevention of decarburization are all

TABLE 1.16
Nominal Seebeck Coefficient (μV/°C) for Standard Thermocouples

Temperature (°C)	Thermocouple Type						
	E	J	K	R	S	T	B
−190	27.3	24.2	17.1	—	—	17.1	—
−100	44.8	41.4	30.6	—	—	28.4	—
0	58.5	50.2	39.4	—	—	38.0	—
200	74.5	55.8	40.0	8.8	8.5	53.0	2.0
400	80.0	55.3	42.3	10.5	9.5	—	4.0
600	81.0	58.5	42.6	1.5	10.3	—	6.0
800	78.5	64.3	41.0	12.3	1.0	—	7.7
1000	—	—	39.0	13.0	1.5	—	9.2
1200	—	—	36.5	13.8	12.0	—	90.3
1400	—	—	—	13.8	12.0	—	1.3
1600	—	—	—	—	1.8	—	1.6

critically important to ensure the overall success of the process, and all are integrally related to the proper selection and operation of the heat treatment equipment.

Furnace atmosphere surface reactions will vary with steel chemistry, process temperature and time, and the purity of the atmosphere itself. In some cases an atmosphere will be reactive with a steel surface, and in other cases the same atmosphere will be protective (nonreactive). Table 1.21 through Table 1.24 and Figure 1.79 provide a summary of some of the most common furnace atmospheres used in heat treatment and their properties [70].

In this section, an overview of furnace atmospheres, both primary furnace gases and controlled atmospheres, is provided, followed by classification, composition, properties, and atmosphere generation.

1.6.1 PRIMARY FURNACE GASES

1.6.1.1 Nitrogen

Nitrogen (N_2), an inert diatomic gas*, is the primary component (78.1%) of atmospheric air as shown in Table 1.25. The remaining components of air are oxygen (20.9%) and other gases in much lower concentrations (<1%). The physical properties of nitrogen and other gases used in furnace atmosphere preparation are summarized in Table 1.26A through Table 1.26D.

Nitrogen is considered to be chemically inert and is used as a carrier gas for reactive furnace atmospheres, for purging furnaces, and in other processes requiring inert gases. However, at high temperature, nitrogen may not be compatible with certain metals such as molybdenum, chromium, titanium, and columbium [80].

---

*The term inert gas is often misleading to both heat treaters and furnace designers. It is incorrectly assumed that furnace atmospheres of nitrogen will not undergo any reactions with steel surfaces. However, steel will be decarburized if it is held for extended periods of time at elevated temperature with a nitrogen atmosphere. To make atmospheres nonreactive and neutral to steel, an enriching gas such as methane or propane that will generate the same C_p as steel must be used with nitrogen. (These will be discussed subsequently.) If nonferrous materials are used, then a reducing gas such as hydrogen is required, but not more than 10%.

TABLE 1.17
Required Voltmeter Sensitivity

Thermocouple Type	Seebeck Coefficient ($\mu V/°C$ at 20°C)	Voltmeter Sensitivity for 0.1°C (μV)
E	62	6.2
J	51	5.1
K	40	4.0
R	7	0.7
S	7	0.7
T	40	4.0

Source: From Anon., *The Temperature Handbook*, Vol. 28, Omega Engineering Corp., Stamford, CT.

Nitrogen of high purity (<1 ppm O_2) can be produced by the oxidation of ammonia or in somewhat less pure form by the combustion of hydrocarbons in insufficient air (which removes O_2) and subsequent purification of residual hydrogen and methane [82]. Table 1.27 illustrates the purity of nitrogen obtained by these processes.

High-purity nitrogen and other gases are also obtained by air liquefaction. A schematic for this process is shown in Figure 1.80. The nitrogen thus obtained can be subsequently purified by palladium-catalyzed hydrogen reduction of the residual oxygen, typically present at a level of ~0.2%. Alternatively, residual oxygen can be removed by high-temperature reaction with copper to form copper oxide, which eliminates the possibility of water vapor contamination of the combustion process [81].

Residual water vapor, which is present as a by-product of this process, must be removed both to minimize vaporization (flash-off) upon expansion, as shown in Figure 1.81, and to ensure subsequent control of reactive gas chemistry in processes such as carburization.

TABLE 1.18
Standard Letter-Designated Thermocouples

Type	Thermoelements	Typical Alloy	Base Composition	Applications	
				Atmosphere	Max Temperature
J	JP	Iron	Fe	Oxidizing	760°C
	JN	Constantan (J)	44Ni/55Cu	Reducing	1000°F
K	KP	Chromel	90N/9Cr	Oxidizing	1260°C
	KN	Alumel	94Ni/Al, Mn, Si	Inert	2300°F
T	TP	Copper	OFHC/Cu	Oxidizing	370°C
	TN	Constantan (T)	44Ni/55Cr	Reducing	700°F
E	EP	Chromel	90Ni/9Cr	Oxidizing	870°C
	EN	Constantan (T)	44Ni/55Cr	Inert	1600°F
N	NP	Nicrosil	Ni/14.2Cr 1.4Si	Oxidizing	1260°C
	NN	Nisil	Ni/4Si/15Mg	Inert	
R	RP	Pt/Rh	87Pt/13Rh	Oxidizing	1480°C
	RN	Pt	Pt	Inert	2700°F
S	SP	Pt/Rh	90Pt/10Rh	Oxidizing	1480°C
	SN	Pt	Pt	Inert	2700°F
B	BP	Pt/Rh	70Pt/30Rh	Oxidizing	1700°C
	BN	Pt/Rh	94Pt/6Rh	Inert and vacuum	3100°F

TABLE 1.19
NBS Thermocouple Polynomial Coefficients (av)

Type E Ni-10%Cr (+)vs. Constantan (−) −100–1000°C	Type J Fe (+) vs. Constantan (−) 0°–760°C	Type K Ni-10%Cr (+) vs. Ni-5% (−) 0° ~ 1370°C	Type R Pt-Rh (+) vs. Pt (−) 0°–1000°C	Type S Pt-10% Rh (+) vs. Pt (−) 0°–1750°C	Type T Cu (+) vs. Constantan (−) 160 ± 400°C
n ±0.5°C	±0.1°C	±0.7°C	±0.5°C	±1°C	±0.5°C
0 0.104967248	−0.048868252	0.226584602	0.26362917	0.927763167	0.100860910
1 17189.45282	19873.14503	24152.10900	179075.491	169526.5150	25727.94369
2 −282639.0850	−218614.5353	67233.4248	−48840341.37	−31568363.94	−767345.8295
3 12695339.5	11569199.78	2210340.682	$1.90002E + 10$	8990730663	78025595.81
4 −448703084.6	−264917531.4	−860963914.9	$-4.82704E + 12$	$-1.635565E + 12$	−9247486589
5 $1.10866\,E + 10$	2018441314	$4.83506E + 10$	$7.62091E + 14$	$1.88027E + 14$	$6.97688E + 11$
6 $-1.76807\,E + 11$		$-1.18452E + 12$	$-7.20026E + 16$	$-1.37241E + 16$	$-2.66192E + 13$
7 $1.71842\,E + 12$		$1.38690E + 13$	$3.71496E + 18$	$6.17501E + 17$	$3.94078E + 14$
8 $-9.19278\,E + 12$		$-6.37708E + 13$	$-8.03104E + 19$	$-1.56105E + 19$	
9 $2.06132\,E + 13$				$1.69535\,E + 20$	

Nitrogen can also be obtained by membrane separation from air based on the selective permeability of a composite membrane fiber. In this process, atmospheric air is filtered, compressed, and cooled and then passed through an air-separation membrane as shown

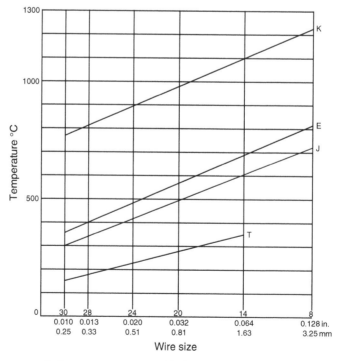

FIGURE 1.75 Recommended upper temperature limits for Type K, E, J, and T thermocouples of various sizes. (From Anon., *Manual on the Use of Thermocouples in Temperature Measurement*, ASTM STP 470 B, American Society for Testing and Materials, Philadelphia, PA, 1968.)

TABLE 1.20
Thermocouple Response Time (s) Variation with Wire and Heat Transfer Medium

Wire Size (in.)	Still Air (800°F/100°F)	Air at 60 ft/s (800°F/100°F)	Still Water (200°F/100°F)
0.001	0.05	0.004	0.002
0.005	1.0	0.08	0.04
0.015	10.0	0.80	0.40
0.032	40.0	3.2	1.6

Source: From Anon., *The Temperature Handbook*, Vol. 28, Omega Engineering Corp., Stamford, CT.

in Figure 1.82 [83,84]. Oxygen, carbon dioxide, and water vapor permeate the hollow membrane fibers and are then vented at low pressure to the atmosphere. Nitrogen is then stored in the desired form.

Liquid nitrogen is also of value for its refrigeration or cooling capabilities as shown in Figure 1.83 [81]. Liquid nitrogen (LN_2) processes are summarized in Table 1.28. Liquid nitrogen is subsequently vaporized to a gas before use [85]. On-site membrane separation system are available from various suppliers in various gas separation capacities.

1.6.1.2 Hydrogen

In addition to being used as a quenchant, hydrogen is a highly reducing atmosphere that is used both for preventing steel oxidation and for oxide reduction according to the surface reactions

$$H_2 + FeO \rightarrow H_2O + 3Fe$$

and

$$H_2 + Fe_3O_4 \rightarrow H_2O + 3FeO$$

The thermal properties of hydrogen, relative to other heat treatment atmospheres, are summarized in Table 1.26A through 1.26D. At temperatures greater than 1300°F (700°C), hydrogen may cause decarburization to form methane by reaction with carbon:

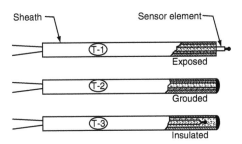

FIGURE 1.76 Common thermocouple junction styles.

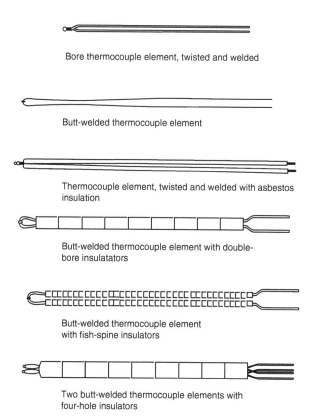

FIGURE 1.77 Thermocouple element assemblies. (From Anon., *Manual on the Use of Thermocouples in Temperature Measurement*, ASTM STP 470 B, American Society for Testing and Materials, Philadelphia, PA, 1968.)

$$C + 2H_2 \leftrightharpoons CH_4$$

In some cases, hydrogen may be adsorbed by the metal at elevated temperatures, causing hydrogen embrittlement [82].

Hydrogen is potentially an extremely explosive and flammable gas. However, if proper safety precautions are followed, it can be used safely in heat treatment.

Hydrogen can be produced by electrolysis of water:

$$2H_2O \rightarrow 2H_2 + O_2$$

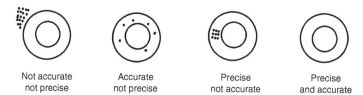

FIGURE 1.78 Illustration of accuracy vs. precision. (From T.P. Wang, *Heat Treating: Equipment and Process* (G.E. Totten and R.A. Wallis, Eds.), Proc. 1994 Conf., ASTM International Materials Park, OH, 1994, pp. 171–174.)

TABLE 1.21
Heat Content of Various Gases above 77°F (Btu/ft³ at 60°F and 30 in. Hg)

Gas Temperature (°F)	CO	CO₂	CH₄	H₂	N₂	O₂	H₂O	AXᵃ	DXᵇ	DXᶜ
100	0.4	0.6	0.5	0.4	0.4	0.4	0.5	0.4	0.4	0.4
200	2.3	3.0	2.9	2.3	2.3	2.3	2.6	2.3	2.3	2.3
300	4.1	5.6	5.4	4.1	4.1	4.2	4.8	4.1	4.2	4.3
400	6.0	8.3	8.2	6.0	6.0	6.1	7.0	6.0	6.1	6.2
500	7.9	1.2	1.1	7.8	7.9	8.1	9.2	7.8	8.0	8.2
600	9.8	14.1	14.3	9.7	9.7	10.1	1.5	9.7	10.0	10.2
700	1.7	17.1	17.7	1.5	1.7	12.2	13.8	1.6	1.9	12.2
800	13.7	20.2	21.3	13.4	13.6	14.2	16.1	13.4	13.9	14.2
900	15.7	23.4	25.0	15.2	15.5	16.3	18.5	15.3	15.9	16.3
1000	17.7	26.6	29.0	17.1	17.5	18.5	21.0	17.2	18.0	18.4
1100	19.7	29.9								
1200	21.8	33.2	37.4	20.9	21.6	22.8	26.0	21.0	22.1	22.7
1300	23.8	36.6	41.9	22.8	23.6	25.0	28.5	23.0	24.2	24.9
1400	25.9	40.1	46.5	24.7	25.7	27.2	31.2	24.9	26.3	27.1
1500	28.1	43.5	51.2	26.6	27.8	29.4	33.8	26.9	28.5	29.3
1600	30.2	47.1	56.1	28.5	29.9	31.6	36.5	28.9	30.7	31.6
1700	32.3	50.6	61.1	30.5	32.0	33.9	39.3	30.9	32.8	33.8
1800	34.5	54.2	66.2	32.5	34.1	36.1	42.1	32.9	35.0	36.1
1900	36.7	57.8	71.4	34.5	36.3	38.4	44.9	34.9	37.2	38.4
2000	38.9	61.4	76.7	36.5	38.4	40.7	47.8	37.0	39.5	40.7
2100	41.1	65.0	82.1	38.5	40.6	43.0	50.7	39.0	41.7	43.0
2200	43.3	68.7	87.6	40.5	42.8	45.3	53.6	41.1	44.0	45.4
2300	45.5	72.4	93.2	42.5	45.0	47.6	56.6	43.2	46.2	47.7
2400	47.8	76.1	98.9	44.6	47.2	49.9	59.6	45.3	48.5	50.1
2500	50.0	79.8	104.7	46.7	49.4	52.2	62.7	47.4	50.8	52.5
2600	52.3	83.6	110.5	48.7	51.7	54.6	65.7	49.5	53.1	54.8
2700	54.5	87.3	116.4	50.8	53.9	56.9	68.9	51.6	55.4	57.2
2800	56.8	91.1	122.4	53.0	56.2	59.3	72.0	53.8	57.7	59.6
2900	59.1	94.9	128.4	55.1	58.4	61.6	75.2	55.9	60.0	62.0
3000	61.3	98.7	134.5	57.2	60.7	64.0	78.3	58.1	62.4	64.5
3100	63.6	102.5	140.7	59.4	63.0	66.4	81.6	60.3	64.7	66.9
3200	65.9	106.3	146.9	61.5	65.2	68.8	84.8	62.4	67.1	69.3
3300	68.2	110.1	153.1	63.7	67.5	71.2	88.1	64.6	69.4	71.7
3400	70.5	114.0	159.4	65.9	69.8	73.6	91.4	66.8	71.8	74.2
3500	72.8	117.8	165.7	68.0	72.1	76.0	94.7	69.1	74.2	76.6
3600	75.2	121.7	172.1	70.2	74.4	78.4	98.0	71.3	76.5	79.1
3700	77.5	125.6	178.5	72.5	76.7	80.9	101.4	73.5	78.9	81.5
3800	79.8	129.4	185.0	74.7	79.09	83.3	104.7	75.8	81.3	84.0
3900	82.1	133.3	191.5	76.9	81.3	85.8	108.1	78.0	83.7	86.5
4000	84.4	137.2	198.0	79.1	83.6	88.2	11.5	80.3	86.1	88.9
4100	86.8	141.1	204.6	81.4	85.9	90.7	115.0	82.5	88.5	91.4
4200	89.1	145.0	21.1	83.7	88.3	93.2	118.4	84.8	90.9	93.9
4300	91.4	148.9	217.8	85.9	90.6	95.6	121.9	87.1	93.3	96.4
4400	93.8	152.8	224.4	88.2	92.9	98.1	125.3	89.4	95.7	98.9
4500	96.1	156.7	231.1	90.5	95.2	100.6	128.8	91.7	98.1	101.3
4600	98.5	160.6	237.8	92.8	97.6	103.1	132.3	94.0	100.5	103.8
4700	100.8	164.5	244.5	95.1	99.9	105.6	135.9	96.3	102.9	106.3

Continued

TABLE 1.21 (Continued)
Heat Content of Various Gases above 77°F (Btu/ft^3 at 60°F and 30 in. Hg)

Gas Temperature (°F)	CO	CO$_2$	CH$_4$	H$_2$	N$_2$	O$_2$	H$_2$O	AX[a]	DX[b]	DX[c]
4800	103.2	168.5	251.2	97.4	102.2	108.1	139.4	98.6	105.4	108.8
4900	105.5	172.4	258.0	99.7	104.6	110.7	142.9	100.9	107.8	11.3
5000	107.9	176.3	254.8	102.0	106.9	113.2	146.5	103.2	110.2	113.8

[a]AX = 75.0% H$_2$, 25.0% N$_2$.
[b]12.0% H$_2$, 72.8% N$_2$.
[c]1.0% H$_2$, 88.0% N$_2$.
Source: From *Atmospheres for Heat Treating Equipment*, Brochure, Surface Combustion, Inc., Maumee, OH.

Hydrogen produced in this way must be further purified to remove traces of contaminants, CO$_2$ and especially O$_2$ that may be present. The dew point of hydrogen is dependent on the degree of contamination as shown in Table 1.29.

The highest purity hydrogen is prepared by ammonia dissociation:

$$2NH_3 \rightarrow 3H_2 + N_2$$

The most common impurities in hydrogen are O$_2$ and H$_2$O. Purification is performed by filtration over palladium, which traps all gases except hydrogen, the smallest molecule. Residual water vapor can be removed by passing hydrogen through either silica gel or a molecular sieve column [82].

1.6.1.3 Carbon Monoxide

Carbon monoxide is also considered to be a reducing gas as it may reduce iron oxide:

$$CO + FeO \leftrightharpoons Fe + CO_2$$

Although CO is a reducing atmosphere, it is not as good a reducing agent as hydrogen. The thermal properties of CO are given in Table 1.26A through Table 1.26D [78]. The preparation of CO is discussed later.

1.6.1.4 Carbon Dioxide

Carbon dioxide is a mildly oxidizing gas. It will form oxides upon reaction with iron at elevated temperatures. When the temperature is greater than 1030°F (540°C), FeO is formed [86]:

$$Fe + CO_2 \leftrightharpoons FeO + CO$$

When the temperature is less than 1030°F (540°C),

$$3FeO + CO_2 \leftrightharpoons Fe_3O_4 + CO$$

Decarburization may also result from the reaction of CO$_2$ with carbides of iron or free carbon [86]:

TABLE 1.22
Characteristics of Simple Gases

Simple Gases and Compound	Critical Temperature (°F)	Critical Pressure (psia)	Inflammability Lower	Inflammability Upper	Ignition Temperature (°F)	Combustion Velocity of Maximum Speed Mixture (ft/s)	Toxicity — Maximum Amt. Inhaled for 1 h Without Serious Disturbance (ppm)	Toxicity — Dangerous in 30 min 1 h (ppm)	Toxicity — Rapidly Fatal (ppm)	Solubility H₂O at 60°F, 30 in Hg	Thermal Conductivity [Btu/(ft²°F in °s)]	Specific Gravity of the Liquid at 60°F	Heat of Vaporization at 60°F (Btu/lb)
H_2	−400	188	4.1	74	1,076–1,094	8.2	—	Simple asphyxiant	—	0.0_3167	3.05×10^{-4}	—	—
O_2	−181	731	—	—	—	—	—	—	—	0.0_49	4.47×10^{-5}	—	—
N_2	−233	492	—	—	—	—	—	Simple asphyxiant	—	0.0_422	4.38×10^{-5}	—	—
CO	−218	515	12.5	74	1,191–1,216	1.6	1,000–1,200	1500–2000	4,000	Very slight	4.12×10^{-5}	—	—
CO_2	88	1,073	—	—	—	—	—	5–7% respiratory stimulant	—	0.090	2.63×10^{-5}	—	—
CH_4	−116	673	5.3	14.0	1,200–1,382	1.2	—	Simple asphyxiant	—	—	5.64×10^{-5}	—	—
C_2H_6	90	717	3.2	12.5	968–1,166	—	—	Simple asphyxiant	—	—	3.46×10^{-5}	0.38	223
C_3H_8	204	632	2.4	9.5	~965	1.03	—	Anesthetic	—	—	—	0.51	210
$n\text{-}C_4H_{10}$	308	529	1.9	8.5	~930	—	—	Anesthetic	—	—	—	0.58	183
$iso\text{-}C_4H_{10}$	273	544	—	—	—	—	—	—	—	—	—	0.56	166
$n\text{-}C_5H_{12}$	387	485	1.4	8.0	~890	—	—	Anesthetic, convulsive, irritant	—	—	223×10^{-5}	0.63	159
$iso\text{-}C_5H_{12}$	370	482	—	—	—	—	—	Anesthetic, convulsive, irritant	—	—	—	—	153
C_6H_{14}	455	434	—	—	—	—	—	Anesthetic, convulsive, irritant	—	—	1.98×10^{-5}	0.66	143
C_2H_4	49	748	3.3	34	1,000–1,020	2.1	—	Simple asphyxiant and anesthetic	—	—	3.14×10^{-5}	—	—
C_3H_6	198	662	2.2	10	—	—	—	Anesthetic	—	—	—	—	—
C_4H_8	—	—	1.7	9	—	—	—	Anesthetic	—	—	—	—	—
C_2H_2	97	911	2.5	80	763–824	4.1	—	Simple asphyxiant and anesthetic	—	—	332×10^{-5}	—	—
C_6H_6	551	701	1.4	8.0	1,364	—	3,100–4,700	—	19,000	—	1.60×10^{-5}	0.88	—
C_7H_8	609	612	1.3	6.75	1,490	—	3,100–4,700	—	19,000	—	—	—	—
C_8H_{10}	—	—	—	—	—	—	3,100–4,700	—	19,000	—	—	—	—
$C_{10}H_8$	—	—	—	—	—	—	—	—	—	—	—	—	—
NH_3	270	1,639	16	27	—	—	300–500	—	5,000–10,000	0.612	384×10^{-5}	—	—
H_2S	212	1,307	—	—	—	—	200–300	500–700	1,000–3,000	0.00466	2.30×10^{-5}	—	—
H_2O	706	3,226	—	—	—	—	—	—	—	—	4.17×10^{-5}	—	1058
Air	−285	547	—	—	—	—	—	—	—	—	4.28×10^{-5}	—	—

Source: From Protective Atmospheres and Analysis Curves, Brochure, Electric Furnace Company, Salem, OH.

TABLE 1.23
Proprties of Typical Commercial Gases

Constituents of Gas (% v/v)

No.	Gas	CO₂	O₂	N₂	CO	H₂	CH₄	C₂H₄	Illuminants C₂H₄	Illuminants C₂H₄	Spec. grav.	Air ref. for Comb. of 1 ft³ gas (ft³)	Btu/ft³ Gross	Btu/ft³ Net	H₂O	CO₂	N₂	Total	Ultimate % CO₂	Btu (net)/ft³ Prod. of Comb.	Flame Temp. n excess a (°F)
1	Natural gas (Birmingham)	—	—	5.0	—	—	90.0	5.0	—	—	0.60	9.41	1002	904	2.02	1.00	7.48	10.50	1.8	86.0	3565
2	Natural gas (Pittsburgh)	—	—	0.8	—	—	83.4	15.8	—	—	0.61	10.58	1129	1021	2.22	1.15	8.37	1.73	12.1	87.0	3562
3	Natural gas (South California)	0.7	—	0.5	—	—	84.0	14.8	—	—	0.64	10.47	1116	1009	2.20	1.14	8.28	1.62	12.1	87.0	3550
4	Natural gas (Los Angeles)	6.5	—	—	—	—	77.5	16.0	—	—	0.70	10.05	1073	971	2.10	1.16	7.94	1.20	12.7	86.7	3550
5	Natural gas (Kansas City)	0.8	—	8.4	—	—	84.1	6.7	—	—	0.63	9.13	974	879	1.95	0.98	7.30	10.23	1.9	86.0	3535
6	Reformed natural gas	1.4	0.2	2.9	9.7	46.6	37.1	—	1.3	(C₃H₆ 0.8)	0.41	5.22	599	536	1.30	0.53	4.16	5.59	1.3	89.6	3615
7	Mixed natural and water gas	4.4	2.1	4.7	25.5	35.1	23.1	4.7	0.2	0.2	0.61	4.43	525	477	1.01	0.64	3.55	5.20	15.3	91.7	3630
8	Coke oven gas	2.2	0.8	8.1	6.3	46.5	32.1	—	3.5	0.5	0.44	4.99	574	514	1.25	0.51	4.02	5.78	1.2	87.0	3610
9	Coal gas (continuous verticals)	3.0	0.2	4.4	10.9	54.5	24.2	—	1.5	1.3	0.42	4.53	532	477	1.15	0.49	3.62	5.26	1.9	90.7	3645
10	Coal gas (inclined retorts)	1.7	0.8	8.1	7.3	49.5	29.2	—	0.4	3.0	0.47	5.23	599	540	1.23	0.57	4.21	6.01	1.9	89.9	3660
11	Coal gas (intermittent verticals)	1.7	0.5	8.2	6.9	49.7	29.9	—	3.0	0.1	0.41	4.64	540	482	1.21	0.45	3.75	5.41	10.7	89.0	3610
12	Coal gas (horizontal retorts)	2.4	0.75	1.35	7.35	47.95	27.15	—	1.32	1.73	0.47	4.68	542	486	1.15	0.50	3.81	5.46	1.6	89.0	3600

Products of Combustion per Cubic Foot of Gas (ft³): H₂O, CO₂, N₂, Total

#	Gas	C1	C2	C3	C4	C5	C6	C7	C8	C9	C10	C11	C12	C13	C14	C15	C16	C17	C18	C19
13	Mixed coke oven and carbureted water gas	3.4	0.3	12.0	17.4	36.8	24.9	3.7	1.5	0.58	4.71	545	495	1.04	0.62	3.85	5.51	13.9	90.0	3630
14	Mixed coal, coke oven, and carbureted water gas	1.8	1.6	13.6	9.0	42.6	28.0	2.4	1.0	0.50	4.52	528	475	1.11	0.50	3.71	5.32	1.8	89.3	3640
15	Carbureted water gas	3.0	0.5	2.9	34.0	40.5	10.2	6.1	2.8	0.63	4.60	550	508	0.87	0.76	3.66	5.29	17.2	96.2	3725
16	Carbureted water gas	4.3	0.7	6.5	32.0	34.0	15.5	4.7	2.3	0.67	4.51	534	493	0.75	0.86	3.63	5.24	17.1	94.2	3700
17	Carbureted water gas (low gravity)	2.8	1.0	5.1	21.0	47.5	15.0	5.2	2.4	0.54	4.61	549	501	0.98	0.64	3.70	5.31	14.7	94.3	9690
18	Water gas (coke)	5.4	0.7	8.3	37.0	47.3	1.3	—	—	0.57	2.10	287	262	0.53	0.44	1.74	2.71	20.1	96.6	3670
19	Water gas (bituminous)	5.5	0.9	27.6	28.2	32.5	4.6	0.4	0.3	0.70	2.01	261	239	0.47	0.41	1.86	2.74	18.0	87.2	3510
20	Oil gas (Pacific coast)	4.7	0.3	3.6	12.7	48.6	26.3	2.7	1.1	0.47	4.73	551	496	1.15	0.56	3.77	5.48	12.9	90.5	3630
21	Producer gas (buckwheat anthracite)	8.0	0.1	50.0	23.2	17.7	1.0	—	—	0.86	1.06	143	133	0.22	0.32	1.34	1.88	19.4	70.5	3040
22	Producer gas (bituminous)	4.5	0.6	50.9	27.0	14.0	3.0	—	—	0.86	1.23	163	153	0.23	0.35	1.48	2.06	18.9	74.6	3175
23	Producer gas (0.6 lb steam/lb coke)	6.4	—	52.8	27.1	13.3	0.4	—	—	0.88	1.00	135	128	0.17	0.34	1.32	1.82	20.5	70.3	3010
24	Blast furnace gas	1.5	—	60.0	27.5	1.0	—	—	—	1.02	0.68	92	92	0.02	0.39	1.14	1.54	25.5	59.5	2650
25	Commercial butane	—	—	C_4H_{10}(93.0)	($C_3H_8$7.0)				1.95	30.47	3225	2977	4.93	3.93	24.07	32.93	14.0	90.5	3640	—
26	Commercial propane	—	—	($C_3H_8$100.0)					1.52	23.82	2572	2371	4.17	3.00	18.82	25.99	13.7	91.2	3660	—

Source: From *Protective Atmospheres and Analysis Curves*, Brochure, Electric Furnace Company, Salem, OH.

TABLE 1.24
Gas Combustion Constants

No.	Gas	Formula	Molecular Weight	lb/ft^3	ft^3/lb	Specific Gravity (Air = 1000)	Heat of Combustion Btu/ft^3 Gross	Net	Btu/lb Gross	Net
1	Carbon	C	12.000	—	—	—	—	—	14,140	14,140
2	Hydrogen	H$_2$	2.015	0.005327	187.723	0.06959	323.8	275.1	61,100	51,643
3	Oxygen	O$_2$	32.000	0.08461	11.819	1.1053	—	—	—	—
4	Nitrogen (atmos.)	N$_2$	28.016	0.07439	13.443	0.9718	—	—	—	—
5	Carbon monoxide	CO	28.000	0.07404	13.506	0.9672	323.5	323.5	4,369	4,369
6	Carbon dioxide	CO$_2$	44.000	0.1170	8.548	1.5282	—	—	—	—
Paraffin series C$_n$H$_{2n+2}$										
7	Methane	CH$_4$	16.031	0.04243	23.565	0.5543	1014.7	913.8	2,3912	21,533
8	Ethane	C$_2$H$_6$	30.046	0.08029	12.455	1.04882	1781	1631	22,215	20,312
9	Propane	C$_3$H$_8$	44.062	0.1196	8.365	1.5617	2572	2371	21,564	19,834
10	Isobutane	C$_2$H$_{10}$	58.077	0.1582	6.321	2.06654	3251	2999	21,247	19,606
11	*n*-Butane	C$_4$H$_{10}$	58.077	0.1582	6.321	2.06654	3353	3102	21,247	19,606
12	*n*-Pentane	C$_5$H$_{12}$	72.092	0.1904	5.252	2.4872	3981	3679	20,908	19,322
13	*n*-Hexane	C$_6$H$_{14}$	86.107	0.2274	4.398	2.9704	4667	4315	20,526	18,976
Olefin series C$_n$H$_{2n}$										
14	Ethylene	C$_2$H$_4$	28.031	0.07456	13.412	0.9740	631	1530	21,884	20,525
15	Propylene	C$_3$H$_6$	42.046	0.1110	9.007	1.4504	2336	2185	21,042	19,683
16	Butylene	C$_4$H$_8$	56.062	0.1480	6.756	1.9336	3135	2884	20,840	19,481
17	Acetylene	C$_2$H$_2$	26.015	0.06971	14.344	0.9107	1503	1453	21,572	20,840
Aromatic series C$_n$H$_{2n-6}$										
18	Benzene	C$_6$H$_6$	78.046	0.2060	4.852	2.6920	3741	3590	18,150	17,418
19	Toluene	C$_7$H$_8$	92.062	0.2431	4.113	3.1760	4408	4206	18,129	17,301
20	Xylene	C$_8$H$_{10}$	106.077	—	—	—	5155	—	18,410	—
21	Naphthalene	C$_{10}$H$_8$	128.062	—	—	—	5589	—	17,298	—
Miscellaneous gases										
22	Ammonia	NH$_3$	17.031	0.04563	21.914	0.5961	433	—	9,598	—
23	Hydrogen sulfide	H$_2$S	34.080	0.09109	10.979	1.189	672	—	7,479	—
24	Sulfur dioxide	SO$_2$	64.06	0.1733	5.770	2.264	—	—	—	—
25	Water vapor	H$_2$O	18.015	0.04758	21.017	0.6215	—	—	—	—
26	Air	—	28.9	0.07655	13.063	1.0000	—	—	—	—

$$Fe_3C + CO_2 \leftrightharpoons 3Fe + 2CO$$

$$C + CO_2 \leftrightharpoons 2CO$$

The properties of CO_2 are listed in Table 1.26. The generation of CO_2 is discussed in a subsequent section.

1.6.1.5 Argon and Helium

Helium and argon are also considered to be inert gases for heat treatment processes because they will not undergo gas–solid reactions, even at high temperatures.

Argon is a significant component of air as shown in Table 1.25 and is obtained in high purity by an air separation process. High-purity argon (>99.999% Ar) contains less than 2 ppm O_2 and less than 10 ppm N_2 and has a dew point of $-110°C$.

| Vol. Gas (ft³)/ft³ | | | | | | Mass Gas (lb)/lb | | | | | |
| Required for Combustion | | | Flue Products | | | Required for Combustion | | | Flue Products | | |
O₂	N₂	Air	CO₂	H₂O	N₂	O₂	N₂	Air	CO₂	H₂O	N₂
—	—	—	—	—	—	2.667	8.873	11.540	3.667	—	8.873
0.5	1.882	2.382	—	1.0	1.882	7.939	26.414	34.353	—	8.939	26.414
—	—	—	—	—	—	—	—	—	—	—	—
0.5	1.882	2.382	1.0	—	1.882	0.571	1.900	2.471	1.571	—	1.900
—	—	—	—	—	—	—	—	—	—	—	—
2.0	7.528	9.528	1.0	2.0	7.528	3.992	13.282	17.274	2.745	2.248	13.282
3.5	13.175	16.675	2.0	3.0	13.175	3.728	12.404	16.132	2.929	1.799	12.404
5.0	18.821	23.821	3.0	4.0	18.821	3.631	12.081	15.712	2.996	1.635	12.081
6.5	24.467	30.967	4.0	5.0	24.467	3.581	11.914	15.495	3.030	1.551	11.914'
6.5	24.467	30.967	4.0	5.0	24.467	3.581	11.914	15.495	3.030	1.551	11.914
8.0	30.114	38.114	5.0	6.0	30.114	3.551	11.815	15.366	3.052	1.499	11.815
9.5	35.760	45.260	6.0	7.0	35.760	3.530	11.745	15.275	3.067	1.465	11.745
3.0	11.293	14.293	2.0	2.0	11.293	3.425	11.935	14.820	3.139	1.285	11.395
4.5	16.939	21.439	3.0	3.0	16.939	3.425	11.395	14.820	3.139	1.285	11.395
6.0	22.585	28.585	4.0	4.0	22.585	3.425	11.395	14.820	3.139	1.285	11.395
2.5	9.411	11.911	2.0	1.0	9.411	3.075	10.231	13.306	3.383	0.692	10.231
7.5	28.232	35.732	6.0	3.0	28.232	3.075	10.231	13.306	3.383	0.692	10.231
9.0	33.878	42.887	7.0	4.0	33.878	3.128	10.407	13.535	3.346	0.783	10.407
10.5	39.524	50.024	8.0	5.0	39.524	3.168	10.540	13.708	3.318	0.849	10.540
12.0	45.170	57.170	10.0	4.0	45.170	2.999	9.978	12.977	3.436	0.563	9.978
—	—	—	—	—	—	—	—	—	—	—	—
1.5	5.646	7.146	SO₂ = 1.0	1.0	5.646	1.408	4.685	6.093	SO₂ = 1.880	0.529	4.085
—	—	—	—	—	—	—	—	—	—	—	—
—	—	—	—	—	—	—	—	—	—	—	—
—	—	—	—	—	—	—	—	—	—	—	—

Helium concentration in air is insignificant, too little for air to be a commercial source of this gas. Instead, natural gas, which contains 5–8% helium, is the commercial source of helium. Air liquefaction is used to separate the helium, which has a dew point of $-100°C$. Physical properties of argon and helium are provided in Table 1.30.

1.6.1.6 Dissociated Ammonia

From ammonia dissociation, which occurs at temperatures of more than 300°C in the presence of a catalyst such as Fe or Ni, it is possible to obtain mixtures of hydrogen and nitrogen that are free of oxygen contamination (see Figure 1.84) [79]. This may be a critically important requirement for some heat treatment processes.

$$2NH_3 \rightarrow N_2 + 3H_2$$

The relative concentrations of nitrogen and hydrogen may be varied by subsequent burning of the hydrogen. The most common atmospheres obtained by ammonia dissociation are summarized in Table 1.31.

TABLE 1.25
Composition of Atmospheric Air

Gaseous Component	Concentration, Dry Basis	
	vol%	ppm
Fixed		
Nitrogen (N_2)	78.084	—
Oxygen (O_2)	20.9476	—
Argon (Ar)	0.934	—
Neon (Ne)	—	18.18
Helium (He)	—	5.24
Krypton (Kr)	—	1.14
Xenon (Xe)	—	0.087
Variable		
Carbon dioxide (CO_2)	—	30–400
Nitrous oxide (N_2O)	—	0.5
Nitrogen dioxide (NO_2)	—	0–0.22
Water (H_2O)	1.25[a]	—
Hydrogen (H_2)	—	0.5 Type
Carbon monoxide (CO)	—	1 Type
Methane (CH_4)	—	2 Type
Ethane (C_2H_6)	—	<0.1 Type
Other hydrocarbons (C_nH_{2n+2})	—	<0.1 Type

[a]The composition of water in atmosphere air can be highly variable ranging from 0.1- > 4 %.
However, a "typical" value is approximately 1.25%
Source: From *Bulk Gases for the Electronics Industry*, Brochure, Praxair Inc., Chicago, IL.

Residual ammonia, typically less than 0.05%, is the primary impurity in dissociated ammonia atmospheres [81]. Residual ammonia is removed using the same methods as those used for water vapor removal. Refrigeration or adsorption in regenerative dryers are the most efficient.

1.6.1.7 Steam

Water vapor (steam) is also an important component in heat treating. As originally reported by Barff [87], steam will react with steel at 650–1200°F (343–650°C) to produce a blueing effect, which imparts a wear-resistant and oxidation-resistant surface furnish. This is due to the formation of either Fe_2O_3, Fe_3O_4, or FeO, depending on the surface temperature of the steel and the ratio of water vapor pressure to hydrogen pressure in the atmosphere as illustrated in Figure 1.85.

The concentration of water vapor is quantified by the dew point, which is the temperature at which a gas is saturated with water vapor (100% relative humidity) [86]. The relationship between dew point and atmospheric temperature is shown in Figure 1.86. As discussed previously, the concentrations of CO, CO_2, H_2O, and H_2 in a furnace atmosphere are interdependent as shown by the water gas reaction:

$$CO + H_2O \leftrightharpoons CO_2 + H_2$$

The equilibrium constant for this process, which defines the actual concentration of these gases in the furnace atmosphere, is written as

TABLE 1.26A
Viscosity (μ) of Gases[a] [μlb/(ft h)]

Temperature (°F)	Air	CO	CO_2	H_2	N_2	O_2	Steam	DX (Lean)	DX (Rich)	RX
100	0.0462	0.0432	0.0379	0.0223	0.0440	0.052	—	0.0432	0.0430	0.0427
200	0.0520	0.0487	0.0439	0.0249	0.0496	0.059	—	0.0488	0.0485	0.0480
300	0.0575	0.054	0.0497	0.0273	0.055	0.066	0.0359	0.0544	0.0539	0.0533
400	0.0626	0.059	0.055	0.0297	0.060	0.072	0.0408	0.0590	0.0587	0.0577
500	0.0675	0.063	0.060	0.0319	0.064	0.077	0.0455	0.0636	0.0635	0.0621
600	0.0722	0.068	0.065	0.0341	0.069	0.082	0.051	0.0681	0.0680	0.0665
700	0.767	0.072	0.070	0.0361	0.073	0.087	0.056	0.0727	0.0725	0.0708
800	0.0810	0.076	0.075	0.0380	0.077	0.092	0.061	0.0768	0.0765	0.0745
900	0.0852	0.080	0.079	0.0399	0.081	0.096	0.065	0.0809	0.0804	0.0784
1000	0.0892	0.083	0.083	0.0419	0.085	0.0100	0.069	0.0849	0.0844	0.0821
1100	0.0932	0.086	0.087	0.0438	0.088	0.105	0.073	0.0883	0.0878	0.0854
1200	0.0970	0.089	0.091	0.0458	0.092	0.109	0.077	0.0917	0.0912	0.0887
1300	0.101	0.093	0.095	0.0477	0.095	0.113	0.081	0.0952	0.0946	0.0921
1400	0.104	0.096	0.099	0.0496	0.099	0.117	0.085	0.0983	0.0977	0.0951
1500	0.108	0.099	0.103	0.051	0.101	0.120	0.089	0.1014	0.1008	0.0981
1600	0.111	0.102	0.107	0.053	0.104	0.123	0.093	0.1046	0.1039	0.1011
1700	0.115	0.105	0.110	0.055	0.108	0.126	0.097	0.1079	0.1072	0.1045
1800	0.118	0.108	0.113	0.056	0.111	0.128	0.101	0.1112	0.1104	0.1071
1900	0.121	0.111	0.116	0.058	0.114	0.130	0.105	0.1145	0.1137	0.1105
2000	0.124	0.114	0.119	0.059	0.117	0.132	0.109	0.1178	0.1170	0.1138

[a]DX, RX defined in Table 1.33.
Source: From *Atmospheres for Heat Treating Equipment*, Brochure, Surface Combustion, Inc., Maumee, OH.

TABLE 1.26B
Thermal Conductivity (k) of Gases [klb/(ft h)]

Temperature (°F)	Air	CO	CO_2	H_2	N_2	O_2	Steam	DX (Lean)	DX (Rich)	RX
100	0.0514	0.0142	0.0101	0.109	0.0151	0.0157	—	0.0150	0.0212	0.0404
200	0.0174	0.160	0.0125	0.122	0.0170	0.0180	—	0.0171	0.0239	0.0455
300	0.0193	0.0178	0.0150	0.135	0.0189	0.0203	0.0171	0.0191	0.0267	0.0506
400	0.0212	0.0196	0.0174	0.0146	0.0207	0.0225	0.0200	0.0210	0.0292	0.0549
500	0.0231	0.0214	0.0198	0.157	0.0225	0.0246	0.0228	0.0229	0.0318	0.0592
600	0.0250	0.0231	0.0222	0.0168	0.0242	0.0265	0.0257	0.0247	0.0342	0.0634
700	0.0268	0.0248	0.0246	0.178	0.0259	0.0283	0.0288	0.0266	0.0366	0.0677
800	0.0286	0.0264	0.0270	0.188	0.0275	0.0301	0.0321	0.0283	0.0388	0.0716
900	0.0303	0.0279	0.0294	0.198	0.0290	0.0319	0.0355	0.0300	0.0411	0.0754
1000	0.0319	0.0294	0.0317	0.208	0.0305	0.0337	0.0388	0.0317	0.0433	0.0793
1100	0.0336	0.0309	0.0339	0.219	0.0319	0.0354	0.0422	0.0333	0.0454	0.0832
1200	0.0353	0.0324	0.0360	0.0229	0.0334	0.0370	0.0457	0.0349	0.0475	0.0870
1300	0.0369	0.0339	0.0380	0.240	0.0349	0.0386	0.0494	0.0365	0.0497	0.0909
1400	0.0385	0.0353	0.0399	0.250	0.0364	0.0401	0.053	0.0381	0.0518	0.0946
1500	0.0400	0.0367	0.0418	0.260	0.0379	0.0404	0.057	0.0397	0.0538	0.0983
1600	0.0415	0.0381	0.0436	0.270	0.0394	0.0425	0.061	0.0413	0.0559	0.1020
1700	0.0430	0.0395	0.0453	0.280	0.0409	0.0436	0.064	0.0428	0.0579	0.1055
1800	0.0444	0.0408	0.0469	0.289	0.0423	0.0446	0.068	0.0443	0.0599	0.1090
1900	0.0458	0.0420	0.0484	0.298	0.0437	0.0456	0.072	0.0458	0.0619	0.1126
2000	0.0471	0.0431	0.050	0.307	0.0450	0.0466	0.076	0.0473	0.0638	0.1161

Source: From *Atmospheres for Heat Treating Equipment*, Brochure, Surface Combustion, Inc., Maumee, OH.

TABLE 1.26C
Specific Heat (C_p) of Gases [Btu/(lb/ft °F)]

Temperature (°F)	Air	CO	CO_2	H_2	N_2	O_2	DX Steam	DX (Lean)	DX (Rich)	RX
100	0.240	0.249	0.203	3.42	0.249	0.220	—	0.2449	0.2756	0.3938
200	0.241	0.250	0.216	3.44	0.249	0.223	—	0.2468	0.2770	0.3948
300	0.243	0.251	0.227	3.45	0.250	0.226	0.456	0.2487	0.2784	0.3957
400	0.245	0.253	0.237	3.46	0.252	0.230	0.462	0.2515	0.2808	0.3977
500	0.247	0.256	0.247	3.47	0.254	0.234	0.470	0.2542	0.2832	0.3996
600	0.250	0.259	0.256	3.48	0.256	0.239	0.477	0.2572	0.2862	0.4032
700	0.253	0.263	0.263	3.49	0.259	0.243	0.485	0.2602	0.2893	0.4069
800	0.256	0.266	0.269	3.49	0.262	0.246	0.494	0.2637	0.2927	0.4101
900	0.259	0.270	0.275	3.50	0.265	0.249	0.50	0.2672	0.2962	0.4133
1000	0.262	0.273	0.280	3.51	0.269	0.252	0.51	0.2708	0.2996	0.4164
1100	0.265	0.276	0.284	3.53	0.272	0.255	0.52	0.2739	0.3028	0.4201
1200	0.268	0.279	0.288	3.55	0.275	0.257	0.53	0.2770	0.3060	0.4238
1300	0.271	0.282	0.292	3.57	0.278	0.259	0.54	0.2800	0.3091	0.4276
1400	0.274	0.284	0.295	3.59	0.280	0.261	0.55	0.2831	0.3124	0.4314
1500	0.276	0.287	0.298	3.62	0.282	0.263	0.56	0.2862	0.3157	0.4352
1600	0.278	0.290	0.301	3.64	0.285	0.265	0.56	0.2893	0.3190	0.4390
1700	0.280	0.292	0.303	3.67	0.288	0.266	0.57	0.2915	0.3214	0.4424
1800	0.282	0.294	0.305	3.70	0.290	0.268	0.58	0.2938	0.3238	0.4457
1900	0.284	0.296	0.307	3.73	0.292	0.269	0.59	0.2960	0.3263	0.4491
2000	0.286	0.298	0.309	3.76	0.294	0.270	0.60	0.2982	0.3287	0.4524

Source: From *Atmospheres for Heat Treating Equipment*, Brochure, Surface Combustion, Inc., Maumee, OH.

TABLE 1.26D
Prandtl Number of Gases

Temperature (°F)	Air	CO	CO_2	H_2	N_2	O_2	DX Steam	DX (Lean)	DX (Rich)	RX
100	0.72	0.76	0.76	0.70	0.73	0.73	—	0.71	0.56	0.42
200	0.72	0.76	0.76	0.70	0.73	0.73	—	0.71	0.56	0.42
300	0.72	0.76	0.75	0.70	0.73	0.73	0.95	0.71	0.56	0.42
400	0.72	0.76	0.75	0.70	0.73	0.74	0.96	0.71	0.56	0.42
500	0.72	0.76	0.75	0.70	0.73	0.74	0.94	0.71	0.57	0.42
600	0.72	0.76	0.75	0.70	0.73	0.74	0.94	0.71	0.57	0.42
700	0.72	0.76	0.75	0.71	0.73	0.75	0.93	0.71	0.57	0.43
800	0.72	0.77	0.75	0.71	0.73	0.75	0.92	0.72	0.58	0.43
900	0.72	0.77	0.74	0.71	0.75	0.75	0.91	0.72	0.58	0.43
1000	0.73	0.77	0.73	0.71	0.74	0.75	0.91	0.73	0.58	0.43
1100	0.73	0.77	0.73	0.71	0.75	0.75	0.90	0.73	0.59	0.43
1200	0.74	0.77	0.73	0.71	0.75	0.75	0.88	0.73	0.59	0.43
1300	0.74	0.77	0.73	0.71	0.75	0.75	0.88	0.73	0.59	0.43
1400	0.74	0.77	0.73	0.72	0.76	0.75	0.87	0.73	0.59	0.43
1500	0.74	0.77	0.73	0.72	0.75	0.75	0.87	0.73	0.59	0.43
1600	0.74	0.78	0.74	0.72	0.75	0.76	0.87	0.73	0.59	0.44
1700	0.75	0.78	0.73	0.72	0.76	0.76	0.87	0.74	0.59	0.44
1800	0.75	0.78	0.73	0.72	0.76	0.77	0.87	0.74	0.60	0.44
1900	0.75	0.78	0.73	0.72	0.76	0.77	0.87	0.74	0.60	0.44
2000	0.75	0.79	0.74	0.72	0.77	0.77	0.87	0.74	0.60	0.44

Source: From *Atmospheres for Heat Treating Equipment*, Brochure, Surface Combustion, Inc., Maumee, OH.

TABLE 1.27
Impurities in Pure Nitrogen

Source of Nitrogen	Impurity					
	O_2	CO_2	CO	H_2	CH_4	Dew Point (°C)
Decomposition of air	0.0001	0.0005	0.001	0.001	0.001	−70
Hydrocarbon combustion	0.0001	0.005	0.001	0.002	0.002	−65

$$K = \frac{p\,H_2 \times p\,CO_2}{p\,CO \times p\,H_2O}$$

where p is the partial pressure of the gas shown.

The temperature dependence of the equilibrium constant is shown in Figure 1.87.

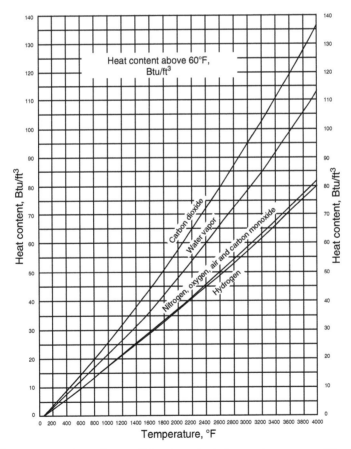

FIGURE 1.79 Temperature dependence of heat content for various gases. (From D. Schwalm, *The Directory of Industrial Heat Processing and Combustion Equipment: United States Manufacturers, 1981–1982*, Energy edition, Published for Industrial Heating Equipment Association by Information Clearing House, pp. 147–153.)

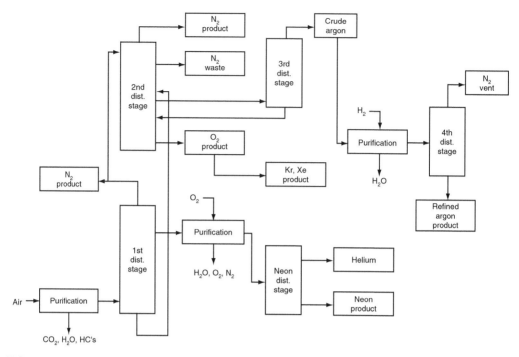

FIGURE 1.80 Process for gas production by air separation. (From *Bulk Gases for the Electronics Industry*, Brochure, Praxair Inc., Chicago, IL.)

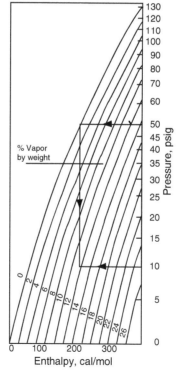

FIGURE 1.81 Example of flash-off with constant-enthalpy throttling of liquid nitrogen. (From *Bulk Gases for the Electronics Industry*, Brochure, Praxair Inc., Chicago, IL.)

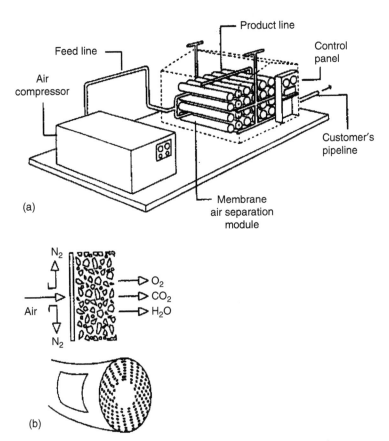

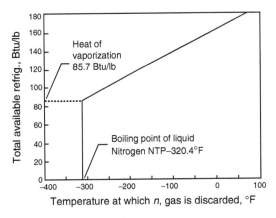

FIGURE 1.82 Gas production by membrane separation. (a) Commercial membrane separation unit; (b) schematic of membrane separation. (Courtesy of Praxair Inc.)

FIGURE 1.83 Refrigeration capacity in liquid nitrogen.

TABLE 1.28
Cooling Chamber Selection Chart for LN$_2$ Processes

Cooling Application	LN$_2$ Immersion Bath	LN$_2$ Cooled Glycol Bath	LN$_2$ Fluid Bed	Indirect Cold Gas Cooling	Direct Cold Gas Cooling
Rapid cooling to 40°F		×	×		
Slow cooling to 40°C			×	×	×
Rapid cooling to 200°F			×		×
Slow cooling to 200°F			×	×	×
Rapid cooling to 320°F	×				×
Slow cooling to 320°F					×
Constant-temperature soaking to −40°F		×	×	×	
Constant soaking to −200°F			×	×	×
Shrink fitting (cool to −320°F)	×		×		×
Metallurgical treatment (−140°F)			×	×	×
Metallurgical treatment (−320°F)	×				×

Source: From *Bulk Gases for the Electronics Industry*, Brochure, Praxair Inc., Chicago, IL.

1.6.1.8 Hydrocarbons

The hydrocarbons used to generate heat treatment atmospheres are most often derived from methane (CH$_4$), propane (C$_3$H$_8$), and natural gas. Natural gas contains approximately 85% methane. The combustion of these gases provides the carbon required for heat treatment processes.

The combustion of these gases and that of other potential sources of carbon result in different furnace efficiencies and require different control procedures. For example, consider the stoichiometry of the oxidation of methane and that of propane, hydrocarbons that differ by only two carbons.

For methane:

$$2CH_4 + O_2 \leftrightharpoons 2CO + 4H_2$$

TABLE 1.29
Impurities of Pure Hydrogen

Method	Impurities (%)			Dew Point (°C)
	CO$_2$	O$_2$	CO$_2$ O$_2$	
Electrolysis of H$_2$	<0.003	<0.0001	<0.0004	−65
Electrolysis of H$_2$	<0.03	<0.0001	<0.0301	−18
Diffusing through palladium	—	—	<0.0001	−80

Source: From R. Nemenyi and G. Bennett, *Controlled Atmospheres for Heat Treatment*, Franklin Book Co., 1995, pp. 22–1022.

TABLE 1.30
Physical Properties of Inert Gas Protective Atmospheres

Gas	Specific Gravity	Thermal Conductivity Relative to Air	Thermal Content (Btu/ft³)
Nitrogen	0.972	0.999	0
Argon	1.379	0.745	0
Helium	1.137	6.217	0

Source: From *Bulk Gases for the Electronics Industry*, Brochure, Praxair Inc., Chicago, IL.

For propane:

$$2C_3H_8 + 3O_2 \leftrightharpoons 6CO + 8H_2$$

Propane provides more CO and less H_2 than methane. In addition, propane may contain small quantities of propylene or butylene, which may lead to soot formation [100].

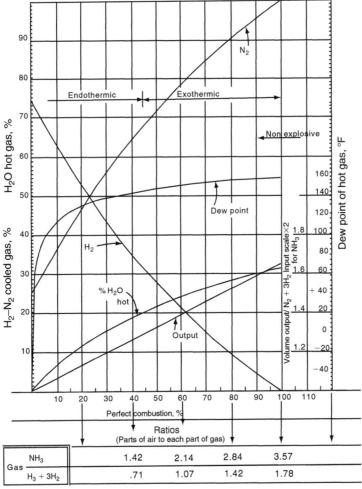

FIGURE 1.84 Generation of nitrogen based atmospheres (From *Protective Atmospheres and Analysis Curves*, Brochure, Electric Furnace Company, Salem, OH.)

TABLE 1.31
Atmospheres Produced by Dissociation of Ammonia

AGA Class	Atmosphere	Composition (%)	
		H_2	N_2
601	Dissociated ammonia	75	25
621	Substantial combustion	1	99
622	Partial combustion	1–20	99–80

Source: From R. Nemenyi and G. Bennett, *Controlled Atmospheres for Heat Treatment*, Franklin Book Co., 1995, pp. 22–1022.

Properties of various hydrocarbons are provided in Table 1.22 and Table 1.23, and gas combustion constants are given in Table 1.24 [79].

1.6.2 CLASSIFICATION

Heat treatment furnace atmospheres have been classified by the American Gas Association (AGA) and are summarized in Table 1.32. Selected AGA and European gas classifications are compared in Table 1.33. The applicability of some of these gases to particular types of heat treatment processes is illustrated in Table 1.34 and Table 1.35.

1.6.2.1 Protective Atmospheres and Gas Generation

1.6.2.1.1 Exothermic Gas Generators

Exothermic (exo) gas, as shown in Table 1.32 and Table 1.33, is essentially a mixture of the reducing gases CO and H_2 and the oxidizing gases CO_2 and water, with nitrogen as an inert carrier gas. Exogas is prepared by the combustion of a hydrocarbon such as methane or propane in a deficiency of air. Lean exogas contains larger quantities of CO_2 and smaller quantities of CO and H_2 than rich exogas. The composition of exogas with respect to the air/gas ratio is shown in the combustion chart of Figure 1.88.

Exogas is used as a protective atmosphere to prevent decarburization, scaling, and other undesirable surface reactions. To achieve the best results, the dew point must be minimized. This is accomplished by drying the gas by water-cooled condensation or refrigeration or by using a molecular sieve adsorbent (see discussion below). The relative effectiveness of these forms of moisture removal is illustrated in Figure 1.89.

A schematic of an exogas generator is shown in Figure 1.90. The hydrocarbon–air fuel mixture is metered into the water-cooled combustion chamber, which contains a catalyst. The combustion products, which include CO, CO_2, H_2, and H_2O (and in some cases residual hydrocarbon gas), are then passed through a water-cooled condenser to remove most of the water vapor. The exogas is further dried either by refrigeration or by using an adsorbent such as activated alumina, activated silica, or a 3–4 Å molecular sieve. The dried exogas is then piped directly to the furnace for use. An example of a commercial exogas generator is shown in Figure 1.91.

1.6.2.1.2 Molecular Sieves

Activated alumina, activated silica, and activated carbon are common adsorbents with high surface areas. The term activation refers to either vacuum or thermal surface desorption

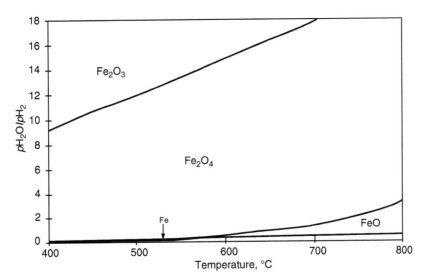

FIGURE 1.85 Effect of hydrogen/water ratio as a function of temperature on the oxidation of iron. (From J. Morris, *Heat Treat. Met.* 2:33–37, 1989.)

freeing the surface-active adsorption sites for subsequent adsorption. Although these materials possess porosity, it is not uniform.

Molecular sieves are aluminosilicates of elements of group I (potassium and sodium) and group II (magnesium, calcium, and barium). These crystalline structures of Al_3O_4 and SiO_4 which are interconnected through oxygen atoms through shared bonding to the metal cations. They possess uniform microporosity with pore sizes varying from 3 to 10 Å, depending on the particular zeolite. Table 1.36 lists commonly available commercial zeolites and their pore sizes.

Zeolites characteristically possess very high internal surface area relative to their outside surface area. Molecular separation is based primarily on zeolite pore size; molecules of dimensions larger than the zeolite pore size are excluded. Table 1.37 and Table 1.38 provide

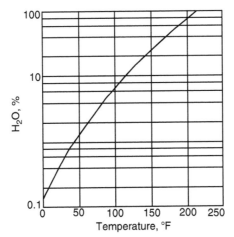

FIGURE 1.86 Dew-point curve. (From J.A. Zahniser, *Furnace Atmospheres*, ASM International, Materials Park, OH.)

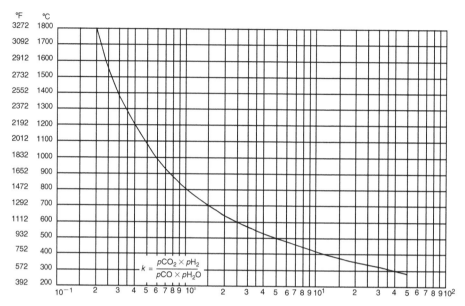

FIGURE 1.87 Water–gas equilibrium temperature as a function of temperature. (From J.A. Zahniser, *Furnace Atmospheres*, ASM International, Materials Park, OH.)

various physical constants, including molecular size, for various adsorbate gases.

Molecular sieves provide excellent alternatives to other activated solid substrates such as silica gel and activated alumina for atmosphere dehydration. Although all of the Type A molecular sieves (Table 1.36) readily absorb water, Type 3A (3 Å) molecular sieves are the

TABLE 1.32
AGA Classification of Heat Treatment Furnace Atmospheres

AGA Class	Atmospheres Designation	Notes
100	Exothermic base	These atmospheres are derived from the exothermic reaction of partial or complete combustion of various air–fuel mixtures and may contain various amounts of ammonia or H_2O
200	Prepared nitrogen base	Class 100 with most of the CO_2 and H_2O vapor removed
300	Endothermic base	These are the endothermic reaction products of various gas–fuel mixtures over a catalyst
400	Charcoal base	Formed by passing air through heated, incandescent charcoal. The desired gases are removed at the maximum temperature zone and purified
500	Exothermic–endothermic	A mixture of air and fuel undergoes complete combustion, and most of the water vapor is removed. The CO is then endothermically formed from the CO_2
600	Ammonia base	Any atmosphere derived from ammonia; may include ammonia or various forms of dissociated ammonia with water vapor and residual ammonia removed

Source: From F.E. Vandaveer and C.G. Segeley, Prepared atmospheres, in *Gas Engineering Handbook*, Industrial Press, New York, 1965, pp. 12/278–12/289.

TABLE 1.33
European and AGA Classification of Protective Atmospheres

Atmosphere	Classification European	AGA	Dew Point (°F)	Typical Gas Composition (% v/v)					
				CO_2	CO	H_2	CH_4	H_2O	N_2
Exothermic	DX, inert	101	100	10.4	0.5	0.5	0.0	6.5	82.1
			40	1.0	0.5	0.5	0.0	0.8	87.2
			100	4.7	9.4	9.4	0.4	6.5	69.6
	DX, rich	102	40	5.0	10.0	10.0	0.4	0.8	73.8
	NX	201	−40	0.05	0.5	0.5	0.0	0.0	98.95
	HNX	223–224 302	−40	0.05	0.05	3.0 –10.0	0.0	0.0	96.9–89.8
Endothermic	RX		30–45	0.1	20.7	40.6	0.4	0.3	37.9
	SRX	323	100	5.2	16.8	71.1	0.4	6.5	0.0
	ASRX	323	100	3.3	18.7	65.4	0.4	6.5	5.7
	HX	325	−40	0.05–0.0	4.55–0.0	95.0–100.0	0.4–0.0	0.0	0.0
Dissociated ammonia	AX	601	−40	0.0	0.0	75.0	0.0	0.0	25.0

Source: From F.E. Vandaveer and C.G. Segeley, *Gas Engineering Handbook*, Industrial Press, New York, 1965, pp. 12/278–12/289.

preferred zeolite for dehydration of hydrocarbon gases [90]. The 3 Å pore size excludes other hydrocarbons such as ethylene and propylene that may catalytically undergo secondary polymerization reactions on the zeolite bed.

Gas adsorption and desorption on molecular sieves are both pressure- and temperature-sensitive and are the basis for the commercial highly efficient engineering practices called pressure swing and temperature (or thermal) swing separations. These separation practices are illustrated schematically in Figure 1.92. For temperature swing processes, adsorption occurs at lower temperatures and desorption at higher temperatures. Pressure swing adsorption is assumed to be an isothermal process, and adsorption occurs at higher pressures and desorption at lower pressures.

TABLE 1.34
Composition and Dew Point of Selected Protective Atmospheres

AGA Class	Typical Atmosphere	Composition (%)					Dew Point (°F)
		CO	CO_2	H_2	N_2	CH_4	
101	Exothermic (lean)	1.5	10.5	1.2	86.8	—	40[a]
102	Exothermic (rich)	10.5	5.0	12.5	71.5	0.5	40[a]
201	Prepared nitrogen (lean)	1.7	—	1.2	97.1	—	−40
202	Prepared nitrogen (rich)	1.0	—	13.2	75.3	—	−40
301	Endothermic (lean)	19.6	0.4	34.6	45.1	0.3	+50
302	Endothermic (rich)	20.7	—	38.7	39.8	0.8	0 to −5
601	Dissociated ammonia	—	—	25	75	—	−60

[a]If tap water cooling, dew point is room temperature and is reduced to 40°F using −50°F refrigeration.

TABLE 1.35
Selected Heat Treatment Processes Using AGA Classified Atmosphere

AGA Class	Atmosphere	Heat Treatment Process
101	Lean exothermic	Forms an oxide coating on steel
201	Lean prepared nitrogen	Neutral heating can cause decarburization of steel
202	Rich prepared nitrogen	Annealing and brazing of stainless steel
301	Lean endothermic	Clean hardening
302	Rich endothermic	Carburizing (rich endogas is not usually used; lean endogas and enrichment gas addition is the preferred method)
501	Lean exothermic–endothermic	Clean hardening
502	Rich exothermic–endothermic	Carburizing
601	Dissociated ammonia	Brazing; sintering
621	Lean combusted ammonia	Neutral hardening

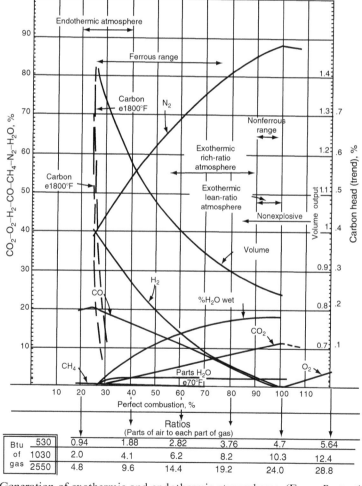

FIGURE 1.88 Generation of exothermic and endothermic atmospheres. (From *Protective Atmospheres and Analysis Curves*, Brochure, Electric Furnace Company, Salem, OH.)

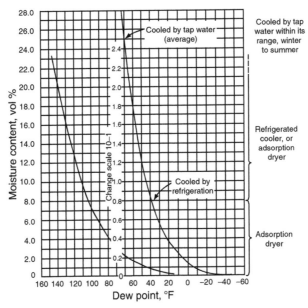

FIGURE 1.89 Effect of cooling temperature on moisture removal. (From A.G. Hotchkiss and H.M. Weber, *Protective Atmospheres*, Wiley, New York, 1963.)

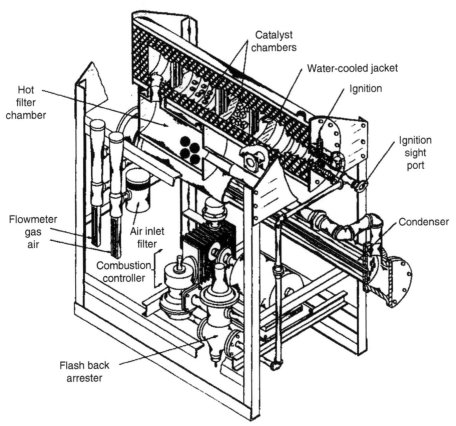

FIGURE 1.90 Schematic illustration of an exothermic generator. (Courtesy of C.I. Hayes, Inc.)

(a)

(b)

FIGURE 1.91 Exothermic generators: (a) (Courtesy of Can-Eng Furnaces, www.can-eng.com.) (b) (Courtesy of AFC-Holcroft, LLC, www.afc-holcroft.com.)

TABLE 1.36
Molecular Sieves

Product Type[a]	Major Cation	Pore Size (Å)	Water Capacity (wt%)
3A	K^+	3	20
4A	Na^+	4	22
5A	Ca^{2+}	5	21.5
10X	Ca^{2+}	8	31.6

[a]*Source*: From D.W. Breck, *Zeolite Molecular Sieves—Structure, Chemistry and Use*, Krieger, Malabar, FL, 1984.

1.6.2.1.3 Monogas (Prepared Nitrogen) Generators
Monogas or prepared nitrogen atmospheres are nitrogen-rich gases (>90%) such as those listed in Table 1.34 that are obtained by the combustion of a hydrocarbon in the presence of a slight deficiency of air using the exogas generator discussed above. The residual CO_2 may be removed (to <0.05%) using a 4–5 Å molecular sieve [26,27]. A molecular sieve of this pore size will remove both residual water vapor and CO_2. (Previously, methanolamine gas scrubbers, which may reduce CO_2 content to 0.05%, were used, but this technology is rarely if ever used today.)

For optimal results in preventing decarburization, it is often important to use a dry, CO_2-free monogas protective atmosphere. This is illustrated using Gonser's curves in Figure 1.93, where increasing CO_2 concentration in the exogas resulted in a corresponding increase in decarburization.

1.6.2.1.4 Endothermic Gas Generators
As shown in the combustion chart of Figure 1.88, endothermic atmospheres are prepared by the combustion of richer mixtures of hydrocarbons in air than those used for the preparation of exogases [79]. Generally, the air/gas ratio is selected to favor the formation of $CO_2 + H_2$

TABLE 1.37
Physical Constants of Adsorbate Gases

Gas	Boiling Point (°C)	Critical Temperature (°C)	Polarizability (A³)	Ionization Potential (V)	Length (Å)	Width (Å)	Kinetic Diameter[a]
Argon	−187.8	−122.4	1.6	15.7	1.92	1.92	3.4
Oxygen	−183.0	−118.8	1.2	12.5	2.0	1.4	3.46
Nitrogen	−195.8	−147.1	1.4	15.5	2.1	1.5	3.64
Methane	−161.4	−82.5	2.6	14.5	2.0	2.0	3.8
CO	−192.0	−139	1.6	14.3	2.1	1.8	3.76
Ethylene	−103.7	9.7	3.5	12.2	2.5	2.2	3.9
Ethane	−88.6	32.1	3.9	12.8	2.6	2.5	3.8
CO₂	−78.5	31.1	1.9	14.4	2.6	1.8	3.3
Propylene	−47.6	92.0	3.5	12.2	3.4	2.2	—
Propane	−42.3	96.8	5.0	12.8	3.3	2.5	4.3

[a]The kinetic diameter is the intermolecular distance of closest approach for two molecules colliding with zero potential energy.

Source: From D.W. Breck, *Zeolite Molecular Sieves—Structure, Chemistry and Use*, Krieger, Malabar, FL, 1984.

TABLE 1.38
Pressure and Temperature Sensitivity of Gas Adsorption on Molecular Sieve[a]

Gas	Temperature (K)	Pressure (torr)	x/m	Pressure (torr)	x/m	Pressure (torr)	x/m
Argon	77	100	<0.01				
	195	100	5	300	15	700	30
	273	100	1	300	1.5	700	3.7
Oxygen	90	0.2	0.11	1	0.17	700	0.26
	195	40	0.003	150	0.01	700	0.044
	195	100	6	300	18	700	34
Nitrogen	97	700	<0.01				
	195	100	0.05	300	0.085	700	0.115
	195	100	30	300	42	700	49
Water	298	0.025	0.16	0.1	0.20	4	0.25
	373	1	0.06	4	0.13	12	0.17
Ammonia	298	3	0.090	10	0.11	100	0.15
CO_2	198	10	0.25	700	0.30		
	298	2	0.070	10	0.12	100	0.165
	423	100	0.034	700	0.105		
CO	198	15	0.070	100	0.091	700	0.11
	273	150	0.024	700	0.055		
	273	150	0.007	700	0.02		

[a] x/m is grams of gas per gram of dehydrated crystalline zeolite.
Source: From D.W. Breck, *Zeolite Molecular Sieves—Structure, Chemistry and Use*, Krieger, Malabar, FL, 1984.

and to be insufficient to form large amounts of CO_2 and H_2O. Endogases contain much higher concentrations of CO and H_2 than exogases and exhibit higher dew points, as shown in Table 1.33. The relationship between the air/gas ratio and dew point for endogas is shown in Figure 1.94. Nitrogen is used as the inert carrier gas.

Since very small amounts of air are used, a catalyst and heat are required to facilitate combustion as shown in the schematic of an endogas generator in Figure 1.95. The endogas is cooled immediately upon departure from the externally heated combustion chamber and before

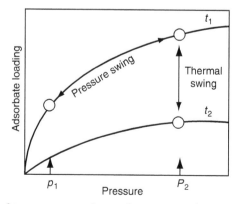

FIGURE 1.92 Illustration of temperature swing and pressure swing separation cycles. (From D.W. Breck, *Zeolite Molecular Sieves—Structure, Chemistry and Use*, Krieger, Malabar, FL, 1984.)

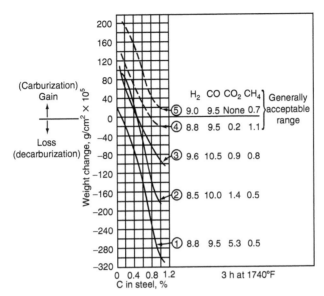

FIGURE 1.93 Gonser's curves illustrating decarburization with increasing CO_2 content of furnace atmosphere. (From B.W. Gonser, *Ind. Heat.* 6(12):1123–1134, 1939.)

piped to the furnace for use. Rapid cooling is necessary to inhibit the formation of CO_2 and soot (carbon):

$$2CO \rightleftharpoons CO_2 + C$$

The air/gas ratio required for endogas formation will vary with the source of combustion gas as shown in Table 1.39. An example of an endogas generator is provided in Figure 1.96.

1.6.2.1.5 Ammonia Dissociators (Crackers)
Ammonia dissociation (cracking) is the preferred method for preparing the highest purity, oxygen-free nitrogen. This reaction is conducted in an ammonia dissociator such as the one illustrated in Figure 1.97.

Ammonia dissociators typically have three essential parts: an instrument panel, a vaporizer, and a reaction chamber. Liquid ammonia is vaporized in a heat exchanger and is fed

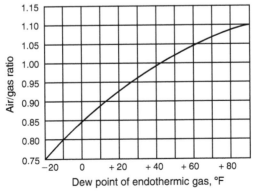

FIGURE 1.94 Effect of air/gas ratio on dew point. (From A.G. Hotchkiss and H.M. Weber, *Protective Atmospheres*, Wiley, New York, 1963.)

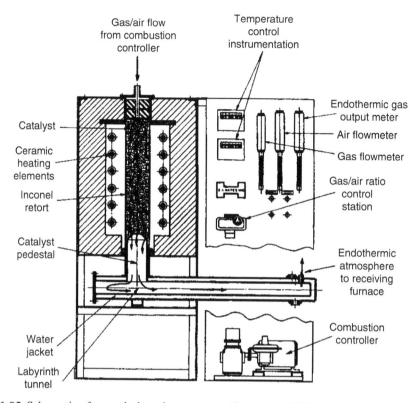

FIGURE 1.95 Schematic of an endothermic generator. (Courtesy of C.I. Hayes, Inc.)

directly to the dissociator, which is a heat-resistant, electrically heated coiled pipe surrounded by refractory material. Hot dissociated ammonia exiting the dissociator is used to heat liquid ammonia entering the heat exchanger and is then piped to the furnace if no further purification is required. The control panel permits control of the gas flows. Gaseous ammonia from cylinders can also be used.

1.6.3 FURNACE ZONING

Traditionally, continuous heat treatment furnaces that use protective atmospheres such as exogas, endogas, and dissociated ammonia were open-ended with an inlet, hot zone, cooling

TABLE 1.39
Air/Gas Ratio for Various Hydrocarbon Sources

Fuel Source	Air/Gas Ratio
Propane	7.0
Butane	9.5
Natural gas[a]	2.5

[a]Natural gas typically contains 85% methane.

Source: From R. Nemenyi and G. Bennett, *Controlled Atmospheres for Heat Treatment*, Franklin Book Co., 1995, pp. 22–1022.

zone, and exit (as shown in Figure 1.98) [92]. This was a relatively inefficient and limiting process because a single reducing or nonoxidizing atmosphere must be used, the atmosphere is typically fed into the hot zone, flammable atmospheres are burned at the ends of the furnace, and a positive pressure is used to prevent atmospheric air from entering the furnace [92].

There is an increasing trend to use zoning to optimize atmosphere in different zones within the furnace rather than using a control in a continuous furnace. Zoning permits the controlled use of different atmospheres. Some advantages of zoning are that it improves atmosphere control throughout the furnace, reduces the amount of gas readded, and improves atmospheric flow. A schematic of a typical continuous furnace using zoning is shown in Figure 1.99.

Significant improvements in zoned temperature control and energy savings are achieved with the use of furnace curtains. Furnace curtains can be fabricated to withstand temperatures up to 200°F. Suitable materials include steel or silica-based or nonasbestos textiles [93].

1.6.4 *IN SITU* ATMOSPHERE GENERATION

Volatile organic liquids or hydrocarbons such as methanol, acetone, isopropanol, and ethyl acetate are used for *in situ* furnace atmosphere generation [94–98]. Other liquids that have been tried include higher alcohols, butane, diesel oil, and kerosene [97]. Comparisons of the CO and CH_4 content of *in situ* generated atmospheres from selected volatile organic liquids are shown in Figure 1.100 and Figure 1.101, and the reactive carbon availability of various fuels is summarized in Table 1.40.

Atmosphere generation occurs either by direct injection of the liquid into the furnace, usually onto a hot surface such as electric heating elements or gas-fired radiant tube heaters, or by external furnace atmosphere preparation. An illustration of an external nitrogen–methanol generator is provided in Figure 1.102. Furnace requirements for this process are [97]:

1. Furnace must be able to supply the additional energy required for *in situ* atmosphere generation.
2. Gas reactions with air must be exothermic.
3. Flow must be strong and directed.
4. Gases entering the furnace must be thoroughly mixed and directed to the hot dissociating surface.

Similarly, there are some general guidelines for selection of the fuel source for *in situ* atmosphere generation [82]:

1. Fuel cannot contain solid material.
2. Components of the fuel mixture must be mutually compatible, and the mixture must have a freezing point above even the coldest use temperature.
3. Liquids with flash points greater than methanol should be avoided.
4. Use of water-immiscible liquids may be undesirable from the standpoint of fire safety.
5. Thermal decomposition should not produce tars or soot.
6. Liquids should not be corrosive or toxic.
7. No liquid should be introduced into the furnace at a temperature below 750°C because of explosion hazards.

1.6.4.1 Nitrogen–Methanol

The most common fuel mixture used for *in situ* generation of furnace atmospheres is nitrogen–methanol. High-purity nitrogen may be generated on site with a membrane separator such as the one shown in Figure 1.82 and may be stored as liquid nitrogen or used as generated.

(a)

(b)

FIGURE 1.96 (a) Endothermic generator. (Courtesy of AFC-Holcroft, LLC, www.afc-holcroft.com.)
(b) Dual retort endothermic generator. (Courtesy of Can-Eng Furnaces, www.can-eng.com.)

FIGURE 1.97 Drever nitrogen generator. (Courtesy of Drever Company.)

Figure 1.103 presents a graph of the free energy released by the thermal decomposition of methanol according to the following furnace atmosphere reactions [98]:

$$CH_3OH \rightleftharpoons C + 2H_2 \tag{1.1}$$

$$2CH_3OH \rightleftharpoons CH_4 + C + 2H_2O \tag{1.2}$$

$$CH_3OH \rightleftharpoons C + H_2 + H_2O \tag{1.3}$$

$$2CH_3OH \rightleftharpoons CH_4 + CO_2 + 2H_2 \tag{1.4}$$

$$2CH_3OH \rightleftharpoons C + CO_2 + 4H_2 \tag{1.5}$$

The overall furnace atmosphere composition is determined by the water–gas equilibrium

$$CO_2 + H_2 \rightleftharpoons CO + H_2O$$

The temperature dependence of methanol dissociation is shown in Figure 1.104 [96,99]. Methanol dissociation may be performed either in the furnace chamber or in an external dissociator. To facilitate more complete decomposition, external liquid methanol dissociators may be used that employ catalysts such as copper and zinc and somewhat lower dissociation temperatures [99].

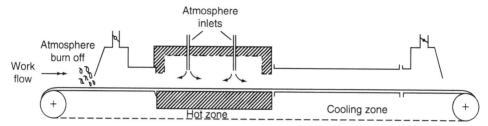

FIGURE 1.98 Illustration of the use of a conventional atmosphere in a continuous furnace. (From M.J. Hill, The efficient use of atmospheres in a continuous furnace using the concept of zoning, Int. Foundry Heat Treat. Conf., Johannesburg, S. Africa, Vol. 2, 1985, pp. 1–30.)

A comparison of the typical composition of atmospheres attainable with nitrogen–methanol mixes and endogases from natural gas (which contains approximately 85% methane) or propane is shown in Table 1.41.

The delivery of a nitrogen–methanol mixture to the furnace in a direct feed system is shown in Figure 1.105. Proper flow control is critical to the success of the process.

The conversion from conventional endothermic gas to nitrogen–methanol is relatively straightforward. The nitrogen–methanol injector pipe is introduced directly into the endogas entry port [95]. The injector atomizes the nitrogen–methanol mixture and directs it to the heated dissociation surface. When high dew points and poor control are encountered, it is usually due to improper injection. Praxair [95] gives some additional guidelines:

1. To ensure safe operation, the inert gas purge line should enter the furnace through a separate furnace port to minimize the possibility of plugging.
2. A methanol pipe that is oversized or excessively long increases the possibility of liquid gas separation (Figure 1.106) and will lead to furnace puffing as shown in Figure 1.107.

1.7 ATMOSPHERE SENSORS

The primary constituents of furnace atmospheres are carbon monoxide (CO), carbon dioxide (CO_2), nitrogen (N_2), hydrogen (H_2), water (H_2O), and hydrocarbon gases such as methane (CH_4). The ratio of these gases with respect to each other is critical to prevent decarburization, oxidation, hydrogen embrittlement, surface blueing due to the low-temperature (400–700°F; 205–370°C) reaction of water vapor with steel, and soot formation. Therefore, it is necessary to use various sensing devices to accurately monitor and control the concentration of these gases

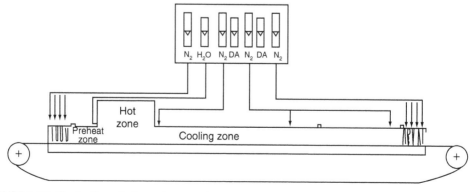

FIGURE 1.99 Illustration of atmosphere zoning.

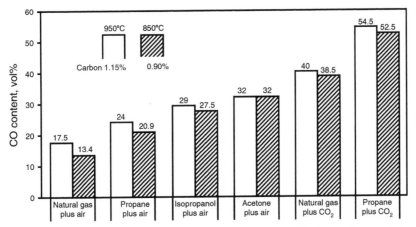

FIGURE 1.100 CO content of selected gas mixtures at 950°C and 850°C. (From B. Edenhofer, Progress in the technology and applications of *in situ* atmosphere production in hardening and case-hardening furnaces, Proc. 2nd Int. Conf. on Carburizing and Nitriding with Atmospheres, December 6–8, 1995, Cleveland, pp. 37–42.)

throughout the heat treatment cycle. A summary of the various options to analyze heat-treating atmospheres is provided in Table 1.42.

The objective of the remainder of this section is to provide an overview of the various measurement techniques that can be used to determine gas composition.

1.7.1 ORSAT ANALYZER

The analytical procedure followed with the Orsat analyzer is based on the relative gas adsorptivity of different gases in selective reagents. A sample of the gaseous atmosphere is passed through a series of liquids, each specific to one of the various gases analyzed: CO_2, CO, O_2, and H_2 (see Table 1.43). The concentration of nitrogen is obtained by difference [82]. The increase in volume is proportional to the relative volume of the gas in the atmosphere analyzed.

Orsat analysis is not commonly used today. It has been replaced by significantly faster and more sensitive methods such as gas chromatography, and dew-point and infrared analyzers [82].

1.7.2 GAS CHROMATOGRAPHY

Gas chromatography is a relatively fast and accurate method for measuring the concentration of the gaseous components in a heat treatment atmosphere [82,101–105]. Gas chromatographic analysis is conducted by injecting the gas to be analyzed onto a column with a carrier gas, typically helium or argon [106]. The separation is primarily dependent on how strongly the adsorbed gases adhere to the column solid support. Gases, as they elute from the column, are detected by flame ionization, thermal conductivity, or some other suitable detector. The more weakly adsorbed gases will elute first, and the most strongly adsorbed gases will elute last. Care should be taken when molecular sieves are used as the solid support because an error due to coelution of argon, which is present in combustion air, with oxygen may inflate the oxygen concentration by approximately 2% [101]. An illustrative gas chromatogram is provided in Figure 1.108.

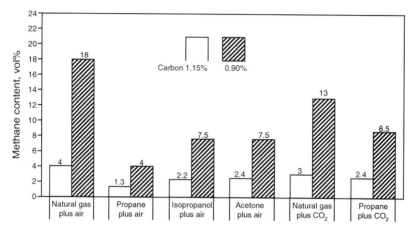

FIGURE 1.101 Average Methane content of selected gas mixtures at 950°C and 850°C. (From B. Edenhofer, Progress in the technology and applications of *in situ* atmosphere production in hardening and case-hardening furnaces, Proc. 2nd Int. Conf. on Carburizing and Nitriding with Atmospheres, December 6–8, 1995, Cleveland, pp. 37–42.)

1.7.3 THERMAL CONDUCTIVITY

The thermal conductivity of a gas is dependent on its composition, as illustrated by the data presented in Table 1.44. The thermal conductivity of a gas is measured electrically using a Pt or Pt-Rh wire in a heated chamber. The thermal conductivity is determined by the amount the wire is cooled in the presence of the gas relative to the cooling provided by a known gas at the same temperature.

TABLE 1.40
Primary and Secondary Carbon Availability and the Carbon Transfer Coefficient for Various Carburizing Methods at 925°C

Carburizing Method	Primary Carbon Availability[a] (g C/nm³)	Secondary Carbon Availability[b] (g C/nm³)	Carbon Transfer Coefficient[c] (N m/s)
Endothermic generator, propane	27	0.5	120
Methanol, generator gas	28	1.1	280
Endothermic–exothermic processor gas	27	0.5	120
Propane processor gas	≤400	0.5–1.5	—
In situ air–propane	25–30	0.5	120
In situ methanol–acetone	3–268	0.8–1.1	250
In situ methanol–ethyl acetate	3–178	1.1–1.2	280
Methanol–water–propane	~28	1.1	280
Nitrogen–methanol	~27	0.5	130

[a]Primary carbon availability is the amount of carbon that $1\,m^3$ (NTP) of a gas carburizer can supply before the carbon potential is reduced to 1.0% C.

[b]Secondary carbon availability is the amount of carbon absorbed at the steel surface by $1\,m^3$ (NTP) of a gas carburizer while its carbon potential is reduced from 0.9 to 1.0% C.

[c]Carbon transfer coefficient describes the kinetics of carbon flow from the gas phase near the steel surface to the steel surface.

FIGURE 1.102 Methanol dissociator. (Courtesy of Surface Combustion, Inc.)

1.7.4 OXYGEN SENSORS

1.7.4.1 Paramagnetic Oxygen Analyzers

A unique feature of oxygen, relative to the other gaseous constituents of heat treatment atmosphere gases, is its magnetic susceptibility [82,107]. This is illustrated in Table 1.45.

Three types of instruments are used to measure the paramagnetic properties of oxygen, magnetodynamic (paramagnetic), thermomagnetic, and magnetopneumatic oxygen analyzers. A schematic of the magnetodynamic oxygen analyzer is shown in Figure 1.109. When the gas to be analyzed is introduced into the instrument, oxygen is attracted to the

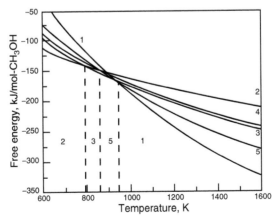

FIGURE 1.103 Free energy released by methanol dissociation reactions (see Equation 1.1 through Equation 1.5).

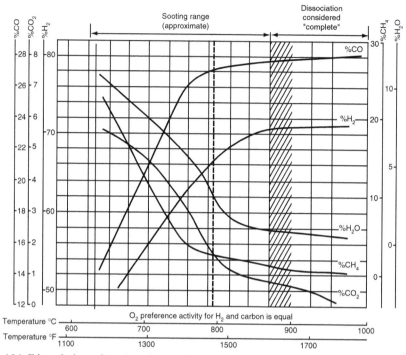

FIGURE 1.104 Dissociation of methanol at various temperatures.

point of greatest magnetic susceptibility, the dumbbell area. The magnitude of displacement of the dumbbell by the oxygen gas is proportional to the amount of oxygen present in the gas.

The operation of the thermomagnetic analyzer is based on the temperature dependence of the magnetic susceptibility c of oxygen [82]:

$$cT = Z$$

where T is the absolute temperature and Z is the Curie constant.

The thermomagnetic analyzer consists of two tubes for gas entry that are connected by a heated crossover tube with an electric heating filament at each end. Each filament is attached to a Wheatstone bridge, so there is a magnetic field surrounding one of the heating filaments. The gas is then introduced equally to both tubes. Because of its magnetic susceptibility,

TABLE 1.41

Comparison of the Composition Ranges Attainable with Nitrogen–Methanol and Endothermic Gas

Analysis	Nitrogen–Methanol	Endothermic Gas from Natural Gas	Endothermic Gas from Propane
% CO	15–20	19.8	23.8
% H_2	35–45	40.4	31.2
% CO_2	0.4	0.3	0.3
% CH_4	0.3	0.5	0.1
N_2	Remainder	Remainder	Remainder

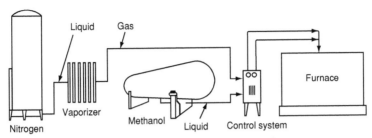

FIGURE 1.105 Schematic of a nitrogen–methanol atmosphere supply system.

oxygen in the gas is attracted to the magnet where it is heated by the electric filament. As the sample is heated, the magnetic susceptibility decreases until it is displaced by a cooler and more magnetic oxygen from the gas.

This mechanism creates a magnetic wind, which cools the filaments. The difference in the resistance of the two wire filaments is proportional to the amount of oxygen gas that is present [107].

The third type of instrumentation that may be used, which is based on the paramagnetic properties of oxygen, is the magnetopneumatic oxygen analyzer. This instrument uses a non-homogeneous magnetic field, and the oxygen in the gas is attracted toward the pole of the greatest magnetic strength. A reference gas of known oxygen content is introduced into the unknown sample and is also analyzed separately, creating a pressure differential. The pressure differential is balanced by a measured flow from the unknown sample gas that is proportional to the differential pressure and is also proportional to the oxygen content of the unknown gas [101].

These types of instruments have not found widespread use in the heat treatment industry because of the following reasons [101]:

1. The magnetodynamic oxygen analyzer is too delicate an instrument for use in rugged heat treatment shop environments.
2. The accuracy of the thermomagnetic oxygen analyzer may be unstable and unreliable when used with high-temperature gases.
3. The magnetopneumatic oxygen analyzer is very sensitive to vibrations, which are often present in the heat treatment shop environment.

1.7.4.2 Electrochemical Oxygen Analyzers

It is also possible to measure the oxygen content of a gas by measuring the electrochemical potential (or voltage, emf) for electrons to flow between platinum electrodes in a known reference cell and an unknown cell [107–113]. The two cells are separated by an electrolyte

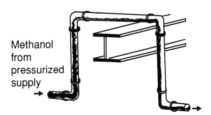

FIGURE 1.106 Elevated piping handling methanol from pressurized supply can accumulate nitrogen with resultant flow irregularities. (From *Installation of Nitrogen Methanol Atmosphere Systems*, Brochure, Praxair Inc., Chicago, IL.)

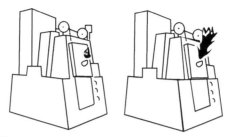

FIGURE 1.107 Furnace puffing due to segregation of liquid and gas in piping. (From *Installation of Nitrogen Methanol Atmosphere Systems*, Brochure, Praxair Inc., Chicago, IL.)

that permits the transfer of electrons between the two half-cells (cathode and anode). The most common electrolyte used for oxygen ions is zirconium oxide stabilized with calcium oxide ($ZrO_2 + 4\%$ CaO).

The electrochemical cell that is formed is represented as [82]:

$$\begin{array}{c|c|c} Pt: O_2 & ZrO_2 & O_2: Pt \\ \text{(unknown)} & \text{(electrolyte for oxygen ions)} & \text{(unknown)} \end{array}$$

The emf potential (or voltage) between the two half-cell reactions is measured and is proportional to the electron transfer between the two cells. The potential can be calculated by the well-known Nernst equation [82]:

$$E = \frac{RT}{nF} \ln \left[\frac{O_2 I}{O_2 II} \right]$$

where R is the gas constant, T is the absolute temperature, n is the number of electrons transferred ($n = 4$ for $O_2 \rightarrow 2O^{-2}$), F is the Faraday constant, and $O_2 I$ and $O_2 II$ are the reactivities of oxygen at either side of the electrolyte.

TABLE 1.42
Gas Composition Measurement Options

Measurement System	Advantages	Disadvantages
O_2 only	Most applicable because only one instrument is needed and low O_2 content indicates proper air/fuel ratio	Substoichiometric conditions are not defined. Inefficient combustion cannot be detected
$O_2 +$ total combustibles	Preferable because there is an instrument that will measure both O_2 and total combustibles. This defines total range	
$O_2 + CO_2$	Defines entire combustion range	Two instruments required
$O_2 + CO$	Defines entire combustion range	Two instruments required
CO_2 only	One instrument required	Particular information on equilibrium flue gas products should be available
Complete gas analysis	Stoichiometric composition gives better information for troubleshooting	More complex. Requires carrier gas and standard gas

Source: From T.J. Schultz, Portable flue gas analyzers for industrial furnaces, Inf. Letter No. 153, Catalog No. C10877, American Gas Association, Arlington, VA.

TABLE 1.43
Liquid Adsorbents for Heat Treatment Atmosphere for Orsat Analysis

Gas	Liquid Adsorbent
CO_2	Aqueous potassium hydroxide
CO	Ammoniacal cupric chloride
O_2	Alkaline pyrogallol

By combining constants, the Nernst equation for the oxygen probe can be written as [101]:

$$\text{emf} = 0.0496T \log \left[\frac{P_{O_2}(\text{ref})}{P_{O_2}} \right]$$

A schematic of the electrochemical oxygen probe is provided in Figure 1.110. The corresponding chemical reactions are

Anode:

$$2O^{2-} \rightarrow 4e + O_2(P_{O_2}\text{furnace})$$

Cathode:

$$O_2(P_{O_2} \text{ air}) + 4e^- \rightarrow 2O^{2-}$$

Sum:

$$O_2(P_{O_2}\text{air}) \rightarrow O_2(P_{O_2}\text{furnace})$$

The relationship between the emf for the oxygen probe and the carbon potential is shown in Figure 1.11.

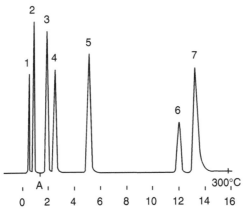

FIGURE 1.108 Gas chromatographic analysis of furnace atmospheres. (From D. Schwalm, *The Directory of Industrial Heat Processing and Combustion Equipment: United States Manufacturers, 1981–1982,* Energy edition, Published for Industrial Heating Equipment Association by Information Clearing House, pp. 147–153.)

TABLE 1.44
Thermal Conductivity of Heat Treatment Gases

Gas	Relative Thermal Conductivity	Gas	Relative Thermal Conductivity
Air (100°C)	100	CO	97
H_2	715	NH_3	90
CH_4	150	CO_2	68
O_2	102	H_2S	54
N_2	99	SO_2	35

Source: From R. Nemenyi and G. Bennett, *Controlled Atmospheres for Heat Treatment*, Franklin Book Co., 1995, pp. 22–1022.

An alternative electrochemical oxygen sensor is the nondepleting coulometric sensor illustrated in Figure 1.112. The advantage of the coulometric sensor is that the sensor does not require replacement. An analysis of the oxygen concentration occurs in an electrochemical cell where potassium hydroxide is used as the electrolyte. The half-cell reactions, which are driven by an external emf of 1.3 V, are [107]:

Cathode:

$$O_2 + 2H_2O + 4e^- \rightarrow 4OH^-$$

Anode:

$$4OH^- \rightarrow O_2 + 2H_2O + 4e^-$$

The following recommendations have been made for oxygen sensor installation [108]:

1. Sensor should be installed in a separate protective tube.
2. Sensor should be installed to monitor the gas stream after it has passed through the load.

TABLE 1.45
Relative Magnetic Susceptibility of Various Gases

Gas	Relative Magnetic Susceptibility
Oxygen (O_2)	100
Nitrogen oxide (NO)	43
Nitrogen dioxide (NO_2)	28
Hydrogen (H_2)	0.24
Carbon monoxide (CO)	0.01
Nitrogen (N_2)	0.00
Methane (CH_4)	−0.20
Argon (Ar)	−0.22
Ethylene (C_2H_4)	−0.26
Ammonia (NH_3)	−0.26
Carbon dioxide (CO_2)	−0.27

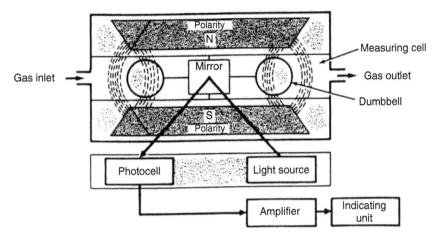

FIGURE 1.109 Schematic illustration of a paramagnetic oxygen analyzer.

3. It is suggested that the sensor be located in the top one third of the heated zone and away from the fan, halfway between the front and back of the heated zone and close to the furnace thermocouple.
4. Sensor should be mounted to permit easy access.
5. A good source of reference air must be available, preferably at a flow rate of 0.5–2.0 scfh.

Some of the most common sources of oxygen probe errors are as follows [110,111]:

1. Improper operating conditions due to probe deterioration. For example, soot formation, which is due to the operation of the furnace outside of the austenite phase field shown in Figure 1.111, will cause carbon to be deposited on the probe, leading to an incorrect response.
2. Alteration of $CO + CO_2$ content in the furnace atmosphere. As shown in Table 1.46, there will be a significant error in the carbon content if CO varies outside the target concentration.
3. Electrolyte failure may develop from cracks within the tube, making it possible for reference air to leak into the sample, leading to an incorrect probe response.

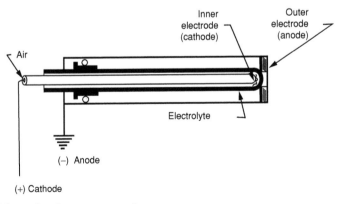

FIGURE 1.110 Schematic of an oxygen probe.

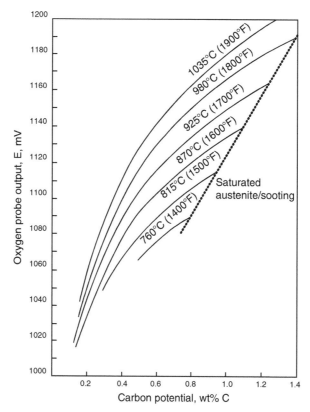

FIGURE 1.111 Isothermal relationship between oxygen probe output and carbon potential. (From R.W. Blumenthal, A technical presentation of the factors affecting the accuracy of carbon/oxygen probes, Proc. 2nd Int. Conf. on Carbonizing and Nitriding with Atmospheres, December 6–8, 1995, Cleveland, OH, pp. 17–22.)

4. Platinum or nickel wire used in the probe assembly may catalytically degrade the CH_4 gas into CO, producing erroneous results.
5. If the reference air sample that is necessary for correct operation of the probe becomes contaminated with furnace atmosphere gases, the response will be erroneous.
6. Reference air may leak out of the probe due to probe cracking and other flaws.

The detection of most of these problems is aided by the use of a second independent measurement of the furnace atmosphere [111].

1.7.4.3 Infrared Sensors

Infrared spectrophotometry provides an indirect measure of various atmospheric gases including CH_4, CO, and CO_2. This method involves passing infrared light through a sample of the gas as shown in Figure 1.113. The wavelength of infrared light characteristically varies from 0.75 to 300 μm, although most applications are performed in the region of 2.5–5 μm. The particular wavelength of infrared light selected depends on the gas analyzed. Figure 1.114 illustrates the infrared spectrum of a CO–CO_2–H_2O mix. Figure 1.115 illustrates the infrared spectrum for CO, CO_2, NH_3, and CH_4.

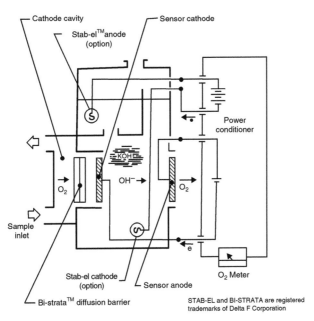

FIGURE 1.112 Coulometric sensor for oxygen probe. (From *Guide to the Selection of Oxygen Analyzers*, Brochure, Delta F Corporation, Woburn, MA.)

When the proper wavelength of light is passed through the gas, it is absorbed, and the amount of absorption is defined by Beer's law,

$$A = abc$$

where A is the absorption of light, a is the absorptivity (a characteristic value for each gas), b is the path length of light through the sample, and c is the concentration of the infrared-active material. Since the path length b is fixed and a is a constant for the gas analyzed, the amount of light absorbed is proportional to the concentration.

Infrared analyzers are used continuously in the system with gas sampling rates of approximately 1 min [82,106,115,116]. They have the advantage of accuracy and speed. However, infrared analyzers are relatively expensive, and significant damage to the optics in the infrared detector may be caused by contamination.

1.7.4.4 Dew-Point Analyzers

Another indirect measure of the carbon content of an atmospheric gas is the dew point. The dew point is defined as the temperature at which gas is saturated with water vapor, that is, it has a relative humidity of 100% [86]. The dew point varies with the temperature of the water

TABLE 1.46
Effect of CO Content on Carbon Potential for Constant Oxygen Probe Output of 1149 mV

	CO Content (%)					
	15	17	19	20	21	23
wt% carbon	0.80	0.88	0.95	1.0	1.05	1.11

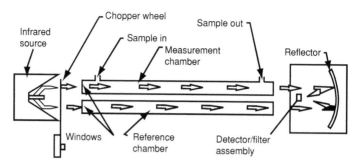

FIGURE 1.113 Nondispersive infrared detector (NDIR) for gas analysis. (From data provided by Teledyne Brown Engineering Analytical Instruments, City of Industry, CA.)

vapor as shown in Figure 1.116. A summary of dew-point variation with temperature and moisture is provided in Table 1.47

There are four principal types of apparatus for measuring dew point: (1) dew cup, (2) chilled mirror (Peltier effect), (3) lithium chloride cell (electrode), and (4) adiabatic expansion [82,106].

1.7.4.4.1 Dew Cup Apparatus

Although relatively inaccurate and unsuitable for continuous automated atmosphere control, the simplest method of measuring the dew point is with the dew cup apparatus shown in Figure 1.117 [82]. This method involves cooling a chromium-plated copper cup within a glass enclosure containing a dry ice–alcohol bath. The gas to be measured enters the instrument, and the cup is cooled until condensation is observed on its surface. The temperature at which this occurs is the dew point.

1.7.4.4.2 Chilled Mirror—The Peltier Effect

A method of measuring the dew point that is based on the Peltier effect is shown schematically in Figure 1.118. The Peltier effect refers to the difference in the current intensity between the two ends of a wire where one end is at the temperature of a reference cold junction and the other is at a higher temperature.

When a chilled mirror is used, the surface of the metal mirror is scanned photoelectrically while it is cooled in the gas to be analyzed. At the dew point, water vapor forms on the mirror. A photoelectric cell will direct the intensity of light reflected from the surface of the mirror, which is dependent on the amount of moisture in the gas. A schematic of a continuous chilled mirror detector is shown in Figure 1.119.

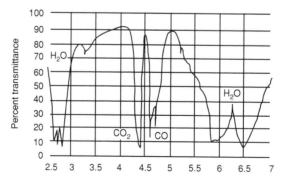

FIGURE 1.114 Infrared spectral frequencies for heat treatment gases. (From data provided by Horiba Instruments Inc., Irvine, CA.)

1.7.4.4.3 Measurement by Lithium Chloride Cell (Electrode)

The lithium chloride cell is the most widely used method for automatic measurement of dew point [82]. This method is based on the variation in the conductivity of hydroscopic lithium chloride in the presence of water vapor. An example of a lithium chloride cell is shown in Figure 1.120. Water vapor from the gas atmosphere is absorbed by the lithium chloride, creating a conductive electrolyte between the platinum wire electrodes. This process produces a temperature rise until equilibrium is attained. The equilibrium temperature is proportional to the dew point.

1.7.5 ADIABATIC EXPANSION

When a gas cools adiabatically, the dew point is achieved only for a particular pressure drop, temperature, and moisture content. The temperature at which this occurs corresponds to the dew point.

1.7.6 CARBON RESISTANCE GAUGE

The electrical resistance of steel varies with carbon content as shown in Figure 1.121. Instruments based on the measurements of continuous iron–nickel alloy wire expanded to the atmosphere gases available.

1.7.7 WEIGHT MEASUREMENT OF EQUILIBRIUM SHIM STOCK

Blumenthal and Hlasny [118] describe a method for calibration of carbon sensors that uses a test for true carbon potential, which is measured by equilibrating an AISI 1010 steel shim sample, 0.003 in. thick and 2.5×3 in. in area in a furnace atmosphere and then determining the carbon content of the shim by weight gain (or by chemical analysis). The shim-holding device is illustrated in Figure 1.122. The test procedure is as follows [118]:

1. Wearing rubber gloves, clean the shim test specimen with acetone and weigh it on a balance to the nearest 0.1 mg.
2. Roll the specimen into a cylinder approximately 3/4 in. in diameter, and insert it into the shim holder as shown in Figure 1.122.

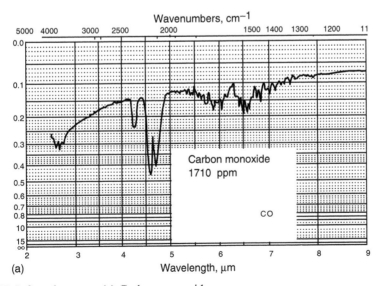

(a)

FIGURE 1.115 Infrared spectra. (a) Carbon monoxide;

(Continued)

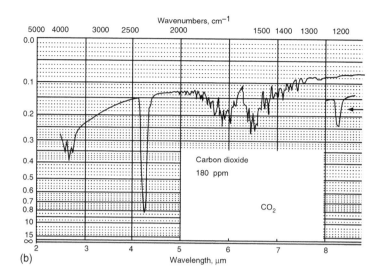

(b)

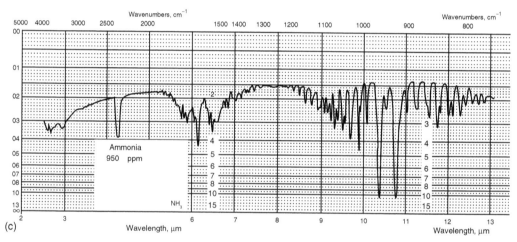

(c)

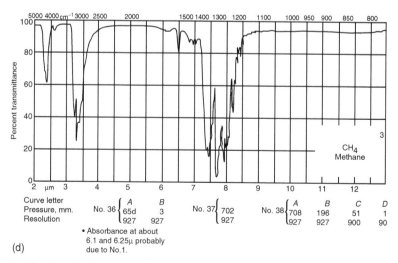

(d)

FIGURE 1.115 (Continued) (b) carbon dioxide; (c) ammonia; (d) methane. (From H.W. Bond, *Mat. Sci. Forum, 102–104*:831–838, 1992.)

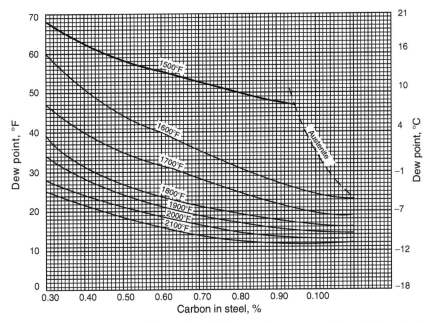

FIGURE 1.116 Correlation of dew point and equilibrium carbon content for endothermic atmosphere as a function of temperature. (From J.A. Zahniser, *Furnace Atmospheres*, ASM International, Materials Park, OH.)

3. Be sure the furnace is operating above 1600°F (871°C) and that there is a load in it. Close the doors with the atmosphere circulation fan running; usually 30 min is sufficient to attain equilibrium.
4. Remove the plug from the sample side of the furnace, and open the gate valve (see Figure 1.122). Insert the shim holder to the same depth as the carbon–oxygen sensor, usually approximately 20 in.
5. Leave the shim specimen in the furnace for 30 min.
6. Record the weight percent carbon from the carbon controller, average furnace temperature, and average probe millivolt output during the test.
7. Move the shim-holder cartridge from the furnace to the cooling chamber.
8. Wearing rubber gloves, remove the steel shim from the shim-holder cartridge.
9. Clean the shim with acetone.
10. Reweigh the shim to the nearest 0.1 mg. Calculate the carbon potential:

$$\text{Carbon potential} = \frac{\text{final weight} - \text{original weight}}{\text{final weight}} \times 100 + \text{wt\% C}$$

where wt% C is the original weight percent carbon content.

1.8 REFRACTORY MATERIALS

Furnace refractories are materials that must withstand severe or destructive high-temperature service conditions and resist attack by chemical substances, exposure to thermal shock, physical impact, and other abuse [119]. Heating efficiency and temperature uniformity within the furnace depend on the type of refractory material used to insulate the furnace. Examples of refractory materials include metal oxides such as MgO, SiO_2, and nonmetallic compounds

TABLE 1.47
Moisture Conversion Table

Dew Point		Vapor Pressure (Water/Ice in Equilibrium) (mmHg)	Water Content (ppm) on Volume Basis at 760 mmHg	Relative Humidity at 70°F (%)	Moisture (ppm) on Weight Basis in Air
°C	°F				
−110	−166	0.0000010	0.00132	0.0000053	0.00082
−108	−162	0.0000018	0.00237	0.0000096	0.0015
−106	−159	0.0000028	0.00368	0.000015	0.0023
−104	−155	0.0000043	0.00566	0.000023	0.0035
−102	−152	0.0000065	0.00855	0.000035	0.0053
−100	−148	0.0000099	0.0130	0.000053	0.0081
−98	−144	0.000015	0.0197	0.000080	0.012
−96	−141	0.00002	0.0289	0.00012	0.018
−94	−137	0.000033	0.0434	0.0018	0.027
−92	−134	0.00048	0.0632	0.000026	0.039
−90	−130	0.000070	0.0921	0.00037	0.057
−88	−126	0.00010	0.132	0.00054	0.082
−86	−123	0.00014	0.184	0.00075	0.11
−84	−119	0.00020	0.263	0.00107	0.16
−82	−116	0.00029	0.382	0.00155	0.24
−80	−112	0.00040	0.562	0.00214	0.33
−78	−108	0.00056	0.737	0.00300	0.46
−76	−105	0.00077	1.01	0.00410	0.63
−74	−101	0.00105	1.38	0.00559	0.86
−72	−98	0.00143	1.88	0.00762	1.17
−70	−94	0.00194	2.55	0.0104	1.58
−68	−90	0.00261	3.43	0.0140	2.13
−66	−87	0.00349	4.59	0.0187	2.84
−64	−83	0.00464	6.11	0.0248	3.79
−62	−80	0.00614	8.08	0.0328	5.01
−60	−76	0.00808	10.6	0.0430	6.59
−58	−72	0.0106	13.9	0.0565	8.63
−56	−69	0.0138	18.2	0.0735	1.3
−54	−65	0.0178	23.4	0.0948	14.5
−52	−62	0.0230	30.3	0.123	18.8
−50	−58	0.0296	38.8	0.157	24.1
−48	−54	0.0378	49.7	0.202	30.9
−46	−51	0.481	63.3	0.257	39.3
−44	−47	0.0609	80.0	0.325	49.7
−42	−44	0.0768	101	0.410	62.7
−40	−40	0.0966	127	0.516	78.9
−38	−36	0.1209	159	0.644	98.6
−36	−33	0.1507	198	0.804	122.9
−34	−29	0.1873	246	1.00	152
−32	−26	0.2318	305	1.24	189
−30	−22	0.2859	376	1.52	234
−28	−18	0.351	462	1.88	287
−26	−15	0.430	566	2.30	351
−24	−11	0.526	692	2.81	430
−22	−8	0.640	842	3.41	523
−20	−4	0.776	1,020	4.13	633
−18	−0	0.939	1,240	5.00	770

TABLE 1.47 (continued)
Moisture Conversion Table

Dew Point		Vapor Pressure (Water/Ice in Equilibrium) (mmHg)	Water Content (ppm) on Volume Basis at 760 mmHg	Relative Humidity at 70°F (%)	Moisture (ppm) on Weight Basis in Air
°C	°F				
−16	+3	1.132	1,490	6.03	925
−14	+7	1.361	1,790	7.25	1,110
−12	+10	1.632	2,150	8.69	1,335
−10	+14	1.950	2,570	10.4	1,596
−8	+18	2.326	3,060	12.4	1,900
−6	+21	2.765	3,640	14.7	2,260
−4	+25	3.280	4,320	17.5	2,680
−2	+28	3.880	5,100	20.7	3,170
0	+32	4.579	6,020	24.4	3,640
+2	+36	5.294	6,970	28.2	4,330
+4	+39	6.101	8,030	2.5	4,990
+6	+43	7.013	9,230	37.4	5,730
+8	+46	8.045	10,590	42.9	6,580
+10	+50	9.029	12,120	49.1	7,530
+12	+54	10.52	13,840	56.1	8,600
+14	+57	1.99	15,780	63.9	9,800
+16	+61	13.63	17,930	72.6	11,140
+18	+64	15.48	20,370	82.5	12,650
+20	+68	17.54	23,080	93.5	14,330
+22	+7135	19.827	26,088		16,699
+24	+75	22.377	29,443		18,847
+26	+79	25.209	33,169		21,232
+28	+82	28.349	37,301		23,877
+30	+86	31.824	41,874		26,804

Source: From R.W. Blumenthal and A.T. Melville, Hot gas measuring probe, U.S. Patent 4,588,493 (May 13, 1986).

such as SiC and SiN. Other materials that are used as refractories include quartz, bauxite, fireclay, chromite ore, magnesite, and zirconium oxide [119]. Refractory materials are available in various shapes such as brick and in bulk form.

Traditionally, the most common lining used in the heat treatment furnaces has been firebrick. Although firebrick continues to be used, materials with significantly better chemical properties such as ceramics and fiber are becoming popular.

The objective of this section is to provide a general overview of refractory materials and their classification.

1.8.1 Refractory Classification

1.8.1.1 Magnesium Compositions

Refractory materials are available in five classes of acid-resistant composition [119]:

1. Magnesia or magnesite (MgO).
2. Magnesia in combination with chromium-containing materials such as Cr_2O_3 [62]. If chromium is the major component, then it is referred to as chrome-magnesite. If magnesia is the predominant component, then it is referred to as magnesite-chrome.

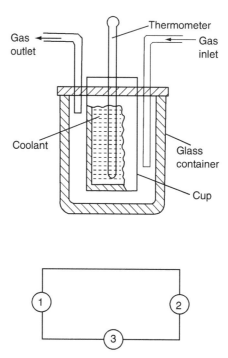

FIGURE 1.117 Dew cup instrument. (From R. Nemenyi and G. Bennett, *Controlled Atmospheres for Heat Treatment*, Franklin Book Co., 1995, pp. 22–1022.)

3. Magnesia in combination with spinel (magnesium aluminum silicate).
4. Magnesia in combination with 2.5 or 4.5% carbon, where carbon is in the form of pitch, which bonds refractory aggregates.
5. Dolomite, which is composed of approximately equal amounts of $MgCO_3$ and $CaCO_3$.

1.8.1.2 Compositions Containing Aluminum Oxide

Another class of refractory compositions is based on high alumina (Al_2O_3) materials that contain more than 47.5% Al_2O_3. There are a number of these materials [82,119].

1. Mullite brick, which is compositionally 71.8% Al_2O_3 and 28% SiO_2. This material has excellent volume stability and strength at high temperatures.
2. Chemically bonded, normally phosphate-bonded, brick that reacts to form aluminum orthophosphate ($AlPO_4$).

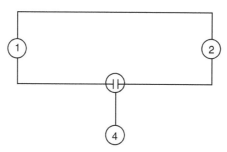

FIGURE 1.118 Dew-point measurement by the Peltier effect. (From R. Nemenyi and G. Bennett, *Controlled Atmospheres for Heat Treatment*, Franklin Book Co., 1995, pp. 22–1022.)

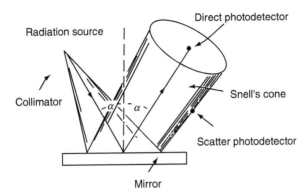

FIGURE 1.119 Continuous on-line dew-point monitor. (From W. Cole, *Heat Treating*, March 1993, 19–20.)

3. Alumina-chrome brick, which is composed of a solid solution of chromium oxide (Cr_2O_3) and a high-purity alumina.
4. Alumina-carbon brick, which is a resin-bonded high alumina graphite-containing composition.

1.8.1.3 Fireclay Compositions

Compositionally, fireclay is a hydrated aluminum silicate ($Al_2O_3 \cdot 2SiO_2 \cdot 2H_2O$). After dehydration at elevated temperatures, the residual material should contain 45.9% alumina and 54.1% silica. There are five standard ASTM classifications of firebrick [119]:

1. Super-duty fireclay contains 40–45% Al_2O_3. This is the refractory classification with the greatest high-temperature volume stability and spalling resistance.
2. High-duty fireclay is slightly less refractory than the super-duty type but is still spalling- or slag-resistant.
3. Medium-duty fireclay is suitable for moderately severe applications.
4. Low-duty fireclay is suitable for moderate temperature application only.
5. Semisilica brick contains only 15–18% alumina. It exhibits good high-temperature load-bearing capacity and volume stability.

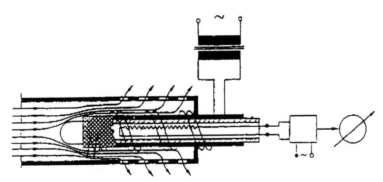

FIGURE 1.120 Lithium chloride dew-point probe. (From M.J. Hill, The efficient use of atmospheres in a continuous furnace using the concept of zoning, Int. Foundry Heat Treat. Conf., Johannesburg, S. Africa, Vol. 2, 1985, pp. 1–30.)

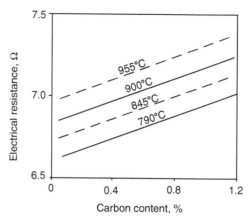

FIGURE 1.121 Electrical resistance of steel vs. carbon content at various temperatures. (From R. Nemenyi and G. Bennett, *Controlled Atmospheres for Heat Treatment*, Franklin Book Co., 1995, pp. 22–1022.)

1.8.1.4 Silica Refractories

Silica refractories are used for high-temperature (3080–3110°F) melting applications. They are capable of withstanding pressures up to 50 psi and are chemically resistant to acid slag. They do not undergo spalling at temperatures above 1200°F. There are two ASTM classifications of silica firebrick: Type A and Type B.

1.8.1.5 Monolithic Refractories

Monolithic refractories are special mixes or blends of dry granular or cohesive plastic materials that form a virtually joint-free lining [119]. They are classified as plastic refractories [120]; ramming mixes, which are denser and stronger than plastic refractories; gunning mixes, which may be pneumatically bonded to a furnace wall; and castables, or refractory concrete. Plastic refractories are prepared from fireclay, high-alumina graphite, alumina-chrome, and other materials.

Numerous other materials are developed, for example high-alumina fiber [121], glass fiber [122], and microporous-insulating refractories [123], to provide greater resistance to slag attack, low thermal expansion, increased resistance to spalling, reduced impurities, improved hot strength, acceptable creep resistance, and improved resistance to thermal shock and corrosion [119].

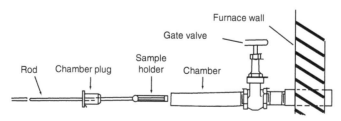

FIGURE 1.122 Apparatus for determination of carbon content by shim-stock exposure and measurement. (From R.N. Blumenthal and R. Hlasny, *Heat Treating*, August 1991.)

1.8.2 DESIGN PROPERTIES

The dimensional change of a material with respect to changes in temperature is described by

$$\frac{\Delta L_0}{L_0} = \alpha(T - T_0)$$

where α is the coefficient of linear expansion. Table 1.48 provides viral coefficients to calculate the percent of linear expansion with respect to its length at 20°C (293 K) for various materials.

The value of $\Delta L_0/L_0$ is of considerable value when estimating the impact of a known temperature change (ΔT) across a wall [(hot face temperature) − (cold face temperature)] and refractory deformation. If the modulus of elasticity (E) is known, it is possible to calculate the stress σ from

$$\sigma = E\alpha(T_2 - T_1)$$

Thermal conductivity δ is calculated from the heat capacity C_p:

$$\delta = k/C_p\rho$$

where k is the thermal conductivity, δ is thermal diffusivity, and ρ is 123 density.

Figure 1.123 and Table 1.49 provide heat capacity data as a function of temperature for various simple refractory compositions [124]. Thermal conductivity as a function of temperature is illustrated in Figure 1.124 [124]. Thermal conductivity data for firebrick and alumina are provided in Figure 1.125 and Figure 1.126, respectively [124].

For firebrick, the thermal conductivity k is dependent on bulk density (BD) [124]:

$$k_{500°F} \text{ [Btu in./(h ft}^2°\text{F)]} = 0.03455\text{BD (lb/ft}^3) - 0.2545$$

$$k_{260°C} \text{ [W/(mK)]} = 0.3110\rho_b \text{ (g/cm}^3) - 0.0367$$

For insulating castables, the thermal conductivity may be calculated from [124]

$$k_{500°F} \text{ [Btu in./(h ft}^2 °\text{F)]} = 0.04375\text{BD (lb/ft}^3) - 0.4250$$

$$k_{260°C} \text{ [W/(mK)]} = 0.3939\rho_b \text{ (g/cm}^3) - 0.0613$$

1.8.3 FURNACE REFRACTORY INSTALLATION

Traditionally, firebrick was used to reline heat treatment furnaces. Today, refractory materials can be installed in a number of ways. Furnaces may be lined with brick or refractory blankets, or they may be lined with refractory modules as shown in Figure 1.127.

1.9 FANS

A fan, blower, or compressor may be selected to move air or gas. They differ from each other in their operational pressure as shown in Table 1.50 [125].

This section presents an overview of the selection and operation of a fan that can be used in heat treatment equipment.

TABLE 1.48
Virial Coefficient for Linear Thermal Expansion of Selected Solid Materials

$$100\Delta L_t/L_{293} = A + B\,(10^{-4}T) + C\,(10^{-4}T)^2 + D\,(10^{-4}T)^3 \quad (T, K)$$

Material	MP (K)	A	B	C	D	Note
Al_2O_3 (hex.)	2327	-0.180	$+5.494$	$+22.520$	-22.940	1
C_2O	3200	-0.321	$+10.590$	$+13.100$	-14.050	
Cr_2O_3 (hex.)	2603	-0.280	$+10.380$	-31.220	$+106.200$	2
Fe_2O_3 (trig.)	1838	-2.537	$+7.300$	$+49.640$	-114.000	3
MgO	3125	-0.326	$+10.400$	$+25.810$	-28.340	
SiO_2 (lo qtz.)	~873 (tr.)	-0.236	$+6.912$	$+0.556$	$+1312.00$	4
SiO_2 (hi qtz.)	1743 (tr.)	$+1.040$	$+0.068$	$+1.660$	$+18.000$	Est.
SiO_2 (vitr.)	~1273 (cryst.)	-0.015	$+0.397$	$+4.666$	-34.460	
ZrO_2 (monocl.)	2988	-0.314	$+13.040$	-90.920	$+408.400$	5
$Al_6Si_2O_{13}$	2193	-0.0929	$+2.580$	$+21.530$	-45.720	
$CaAl_2O_4$	1873	-0.107	$+2.578$	$+39.680$	-90.770	
Ca_2SiO_4	2403	-0.345	$+1.260$	$+16.560$	$+27.330$	
$MgAl_2O_4$	2408	-0.183	$+5.456$	$+28.060$	-41.810	
$Mg_2Al_4\,Si_5O_{18}$	~1773	$+0.00911$	-0.912	$+20.640$	-3.921	6
$MgCr_2O_4$	2673	-0.176	$+5.822$	$+55.80$	$+23.360$	
$MgFe_2O_4$	2023	-0.218	$+6.003$	$+52.560$	-94.040	
Mg_2SiO_4	2183	-0.238	$+7.166$	$+33.810$	-37.970	
Mg_2TiO_4	~2100	-0.249	$+8.294$	$+4.074$	$+94.300$	
$ZrSiO_4$	2673	-0.136	$+5.337$	-30.420	$+209.400$	
AIN (hex.)	~2500	-0.0809	$+1.806$	$+31.760$	-72.560	
B_4C (rhomb.)	2623	-0.114	$+3.523$	$+12.660$	-5.085	8
BN	~3273 (sub.)	-0.00133	-1.278	$+49.110$	-86.350	
SiC	~2923 (dec.)	-0.0991	$+2.970$	$+13.880$	-15.480	
TiC	~3410	-0.177	$+5.710$	$+1.740$	$+2.412$	
C (graphite)	~3900	-0.0550	$+1.552$	$+12.050$	-10.330	9
C (graphite)	~3900	-0.1580	$+5.561$	-8.850	$+35.550$	9
C (vir.)	~2700 (cryst.)	-0.890	$+3.015$	$+1.285$	$+17.240$	
Fe	1800	-0.289	$+7.350$	$+93.300$	-314.000	10

Notes:

1. Cryst. exp. $c/a \sim 1.1$.
2. Cryst. exp. $a/c \sim 1.3$.
3. Cryst. exp. $a/c \sim 1.26$.
4. Cryst. exp. $a/c \sim 1.58$.
5. Cryst. exp. $c/b \sim 2.5$.
6. Cordierite refractory.
7. Cryst. exp. $a/c \sim 1.18$.
8. Cryst exp. $\sim$ isotrop.
9. Grade ATJ, parallel and perpendicular to the textural grain, respectively. Cryst. exp. $c/a \sim 10$.
10. Numerous steels and SS agree with Fe within $\sim \pm 15\%$.

1.9.1 Calculation of Fan Performance (The Fan Laws)

The work done by a fan is related to the volume change of the gas moved due to compression (see Figure 1.128) and is calculated from

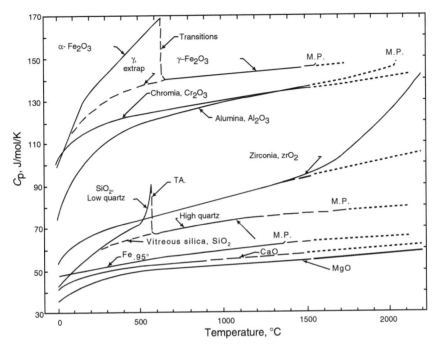

FIGURE 1.123 Heat capacity of various refractory materials.

$$\text{Work} = \Delta p \times V$$

where Δp is the pressure increase across the fan and V is the volume of the gas moved. The pressure difference (Δp) is referred to as head and is usually quantified in inches of water.

Static pressure (SP) is the pressure on the walls, ducts, and piping. The velocity pressure required to move the gas is the difference between total pressure and the static head.

Air horsepower (Ahp) output is defined as

$$\text{Ahp} = \frac{VH}{6356}$$

where V is the volumetric flow through the fan (ft^3/min) and H is the head (Δp) across the fan. The head may be static (H_s) or total (H_t):

$$H_t = H_s + H_s$$

Static air horsepower (Ahp$_s$) is the resistance that must be overcome and is related to static head. Total air horsepower (Ahp$_t$) or power output is based on total head.

The static and mechanical efficiency is calculated from

$$E_s = \frac{\text{Ahp}_s}{\text{power input}}$$

$$E_t = \frac{\text{Ahp}_t}{\text{power input}}$$

TABLE 1.49
Molal Heat Capacity of Solid Substances [C_p in J/(mol K); T in K]

$$C_p = a + b\,(10^{-3}T) + c\,(10^{-3}T)^2 + d/(10^{-3}T)$$

	Formula	a	b	c	d	Note
Al_2O_3	101.96	+154.96	−16.168	+7.120	−20.817	
CaO	56.08	+57.68	−1.324	+1.560	−4.418	
Cr_2O_3	152.00	+137.14	−3.568	+3.120	−9.585	
Fe_2O_3	159.70	+176.70	−24.000	+7.200	−19.843	1
$Fe_{0.95}O$	69.06	+45.31	+14.900	−2.600	−0.374	2
MgO	40.31	+63.24	−7.632	+2.880	−7.263	
SiO_2	60.09	+77.09	+3.384	−0.160	−10.558	3
ZrO_2	123.32	+60.88	+22.320	−1.600	−3.370	
$Al_6Si_2O_{13}$	426.06	+607.83	−42.752	+22.880	−81.020	4
$CaAl_2O_4$	158.04	+236.00	−39.240	+14.360	−32.120	
Ca_2SiO_4	172.25	+182.73	+3.768	+1.580	−17.372	
$CaTiO_3$	135.96	+149.08	−2.916	−0.360	−15.051	
$MgAl_2O_4$	142.27	+233.16	−39.244	+14.360	−32.124	4
$MgCr_2O_4$	192.31	+212.76	−18.404	+8.360	−24.261	
Mg_2SiO_4	140.71	+191.48	−4.236	+8.040	−21.790	5
$MgTiO_3$	120.19	+145.39	−3.128	+3.520	−15.949	
$ZrSiO_4$	183.31	+179.49	−21.380	+12.200	−22.838	
AlN	40.99	+61.68	−7.344	+2.560	−8.911	
B_4C	55.25	+35.55	+86.508	−16.920	−3.145	
BN	24.82	+37.02	+19.324	−5.560	−6.814	
SiC	40.10	+55.08	−0.876	+2.040	−8.312	
TiC	59.89	+65.44	−7.964	+2.760	−8.911	
C (graph.)	12.01	+18.37	+8.592	−2.480	−3.969	6

Notes:
1. Constants fit γ-Fe_2O_3 Tr. ~630°C; mp 1565°C.
2. Melting point 1369°C.
3. Constants fit high-quartz. Tr. ~577°C; mp 1723°C.
4. Constants estimated as sum of $CaO + Al_2O_3$ and $MgO + Al_2O_3$.
5. Melting point 1910°C.
6. No reference nongraphitic carbon exists.

Head and horsepower vary inversely with absolute gas temperature T and absolute gas pressure P. This is calculated from

$$\text{Corrected head } (H_a) = H_b \frac{P_a T_b}{P_b T_a}$$

$$\text{Corrected horsepower } (\text{Hp}_a) = \text{Hp}_b \frac{P_a T_b}{P_b T_a}$$

where the subscript a indicates after and b indicates before.

The interrelationships of head, efficiency, volume, and horsepower are shown in Figure 1.129. A system resistance curve relates fan pressure and horsepower to flow as shown in Figure 1.130. The fan supply equals the demand at the intersection point. To conserve energy,

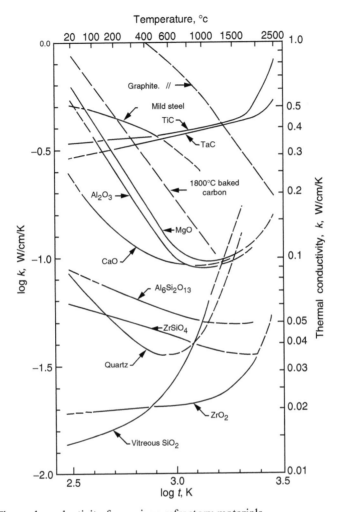

FIGURE 1.124 Thermal conductivity for various refractory materials.

fan speed can be varied to match output pressure and system resistance for a given flow as shown in Figure 1.131. From these curves, it is evident that fan capacity Q, pressure P, and horsepower (hp) vary with speed and can be calculated as

$$\frac{Q_1}{Q_2} = \frac{\text{rpm}_1}{\text{rpm}_2}$$

$$\frac{P_1}{P_2} = \left(\frac{\text{rpm}_1}{\text{rpm}_2}\right)^2$$

$$\frac{\text{hp}_1}{\text{hp}_2} = \left(\frac{\text{rpm}_1}{\text{rpm}_2}\right)^3$$

Fan capacity and speed vary with the square root of pressure:

$$\frac{\mathrm{rpm}_1}{\mathrm{rpm}_2} = \frac{Q_1}{Q_2} = \left(\frac{P_1}{P_2}\right)^{1/2}$$

Horsepower varies with pressure:

$$\frac{\mathrm{hp}_1}{\mathrm{hp}_2} = \left(\frac{P_1}{P_2}\right)^{3/2}$$

At constant pressure and density, for geometrically similar fans the capacity power and speed are calculated from the wheel diameter (dia):

$$\frac{Q_1}{Q_2} = \frac{\mathrm{hp}_1}{\mathrm{hp}_2} = \left(\frac{\mathrm{dia}_1}{\mathrm{dia}_2}\right)^2$$

$$\frac{\mathrm{rpm}_1}{\mathrm{rpm}_2} = \frac{\mathrm{dia}_1}{\mathrm{dia}_2}$$

If both the speed and diameter change, then

$$\frac{Q_1}{Q_2} = \frac{\mathrm{rpm}_1}{\mathrm{rpm}_2} \times \left(\frac{\mathrm{dia}_1}{\mathrm{dia}_2}\right)^3$$

$$\frac{P_1}{P_2} = \left(\frac{\mathrm{rpm}_1}{\mathrm{rpm}_2}\right)^2 \times \left(\frac{\mathrm{dia}_1}{\mathrm{dia}_2}\right)^2$$

$$\frac{\mathrm{hp}_1}{\mathrm{hp}_2} = \left(\frac{\mathrm{rpm}_1}{\mathrm{rpm}_2}\right)^3 \times \left(\frac{\mathrm{dia}_1}{\mathrm{dia}_2}\right)^5$$

$$\frac{\mathrm{hp}_1}{\mathrm{hp}_2} = \frac{Q_1}{Q_2} \times \frac{P_1}{P_2}$$

$$\frac{\mathrm{rpm}_1}{\mathrm{rpm}_2} = \frac{Q_1}{Q_2} = \frac{\mathrm{hp}_1}{\mathrm{hp}_2} = \left(\frac{d_2}{d_1}\right)^{1/2}$$

and

$$\frac{\mathrm{rpm}_1}{\mathrm{rpm}_2} = \frac{Q_1}{Q_2} = \frac{\mathrm{hp}_1}{\mathrm{hp}_2} = \left(\frac{b_2}{b_1}\right)^{1/2} = \left(\frac{T_1}{T_2}\right)^{1/2}$$

where d is the density of the gas transferred, b is the barometric pressure, and T is the absolute temperature.

At constant speed and capacity,

$$\frac{\mathrm{hp}_1}{\mathrm{hp}_2} = \frac{p_1}{p_2} = \frac{d_1}{d_2} = \frac{b_1}{b_2} = \frac{T_2}{T_1}$$

Transferring a constant amount of gas results in

$$\frac{Q_1}{Q_2} = \frac{\mathrm{rpm}_1}{\mathrm{rpm}_2} = \frac{P_1}{P_2} = \frac{d_2}{d_1} = \frac{b_2}{b_1} = \frac{T_1}{T_2}$$

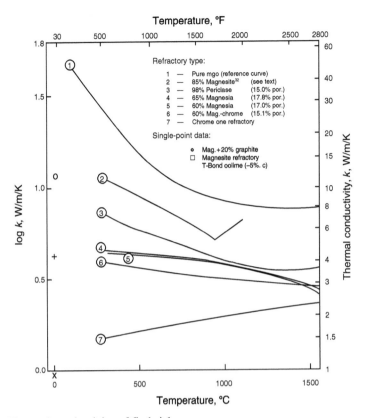

FIGURE 1.125 Thermal conductivity of firebrick.

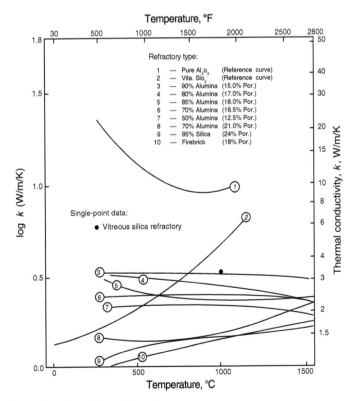

FIGURE 1.126 Thermal conductivity of alumina.

FIGURE 1.127 Car-bottom furnace lined with power-lock® modules. (Courtesy of The Carborundum Company, Fibers Division.)

$$\frac{hp_1}{hp_2} = \left(\frac{d_2}{d_1}\right)^2$$

$$\frac{hp_1}{hp_2} = \left(\frac{b_2}{b_1}\right)^2 = \left(\frac{T_1}{T_2}\right)^2$$

If both the temperature and pressure vary, then

$$\frac{Q_1}{Q_2} = \frac{rpm_1}{rpm_2} = \left(\frac{P_1}{P_2} \times \frac{T_1}{T_2}\right)^{1/2}$$

$$\frac{hp_1}{hp_2} = \left(\frac{P_1^3}{P_2^3} \times \frac{T_1}{T_2}\right)^{1/2}$$

TABLE 1.50
Operational Pressures of Fans, Blowers, and Compressors

Machine	Operational Pressure (psi)
Fan	<1
Blower	<50
Compressor	>35

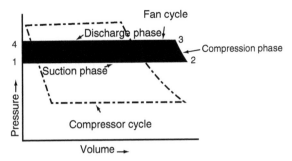

FIGURE 1.128 Illustration of a fan pressure cycle.

For geometrically similar fans, the flow, speed, and head are related by specific speed and specific diameter [125]. Specific speed (N_s) is that rpm at which a fan would operate if reduced proportionately in size so that it delivers 1 ft^3/min of air at standard conditions against a 1 in. H$_2$O SP [125].

$$N_s = \frac{\text{rpm}(\text{ft}^3/\text{min})^{1/2}}{\text{SP}^{3/4}}$$

The specific diameter D_s is the fan diameter D required to deliver 1 ft^3/min standard air against 1 in. H$_2$O SP at a given speed

$$D_s = \frac{D(\text{SP})^{1/4}}{(\text{ft}^3/\text{min})^{1/2}}$$

1.9.2 FAN SELECTION

The first step in selecting a fan is to determine the system requirements. First determine the amount of gas to be moved, making appropriate corrections for temperature, density, and barometric pressure. If multiple fans are to be sued, divide the total volume by the number of fans.

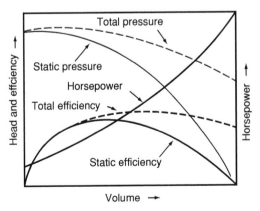

FIGURE 1.129 Fan performance curve. (From R.J. Aberbach, *Power*, New York, NY.)

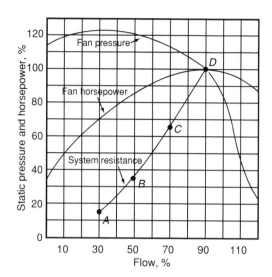

FIGURE 1.130 System resistance curve.

The second step is to determine SP where the fan will operate. This is done by plotting specific speed vs. specific diameter and static efficiency for the number of fans to be considered as shown in Figure 1.132 [125].

The flow (ft^3/min), SP, and specific speed (N_s) can be calculated from the nomogram shown in Figure 1.133 from the N_s. The specific diameter (D_s) is determined by calculation, and the type of fan required is identified from a graph such as Figure 1.132.

When selecting fans, consideration must be given to whether they will be installed in series or in parallel. When installed in series, each fan handles the same amount of gas and the volumetric capacity varies with the density of the gas. When installed in parallel, each fan handles only part of the total volume of gas.

When determining the system resistance, one must account for flow losses due to duct resistance and diameter (Figure 1.134), shape (Figure 1.135), and frictional losses in elbows in the duct system (Figure 1.136) [125].

1.9.3 FLOW CALIBRATION

The flow rate of the fan system can be determined after initial installation or at any time by placing a pitot tube in a straight length of duct at approximately 10 diameter lengths from either the inlet or the outlet side of the fan. Velocity pressure (H_v) is then determined at 20 different positions in the duct as shown in Figure 1.137. The square root of each velocity is determined and then averaged for the 20 readings.

The gas density (D) is determined by correcting for temperature and pressure

$$D \ (\text{lb/ft}^3) = 0.075 \left[\frac{530}{460 + \text{local temperature } (^\circ\text{F})} \right] \left[\frac{\text{barometer reading}}{29.92} \right]$$

The ratio of standard to actual gas (air) density (K) is determined as

$$K = \frac{0.075}{D}$$

The average air velocity V (ft/min) is calculated from

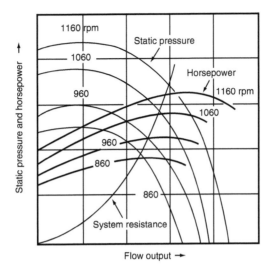

FIGURE 1.131 System resistance curve variation with fan speed.

$$V = 4000\left(\sqrt{H_\text{v}}\right)_\text{ave}$$

The volume flow rate is determined by multiplying the linear flow rate by the cross-sectional area of the duct.

1.10 FIXTURE MATERIALS

Heat treatment conditions such as corrosive and reactive atmospheres and high temperatures provide an especially harsh environment for furnace fixtures: baskets, trays, chains, conveyor belts, and radiant tubes. This section provides an overview of typical heat-resistant alloys that are used for fixture construction.

1.10.1 COMMON HIGH-TEMPERATURE ALLOYS

The most common heat-resistant alloy materials used for furnace fixture construction are wrought and cast materials of Fe–Cr–Ni, Fe–Ni–Cr, and Ni-based alloys. Data on the composition of selected alloys are provided in Table 1.51.

A number of alloying elements are used for heat-resistant materials [127]:

Chromium (Cr) improves oxidation resistance at temperatures below 950°C, improves resistance to sulfidation and carburization, exhibits poor nitriding resistance, increases high-temperature resistance.

Silicon (Si) improves resistance to sulfidation, nitriding, oxidation, and carburizing. In synergy with Cr, Si improves scale resilience.

Molybdenum (Mo) improves high-temperature resistance and creep strength and reduces oxidation resistance.

Nickel (Ni) improves resistance to carburization and nitriding and reduces sulfidation resistance.

Tungsten (W) has effects similar to those of Mo.

Carbon (C) improves strength, improves nitriding and carburizing resistance, reduces resistance to oxidation.

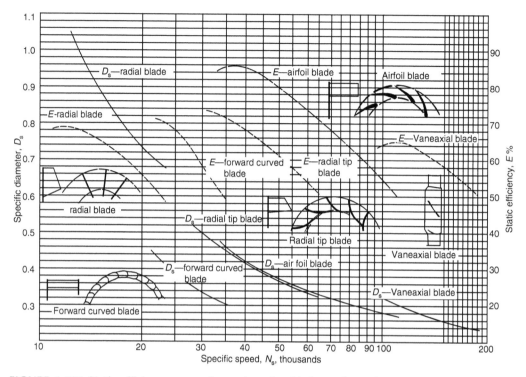

FIGURE 1.132 Static efficiency curves for various fan blade configurations. (From R.J. Aberbach, *Power*, New York, NY.)

Yttrium (Y) and rhenium (Re) improve resistance to spalling, oxidation, carburizing, and sulfidation.

Aluminum (Al) acts independently of and synergistically with Cr to improve oxidation and sulfidation resistance and reduces nitriding resistance.

Titanium (Ti) imparts poor nitriding resistance.

Niobium (Nb) improves creep strength.

Manganese (Mn) improves high-temperature strength and creep but gives poor oxidation and nitriding resistance.

Cobalt (Co) improves sulfidation resistance.

High-temperature alloys should be selected with regard to resistance to combustion products (see Figure 1.138 and Figure 1.139), oxidation (Table 1.52), carburization (Table 1.53), nitriding (Table 1.54), molten salts (Table 1.55), and other severe environments encountered in the particular heat treatment of interest [126].

1.11 PARTS WASHING

1.11.1 WASHING PROCESSES

Many parts enter the heat treatment shop with residual lubricants, coolants, or corrosion inhibitor films on the surface. These surface contaminants may contain graphite, molybdenum, sulfide (MoS_2), or compounds of chlorine, sulfur, silicon, phosphorus, or boron. Their presence on the surface of parts to be heat treated may prevent uniform atmosphere diffusion or heat transfer (on heating or cooling) and lead to soft spotting or increased distortion before the heat treatment operation [129]. Degreasing may be performed before

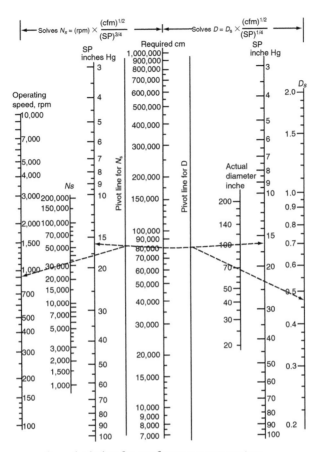

FIGURE 1.133 Nomogram for calculating fan performance parameters.

or after the heat treatment process, or both. A schematic illustrating the process is shown in Figure 1.139.

Until recently, one of the most common parts cleaning operations was vapor degreasing. The solvents that have been used for vapor degreasing include trichloroethane, carbon tetrachloride, trichloroethylene, percholoroethylene, and tetrachloroethane (see Figure 1.140) [129,131]. Cleaning with such solvents is expensive, in many cases toxic; they are often pollutants, and their use is accompanied by a high disposal cost.

An alternative to vapor degreasing using a chlorohydrocarbon is to use a hydrocarbon solvent. Hydrocarbon solvents for these processes typically exhibit flash points of more than 70°C. The benefits of hydrocarbon-based parts cleaning processes include the following [132]:

1. They have no detrimental impact on the ozone layer.
2. They provide high solvent efficiency.
3. Petroleum solvents are noncorrosive.
4. Hydrocarbon solvents are relatively nontoxic and easy to handle.
5. The solvents are reusable.
6. There is no need for drainage or exhaust gas facilities.

Currently, systems employ a combination of vapor degreasing with hydrocarbon solvents and vacuum drying [132]. The relationship between the vapor pressure and the temperature of

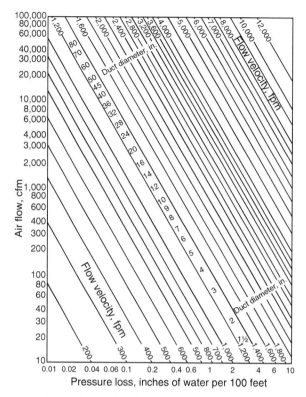

FIGURE 1.134 Fan pressure loss with varying flow rates and duct diameters.

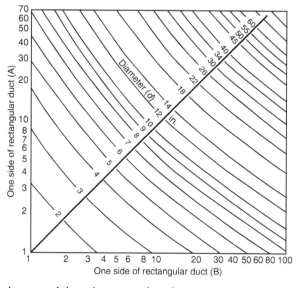

FIGURE 1.135 Equivalent round duct size conversion chart.

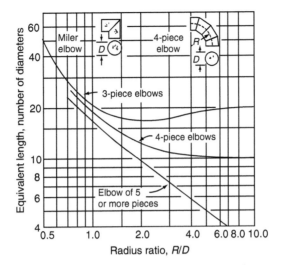

FIGURE 1.136 Equivalent length chart for ducts with elbows.

a typical hydrocarbon solvent used for this process and more traditional halogenated hydro-carbons is shown in Figure 1.140.

Currently, the most commonly used alternatives to solvent-based vapor degreasing are a water-based washing system. These systems include the following [129,131]:

Emulsion cleaners. An organic solvent is emulsified in water that contains various addi-tives such as surfactants to aid in the surface soil removal process.

Semiaqueous cleaners. Parts may be cleaned using solvents such as terpenes, esters, or a blend of hydrocarbons. Residual solvents are removed as an emulsion by subsequent cleaning with water or surfactant.

Alkaline cleaners. This process uses alkaline metal salts (also known as detergent builders) (see Table 1.56) that include a mixture of silicates, phosphates, and surfactant. These cleaning solutions are suitable for automated systems and heavily soiled parts.

Neutral cleaners. These cleaning solutions are composed of water and are usually nontoxic surfactants. They are suitable for lightly soiled surfaces and are considerably less alkaline (neutral) than the alkaline cleaners described above.

A schematic of a water-based cleaning process is presented in Figure 1.141.

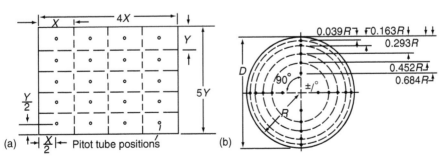

FIGURE 1.137 Recommended pitot placement positions for fan flow calibrations. (a) In rectangular duct; (b) in round duct, keep pitot tube within $\pm 0.5\%$ of R and $\pm 1°$ of indicated positions.

TABLE 1.51
Nominal Compositions of Alloys Used in Laboratory and Field Tests[a]

Alloy (UNS No.)	Composition (%)										
	Ni	Fe	Co	Cr	Mo	W	Al	Ti	Si	C	Other
Cabot No. 214	75	2.5	2.0	16.0	0.5	0.5	4.5	—	—	—	Y (present)
Hastelloy X (N06002)	—	47	—	18.5	1.5	22.0	9.0	0.6	—	—	—
Alloy 600 (N06600)	72	8.0	1.0	15.5	—	—	0.35^b	0.3^b	—	—	—
Inconel 601 (N06601)	Bal	14.0	—	25	—	—	1.4	0.3	—	—	—
Cabot No. 625 (N06625)	Bal	5.0	1.5	9.0	—	0.4	0.4^b	—	—	—	Cb + Ta = 3.5
Haynes No. 230	57	3.0	3.0	22.0	2.0	14.0	0.3	—	—	—	0.03 La
Hastelloy S	67	3.0	2.0	5.5	14.5	1.0	0.25	—	—	—	0.05 La
Waspaloy (N07001)	58	2.0	13.5	19.0	4.3	—	1.5	3.0	—	—	0.05 Zr
RA 333 (N06333)	45	18.0	3.0	25.0	3.0	3.0	—	—	—	—	1.25 Si
Haynes No. 188 (R30188)	22	3.0	39.0	22.0	—	14.0	—	—	—	—	0.07 La
Alloy 800H (N08810)	32.5	44	2.0^b	21	—	—	0.4	0.4	—	—	—
Haynes No. 556	20.0	31	18.0	22	3.0	2.5	0.2	—	—	—	0.8 Ta, 0.02 La
Multimet (R30155)	20.0	30	20.0	21	3.0	25	—	—	—	—	Ch + Ta = 1.0
RA330 (N08330)	35.0	43	—	19	—	—	—	—	—	—	1.25 Si
Type 304 SS (S30400)	9.0	Bal	—	—	—	—	—	—	—	—	2.0 Mn, 1.09 Si
Type 310 S (S31000)	20	—	25	—	—	—	—	—	—	—	2.0 Mn, 1.5 Mn
Type 316 SS (S31600)	12	—	17	2.5	—	—	—	—	—	—	2.0 Mn, 1.0 Si
Type 446 SS (S44600)	—	—	25	—	—	—	—	—	—	—	1.5 Mn, 1.0 Si
RA 85 H (S30615)	14.5	—	—	18.5	—	—	1.0	—	3.5	0.2	
HR 120	37	—	—	25	—	—	0.1	—	0.6	0.05	07 C, 0.2 N
RA 309 (S30908)	13	—	—	23	—	—	—	—	0.8	0.05	—
RA 310 (S31008)	20	—	—	25	—	—	—	—	0.5	0.05	—
RA 446 (S44600)	—	—	—	25	—	—	—	—	0.5	0.05	07 Mn, 0.1 N
253 MA (S30815)	11	—	—	21	—	—	—	—	1.7	0.08	0.04 Ce, 17N
314 (S31400)	20	—	—	25	—	—	—	—	2.2	0.1	—
Alloy 800 (N08800)	31.8	Bal	—	21.4	—	—	0.35	—	0.35	0.04	0.79 Mn, 0.44 Ti
Alloy 520	35.0	Bal	—	21.0	—	—	NA	—	2.0	—	1 Cb
Alloy DS	34.3	Bal	—	18.0	—	—	<0.1	—	2.20	0.03	1.3 Mn, <0.05 Ti
253	11	Bal	—	21	—	—	—	—	1.7	—	N, Ce
DS	36	Bal	—	18	—	—	—	—	2.2	0.06	—
45 TM	Bal	23	—	27	—	—	—	—	2.7	0.08	—
602 CA	Bal	9.5	—	25	—	—	2.1	—	—	0.18	Y, Zr, Ti
X	Bal	18	—	22	—	—	—	—	—	0.10	9 Mo, W, Co
625	Bal	3	—	22	—	—	—	—	—	0.10	9 Mo, 3.5 Cb
617	Bal	1.5	—	22	—	—	1.2	—	—	0.06	9 Mo, 12 Co

SS, stainless steel.

[a]Cabot, Hastelloy, Haynes, and Multimel are registered trademarks of Cabot Corporation; Waspaloy is a trademark of United Technologies. RA is a registered trademark of Rolled Alloys. Inconel is a registered trademark of the INCO family of companies.

[b]Maximum.

Source: From D.E. Fluck, R.B. Herchenroeder, G.Y. Lai, and M.F. Rothman, *Met. Prog. 128*(4):35–40, 1985; J. Kelly, *Heat Treating*, July 1993, 24–27.

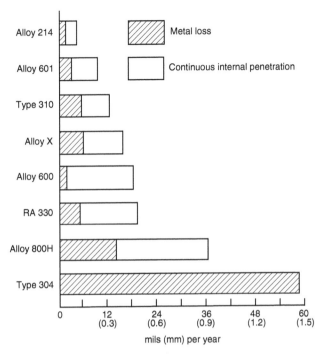

FIGURE 1.138 Alloy attack by combustion by-products.

The mechanistic cleaning process of alkaline and neutral cleaners is illustrated in Figure 1.142. The aqueous surfactant solution penetrates the oil film on the metal surface and solubilizes it by a micellation process. The micellized oil is then removed to the bulk of the aqueous cleaning fluid. The detergent cleaning action is affected by time, temperature, concentration, contamination, and additional additives, including surfactants and defoamants.

1.11.2 EQUIPMENT

Most parts washers used in heat treatment operations are classified as either dunk or spray-and-dunk washers. Both processes enhance the mixing of the cleaning fluid with the part and facilitate

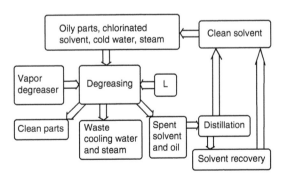

FIGURE 1.139 Anderson diagram for vapor degreasing. (From W.H. Michels, *Heat Treating: Equipment and Processes* (G.E. Totten and R.A. Wallis, Eds.), Proc. 1994 Conf., ASM International, Materials Park, OH, 1994, pp. 383–388.)

TABLE 1.52
Static Oxidation Behavior of Various Alloys—Metal Loss and Continuous Internal Penetration, mils (μm) per side[a]

Alloy	1800°F (980°C)	2000°F (1095°C)	2200°F (1205°C)
Alloy 214	0.2 (5)	0.1 (3)	0.7 (18)
Alloy 600	0.9 (23)	1.6 (41)	8.4 (213)
Alloy X	0.9 (23)	2.7 (69)	(672 h)[b]
Alloy 556	1.1 (28)	2.6 (66)	(168 h)[b]
Type 310	1.1 (28)	2.3 (58)	10.3 (262)
Alloy 800H	1.8 (46)	7.4 (188)	13.6 (345)
Type 446	2.3 (58)	14.5 (368)	(1000 h)[b]
RA 330	4.3 (109)	6.7 (170)	8.3 (211)
Type 304	8.1 (206)	<23.0 (584)	(336 h)[b]
Type 316	14.3 (363)	<69.0 (1753)	(168 h)[b]

[a]Exposure at 1008 h in air. Cooled to room temperature weekly.
[b]Alloy element consumed in test.

subsequent removal of the micellized soil from the part surface. Examples of these processes are shown in Figure 1.143. A photo of a commercial washer is shown in Figure 1.144.

The washing process should [131,133]

Use hot water, 160°F minimum
Continually remove residual oil throughout the cycle
Employ a drying cycle to eliminate the possibility of water contamination of the furnace
Use the correct cleaner for the job and use it correctly

1.12 QUENCH SYSTEM DESIGN

Quenching is one of the most critical processes in the overall heat treatment operation. It is essential to achieve the necessary heat transfer rates and optimal uniformity of the heat transfer

TABLE 1.53
Carburization Resistance of Various Alloys[a]

Alloy	Carbon Absorption (g/cm²)
Alloy 214	0.6
RA 333	1.0
Alloy 800H	1.0
Multimelt	1.3
Alloy 556	1.3
Alloy X	2.5
Alloy 600	2.8
Alloy 625	5.3
Type 310 stainless steel	Heavy localized attack

[a]Carburization conditions: in Ar, 5% H_2, and 5% CH_4 at 1800°F (980°C) for 55 h. $P_{O_2} = 9 \times 10^{-22}$ atm, carbon activity 1.0.

TABLE 1.54
Nitriding Resistance of Selected Alloy[a]

Alloy	Nitrogen Absorption (mg/cm^2)	Depth of Nitrided Layer (μm)
Alloy 230	0.7	30
Alloy 600	0.8	33
Alloy 188	1.2	15
Alloy 214	1.5	38
Alloy X	1.7	38
RA 330	3.9	97
Alloy 800H	4.3	104
Alloy 556	4.9	89
Type 310	7.4	152
Type 304	9.8	213

[a]Tested in ammonia at 1200°F for 1 week.

process if the desired steel transformation products are to be obtained and if thermal gradients are to be minimized [134–136]. Optimizing the overall quenching process is interdependent with quench system design, especially the direction, turbulence, and velocity of quenchant flow [136]. This is true for all quench systems, whether gas or liquid, including high-pressure gas quenching and molten salt, fluidized-bed, brine, oil, and aqueous polymer quenchants.

In this section, an introduction is provided to quench system design criteria with particular focus on vaporizable liquids. The design criteria include: (1) tank sizing, (2) heat exchanger selection, (3) agitator selection, (4) chute design, (5) flood quench systems, and (6) filtration.

1.12.1 QUENCH TANK SIZING

The sizing of the quench system depends on the thermal properties of the metal (Figure 1.145) and on those of the quenchant. Heat (q) transfer rates depend on the heat transfer coefficient of the quenchant film surrounding the heated metal (h), surface area (A), and the difference between the temperature of the metal before quenching (T_i) and the bath temperature (T_b).

TABLE 1.55
Resistance to Molten Chloride Salts at 1550°F (845°C)

Alloy	Metal Loss and Continuous Internal Penetration (mm/side)
Alloy 188	0.69
Multimet alloy	0.76
Alloy X	0.97
Alloy s	1.0
Alloy 556	1.1
Waspaloy alloy	1.7
Alloy 214	1.8
Type 304	1.9
Type 310	2.0
Alloy 600	2.4

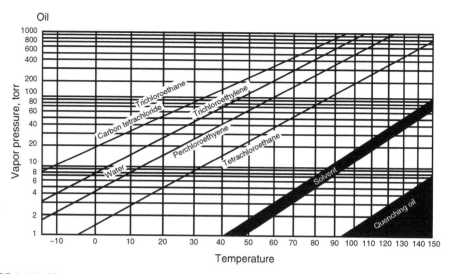

FIGURE 1.140 Vapor pressure–temperature relationship of various cleaning solvents.

$$q = hA(T_i - T_b)$$

The amount of heat transferred from the hot metal to the cooler quenchant can be estimated as [137]:

$$Q = C_{pm} W_m (T_i - T_b)$$

where C_{pm} is the specific heat of the metal, W_m is the weight of the metal, including fixtures such as trays, and T_i and T_b are the initial and final temperatures, respectively, of the metal.

The temperature rise (T_r) of the quenchant can be estimated from [137]:

$$T_r = (T_i - T_b) = Q W_q C_{pq}$$

where C_{pq} is the heat capacity of the quenchant, W_q is the weight of the quenchant, and T_i and T_b are as defined above.

Clearly the temperature increase of the quenchant is dependent on the initial temperature of the metal; the volume, temperature, and heat capacity of the quenchant; and the heat capacity and size of the load that is quenched.

TABLE 1.56
Summary of Alkaline Salts (Builders) Used for Aqueous Cleaning Solutions

Salt Class	pH	Comment
Caustics (NaOH, KOH)	2–14	Clean fats and oil may neutralize acid. *Not safe* for soft metals such as aluminum and zinc
Silicates	11–12.5	Good detergency; attack soft metal and become insoluble at pH < 10
Phosphates	9.5–1.5	Good detergency; safe for soft metal and suitable for hard water
Carbonates	9.0–9.5	Used to provide reserve alkalinity for other builders in products
Borates	8.0–10.5	Used when other more alkaline salts such as phosphates and silicates cannot be used. Provides some rust protection

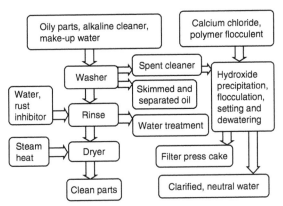

FIGURE 1.141 Schematic of aqueous alkaline washing process. (From W.H. Michels, *Heat Treating: Equipment and Processes* (G.E. Totten and R.A. Wallis, Eds.), Proc. 1994 Conf., ASM International, Materials Park, OH, 1994, pp. 383–388.)

Two general rules have been traditionally and successfully applied to oil and polymer quenchant systems [138,139]:

1. The size of the load (including fixtures) quenched should not exceed 1 lb/gal (0.11 kg/l).
2. The temperature rise of the quenchant should not exceed 10°F (5°C).

It is recommended that mild steel be used for quench tank construction. Cast iron may be used for pumps and agitators [140]. Soft metals such as zinc, lead, magnesium, and galvanized steel should not be used for aqueous polymer quenchants that have a basic pH (pH > 7.0). Copper oil and brass should not be used for mineral oil systems because they promote oxidation [140].

1.12.2 HEAT EXCHANGER SELECTION

There are four primary types of heat exchangers used in quenching systems: mechanical refrigeration, shell-and-tube (or plate-and-frame), evaporative water-cooled spray towers, and air-cooled radiators [134]. For small systems, submerged water cooling pipes and cooling jackets around the quench tank may also be used [140].

Selection of the proper heat exchanger depends on the final temperature of the fluid. The selection criteria are summarized Table 1.57. Examples of air-cooled and evaporative heat exchangers, two of the more commonly used heat exchangers in larger heat treatment processes, are shown schematically in Figure 1.146 and Figure 1.147.

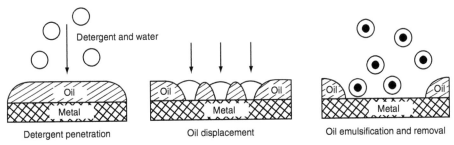

FIGURE 1.142 Detergency mechanism for alkaline washing process. (From *Heat Treating Handbook*, Seco/Warwick Corp., Meadville, PA.)

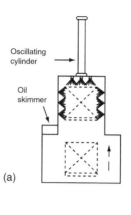

(a)

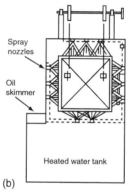

(b)

FIGURE 1.143 Schematics of (a) spray-and-dunk washer; (b) spray washer. (Courtesy of BeaverMatic.)

Shell-and-tube heat exchangers typically contain copper cooling surfaces and are subject to fouling and plugging, particularly when mineral oil quenchants are used. Plate-and-frame heat exchangers have stainless steel plates on the heat exchange surface. Water contamination problems may occur with both of these types of heat exchangers. These can be disastrous for oil quenchants, ultimately leading to fires. This problem can be avoided by pressurizing the quench oil side to prevent water from flowing into the oil. To avoid potential problems in the event of pressurizing pump failure, it is recommended that a pressure-sensing device (140–280 kPa) be installed [134].

Centrifugal pumps are generally recommended. The appropriate pipe size is dependent on the required flow rate as shown in Table 1.58. Correct pipe size is essential to ensure the necessary turbulent flow and to reduce the fouling potential. Maximum efficiency is obtained by removing the heated quenchant from the top of the tank and returning the cooled fluid to the bottom of the tank [140].

1.12.3 AGITATOR SELECTION

There are numerous possible design alternatives to provide the necessary agitation to optimize the heat transfer rates and the uniformity of the quenching process. They may be used either individually or in combination with each other. Some of the more common agitator design options are briefly reviewed here.

1.12.3.1 Sparging

Bubbling air or inert gases into the quenchant, a process known as sparging, may be used either as the sole source of agitation [141,142] or to supplement other forms of agitation such as pump agitation [143]. Air sparging can be readily used for water and brine. It is a relatively

(a)

(b)

FIGURE 1.144 (a) Continuous vapor degreasing line. (Courtesy of Fluid therm Technology P. Ltd., www.fluidtherm.com.) Aqueous alkaline washers (b) Open batch washer.

Continued

(c)

FIGURE 1.144 (continued) (c) Continous batch washer. (Courtesy of Universal Separators, Madison, WI, www.smartskin.com.)

poor choice for other vaporizable quenchants such as oil because the increased presence of air will facilitate oxidative degradation. Although an inert gas such as nitrogen should be used, it is significantly more expensive than air.

Agitation rates provided by gas sparging are dependent on the gas pressure. The pressure required to provide the desired amount of agitation depends on the head of the fluid, frictional losses in the delivery pipe, and the pressure difference required to force the gas through the pipe orifice [142].

Airflow rate is a function of the pipe diameter:

$$Q = 0.327 V D^2$$

where Q is the airflow volume (ft^3/min), V is the linear velocity (ft/min), and D is the pipe diameter (ft).

Pressure losses due to friction (P) can be calculated as

$$P = \frac{Q^2 L}{2690 D^5}$$

where L is the length of the pipe (ft).

TABLE 1.57
Heat Exchanger Selection Guidelines

Final Fluid Temperature	Heat Exchanger Type
<24°C (<75°F)	Mechanical refrigeration
33–45°C (95–110°F)	Evaporation or water-cooled
<45°C (<110°F)	Air-cooled

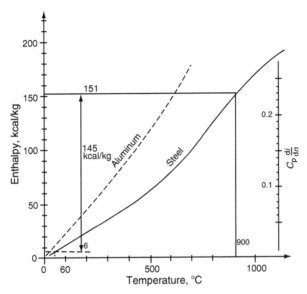

FIGURE 1.145 Heat capacity curves for aluminum and steel.

To determine pressure loss due to friction in flow through a pipe elbow, the equivalent length of straight pipe is first determined by using a table such as Table 1.59.

The rate of airflow (Q) through a pipe orifice of known diameter (in.) is

$$Q = 21.7D^2 C\sqrt{H}$$

where H is the head pressure (in H_2O) and C is an orifice-dependent constant selected from Table 1.60.

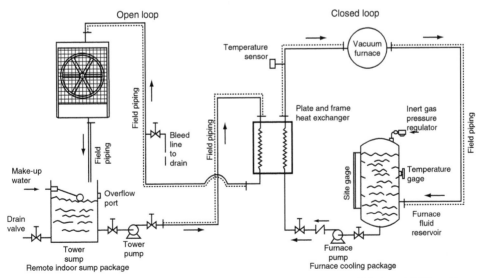

FIGURE 1.146 Example of an evaporative heat exchanger. (Courtesy of Industrial Heat Recovery Equipment.)

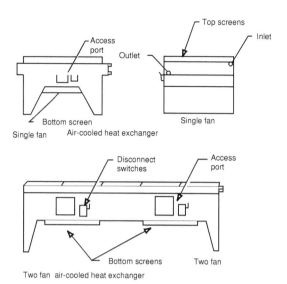

FIGURE 1.147 Example of an air-cooled heat exchanger. (Courtesy of Industrial Heat Recovery Equipment.)

1.12.3.2 Centrifugal Pumps

Centrifugal pump systems can be used to provide quenchant bath agitation and to transfer the fluid to and from the heat exchanger. However, if centrifugal pumps are used, particular attention must be given to the position of the sparge manifolds around the part, or load, that is quenched. This is especially significant because the flow velocity drops rapidly at distances greater than 10 diameters from the discharge point, as shown in Figure 1.148 [134].

A detailed description of centrifugal pump selection, use, and system design is beyond the scope of this text. However, excellent information is provided in Refs. [146–148] and other texts. *Applied Process Design for Chemical and Petrochemical Plants* [147], which is particularly good, is available free of charge.

1.12.3.3 Impeller Agitation

Perhaps the most commonly used source of agitation, especially for larger quench tanks, is the impeller. One of the earliest references to the selection and use of impeller agitation for quench systems was a now-classic paper published by U.S. Steel [144]. The newer impeller mixer technology is discussed in Refs. [145,149].

TABLE 1.58
Heat Exchanger Pipe Sizing Recommendations

Flow Rate		Pipe Size	
gal/min	l/min	in.	mm
50–90	190–340	2	50
90–180	340–680	2.5	63
180–250	680–950	3	75

TABLE 1.59
Equivalent Straight Pipe Length for Elbows

Pipe Elbow Radius		Equivalent Length of Straight Pipe	
in.	mm	in.	mm
0.5	12.7	121.0	3070
0.75	19.0	35.0	889
1.0	25.4	17.5	445
1.25	31.8	12.7	323
1.5	38.1	10.3	262
2.0	50.8	9.0	229
3.0	76.2	8.4	213
5.0	127.0	7.8	198

Source: From G.B. Tatterson, *Fluid Mixing and Gas Dispersion in Agitated Tanks*, McGraw-Hill, New York, 1991.

Many quench tanks are agitated using one or more open-impeller mixers or an impeller encased in a tube for directing the flow. If a draft tube is not used, then flow must be directed by the impeller itself or by flow-directing baffles in the quenching region of the tank.

A common impeller used for open systems is an axial flow impeller such as the marine impeller shown in Figure 1.149b. An axial flow impeller directs flow parallel to the impeller shaft. Axial flow impellers may be top-entering, side-entering, or angled top-entering as shown diagrammatically in Figure 1.150.

Three-blade marine impeller mixers have traditionally been used for quench tank agitation. The mixer power requirements for a marine impeller with a 1.0 pitch ratio operating at 420 rpm are summarized in Table 1.61 [144].

The impeller currently most often recommended for angled top entry or vertical top entry is the airfoil type illustrated in Figure 1.150a. A comparison of the power requirements for a marine and an airflow-type impeller is given in Table 1.62 [145]. The recommended power requirements for an airfoil impeller in both open and draft tube configurations are provided in Table 1.63.

The required power P is proportional to speed N and can be calculated for other output speeds from

$$P \propto N^{4/3}$$

TABLE 1.60
Gas Injection Orifice Constant[a]

Orifice Type	Constant C
Conoidal mouthpiece (contracted vein)	0.97–0.99
Conoidal (converging mouthpiece)	0.90–0.99
Short (cylindrical mouthpiece rounded at inner end)	0.92–0.93
Short (cylindrical mouthpiece)	0.81–0.84
Thin circular plate	0.56–0.59

[a]For equation $Q = 21.7 D^2 C \sqrt{H}$; see text.

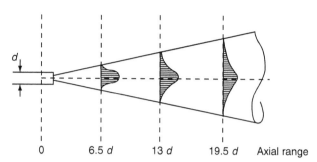

FIGURE 1.148 Flow pattern as a function of nozzle diameter and distance from nozzle for a jet mixer. (From G.E. Totten, C.E. Bates, and N.A. Clinton, in *Handbook of Quenchants and Quenching Technology*, ASM International, Materials Park, OH, 1993, chap. 9.)

The impeller blades used for a side-entry mixer (Figure 1.151c) have a different shape than those used for top-entry mixers (Figure 1.151b) to balance the mechanical forces acting on the impeller in the side-entry mode.

The sizing of the impeller diameter with respect to power is shown in Table 1.64. The required power is taken from Table 1.63.

1.12.3.4 Draft Tubes

A draft tube agitator is used to provide directed airflow in many quenching system designs. A schematic of a draft tube agitator is provided in Figure 1.151.

Marine impellers are normally used for smaller draft tube systems (<24 in.). Airfoil impellers are better performing, lower cost alternatives for larger (>24 in.) systems. When airfoil impellers are used in draft tube applications, a lower tip-chord angle is used than for side-entering mixers to avoid stalling under conditions of high heat resistance.

A draft tube should have the following characteristics [145,149]:

1. A down-pumping operation is used to take advantage of the tank bottom as a flow-directing device.
2. A 30° entrance flow on the draft tube minimizes the entrance head losses and ensures a uniform velocity profile at the inlet.
3. Liquid depth over the draft tube should be at least one-half of the tube diameter to avoid flow loss due to disruption of the impeller inlet velocity profile.
4. Internal flow straightening vanes are used to prevent fluid swirl.
5. Impeller should be inserted into the draft tube a distance equal to at least one-half of the tube diameter.
6. A steady bearing or limit ring is used to protect the impeller from occasional high deflection. A steady bearing is the lower cost alternative but requires maintenance.
7. Impeller requires 1–2 in. (25–50 mm) of radial clearance between the blade tips and the draft tube. When the draft tube must be minimized, an external notch can be used to reduce the draft tube dimensions by 2–3 in. (50–75 mm).

1.12.3.5 Multiple Mixers

Ensuring uniform heat transfer throughout the quench zone often requires multiple mixers. Although there are no simple quantitative prediction methods to determine the necessity of using multiple mixers, there are two general guidelines:

1. If the length/width ratio is 2, a single properly design mixer is usually sufficient.
2. If the length/width ratio is greater than 2, then multiple mixers are recommended.

Possible draft tube and side-entry mixer arrangements for rectangular tanks are shown in Figure 1.152. Top-entry mixers may be arranged differently.

(a)

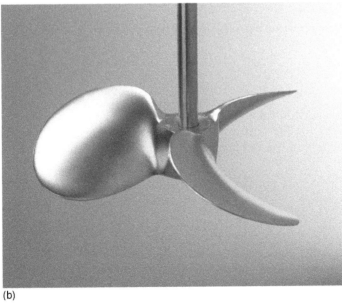

(b)

FIGURE 1.149 Examples of impeller design: (a) lightnin A-310 airfoil-type impeller; (b) lightnin A-310 marine impeller.

Continued

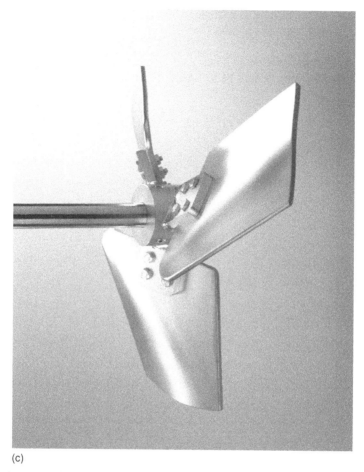

(c)

FIGURE 1.149 (Continued) (c) lightnin A-312 side-entry impeller. (Courtesy of SPX Process Equipment, Lightning Operation.)

To size a mixer used in a multiple arrangement, it is first necessary to determine the total power requirement for the tank using Table 1.63. The impeller sizing for each mixer is determined from Table 1.64, and the power per mixer is determined from the relationship

$$\text{Power per mixer} = \frac{\text{total power}}{\text{number of mixers}}$$

1.12.3.6   Cavitation

Cavitation occurs when the pumping action of the impeller creates localized zones of low pressure, pressure below the vapor pressure of the fluid. Vaporization can lead to erosion of the impeller and unstable operating performance. This should not occur with properly designed systems.

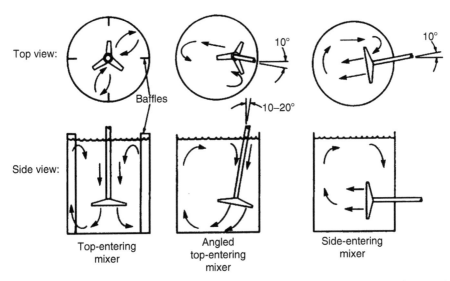

Top view:

Baffles

10°

10°

10–20°

Side view:

Top-entering mixer

Angled top-entering mixer

Side-entering mixer

FIGURE 1.150 Examples of flow patterns and impeller arrangements for top-entering, angled top-entering, and side-entering mixers.

1.12.4 Computational Fluid Dynamics

Computational fluid dynamics (CFD) may provide greater understanding of the details of fluid flow in the quench system and greater insights into how to determine the optimum location and orientation of the part in the quench zone [150,151]. It can also assist in troubleshooting fluid flow problems that can produce distortion or cracking [152].

An example of the use of CFD to model flow in a simple quench tank configuration using a draft tube [149] is shown in Figure 1.153. The CFD model illustration shows that there is a general flow pattern that sweeps across the bottom of the tank from left to right with a rotational element around the axis of the general flow pattern. The flow is relatively quiet in the corners of the tank except where the draft tube is located.

TABLE 1.61
Power Requirements for Propeller Agitation[a]

| Tank Volume | | Power Required | | | |
| | | Standard Quench Oil | | Water or Brine | |
gal	l	hp/gal	kW/l	hp/gal	kW/l
50–800	2,000–3,200	0.005	0.001	0.004	0.0008
800–2,000	3,200–8,000	0.006	0.0012	0.004	0.0008
2,000–3,000	8,000–12,000	0.006	0.0012	0.005	0.001
>3,000	>12,000	0.007	0.0014	0.005	0.001

[a]Agitation at 420 rpm. Marine propeller with 1.0 pitch ratio.
Source: From U.S. Steel, *Improved Quenching of Steel by Propeller Agitation*, 1954.

TABLE 1.62
Equivalent Quench Tank Mixer Size[a]

Impeller	rpm[b]	Standard Quench Oil		Water or Brine	
		hp/gal	kW/l	hp/gal	kW/l
Marine	480	0.007	0.0014	0.005	0.001
Marine	280	0.004	0.0008	0.003	0.0006
Airfoil	280	0.003	0.0006	0.002	0.0004

[a]Based on a 3000 gal tank with an open-impeller mixer providing violent circulation. The recommended power requirement for an airfoil impeller used in a draft tube application is 0.006 hp/gal for a quench oil and 0.0045 hp/gal for water or brine.

[b]The power levels for other output speeds are adjusted using $P \propto N^{4/3}$, where P is the required power level and N is speed.

Similar studies were performed by Garwood et al. [150] and Wallis et al. [151]. Their work also showed that even with considerable surface flow, there were substantial velocity gradients within the tank around the mixers positioned at all four corners of the tank.

These results show that CFD calculations have tremendous potential in the study of existing quench tank flow problems and in assisting the design of new systems.

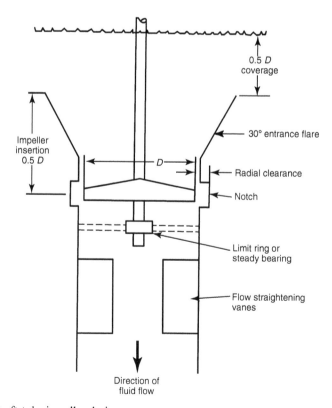

FIGURE 1.151 Draft tube impeller design.

TABLE 1.63
Recommended Quench Tank Mixer Sizes

Mixer Type	Standard Quench Oil		Water or Brine	
	hp/gal	kW/l	hp/gal	kW/l
Open-impeller mixers—top-entry or side-entry airfoil impeller at 280 rpm	0.004	0.0008	0.003	0.0006
Draft tube mixer—airfoil impeller at 280 rpm	0.006	0.0012	0.0045	0.0009

1.12.5 CHUTE-QUENCH DESIGN

Illgner [153] conducted a study of chute quenching in the continuous heat treatment process illustrated in Figure 1.154. The variables affecting quenching in the chute zone include effect of the time to drop through the chute zone, cross-sectional size, and quenchant viscosity. The conclusions of this study were that [153,154]:

1. The quenchant viscosity has only minimal effect on sinking speeds.
2. The position of the part as it enters the quenchant will significantly affect the sinking speed.

TABLE 1.64
Size of Impeller Mixers

Motor[a,b]		Impeller Size[c,d]	
hp	kW	in.	cm
0.25	0.19	13	33.0
0.33	0.25	14	35.6
0.50	0.37	15	38.1
0.75	0.56	16	40.6
1.0	0.75	17	43.2
2.0	1.49	20	50.8
3.0	2.34	22	55.9
5.0	3.73	24	61.0
7.5	5.59	26	66.0
10.0	7.46	28	71.1
15.0	1.19	30	76.2
20.0	14.92	32	81.3
25	18.65	33	83.8

[a]The power requirements were calculated assuming 280 rpm, specific gravity 1.0, and airfoil impeller with Np 0.33. (Airfoil and marine propeller power numbers are nearly identical.)

[b]The shaft horsepower (hp$_s$ is equal to 80% of the motor horsepower (hp$_m$) (0.8 × hp$_m$ – hp$_s$).

[c]These are the power requirements for an open impeller operating at 280 rpm.

[d]When used in a draft tube, the impeller size should be reduced by 3%. Axial flow impellers are used in draft tubes to more closely control the direction of the flow pattern. Draft tube circulators have a higher resistance head that the impeller must pump against, which is due to the fluid friction losses in the draft tube pipe. Velocity head losses also occur at the entrance, exit, and at any angles in the draft tube. The higher head conditions require a slightly different impeller for optimum pumping performance. Impellers in both open- and draft tube configurations are provided in Table 1.58.

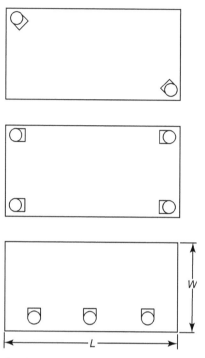

FIGURE 1.152 Examples of draft tube mixer placement for tanks with $L/W > 2$.

3. Even small (16×100 mm) parts do not cool to the M_s transformation temperature in the chute zone.

For optimal quench uniformity, both vigorous agitation and adequate quenchant turn-over in the chute zone are necessary. If the temperature is allowed to vary, it may not be possible to control hardness because quench severity depends on the quenchant temperature.

Quenchant temperature control in the chute zone is especially critical for aqueous poly-alkylene glycol polymer quenchants, which will thermally separate from solution at elevated temperatures, producing a heterogeneous quenching medium. These conditions will result in nonuniform cooling, and the increased thermal gradients may be sufficient to increase distortion or cracking.

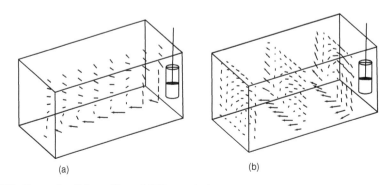

         (a)   (b)

FIGURE 1.153 Example of the utility of CFD analysis to model agitation uniformity in quench tanks. (a) Vectors in xz planes; (b) vectors in yz planes.

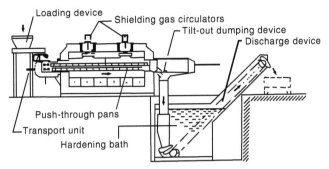

FIGURE 1.154 Illgner continuous chute-quench design. (From K. Illgner, *Harterei-Tech. Mitt. 41*(2):113–120, 1987.)

If thermal separation of the fluid is accompanied by poor agitation and fluid turnover in the chute zone, the polymer may adsorb on the part and subsequently be removed before polymer dissolution occurs. This will lead to a depletion of the polymer in the chute zone. If sufficient polymer is removed with replenishment, the quenchant in the chute zone will become almost entirely water.

Various chute-quench designs have been used to achieve the appropriate agitation and quenchant turnover in the chute zone. The illustrative examples are shown in Figure 1.155 [154]. The systems include (a–c) impeller agitation in the chute zone, (d–f) up-flow and submerged spray-quench designs, (h) the SECO-Warwick Whirl–Away countercurrent pump flow design, and (i) a special roller-track design (which must also be accompanied by some form of agitation such as the submerged spray). A rotating chute design has also been reported (Figure 1.155j) [156]. The dual-belt chute-quench design (immersion time quenching system; ITQS) shown in Figure 1.155g permits variable cooling rates during the quench [155].

A properly designed chute-quench system should incorporate the following features [139]:

1. Sufficient agitation and turnover in the chute zone to provide adequate and uniform heat transfer.
2. Cooling jacket for the chute above the quench zone to prevent oil and water vapor from entering the furnace vestibule. Cooling can be achieved by routing the quenchant returning from the heat exchanger through the chute zone cooling jacket.
3. Fume educator located in the chute zone above the cooling jacket to prevent vapor contamination of the furnace atmosphere.
4. Perforated or screened opening in the chute area to allow heated quenchant to escape during the quench. Solid chutes should never be used.
5. Mesh belt of sufficient porosity and length to permit quenchant agitation around the part to facilitate completion of the quench.

TABLE 1.65
Pressure and Orifice Size for Indirect Spray-Quench Systems

| Type of Spray | Pressure (psig) | Orifice Size | |
		in.	mm
Open	<20	1/8	3.18
Submerged	>40	1/4	6.35

TABLE 1.66
Interrelationship between Hole Cross Section and Flow Rate

Cross Section		Hole Diameter		Flow Rate at 20 psi
in.	mm	in.	mm	(gpm/hole)
0.5	12.7	1/16	1.58	0.33
1.0	25.4	1/8	3.18	1.5

Source: From D.J. Williams, *Met. Heat Treat.*, July/August 1995, pp. 33–37.

1.12.6 FLOOD QUENCH SYSTEMS

Typically either spray or dunk quenching is used for induction hardening [136,138,157]. Polymer quenchants are the most commonly used quenchants for either process [157].

For dunk quenching, it is important that the reservoir be sufficiently large to allow the foam head to dissipate before the quenchant is pumped back into the system [139]. Therefore, the reservoir volume should be at least five to eight times the volume rate of flow. For example, if the flow rate is 10 gal/min, then the reservoir capacity should be 50–80 gal.

One of the most common problems with dunk quenching systems is that the reservoir is undersized. If the reservoir is too small, a mixture of the foam and quenchant will be used to quench the part. This will often lead to increased distortion and cracking. Dunk quench systems also require the use of filters and heat exchangers.

The major factors in the selection and design of spray quenching systems are the material induction hardened, the area to be quenched, and the quenchant. Pressure and orifice size recommendations for polymer quenchants are presented in Table 1.65.

The total flow of the quenchant onto the part depends on the size and number of holes in the spray ring. Generally, the total quench hole area should be a minimum of 5–10% of the area quenched [157]. The size of the holes in the quench ring is a function of the cross-sectional size of the part (Table 1.66).

Mass flow vs. pipe size, effect of inlet/outlet ratio, and hole size on quenching size vs. pressure are summarized in Figure 1.156 through Figure 1.158.

Ideally the rate of quench hole area to surface area of the lines feeding the coil will be 1:1 and no greater than 2:1 [157].

1.12.7 FILTRATION

Quenchants may contain various types of solid contaminants such as sludge and carbon. In addition to causing excessive wear of pumps and seals, scale, and heat exchange fouling, the presence of contaminants may affect the uniformity of the quenching process, resulting in increased distortion and cracking frequency for the lifetime of the bath.

One method of solid contaminant removal is centrifugation. An example of a centrifugal separator is shown in Figure 1.159. Solid contaminants may also be removed by filtration. Filters can be divided into two classes: fixed-pore surface filters and nonfixed-pore depth filters. Fixed-pore surfaces such as mesh or screen filters are suitable for removal of relatively large solids (or small parts). Dimensions of commonly available mesh filters are given in Table 1.67.

Nonfixed-pore depth filters have a higher capacity than mesh filters and permit the separation of smaller particles. They may be constructed from cotton, wool, cellulose, or synthetic materials such as polyester, polyethylene, fiberglass, and Teflon. The best quenching results are achieved using 3–5 μm filtration [158].

Filters can be used in either full-flow or proportional flow mode. For proportional flow filtration, only a portion of the pumped fluid stream is passed through the filter. For full-flow filtration, the entire fluid stream pumped is passed through the filter.

Filter selection is dependent on the flow rate, the fluid composition, and the temperature. The maximum flow rate and allowable pressure will dictate the filter element size that can be

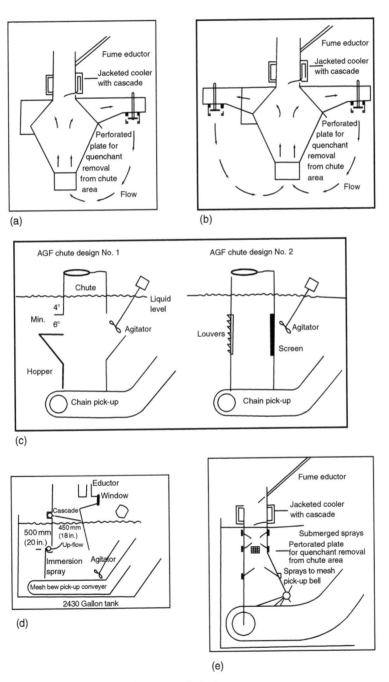

FIGURE 1.155 Examples of possible chute-quench designs.

Continued

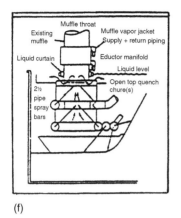

(f)

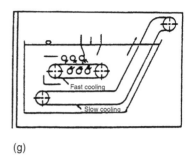

(g)

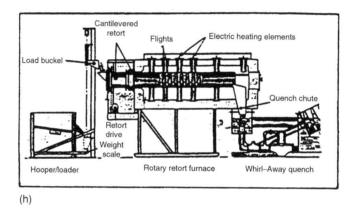

(h)

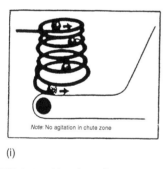

(i)

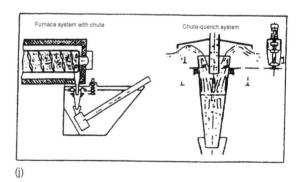

(j)

FIGURE 1.155 (Continued)

used. The filter selected must be compatible with the fluid type at the operating temperature. *Note*: Cellulosic filters should never be used with aqueous quenchants.

1.12.7.1 Membrane Separation

As illustrated in Figure 1.160, microporous membranes are capable of removing various organic and inorganic materials from water [159]. The porosity of these membranes, which are commonly manufactured from cellulose acetate, is selected to separate large molecules

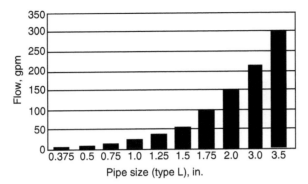

FIGURE 1.156 Correlation of volume flow rate and pipe size at quench flow of 10 ft/s. (From D.J. Williams, *Met. Heat Treat.*, July/August 1995, 33–37.)

such as polymer from smaller molecules such as salt and water. If the porosity is sufficiently small, everything larger than a water molecule can be removed.

Two of the most common membrane separation techniques for the removal of quenchant polymers, salts, and additives from water are reverse osmosis (RO) and ultrafiltration (UF). The primary differences between these methods are the system pressure and molecular size separated (see Table 1.68). RO is more common because it provides the most complete separation.

The separation process is carried out by pumping the quenchant solution through an array of membrane cartridges such as the one illustrated in Figure 1.161. Pure water passes into the core of the filter cartridge, from which it is released to the environment or delivered to a storage tank. The separated polymer is pumped to a storage tank, rediluted to the proper concentration, and reused.

If the aqueous quenchant is contaminated with oil, the oil must be removed before membrane separation. This can be accomplished by liquid–liquid surface coalescence of the micrometer-size oil droplets, which are then separated by density difference as illustrated in Figure 1.162. The agglomerated oil is then collected in a setting tank.

1.12.8 PRESS DIE QUENCHING

Many batch and continuous processes call for the heat treatment of parts, such as gears and rings, for which the maintenance of dimensional tolerances is critically necessary. Various metallurgical and heat treatment process variables will affect the final part dimensions [160,161]:

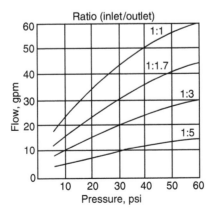

FIGURE 1.157 Correlation of pipe inlet/outlet ratio with flow rate and pressure. (From D.J. Williams, *Met. Heat Treat.*, July/August 1995, 33–37.)

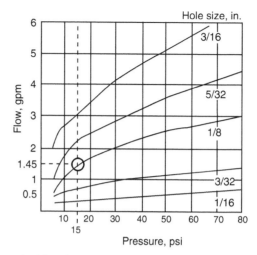

FIGURE 1.158 Correlation of orifice size with flow rate and pressure. (From D.J. Williams, *Met. Heat Treat.*, July/August 1995, 33–37.)

1. Materials: Irregular grain structure (texture, grain size), hardenability, chemical composition
2. Workpieces: Geometry of thickness variation, internal stress (machining/forming)
3. Heat treatment: Thermal stress during heating and cooling, texture transformation (austenitizing/hardening), nonuniform carburizing, exterior stress (loading configuration, part weight)

The variables that may potentially affect heat treatment, particularly quenching, are often the processes that receive the most attention. (For a more complete discussion of this subject, see Chapter 10 of *Steel Heat Treatment: Metallurgy and Technologies*.)

Previously, it was shown that to minimize thermal and transformational gradients, the optimization of quench uniformity is critical. However, this may be insufficient to obtain

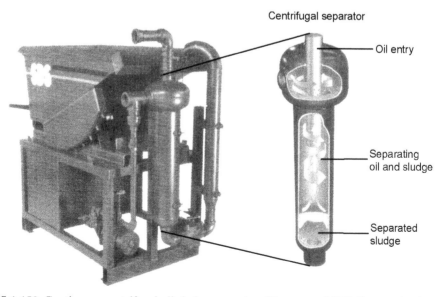

FIGURE 1.159 Continuous centrifugal oil sludge separator. (Courtesy of SBS Corporation.)

TABLE 1.67
Standard Mesh Screen Sizes

		Sieve Opening	
Mesh Per Linear Inch	U.S. Mesh No.	in.	μm
52.36	50	0.0017	297
72.45	70	0.0083	210
101.01	100	0.0059	149
142.86	140	0.0041	105
200.00	200	0.0029	74
270.26	270	0.0021	53
323.00	325	0.0017	44

the desired dimensional tolerances. In these cases, the use of press (die) quenching may be required.

1.12.8.1 Press Quenching Machines

A simplified schematic of a press quenching machine is shown in Figure 1.163 [162,163]. The part to be quenched is placed on a die that is custom manufactured for the particular part processed. The press, which is either hydraulically or pneumatically actuated, closes over the part and die. The part, pressed in the die at the minimum required pressure, is then either submerged in or sprayed with the quenchant. After the quench, the press is opened and the part is removed. (The die-pressing operation also holds the part to prevent distortion during quenching.)

An example of a press quenching machine for a batch production process is illustrated in Figure 1.164. However, press quenching may be part of either a batch or a continuous process [27].

Symmetrical parts may be manufactured to the straightness relationship [160]

$$\text{TIR} = k\frac{l}{d}$$

where TIR is the total indicator reading or straightness, l length (in.), d diameter (in.), and $k = 10^{-4}$.

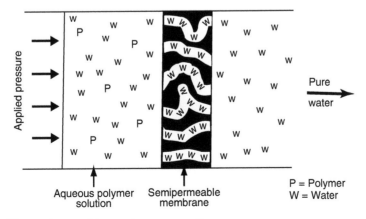

FIGURE 1.160 The semipermeable membrane separation process.

TABLE 1.68
System Pressure and Solute Size Membrane Separation Techniques

Technique	Size of Solute Separated (μm)	Pressure (psig)
Ultrafiltration	0.001–2	—
Reverse osmosis	0.0001–0.001	200–1000

1.13 FURNACE SAFETY

There are numerous examples of industrial accidents that have been caused by improper and unsafe furnace operation [165]. Many of these incidents are related to the unsafe use of furnaces with controlled atmospheres. For example, furnace explosions may occur with sufficient force to literally blow the door off the furnace and propel it across the shop. Unfortunately, these incidents occur with such frequency that it is commonly understood that one should never stand in front of the furnace door.

The objective in this section is to provide an overview and summary of furnace safety precautions. *Note*: The National Fire Protection Agency and the U.S. Occupational Safety and Health Administration, insurance underwriters, and equipment manufacturers all provide guidelines for safe operation of industrial furnaces. The furnace operator should be well versed in these regulations before operating any piece of equipment.

1.13.1 EXPLOSIVE MIXTURES

When combustible gas and air are mixed and the necessary source of ignition, such as a spark, is present, combustion of the mixture will occur. There will be a corresponding volumetric expansion due to the temperature rise of the gaseous combustion by-products that may lead to the rupture of the container in which they are confined (an explosion).

Many furnace atmospheres, when combined with the proper amount of air, are explosive. Some of those mixtures have been noted in Table 1.21. One of the first steps of safety is for the furnace operator to be familiar with the explosive limits of all the gases used in the shop.

When positive ignition is necessary to facilitate combustion of a mixture containing more than 4% combustibles, the minimum temperature for safe operations is 1400°F (760°C).

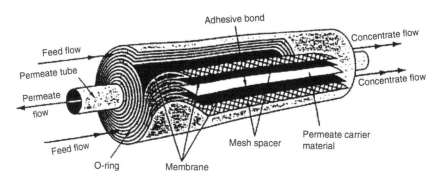

FIGURE 1.161 Typical membrane assembly. (Courtesy of Despatch Industries.)

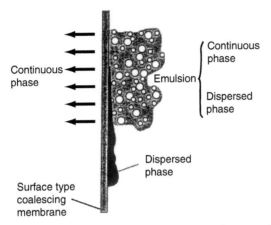

FIGURE 1.162 Schematic of separation by coalescence. (Courtesy of Despatch Industries.)

1.13.2 PURGING

The furnace must be purged of air with an inert gas, usually nitrogen, before a flammable mixture is introduced. It is generally true that there is no residual oxygen if the furnace is purged with a volume of N_2 equal to at least five times the volume of the furnace [164].

1.13.3 SAFETY OF OPERATION TEMPERATURE

There are four variations of furnace temperatures. They include [164]:

1. Entire furnace temperature is above 1400°F (760°C).
2. Part of the furnace is above 1400°F with short cooler zones.
3. Part of the furnace is above 1400°F (760°C) with longer cold zones.
4. The entire furnace is below 1400°F (760°C).

When the furnace temperature is above 1400°F (760°C), combustible gas ignites immediately and is therefore relatively safe. However, if the furnace temperature is less than 1400°F

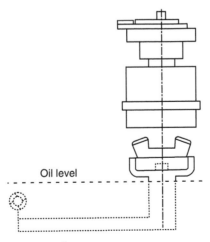

FIGURE 1.163 Schematic of a press quench process.

(a)

(b)

FIGURE 1.164 (a) A press quench machine. (b) Die installation into press quench machine.

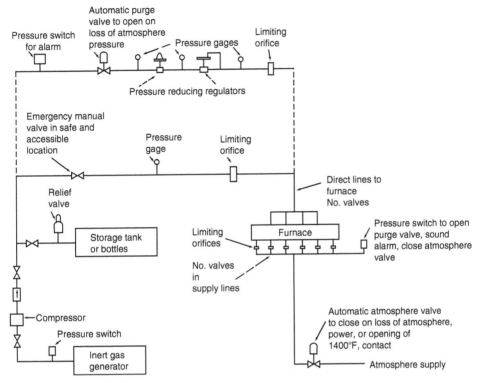

FIGURE 1.165 Inert gas purging system.

(760°C), an explosive situation may result. In such cases, it is recommended that the furnace be purged with an inert gas to remove the combustible gas.

If there are long hot zones, especially at the ends of the furnace, and relatively short cooler zones, the inert purge gas needs to be added only in the cold zones. Care should be taken to ensure that the atmosphere is added only to the hot zones. Batch furnaces below 1400°F

TABLE 1.69
Prevention of Explosions of Furnace Atmospheres during Emergencies

	Emergency Interruption		
	Power	**Fuel Gas**	**Atmosphere**
Protective system	×	×	×
Inert gas storage	×	×	×
Bottled inert gas	×	×	×
Steam purging system	×	×	×
Preventive method			
Motor generator	×		
Manifold generator			×
Natural gas[a]	×		×
Bottle LP fuel gas		×	

[a]Natural gas should be used only at temperatures greater than 1400°F (760°C).

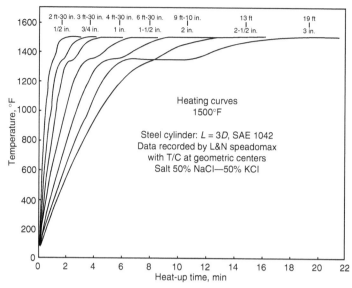

FIGURE 1.166 Steel heating curves with respect to section size. (Courtesy of Ajax Electric Company.)

(760°C) should be purged with the introduction of air when starting cold and at the end of the cooling cycle, only after it is ensured that there can be no ignition source [164].

Large cold zones with short hot zones represent a potentially unsafe condition, and the furnace must be operated only with special precautions such as with the use of an inert gas production and storage system (Figure 1.165).

1.13.4 POWER FAILURES

If a power failure occurs, the atmosphere and heat will be lost, creating unsafe (<1400°F (760°C)) conditions. In such cases [164]:

1. Flammable mixture should be burned out immediately if the burnout procedure is safe.
2. If the furnace is operated below 1400°F, standby storage of an inert gas purge is recommended.

TABLE 1.70
Composition and Recommended Use Temperature Range for Steel Heat-Treating Salt Mixture

Process	$BaCl_2$	$NaCl_2$	KCl	$CaCl_2$	$NaNO_3$	KNO_3	°C	°F	°C	°F
	Composition (%)						Melting Point		Use Temperature	
Austenitizing										
	98–100	—	—	—	—	—	950	1742	1035–1300	1895–2370
	89–90	10–20	—	—	—	—	870	1598	930–1300	1705–2370
Preheating salts										
	70	30	—	—	—	—	335	635	700–1035	1290–1895
	55	20	25	—	—	—	550	1022	590–925	1095–1700
Quench and temper										
	30	20	—	50	—	—	450	842	500–675	930–1250
	—	—	—	—	55–80	20–45	250	482	285–575	545–1065

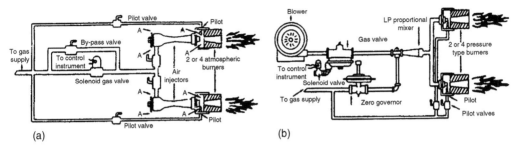

FIGURE 1.167 A gas combustion heating systems. (a) Atmospheric combustion system (A, air); (b) blast-type combustion system. (Courtesy of Ajax Electric Company.)

Some common protective systems are summarized in Table 1.69 [164].

The following protective controls should be installed and interlocked for all atmosphere furnace systems [165]:

1. Atmosphere supply line should have a safety shut-off valve.
2. Operator should be able to visually monitor the atmosphere gas supply to verify gas flow.
3. Temperature of all zones in the furnace should be adequately monitored and interlocked to the atmosphere supply safety shut-off at (or below) 1400°F (760°C).
4. An automatic safety shut-off valve for the flame curtain should be installed and interlocked to prevent opening below 1400°F (760°C).
5. Both audible and visual temperature atmosphere and gas flow alarms should be installed.
6. It should be possible to open the door manually in the event of power failure.

1.14 SALT BATH FURNACE

There are three components to a salt bath furnace: the salt, hardware to melt the salt and control temperature, and a means of reclaiming the salt. This section first provides an overview of the use of molten salts in heat treatment and their selection. The second subsection reviews the furnaces (or salt pots) that are used to provide the necessary temperature

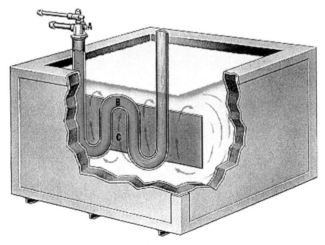

FIGURE 1.168 A gas-fired radiant tube heating/circulation system. (Courtesy of Ajax Electric Company.)

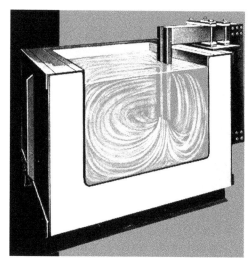

FIGURE 1.169 An over-the-top electrical heating system. (Courtesy of Ajax Electric Company.)

control. The third and fourth subsections deal with salt bath furnace safety and salt contamination. Finally, salt reclamation is briefly discussed.

1.14.1 SALT BATHS

An alternative to heating a metal part in a conventional furnace is to heat it in a liquid such as a molten salt. The advantages of using molten salts for heating include rapid heating, superior temperature uniformity and control, and protection from air, thus eliminating surface oxidation and scale formation. Heating processes where molten salt heating is used [166–169] include neutral hardening, isothermal heat treating, annealing, tempering, descaling, cyanide hardening, carburizing, and brazing.

Hardening of some distortion- and crack-sensitive steels requires cooling the steel at a temperature at or near its martensitic temperature. The high-temperature cooling process (martempering or austempering) can also be performed in a molten salt [167–171]. Figure 1.166 provides typical heat-up times for steel rods of various cross sections.

The use temperature of a salt bath is dependent on the melting point of the salt. Melting points are controlled by blending mixtures, either binary or ternary, of different salts. Table 1.70 lists some typical salt mixtures used for preheating, heating (austenitizing), and

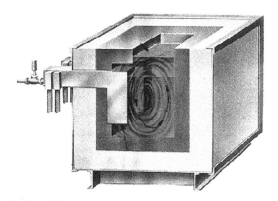

FIGURE 1.170 A through-the-wall heating system. (Courtesy of Ajax Electric Company.)

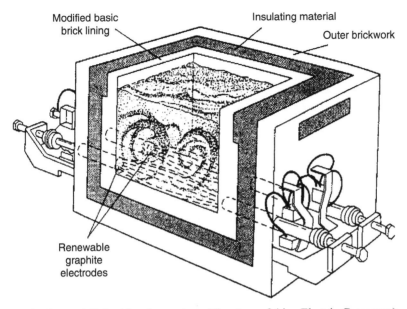

FIGURE 1.171 An internal Calrod heating system. (Courtesy of Ajax Electric Company.)

cooling [167]. (Preheating of crack- or distortion-sensitive parts may be required to minimize thermal shock and to minimize heating time in the higher temperature furnace [167,172].) Generally, higher melting chlorides are used for heating. Lower melting nitrates (and nitrite ternary mixtures discussed in Chapter 9 of *Steel Heat Treatment: Mettalurgy and Technologies*) are used for martempering and austempering.

1.14.2 FURNACE (SALT POT) DESIGN

1.14.2.1 Gas- or Oil-Fired Furnaces

Molten salt furnaces and salt pots can be heated with a gas- or oil-fired radiant tube combustion system as shown in Figure 1.167. The schematic of the heating system illustrated in Figure 1.168 indicates that the pressurized burner fires into a radiant tube, which heats the salt inside the baffled area. The molten salt is subsequently recirculated throughout the furnace zone using mechanical agitators.

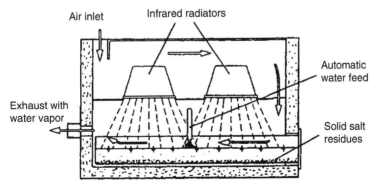

FIGURE 1.172 Schematic illustration of a salt recovery system.

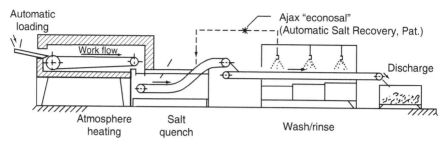

FIGURE 1.173 Example of a continuous salt recovery system. (Courtesy of Ajax Electric Company.)

Gas- or oil-fired burners provide significant energy savings in comparison with electrical heating. Further savings can be achieved by recirculating the hot off-gases to provide additional thermal energy to heat the salt [172].

1.14.2.2 Electrically Heated Furnaces

Molten salt furnaces may also be heated electrically. There are a number of advantages to electrical heating: excellent temperature control ($\pm 5°C$ compared to $\pm 10°C$ for gas-fired heating), ease of changing electrodes, faster heating, and more furnace flexibility.

There are three common designs used for electrical heating of molten salts: over-the-top (see Figure 1.169), through-the-wall (see Figure 1.170), and submerged electrode (Calrod) heating (see Figure 1.171).

1.14.3 SALT BATH FURNACE SAFETY

In addition to the potential hazards of overheating a salt, a number of additional furnace safety concerns have been noted by Laird [172] and Becherer [167].

1. Some mixtures are subject to potential explosive degradation at temperatures above 600°C (1110°F) [173]. However, molten salts can be used safely if the manufacturer's use recommendations and hazard warnings, which are available on the product material safety data sheets (MSDSs), are followed.
2. Care should be taken not to overheat the sidewalls or bottom during start-up to prevent the salt from flowing from the pot.
3. A nitrate–nitrite salt mixture dripping on a hot refractory may present a fire hazard.
4. Chloride salts used at temperatures in excess of 650°C (1200°F), although nontoxic, may produce fumes that require venting from the work area.
5. Flue gases must be vented from the work area.

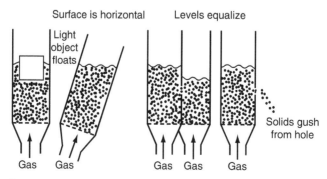

FIGURE 1.174 Typical properties of a fluidized bed mixture. (Courtesy of Quality Heat Treatment Pty. Ltd.)

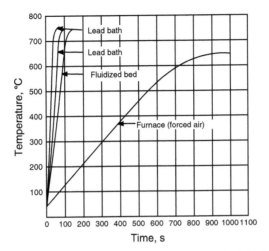

FIGURE 1.175 Comparison of cooling rates achievable with salt and lead baths and fluidized beds for a steel bar 16 mm in diameter. (From *An Introduction to Fluidized Bed Heat Treating*, Brochure, Fluidtherm Technology, Madras, India.)

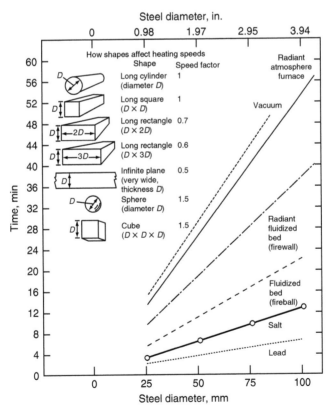

FIGURE 1.176 Recommended heat-up times for various heating media. (From R.W. Reynoldson, *Heat Treatment in Fluidized Bed Furnace*, ASM International, Materials Park, OH, 1993.)

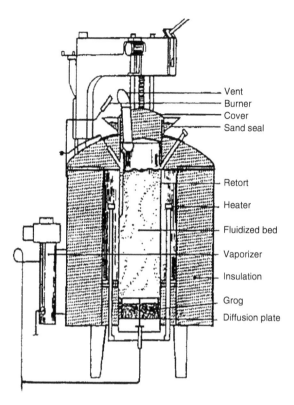

FIGURE 1.177 Schematic illustration of a fluidized bed furnace for methanol carburizing. (Courtesy of Procedyne Corporation.)

1.14.4 SALT CONTAMINATION

A part may be austenitized in a chloride salt pot, withdrawn, and then austempered in molten nitrite salt. In this situation, it is likely that some of the chloride salt will drag out on the heated part and subsequently contaminate the nitrate salt. Since chloride salts are insoluble, a

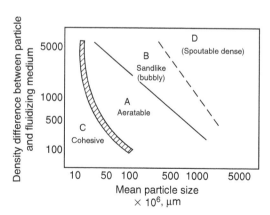

FIGURE 1.178 Geldart chart illustration of effect of selection of solid on fluidized bed properties. (From M.A. Delano and J. Van den Sype, *Heat Treating*, December 1988.)

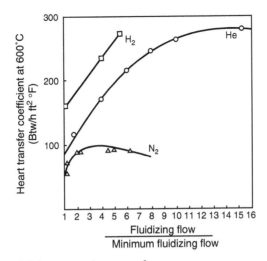

FIGURE 1.179 Effect of fluidizing gas on heat transfer.

chloride sludge will be formed that must be removed by filtration. Conversely, contamination of a chloride salt solution with even 600 ppm of a nitrate salt can result in severe damage to the surface of the heated part.

Chloride salts may become contaminated with soluble oxides and dissolved metals. This will produce an oxidizing environment that may lead to decarburization. When this occurs, the salt solutions must be rectified, sometimes daily. Rectification can be accomplished in one of the two ways. The first method involves the reaction of the soluble oxide to form an insoluble compound. For example, silica or silicon carbide will react with a soluble metal oxide to form an insoluble silicate that can be removed by filtration. A second method of rectification is to react the soluble metal oxides with methyl chloride gas or ammonium chloride, a solid. Either process results in the re-formation of the original chloride salt.

Soluble metals will be reduced at the surface of a graphite rod, thus facilitating their removal.

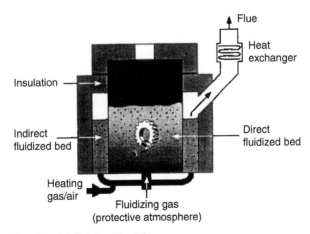

FIGURE 1.180 A gas (or electric) fluidized bed furnace.

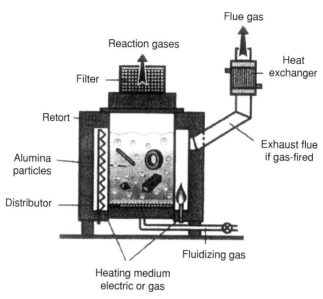

FIGURE 1.181 An externally heated fluidized bed furnace.

1.14.5 SALT RECLAMATION

Increasingly stringent environmental regulations reflect worldwide demand that salt contaminants not be released to the environment. This problem has been successfully addressed by the use of an automatic surface evaporation scheme that permits salvaging the salt from the waste cleaning water and its subsequent reuse [174]. One process used to accomplish this is shown schematically in Figure 1.172 and is a schematic of a continuous salt recovery system for use with a conveyorized wash system.

1.15 FLUIDIZED BED FURNACES

Fluidized bed furnaces employ one of the newer techniques for heat treating. They were first developed in the early 1970s when the oil crisis caused designers to look at more energy-efficient techniques for heating metals. Fluidized beds are formed by passing a gas, often air or nitrogen, through solid particles such as aluminum oxide and silica sand, causing the particles to become microscopically separated from each other and to behave like a bubbling fluid analogous to a boiling liquid. Typical properties are shown in Figure 1.174. The particles in general are inert (with a high melting point) and do not react with metal parts, but act purely as the mechanism to improve heat transfer between the fluidizing gas and the part that is processed.

A broad range of heat transfer rates are possible, depending on bed particle size (particle composition is relatively insignificant), fluidizing gas velocity, choice of fluidizing gas, and bed temperature.

Apart from more energy-efficient than conventional furnaces [175], it was found that the natural physical properties of the basic fluidized bed technique such as temperature uniformity and heat transfer rates similar to those of salt and lead made it a safe and ecologically sound alternatives to molten metal [176,177] and salt baths [178].

It is possible using fluidizing gases such as argon and nitrogen to perform conventional neutral hardening processes such as austenitizing of all ferrous and nonferrous alloys [175]. In addition, the natural characteristics of the fluidized bed allow the normal thermochemical

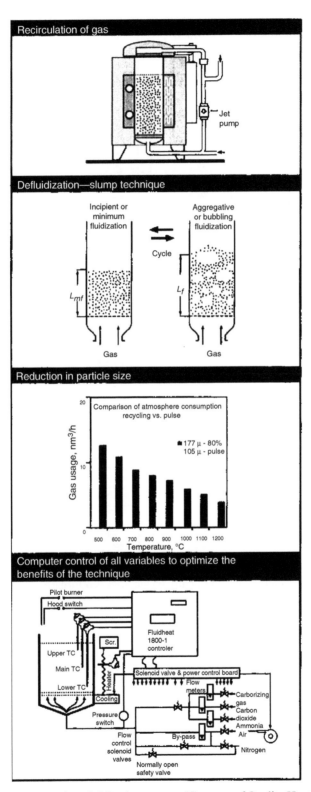

FIGURE 1.182 Procedures to reduce fluidization gas use. (Courtesy of Quality Heat Treatment Pty. Ltd.)

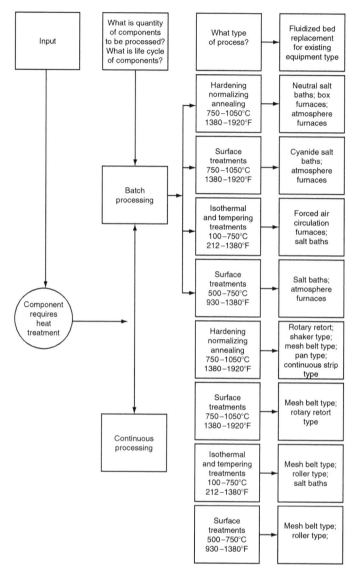

FIGURE 1.183 Decision model for applicability of fluidized bed furnaces. (Courtesy of Quality Heat Treatment Pty. Ltd.)

processes to proceed as well as newly developed processes, e.g., carbide-dispersed carburizing [179], which cannot be performed in other atmosphere-based furnaces. The basic carburizing and nitriding [178] surface treatments can be performed as well as hard surface coating [175] of ferrous materials. The relatively high heat transfer characteristics of the fluidized bed can also be used to cool or quench metal parts after austenitizing or solution treatment [5,6] under certain circumstances [180], providing significant advantages over the more established quenching techniques.

Heating and cooling rates approaching those for salt and lead baths as shown in Figure 1.175 can be achieved depending on temperature and choice of gas. For comparison, recommended heat-up times for various types of heat transfer media are shown in Figure 1.176. Similarly, when the fluidized bed is used for quenching, a broad range of cooling rates are also

FIGURE 1.184 Typical fluidised bed line for semi continuous processing of hot die steel tooling. (Courtesy of Fluidtherm Technology.)

possible, from gas quenching to those attainable with the slower quench oils. The primary focus of this section is placed on the fluidized bed furnace rather than on the associated process technology.

1.15.1 Design of Fluidized Bed Furnaces

An illustrative schematic of a fluidized bed furnace is provided in Figure 1.177. This section discusses the main components of a fluidized bed furnace.

1.15.1.1 Heat Transfer Particles

The characteristic properties of fluidized bed media are based on their position in the Geldart chart as shown in Figure 1.178. The Geldart chart permits characterization of solid media based on their ability to form gas solid fluid. The greatest heat transfer rates occur with Type B media, which include aluminum oxide. Of these media, aluminum oxide provides the best heat transfer capacity, thermal stability, and uniformity [183].

In general, the optimum heat transfer on heating and cooling is achieved with particles in the range of 100–125 μm. If the particle size is less than 100 μm, the uniformity of fluidizing deteriorates, and if the particle size becomes greater than 250 μm, the use of gas becomes uneconomical.

The exception to the choice of aluminum oxide for the fluidized particles is where the particles play a significant role in the thermochemical treatment. For example, ferrovanadium particles are used in the hard surface or vanadium carbide coating of ferrous metals [175]. The choice is based on the coating required and its reaction with the fluidizing gas.

1.15.1.2 Retort

The retort, which contains the particles, is generally fabricated from a high melting point metallic alloy that determines the operating temperature limitations on the bed (i.e., safe working temperature). For some heating methods, e.g., internal gas combustion, refractory containers have been used.

1.15.1.3 Fluidizing Gas Distributor

The design of the distributor, which controls fluidization characteristics, is critical to the function of the bed in that it controls the temperature uniformity and distribution of the gas phase. Many types of distributors have been used, ranging from perforated metal to ceramic filters.

1.15.1.4 Fluidizing Gas

The type of gas used for fluidizing has an effect on the heat transfer rates, particularly at temperatures below 600°C (where radiation effects are minimal). Hydrogen and helium give the highest rates of heat transfer (see Figure 1.179). A mixture of gases used for thermochemical treatments can play a significant part not only in heat transfer but also in controlling the rate of thermochemical reactions.

1.15.1.5 Heating Systems

1.15.1.5.1 External Heating—Electricity or Gas
Externally resistance-heated systems are the most commonly used. For gas, natural draft and flat flame burners with and without heat recovery systems have been developed (Figure 1.180). One recent development has been the use of an externally heated fluidized bed as shown in Figure 1.181, in which internal gas combustion is used to replace burners, resulting in significant increases in heating energy utilization. Externally heated fluidized beds are limited to maximum operating temperatures of 1200°C, which in fact is the safe operating temperature for most heat-resistant alloys. With the development of alloys with higher operating temperatures, fluidized bed furnaces could operate at much higher temperatures.

1.15.1.5.2 Internal Heating—Electricity or Gas
The use of internal or submerged combustion techniques using stoichiometric mixtures of gas and air is very energy-efficient, as is direct resistance heating. However, separation of the control of the fluidizing atmosphere and velocity from the control of heat input makes these techniques more difficult to use than external heating.

1.15.1.6 Retort Support and Casing

The method of construction for supporting the fluidized bed uses conventional high performance insulation and heating element design.

1.15.2 Development of Fluidized Bed Furnaces—Energy and Fluidizing Gas Utilization

As with all furnace designs, the development of the fluidized bed has occurred in discrete stages since the early 1970s. After the initial signs with relatively large particles of aluminum oxide, it was recognized that the use of a column of fluidizing gas was uneconomical. This led to the development of various techniques.

Fluidizing gas recirculation
Smaller particle size (fluidizing gas use is proportional to the square of the diameter of the particle)

Slump techniques combined with controlled fluidization
Computer control

Today these techniques can be combined, making the furnaces (see Figure 1.182) not only energy-efficient in the transfer of heat to the metal but also efficient in the use of fluidizing gases.

1.15.3 APPLICATIONS OF FLUIDIZED BEDS FOR METAL PROCESSING

The applications of fluidized bed furnaces are rapidly increasing in number as the technology becomes more widely accepted. The flowchart in Figure 1.183 can be used in choosing the most suitable type of equipment. In continuous processing, such as wire heating and patenting, fluidized bed furnaces have been used to replace lead and salt baths and should be considered as a viable alternative processing technique for many types of materials.

Figure 1.184 illustrates typical commercial installations of a fluidized bed furnace.

REFERENCES

1. A.N. Kulakov, V.Ya. Lipov, A.P. Potapov, G.K. Rubin, and I.I. Trusova, Modernization of the electric furnace SKZ-8.40.1/9, *Met. Sci. Heat Treat. Met. 8*:551–553 (1981).
2. T.W. Ruffler, Bulk heat treatment of small components, 2nd Int. Congr. Heat Treat. Mater.: 1st Natl. Conf. Metall. Coatings, Florence, Italy, September 20–24, 1982, pp. 597–608.
3. *Heat Treating Furnacers and Ovens*, Brochure, K.H. Hupper & Co. (KHH), South Holland, MI.
4. J.H. Greenberg, *Industrial Thermal Processing Equipment Handbook*, ASM International, Materials Park, OH, 1994.
5. Anon., New heat-treating furnace sets high-performance standards, *Nat. Gas. Ind. Technol.*, Spring 1988, pp. 4–5.
6. Anon., Furnace technology applied to the production of wire, *Wire Ind.*, November 1994, pp. 744–745.
7. E. Ford, The new efficiencies of anti-pollutant furnaces, *Eng. Dig. 38*(1):11–12 (1992).
8. *Sintering Systems*, Product Bulletin No. 901, C.I. Hayes, Cranston, RI.
9. J.P. Hutchinson and S.T. Passano, Control of Lukens 110-in. Plate mill reheat furnaces at Conshohocken, *Iron Steel Eng.*, August 1990, pp. 29–32.
10. K.H. Illgner, Furnace plant for the heat treatment of small mass-produced components, *Wire 39*:246–250 (1989).
11. H. Trojan, Car-bottom furnace of versatile design built in co-operative venture to heat treat gears, *Ind. Heat.*, October 1992, pp. 25–27.
12. T. Banno, Heat treating technology present states and challenges, *Heat and Surface '92*, Kyoto, Japan, November 1992, pp. 449–454.
13. D. Schwalm, Energy saving design concepts for heat treating type furnace, in *The Directory of Industrial Heat Processing and Combustion Equipment: United States Manufacturers, 1981–1982*, Energy edition, Published for Industrial Heating Equipment Association by Information Clearing House, 1111 N. 19th St. Suite 425, Arlington, VA 22209, pp. 147–153.
14. S.N. Piwtorak, Energy conservation in low temperature oven, in *The Directory of Industrial Heat Processing and Combustion Equipment: United States Manufacturers, 1981–1982*, Energy edition, Published for Industrial Heating Equipment Association by Information Clearing House 1111 N. 19th St. Suite 425, Arlington, VA 22209.
15. P. Shefsiek, Gas vs. electricity: another look at relative efficiencies, *Heat Treating*, September 1980, pp. 40–41.
16. M.D. Bullen and J. Bacon, Advances towards the electric heat treatment shop, *Metallurgia 49*(11):518–519, 523–524 (1982).

17. *Kanthal Handbook—Resistance Heating Alloys and Elements for Industrial Furnaces*, Brochure, Kanthal Corporation Heating Systems, Bethel, CT.

18. *Hot Tips for Maximum Performance and Service—Globar Silicon Carbide Electric Heating Elements*, Brochure, The Carborundun Company, Niagara Falls, NY.

19. G.C. Schwartz and R.L. Hexemer, Designing for most effective electric element furnace operation, *Ind. Heat.*, March 1995, pp. 69–72.

20. W. Trinks and M.H. Mawhinney, *Industrial Furnaces*, 4th ed., Vol. 2, Wiley, New York, 1967, pp. 93–11.

21. J. Oare and G. Eklund, Gas-to-electric furnace conversion hikes production, *Metal Prog.*, February 1978, pp. 32–35.

22. *Kanthal Super Electric Heating Elements for Use up to 1900°C*, Brochure, Kanthal Furnace Products, Hallstahammer, Sweden.

23. T. Thomonder, *Some View Points Concerning the Design of Electric Resistance Furnaces*, Brochure, Aktiebolaget Kanthal, Hallstahammer, Sweden.

24. B. Barton, Case study shows natural gas to be cheaper than electricity, *Heat Treating*, February 1981, pp. 36–38.

25. R.A. Andrews, Convert electric furnaces to gas, *Heat Treating*, February 1993, pp. 2–23.

26. *Single-End Recuperative Radiant Tube Combustion Systems*, Brochure, Pyronics Inc., Cleveland, OH.

27. *Fundamentals of Gas Combustion*, 10th ed., 1994, Brochure, Catalog No. XH 03737, American Gas Association, Arlington, VA 22209.

28. R.G. Martinek, *Eclipse Industrial Process Heating Guide*, Brochure, Eclipse Fired Engineering Co., Dan Mills, ON, Canada.

29. *The Prudent Use of Gas—An Industrial Guide to Energy Conservation*, Brochure No. 7782-573, Catalog No. C30070, American Gas Association, Arlington, VA.

30. R.J. Reed, Maintaining and adjusting combustion systems for fuel economy, *J. Heat Treat.* $1(1):93–95$ (1979).

31. *Handbook Supplement 6–57*, Chart 40, North American Manufacturing Co., Cleveland, OH.

32. G. Gross, The use of oxygen in industrial furnace, Int. Foundry Heat Treat. Conf., Vol. 4, Paper No. 6, September 30–October 4, 1985, Johannesburg, South Africa.

33. L.A. Weaver, Systems control for energy conservation, in *The Directory of Industrial Heat Processing and Combustion Equipment: United States Manufacturers, 1981–1982*, Energy edition, Published for Industrial Heating Equipment Association by Information Clearing House, pp. 160–168.

34. S. Lampman, Energy-efficient furnace design and operation, in *ASM Handbook*, Vol. 4, *Heat Treating*, ASM International, Materials Park, OH, 1991, pp. 519–526.

35. W. Trinks and M.H. Mawhinney, *Industrial Furnaces*, 4th ed., Vol. 2, Wiley, New York, 1967, pp. 29–73.

36. *Heat Treating Handbook*, Seco/Warwick Corp., Meadville, PA.

37. K. Keller, Use high-speed burners in heating and heat treating furnaces, *Formage Trait. Met.*, November 1977, pp. 39–44.

38. T. Darroudi, J.R. Hellmann, R.E. Tressler, and L. Gorski, Strength evaluation of reaction-bonded silicon carbide radiant tubes, *J. Am. Ceram. Soc. 75*(12):3445–3451 (1992).

39. D. Hibberd, Recent developments in reheating and heat-treatment furnaces, *Metallurgia*, February 1968, pp. 52–58.

40. D. Hibberd and T. Hallatt, Regenerative firing in the steel industry, *Metallurgia 58*(12):FUS 3-Fus (1991).

41. B. Vinton, Ceramic radiant tube system speeds batch furnace recovery, *Heat Treating*, February 1989, pp. 24–27.

42. Anon., Furnace design moves into the digital age with pulse-firing burner technology, *Nat. Gas. Ind. Technol.*, Summer 1988, pp. 1–2.

43. Anon., High velocity combustion in a heat treatment furnace, *Steelmaker*, September 1968, pp. 28–30.

44. R.A. Wallis, B.C. McMillian, and A.C. Sanderfer, Variable excess air system for controlled firing saves gas while maintaining temperature uniformity in a batch heat treat furnace, *Ind. Heat.*, June 1995, pp. 53–55.

45. M. Shay, Pulse firing—how this "new" combustion technique works, *Heat Treating*, February 1987, pp. 24–25.

46. F.M. Heyn, Heat recovery from waste gases with recuperators, in *The Directory of Industrial Heat Processing and Combustion Equipment: United States Manufacturers, 1981–1982*, Energy edition, Published for the Industrial Heating Equipment Association by Information Clearing House.

47. R.J. Evans, New recuperative-burner system gives MOI a better workhorse, *Heat Treating*, 1984, pp. 18–20.

48. F.J. Bartkowski and K.H. Kohnken, Carbottom furnace retrofitted with burner—ceramic recuperative system to save energy in heat treating casting, *Ind. Heat.*, 1982, pp. 48–50.

49. J.A. Wünning and J.G. Wünning, Burner design for flameless oxidation with low NO-formation even at maximum preheat, *Ind. Heat.*, January 1995, pp. 24–28.

50. W. Trinks, *Industrial Furnaces*, 4th ed., Vol. 1, Wiley, New York, 1950, pp. 220–262.

51. T. Ward and R.J. Webb, Regenerative burners for use in high temperature furnaces, *Inst. Gas Eng. Commun. 1273*, 1985.

52. D.O. Swinder, Gas—the natural winner for heat treatment, *Heat Treat. Steel*, September 1989, pp. 6–25.

53. Anon., The Performance of Impulse-Fired Regenerative Burners on a Small Batch Heat Treatment Furnace, Br. Energy Efficiency Office, Dept. of Energy, Rep. ED/220/332.

54. D. Lupton, Regenerative burners in bloom reheating furnace, *Steel Times Int.*, November 1989, pp. 22–25.

55. Anon., New generation, regenerative burner for continuous slab reheating furnace in Japan to reduce NO_x emission and save energy, *Ind. Heat.*, March 1996, p. 112.

56. Anon., The Application of Regenerative Burners to a Continuous Heat Treatment Furnace, Br. Energy Efficiency Office, Dept. of Energy, Rep. ED/166/191.

57. J.D. Bowers, Regenerative burners slash fuel consumption, *Adv. Mater. Proc.*, March 1990, pp. 63–64.

58. T. Martin, Regenerative ceramic burner technology and utilization, *Ind. Heat.*, November 1988, pp. 12–15.

59. N. Fricker, K.F. Pomfret, and J.D. Waddington, Commun. 1072, Inst. Gas Eng., 44th Annu. Meeting, London, November 1978.

60. V. Paschkis and J. Persson, *Industrial Electric Furnaces and Appliances*, Interscience, New York, 1960, pp. 14–25.

61. D.F. Hibbard, Modern approaches to heat treatment furnace design, *Melt. Mater. 3*(1):22–27 (1987).

62. AFS, *Refractories Manual*, 2nd ed., American Foundrymen's Society, Des Plaines, IL, 1989.

63. W.H. McAdams, *Heat Transmission*, 3rd ed., McGraw-Hill, New York, 1954.

64. J.P. Holman, *Heat Transfer*, 2nd ed., McGraw-Hill, New York, 1968.

65. A.J. Beck, *Heat Treating*, May 1973, pp. 29–32.

66. R.N. Britz, Convection heat treating. Part II. Furnace considerations, *Ind. Heat.*, January 1975, pp. 39–47.

67. D. Nicholson, S. Ruhemann, and R.J. Wingrove, Heat transmission in reheating and heat treatment furnace: some recent developments, in *Heat Treatment of Metals*, Spec. Rep. 95, Iron and Steel Inst., 1966, pp. 173–182.

68. S. Lampman, Energy-efficient furnace design and operation, in *ASM Handbook*, Vol. 4, *Heat Treating*, ASM International, Materials Park, OH, pp. 519–526.

69. M.A. Aronov, J.F. Wallace, and W.A. Ordillas, System for prediction of heat-up and soak times for bulk heat treatment processes, in *Heat Treating: Equipment and Processes* (G.E. Totten and R.A. Wallis, Eds.), Proc. 1994 Conf., ASM International, Materials Park, OH, 1994, pp. 55–61.

70. M. Aronov, J. Wallace, and M. Ordillas, Development of validated system for prediction of heat-up and soak times for bulk heat treatment process for materials, in *Soak Time Determination Manual*, Appendix to Find Report HTN/CT-020112/TR93, Heat Treating Network, Cleveland, OH.

71. Z.S. Tian, M.S. Xu, Y.H. Chui, and J.F. Sui, The thermoelectric effect and theory of the thermocouple, *J. Heat Treat. Met. (China)* 6:42–44 (1994).

72. H.D. Baker, E.A. Ryder, and N.H. Baker, *Temperature Measurement in Engineering*, Vol. 1, Omega Press, Stamford, CT, 1975.

73. Anon., *The Theory and Properties of Thermocouple Elements*, ASTM STP 492, American Society for Testing and Materials, Philadelphia, PA, 1971.

74. Anon., *The Temperature Handbook*, Vol. 28, Omega Engineering Corp., Stamford, CT.

75. T.P. Wang, Thermocouples for special applications, in *Heat Treating: Equipment and Process* (G.E. Totten and R.A. Wallis, Eds.), Proc. 1994 Conf., ASTM International Materials Park, OH, 1994, pp. 171–174.

76. Anon., *Manual on the Use of Thermocouples in Temperature Measurement*, ASTM STP 470 B, American Society for Testing and Materials, Philadelphia, PA, 1968.

77. J. Nanigan, Improving accuracy and response of thermocouples in ovens and furnace, in *Heat Treating: Equipment and Processes* (G.E. Totten and R.A. Wallis, Eds.), Proc. 1994 Conf., ASM International, Materials Park, OH, 1994, pp. 171–174.

78. *Atmospheres for Heat Treating Equipment*, Brochure, Surface Combustion, Inc., Maumee, OH.

79. *Protective Atmospheres and Analysis Curves*, Brochure, Electric Furnace Company, Salem, OH.

80. *Bulk Gases for the Electronics Industry*, Brochure, Praxair Inc., Chicago, IL.

81. A.G. Hotchkiss and H.M. Weber, *Protective Atmospheres*, Wiley, New York, 1963.

82. R. Nemenyi and G. Bennett, *Controlled Atmospheres for Heat Treatment*, Franklin Book Co., Philadelphia, PA, USA 1995, pp. 22–1022.

83. *Praxair Nitrogen® Membrane System*, Brochure, Praxair Inc., Danbury, CT.

84. *Nitrogen Supply Systems to Meet Every Requirement*, Brochure, Praxair Inc., Danbury, CT.

85. *Liquid Nitrogen Cooling Systems for Heat Treaters*, Brochure, Praxair Inc., Chicago, IL.

86. J.A. Zahniser, *Furnace Atmospheres*, ASM International, Materials Park, OH.

87. F.S. Barff, Zinc white as paint, and the treatment of iron for the prevention of corrosion, *J. Soc. Arts 25*:254–260 (1877).

88. J. Morris, The use of water in furnace atmospheres, *Heat Treat. Met. 2*:33–37 (1989).

89. F.E. Vandaveer and C.G. Segeley, Prepared atmospheres, in *Gas Engineering Handbook*, Industrial Press, New York, 1965, pp. 12/278–12/289.

90. D.W. Breck, *Zeolite Molecular Sieves—Structure, Chemistry and Use*, Krieger, Malabar, FL, 1984.

91. B.W. Gonser, The status of prepared atmospheres in the heat treatment of steel, *Ind. Heat. 6*(12):1123–1134 (1939).

92. M.J. Hill, The efficient use of atmospheres in a continuous furnace using the concept of zoning, Int. Foundry Heat Treat. Conf., Johannesburg, S. Africa, Vol. 2, 1985, pp. 1–30.

93. Anon., Furnace curtains save energy, *Metal Heat Treat.*, November/December, 1994, p. 40.

94. W. Olszanski, T. Sobusiak, and T. Trzcialkowski, Carbomix system of controlled carburizing and carbonitriding in pit-type furnaces, 5th Int. Cong. Heat Treat. Mater., Budapest, Hungary, Vol. II, October 20–24, 1986, pp. 1276–1285.

95. *Installation of Nitrogen Methanol Atmosphere Systems*, Brochure, Praxair Inc., Chicago, IL.

96. Data available from Airco Industrial Gases, Murray Hill, NJ.

97. B. Edenhofer, Progress in the technology and applications of *in-situ* atmosphere production in hardening and case-hardening furnaces, Proc. 2nd Int. Conf. on Carburizing and Nitriding with Atmospheres, December 6–8, 1995, Cleveland, pp. 37–42.

98. D.S. Mackenzie, The dissociation of methanol used for neutral hardening of steel, in *Heat Treating: Equipment and Processes* (G.E. Totten and R.A. Wallis, Eds.), Proc. 1994 Conf., ASM International, Materials Park, OH, 1994.

99. M.J. Huber, High temperature methanol dissociation, in *Heat Treating: Equipment and Processes* (G.E. Totten and R.A. Wallis, Eds.), Proc. 1994 Conf., ASM International Materials Park, OH, 1994, pp. 437–440.

100. H. Walton, Atmospheres for the hardening of steel, in *Heat Treating: Equipment and Processes* (G.E. Totten and R.A. Wallis, Eds.), Proc. 1994 Conf., ASM International, Materials Park, OH, 1994, pp. 441–447.

101. T.J. Schultz, Portable flue gas analyzers for industrial furnaces, Inf. Letter No. 153, Catalog No. C10877, American Gas Association, Arlington, VA.

102. S. Yasui, Automatic gas chromatography, *Netsu Shori 26*(3):231–237 (1986).

103. M. Okumiya, Y. Tsunekawa, I. Niimi, M. Harmada, and M. Mabe, Control of the surface carbon content by measurement of retained methane in plasma-carborizing, *Nippon Kinzoku Gakkaishi 55*(1):981–985 (1991).

104. F. Trombetta and M. Caon, Confronto ta analizzatore a raggi infraosse e la sonda ord ossido di friconio nel trattamento. Termico di cementazione e tempra di pari en sinterizzato, *Metall. Stalenana 2*:65–74 (1981).

105. K. Derge, Uberwacbury du Often Atmosphaere mit Hilfe de Gas-Chromatographie, *Giesserei-Praxis 2l*:433–440 (1965).

106. R.K. Singh and C.R. Chakrovorty, Method for atmosphere control of heat treatment furnaces, *Tool Alloy Steels*, April/May 1986, pp. 141–144.

107. *Guide to the Selection of Oxygen Analyzers*, Brochure, Delta F Corporation, Woburn, MA.

108. H.W. Bond, Atmosphere control of heat treating furnace using O_2 sensors: Current standing and the future, *Heat and Surface '92*, Kyoto, Japan, November 17–20, 1992, pp. 479–482.

109. Y.-C. Chen, Automatic control of carbon potential of furnace atmospheres without adding enriched gas, *Metall. Trans. B 24B*:881–888 (1993).

110. R.W. Blumenthal, A technical presentation of the factors affecting the accuracy of carbon/oxygen probes, Proc. 2nd Int. Conf. on Carbonizing and Nitriding with Atmospheres, December 6–8, 1995, Cleveland, OH, pp. 17–22.

111. D.W. McCurdy, Improving the accuracy of oxygen probe control system, in *Heat Treating: Equipment and Processes* (G.E. Totten and R.A. Willis, Eds.), Proc. 1994 Conf., ASM International, Materials Park, OH, 1994, pp. 117–121.

112. H.W. Bond, Oxygen-sensors—a review of their impact on the heat treating industry, *Mat. Sci. Forum, 102–104*:831–838 (1992).

113. R.W. Blumenthal and A.T. Melville, Hot gas measuring probe, U.S. Patent 4,588,493 (May 13, 1986).

114. Data provided by Teledyne Brown Engineering Analytical Instruments, City of Industry, CA.

115. P.C. Prasannan, Carburization of steels—an overview, *Indian J. Eng. Mater. Sci. 1*:221–228 (1994).

116. M.J. Fischer, Distributed infrared atmosphere monitoring system, in *Heat Treating: Equipment and Processes* (G.E. Totten and R.A. Wallis, Eds.), Proc. 1994 Conf., ASM International, Materials Park, OH, 1994, pp. 167–169.

117. W. Cole, Continuous dew point monitor increases accuracy, *Heat Treating*, March 1993, pp. 19–20.

118. R.N. Blumenthal and R. Hlasny, Check out carbon control systems—step by step, *Heat Treating*, August 1991.

119. H. Heine, Refractories revisited: a review and outlook, *Met. Heat Treat.*, March/April 1996, pp. 87–93.

120. R.C. Johnson, Evaluation of state-of-the-art refractory system, *Ind. Heat.*, February 1994, pp. 39–43.

121. J. Dinwoodie, High alumina fiber: manufacture, properties and application in high temperature furnaces, *Ind. Heat.*, February 1996, pp. 40–46.

122. G. Deren and M.A. Rhoa, Favorable properties of high temperature glass fiber insulating material by improved chemistry, *Ind. Heat.*, November 1993, pp. 46–49.

123. Anon., Characteristics and applicability of specially designed microporous insulating refractory, *Ind. Heat.*, August 1995, pp. 47–51.

124. S.C. Carniglia and G.L. Barna, *Handbook of Industrial Refractories Technology*, Noyes, Park Ridge, NJ, 1992.

125. R.J. Aberbach, Fans—a special report, *Power*, New York, NY.

126. D.E. Fluck, R.B. Herchenroeder, G.Y. Lai, and M.F. Rothman, Selecting alloys for heat treatment equipment, *Met. Prog. 128*(4):35–40 (1985).

127. D.C. Agarwal and U. Brill, Material degradation problems in high temperature environments (alloys—alloying elements—solutions), *Ind. Heat.*, October 1994, pp. 55–60.

128. J. Kelly, Heat resistant alloy performance, *Heat Treating*, July 1993, pp. 24–27.

129. T. Cronan, Parts cleaning and its integration into heat treating, *Heat Treating: Equipment and Processes* (G.E. Totten and R.A. Wallis, Eds.), Proc. 1994 Conf., ASM International, Materials Park, OH, 1994, pp. 311–315.

130. W.H. Michels, Pollution prevention analysis of oil and polymer quenching in the heat treatment of steel, in *Heat Treating: Equipment and Processes* (G.E. Totten and R.A. Wallis, Eds.), Proc. 1994 Conf., ASM International, Materials Park, OH, 1994, pp. 383–388.

131. D.B. Lebart, Parts cleaning: alternatives for the heat treat shop, *Metal Heat Treat.*, January/February 1995, pp. 21–24.

132. M. Sugiyama and N. Hiramoto, Vacuum vapor solvent degreasing: an effective alternative for pollution control in thermal processing, *Ind. Heat.*, November 1994, pp. 36–39.

133. L.E. Jones and B. Strebing, Heat Treating: Equipment Maintenance, Safety and Equipment Operations, ASM Int. Short Course, presented April 16–17, 1994, Schaumberg, IL.

134. G.E. Totten, C.E. Bates, and N.A. Clinton, in *Handbook of Quenchants and Quenching Technology*, ASM International, Materials Park, OH, 1993, chap. 9.

135. F. Mayinger, Thermo and fluid-dynamic principles of heat transfer during cooling, in *Theory and Technology of Quenching: A Handbook* (B. Liscic, H.M. Tensi, and W. Luty, Eds.), Springer-Verlag, New York, chap. 3.

136. G.E. Totten, G.M. Webster, and N. Gopinath, Quenching fundamentals—effect of agitation, *Adv. Mater. Process. 149*(2):73–76 (1996).

137. C.E. Bates, G.E. Totten, and R.L. Bremman, in *ASM Handbook*, Vol. 4, *Heat Treating*, ASM International, Materials Park, OH, 1991, pp. 67–120.

138. J.J. Lakin, Use of polymer quenchants in surface heat treatments, *Heat Treat. Met. 9*(3):73–76 (1982).

139. G.E. Totten, K.B. Orszak, L.M. Jarvis, and R.R. Blackwood, How to effectively use polymer quenchants, *Ind. Heat.*, October 1991, pp. 37–41.

140. J. Hasson, Quench system design factors, *Adv. Mater. Process. 148*(3):425S–424U (1995).

141. G.B. Tatterson, *Fluid Mixing and Gas Dispersion in Agitated Tanks*, McGraw-Hill, New York, 1991.

142. H.L. Kauffman, *Chem. Metall. Eng. 37*:177–180 (1930).

143. V.G. Stognei and A.T. Kruk, Optimization of the quenching process of forgings in bubbling tanks, *Met. Sci. Heat Treat. 31*(1–2):69–71 (1989).

144. U.S. Steel, *Improved Quenching of Steel by Propeller Agitation*, 1954.

145. G.E. Totten and K.S. Lally, Proper agitation dictates quench success. Part I, *Heat Treat.*, September 1992, pp. 12–1.

146. I.J. Krassik, W.C. Krutzch, W.H. Fraser, and J.P. Messina, *Pump Handbook*, 2nd ed., McGraw-Hill, New York, 1985.

147. E.E. Ludwig, *Applied Process Design for Chemical and Petrochemical Plants*, 2nd ed., Vol. 1., Gulf Publishing, London, 1984.

148. *Goulds Pump Manual*, Goulds Pumps Inc., Industrial Products Group, Seneca Falls, NY.

149. K.S. Lally and G.E. Totten, Proper agitation dictates quench success. Part 2, *Heat Treat.*, October 1992, pp. 28–31.

150. D.R. Garwood, J.D. Lucas, R.A. Walls, and J. Ward, Modeling of flow distribution in an oil quench tank, *J. Mater. Eng. Perfect. 1*(6):781–787 (1992).

151. R.A. Wallis, D.R. Garwood, and J. Ward, The use of modeling techniques and improved quenching of components, in *Heat Treating: Equipment and Processes* (G.E. Totten and R.A. Wallis, Eds.), Proc. 1994 Conf., ASM International, Materials Park, OH, 1994, pp. 51–54.

152. N. Bogh, Quench tank agitation design using flow modeling, in *Heat Treating: Equipment and Processes* (G.E. Totten and R.A. Wallis, Eds.), Proc. 1994 Conf., ASM International, Materials Park, OH, 1994, pp. 51–54.

153. K. Illgner, Quenching of small parts, *Harterei-Tech. Mitt. 41*(2):113–120 (1987).

154. G.E. Totten, G.M. Webster, R.R. Blackwood, L.M. Jarvis, and T. Narumi, Designing chute quench for continuous furnace heat treating effectively, *Ind. Heat.*, November 1995, pp. 49–52.

155. S.W. Han, S.H. Kang, G.E. Totten, and G.M. Webster, Immersion time quenching, *Adv. Mater. Process.*, September 1995, pp. 42AA–42DD.

156. A.P. Petrukhin, USSR Patent 1,247,424 (1986).

157. D.J. Williams, Quench system for industrious hardening, *Met. Heat Treat.*, July/August 1995, pp. 33–37.

158. V. Srimongkolkul, Is there a need for really clean oil in quenching operations?, *Heat Treating*, December 1990, pp. 27–28.

159. R.D. Howard and G.E. Totten, Membrane separation of polymer quenchants, *Met. Heat Treat.*, September/October 1994, pp. 22–24.

160. R. Kern, Distortion and cracking. II. Distortion from quenching, *Heat Treating*, March 1985, pp. 41–45.

161. D. Grassl, Heat treating furnace system with integrated single-part press quenching, *Heat and Surface '92*, Conf., Proc., Kyoto, Japan, November, 1992, pp. 625–628.

162. L.E. Jones, The fundamentals of gear press quenching, *Ind. Heat.*, April 1995, pp. 54–58.

163. L.E. Jones, The fundamentals of gear press quenching, *Gear Technol.*, March/April 1994, pp. 32–40.

164. Anon., The safe operation of atmosphere furnace, *Metal Prog.*, December 1956, pp. 1–7.

165. R. Ostrowski, Furnace safety, in *ASM Handbook*, Vol. 4, *Heat Treating*, ASM Int., Materials Park, OH, 1991, pp. 657–663.

166. G. Wahl, Development and application of salt baths in the heat treatment of case-hardened steels, Proc. ASM Heat Treating Conf. Carburizing, Processing and Performance, ASM International, Materials Park, OH, 1989, pp. 41–56.

167. B.A. Becherer, Processes and furnace equipment for heat treating of tool steels, in *ASM Handbook*, Vol. 4, *Heat Treating*, ASM International, Materials Park, OH, 1991, pp. 726–733.

168. Q.D. Mehrkam, Salt bath heat treating, Part I, *Tooling Prod.*, June 1967, pp. 1–6.

169. Q.D. Mehrkam, Salt bath heat treating, Part II, *Tooling Prod.*, July 1967, pp. 7–12.

170. R.W. Foreman, Salt-bath quenching, in *Quenching and Distortion Control* (G.E. Totten, Ed.), ASM International, Materials Park, OH, 1992, pp. 87–94.

171. K.S. Sreenivasa Marthy and K.S. Shamanna, Heat treatment salts, *Tool and Alloy Steels*, May 1992, pp. 115–11.

172. W.J. Laird, Salt bath equipment, in *ASM Handbook*, Vol. 4, *Heat Treating*, ASM International, Materials Park, OH, 1991, pp. 475–483.

173. M.A.H. Howes, The Cooling of Steel Shapes in Molten Salt and Hot Oil, Ph.D. thesis, London University, 1959.

174. E.H. Burgdorf, Use and disposal of quenching media—recent developments with respect to environmental regulations, *Quenching and Carburizing*, 3rd Int. IFHT Seminar, Melbourne, Australia, September 1991, pp. 66–77.

175. R.W. Reynoldson, *Heat Treatment in Fluidized Bed Furnace*, ASM International, Materials Park, OH, 1993.

176. R. Branders, Patenting in a fluidized bed, *Wire Ind.*, February 1990, pp. 89–91.

177. H. Lochner, Molten-metal and hydrogen quenching technologies for steel and strip, *Rev. Fr. Metall.*, February 1993, pp. 65–73.

178. W. Krebs, Fluidized-bed furnaces instead of salt baths for the heat treatment of tool steels, *Giesserei* 77(10):337–334(1990).

179. A. Killian, Fluidized bed and QCD carburizing, Materials Research Conference, 1996.

180. R.W. Reynoldson and E. Ninham, Optimizing the performance of die casting tools manufactured from H13 hot work tool steels, Technical Paper, 1996.

181. *An Introduction to Fluidized Bed Heat Treating*, Brochure, Fluidtherm Technology, Madras, India.

182. M.A. Delano and J. Van den Sype, Fluid bed quenching of steels: applications are widening, *Heat Treating*, December 1988.

183. P. Sommer, Quenching in fluidized beds, *Heat Treat. Met.* 2:39–44 (1986).

2 Design of Steel-Intensive Quench Processes

Nikolai I. Kobasko, Wytal S. Morhuniuk,
and Boris K. Ushakov

CONTENTS

2.1 INTRODUCTION

The development of new technology of heat treatment based on the intensification of heat transfer processes during phase transformations and on the existence of the optimal depth of hard layer is of great interest. As is known, in the case of the intensification of the cooling processes for machine parts to be quenched, great thermal and structural stresses appear, which often results in the destruction of the material. For this reason, it is important to investigate current and residual stresses with regard to cooling conditions and character of phase transformations that occur in accordance with CCT or TTT diagrams for the transformation of overcooled austenite.

Here there are results of numerical investigations of current and residual stresses made on computers based on well-known methods, and the main attention is paid to the optimal layer. On the basis of investigations made, common recommendations have been given, which can be used for the improvement of the technology of heat treatment of machine parts and equipment and can be applied in machine construction and other fields of engineering. Besides, the main attention is paid to the maximum compressive stresses which are formed at the surface. These data are based on the work done in 1979–1983, results of which published in 1983 [1]. This chapter also discusses the advantages of using controlled-hardenability steels, the so-called shell hardening. The practical use of controlled-hardenability steels and their great advantages have been provided by Professor Ushakov, which are described in detail in Refs. [2–4]. Additional calculations have been made, which prove the existence of optimal hard layer corresponding to the maximum compressive stresses at the surface.

It should be noted that different authors dealing with intensive quenching at which the surface hard layer is formed introduced different names: shell hardening [5] and through-surface hardening (TSH) [3].

Regularities of intensive quenching using controlled-hardenability steels are also discussed, which are described in detail in Ref. [6]. The optimal hard layer corresponding to maximum compressive stresses at the surface is also related to shell hardening or TSH.

The chapter has the following plan:

1. Study of the effect of the intensification of the cooling process upon the value of residual stresses during quenching. Determination of the notion of optimal depth of hard layer.
2. Investigation of the effect of size upon thermal and stress–strain state of parts to be quenched. Similarity in the distribution of residual stresses.
3. Some recommendations concerning the heat treatment of machine parts.

2.2 MATHEMATICAL MODELS AND METHODS OF CALCULATION OF THERMAL AND STRESS–STRAIN STATE

The coupled equation of nonstationary thermal conductivity, as known, is given in the form [1,7,8]:

$$c\rho \frac{\partial T}{\partial \tau} - \mathrm{div}(\lambda \ \mathrm{grad}\ T) - \sigma_{ij}\dot{\varepsilon}_{ij}^{p} + \sum \rho_{l}l_{l}\dot{\varepsilon}_{l} = 0 \qquad (2.1)$$

with corresponding boundary conditions for film boiling

$$\frac{\partial T}{\partial r} + \frac{\alpha_{f}}{\lambda}(T - T_{s})|_{r=R} = 0 \qquad (2.2)$$

and initial conditions:

$$T(r,0) = T_{0}. \qquad (2.3)$$

The transition from film boiling to nucleate boiling is made when the following is fulfilled:

$$q_{cr2} = \alpha_{f}(T_{per} - T_{s}) \qquad (2.4)$$

where $q_{cr2} = 0.2q_{cr1}$.

At the stage of the nucleate boiling, the boundary conditions have the following form:

$$\frac{\partial T}{\partial r} + \frac{\beta^m}{\lambda_T}(T - T_s)^m|_{r=R} = 0, \tag{2.5}$$

$$T(r,\tau_f) = \varphi(r). \tag{2.6}$$

Finally, at the area of convection heat transfer, the boundary conditions are analogous to those for film boiling.

$$\frac{\partial T}{\partial r} + \frac{\alpha_{cn}}{\lambda_T}(T - T_c) = 0, \; T(r,\tau_{nb}) = \psi(r). \tag{2.7}$$

At intensive quenching, the film boiling is absent and the main process is nucleate boiling and convection; that is, boundary conditions (2.5) and (2.7) are used.

For the determination of current and residual stresses, plasticity theory equations are used, which are presented in detail in Refs. [9–11] and have the following form:

$$\dot{\varepsilon}_{ij} = \dot{\varepsilon}_{ij}^p + \dot{\varepsilon}_{ij}^e + \dot{\varepsilon}_{ij}^T + \dot{\varepsilon}_{ij}^m + \dot{\varepsilon}_{ij}^c, \tag{2.8}$$

where

$$\varepsilon_{ij}^e = \frac{1-\nu}{E}\sigma_{ij} - \frac{\nu}{E}\sigma_{kk}\delta_{ij}; \; \varepsilon_{ij}^T = \alpha_T(T - T_0)\delta_{ij};$$

$$\varepsilon_{ij}^m = \sum_{I=1}^{N} \beta_I \xi_I \delta_{ij},$$

$$\dot{\varepsilon}_{ij}^p = \Lambda \frac{\partial F}{\partial \sigma_{ij}} = \hat{G}\left(\frac{\partial F}{\partial \sigma_{kl}}\dot{\sigma}_{kl} + \frac{\partial F}{\partial T}T + \sum_{I=1}^{N} \frac{\partial F}{\partial \xi_I}\dot{\xi}_I \right)\frac{\partial F}{\partial \sigma_{ij}}.$$

Here $F = F(\sigma_{ij}, \varepsilon^p, k, T, \xi_I)$,

$$\frac{1}{\hat{G}} = -\left(\frac{\partial F}{\partial \varepsilon_{mn}^p} + \frac{\partial F}{\partial k}\sigma_{mn} \right)\frac{\partial F}{\partial \sigma_{mn}}.$$

The creep strain rate is as follows:

$$\dot{\varepsilon}_{ij}^c = \frac{3}{2}A_c^{1/m}\sigma^{(n-m)/m}\varepsilon^{c(m-1)/m}S_{ij}. \tag{2.9}$$

2.2.1 KINETICS OF PHASE TRANSFORMATIONS AND MECHANICAL PROPERTIES OF MATERIAL

Kinetics of phase transformations is described by CCT or TTT diagrams; therefore in our program, when temperature fields were computed at each step by the space and time, results of calculations were compared with thermal-kinetic diagrams with regard to the location, and the corresponding law of phase transformations was chosen at the plane of diagram.

In other words, the program is constructed in such a way that the data entered into computer include the coordinates of all areas of main phase transformations, and for each

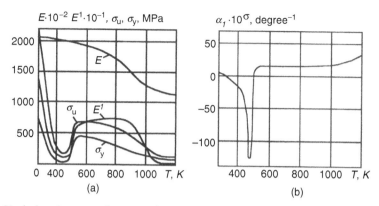

FIGURE 2.1 Variation in strength properties, elastic modulus E, hardening modulus E' (a), and coefficient of linear expansion α_T, ultimate stress σ_u, yield stress σ_y of steel AISI 52100 (ShKh15) with allowance for increase in specific volume as austenite transforms into martensite (b), as functions of temperature.

separate area, the corresponding formulas are provided for the description of the process of the phase transformations. As for this approach, there is a good agreement between data computed and results of experiments.

The regularities of phase transformations are considered in detail in the works of many authors [12–16].

It should be noted that the determination of mechanical steel properties is a quite difficult and serious problem since phases are not stable and the process is fast at high temperatures. Let us note that the coefficient of linear expansion α_T is negative within the martensite range since the expansion of the martensite is observed, which is presented in Figure 2.1b. Within the martensite range, superplasticity is also observed, which is reflected in Figure 2.1a in the form of decline of plasticity. In calculations, these declines result in oscillations and this causes difficulties such as lower precision of numerical calculations.

The phase distribution was calculated on the basis of CCT diagram, presented in Figure 2.2.

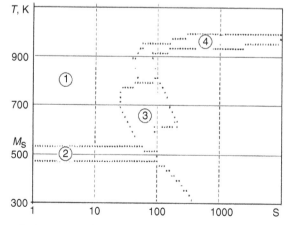

FIGURE 2.2 Diagram of the martensite–austenite transformations for steel 52100: 1—austenite, 2—martensite, 3—intermediary phase, and 4—perlite.

The process of the calculation is divided into time steps. At each time step, a heat conductivity problem is solved. The resulting temperature field determines mechanical characteristics and temperature load for the thermal plasticity problem. The initial conditions for thermal conductivity and thermal plasticity problems are taken from results of solving the corresponding problems at previous steps. Besides, at each time step, the calculation results are compared with the diagram of the supercooled austenite transformations, and new thermal, physical, and mechanical characteristics for the next step are chosen depending on the structural components. In every separate region of the diagram, a specific transformation law is established, if necessary. The calculations are made until the part is cooled.

The time of reaching the maximum compressive stresses at the surface is sought by solving the same problem, in intensive quenching conditions, while the temperature of the core reaches a certain low value. Then, stress distribution fields are analyzed by computer or visually and the time is found when the stress distribution is optimal. After reaching maximum compressive stresses, self-tempering is made by air-cooling and it is conducted until the temperature fields are equalized at cross sections.

2.3   BASIC REGULARITIES OF THE FORMATION OF RESIDUAL STRESSES

The calculation was fulfilled by the thermal and stress–strain states of steel parts on the basis of finite element methods using software "TANDEM-HART" [8].

At each time and space step, the calculation results were compared with the CCT or TTT diagrams of the supercooled austenite transformation, and new thermophysical and mechanical characteristics for the next step were chosen depending on the structural components. The principal block diagram of the computing complex TANDEM-HART is shown in Figure 2.3 [8]. The calculation results are temperature field, material phase composition, migration of points in the volume which is calculated, components of stress and strain tensors, intensities of stress and strain, and the field of the safety factor, which means the relation of stresses for which the material will be destroyed. These values are presented in the form of tables and isometric lines that allow to observe the kinetics of phase changes in the process of heating and cooling.

Using the potentialities of the computing process mentioned above, the current and residual stresses were determined depending on the cooling intensity for cylindrical specimens made of different steel grades. Similar calculations were fulfilled for quenching the parts of complex configuration.

The most important results used at developing the intensive methods of quenching were as follows.

It has been stated [1] that with the quench intensity increase, the residual stresses grow at first, then lower, with the further increase of Biot number and transfer to compression stresses (Figure 2.4).

The dependence of residual stresses on the cooling rate of the specimen core at temperature of 300°C is readily represented.

It appeared that the maximum probability of quench cracks formation and maximum tensile stresses, which depend on the specimen's cooling rate, coincided [17–20].

The absence of quench cracks at quenching alloyed steels under intensive heat transfer can be explained by high compression stresses arising at the surface of the parts being quenched.

The mechanism of arising high compression stresses in the process of intensive heat transfer is described in Refs. [1, 20, 21].

The results obtained were confirmed experimentally by measuring residual stresses at the surface of quenched parts (specimens) with the help of x-ray crystallography technique [22].

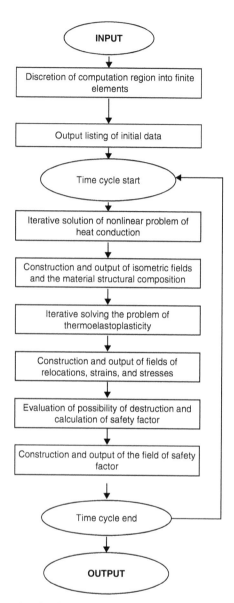

FIGURE 2.3 Chart describing the algorithm of the software package.

The intricate character of the dependence of residual stresses on the cooling rate can be explained by "super-plasticity" and variation of the phase specific volume at phase changes [23]. Under the conditions of high-forced heat transfer (Bi $\rightarrow \infty$), the part surface layer is cooled down initially to the ambient temperature, while the core temperature remains practically constant.

In the process of cooling, the surface layers must compress. However, this process is hampered by a heated and expanded core. That is why compression is balanced out by the surface layer expansion at the moment of superplasticity. The higher the temperature gradient and the part initial temperature are, the greater the surface layer expansion is. At the further cooling, the core is compressed due to which the surface layer begins to shrink towards the center and compression stresses arise in it. At the core cooling, transformation of austenite

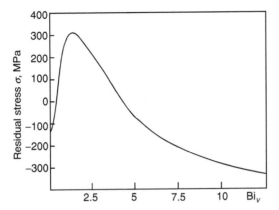

FIGURE 2.4 Residual hoop stresses at the surface of a cylindrical specimen vs. generalized Biot number.

into martensite takes place. The martensite specific volume is higher than that of austenite. For this reason, the core swelling is taking place that causes the surface layer extension at moderate cooling.

At Bi→ ∞, the surface layer is stretched to a maximum and therefore despite swelling, it cannot completely occupy an additional volume formed due to the external layer extension. It is just under the conditions of high-forced heat transfer that compression stresses occur in the surface layer. More detailed information about the calculation results can be found in Refs. [23–25].

2.3.1 MODELING OF RESIDUAL STRESS FORMATION

The mechanism of the formation of compressive stresses on the surface of steel parts is very important to the development of new techniques for thermal strengthening of metals, such as intensive quenching.

The reason why intensive quenching results in high compressive stresses can be explained using a simple mechanical model (Figure 2.5) consisting of a set of segments (1) joined together by springs (2) to form an elastic ring. The segments are placed on a plane surface and connected with rigid threads (3), which pass through a hole (4) in the center of the ring and are attached to the opposite side of the plane surface.

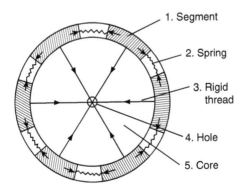

FIGURE 2.5 Mechanical model that can be used to explain the formation of hoop stresses at the surface of a cylindrical specimen during intensive quenching.

Now consider the processes that take place while quenching a cylindrical steel specimen and how they would affect the behavior of the model. Assume that the specimen is being quenched under conditions of intensive cooling. In this case, the cylinder's surface layer is cooled to a certain depth, while the core remains at almost the austenizing temperature and considerably expanded in volume. Let the cooled surface layer correspond to the model's segmented ring.

Because metals contract when cooled, the ring's segments (1) also will contract. The springs (2) will then extend by an amount that corresponds to the increase in tangential tensile stresses. However, when the surface layer is further cooled, austenite transforms to martensite, which has a high specific volume. That is why the cooled layer enlarges or swells.

Now imagine that the segments expand. In this case, resulting compression of the springs corresponds to the appearance of tangential compressive stresses on the surface of the part. With additional time, the temperature of the specimen's core drops, and its diameter decreases. In the model, the core is represented by the smaller blank circle, which is held in tension by the rigid threads.

When the threads are taut, the springs also will compress. The level of hoop compressive stresses will increase until the austenite in the core of the part transforms to martensite. The core will then start to swell because the specific volume of martensite is greater than that of austenite, which causes the compressive stresses to decrease. In the model, this would be reflected by an enlargement of the blank circle and a resulting decrease in the springs' compressive power.

2.3.2 WHY COMPRESSIVE STRESSES REMAIN IN THE CASE OF THROUGH-HARDENING?

At intensive quenching, the temperature of the core almost does not change at the first period of time, and the temperature at the surface instantly drops to the martensite start temperature M_s. At the surface, tensile stresses are formed. At the beginning of martensite transformations, the phenomenon of superplasticity takes place. Due to tensile stresses and superplasticity, the surface stress layer obtains the shape of the part, for example, a cylinder; that is, the surface layer becomes essentially extended. Then, martensite transformations take place, as a result of which the surface layer has increased volume. The core cools to the martensite start temperature, and because of great specific volume of the martensite, the core starts to expand. However, this volume is not enough to fill that initial volume formed by the shell. It looks like the formation of the empty space between the core and shell, and the core pulls this surface layer to itself. In Figure 2.38, showing segments, this process can be illustrated by threads pulling the shell to the core. Due to it, compressive stresses are formed at the surface.

In the case of conventional slow cooling, the difference of temperatures between the surface and core at the time of reaching martensite start temperature is not large. Therefore the initial volume of the shell is not big either. In this case, when the core expands, the volume of the core becomes greater than the initial volume of the shell, and the core expands the surface layer and it causes fracture. It is similar to ice cooling in a bottle, which leads to cracking due to the expansion of the ice. The calculations of the linear elongation factor and changes in the surface layer and volume of the core support this fact. It is accounted that the specific volume of the martensite is greater than those of austenite by 4%.

2.3.3 SIMILARITY IN THE DISTRIBUTION OF RESIDUAL STRESSES

2.3.3.1 Opportunities of Natural and Numerical Modeling of Steel Part Quenching Process

The numerical calculation of current and residual stresses in accordance with the method described above was made for cylindrical bodies of different sizes, i.e., for cylinders having diameter of 6, 40, 50, 60, 80, 150, 200, and 300 mm. Besides, the calculations were made for 45 steel and for cases when CCT diagram is shifted to the right by 20, 100, and 1000 s. This allowed to simulate the quenching process for alloyed steels, where the martensite formation is observed on all cross sections of parts to be quenched.

After making investigations, it has been established that in the case of fulfillment of certain conditions, the distribution of current and residual stresses is similar for cylinders of different sizes. This condition meets the following correlation:

$$\theta = F(\overline{Bi}, \overline{Fo}, r/R), \tag{2.10}$$

where

$$\theta = \frac{T - T_m}{T_0 - T_m}; \quad Bi = \frac{\alpha}{\lambda}R = idem; \quad Fo = \frac{a\tau}{R^2} = idem.$$

For this case, average values of heat conductivity and thermal diffusivity of the material within the range from T_m to T_0 are used. Despite of this, there is a good coincidence of the character of the distribution of current stresses in cylinders of different sizes.

Thus Figure 2.6 represents the results of computations made for a cylinder of 6- and 60-mm diameter. In both, cases the martensite was formed through all cross section of the cylinder, which was fulfilled through the shift of the CCT diagram by 100 s. For the comparison of current stresses, the first time moment was chosen when compressive stresses on the surface of the cylinder to be quenched come to their maximum values. For cylinder of 6-mm diameter, this time was 0.4 s, and for the cylinder of 60-mm diameter, the maximum compressive stresses on the surface are reached after 40 s provided that $Bi = (\alpha/\lambda)R = idem$. The latter was reached due to calculating current and residual stresses for 6-mm-diameter cylinder with $\alpha = 300,000$ W/ m^2 K, and for 60-mm-diameter cylinder, with $\alpha = 30,000$ W/m^2 K. For both cases, $Bi = 45$. Correspondingly, for both cases, maximum compressive stresses were reached at $Fo = 0.24$, i.e., for 6-mm-diameter cylinder at $\tau = 0.4$ s and for 60-mm-diameter at $\tau = 40$ s.

The same values of hoop and tangential values are reached at the same correlation of r/R (Figure 6). Thus hoop stresses for both cases are zero at $r/R = 0.65$; that is, for 6-mm-diameter cylinder, hoop stresses are zero at $r = 1.95$ mm, and for 60-mm-diameter cylinder, hoop stresses are reached at $r = 19.5$ mm correspondingly.

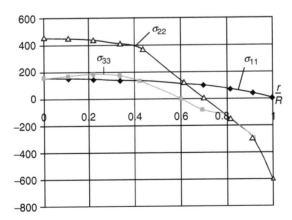

FIGURE 2.6 The distribution of stresses on the cross section of cylindrical sample of diameter of 6 and 60 mm at the time of reaching maximum compressive stresses on the surface ($\varepsilon_1 = 7$; $Fo = 0.7$); 1—sample of 6-mm diameter; 2—sample of 60-mm diameter.

More accurate modeling of the hardening process can be fulfilled with the use of water–air cooling, which allows to change the heat transfer coefficient by a law set in advance. Knowing the cooling conditions, for example, for a turbine rotor, one can make them so that for the rotor and its model, the function $Bi = f(T)$ has the same value. In this case, there will be similarity in the distribution of residual stresses. In practice, it is advisable to investigate the distribution of residual stresses in large-size power machine parts by models made in accordance with theorems of similarity with regard to necessary conditions of cooling and appropriate CCT diagrams.

2.4   REGULARITIES OF CURRENT STRESS DISTRIBUTION

It is of high practical interest to study the character of stress distribution in the case when the end phase through all sections is martensite. Calculations showed that at the initial moment of time on the surface, tensile stresses appear, which, while moving to the core, are changed to compressive ones Figure 2.7. While martensite appears in the surface layer, tensile stresses start to reduce, and while martensite moves to the core of the part, they become compressive.

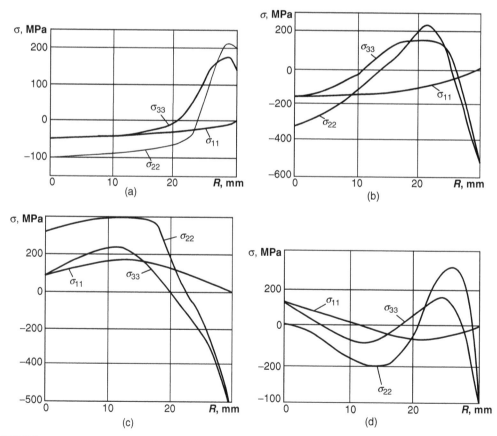

FIGURE 2.7 Stress distribution on the cross section of a cylindrical sample quenched thoroughly of 60-mm diameter at different moments of time ($\alpha = 25,000$ W/m^2 K): (a) 1 s, (b) 10 s; (c) 40 s, (d) 100 s.

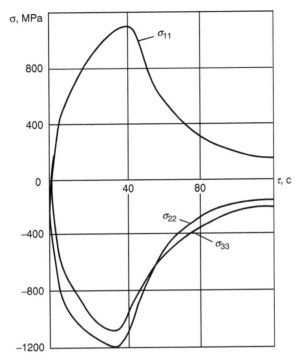

FIGURE 2.8 Stresses on the surface of cylindrical sample (diameter of 60 mm) at through hardening (Bi = 40) vs. time.

In some time, compressive stresses on the surface reach the maximum (Figure 2.8), at that time inside the part tensile stresses are in force. When the part cools completely, depending on the Biot number, tensile or compressive stresses appear.

The characteristic feature of the residual stress distribution in parts quenched thoroughly is that on surface layers, there are always tensile stresses, although on the surface itself, tangential and hoop stresses are compressive. Figure 2.7 shows results of the stress distribution calculation in a cylinder at different moments of time which prove the above-said statement (Figure 2.9).

In the case of thoroughly heating of samples or parts in which the martensite layer is formed after full cooling at a not big depth, a completely different situation is observed. On the surface of part or sample, high compressive tangential and hoop stresses appear, which change gradually to tensile stresses only while moving to inner layers.

It becomes obvious when high-alloy steel parts are quenched into water leading to quench-cracking. However, no cracking is observed for carbon steel parts quenched in the same manner.

In this connection, it is of great practical interest to study the characteristics of stress distribution of parts with different configurations as a function of the conditions of heat transfer and depth of hardness. Analyses of these results permit the development of advanced steel quenching technology.

It is well-known that one of the factors having effect on residual stress distribution is the depth of hard layer. In practice, steel grades of controlled hardenability have become widely used. Parts made out of these steel grades have longer service life for the account of the creation of high compressive stresses on the surface of bodies quenched. For this reason, it is of great practical interest to investigate the regularities of the distribution of current and

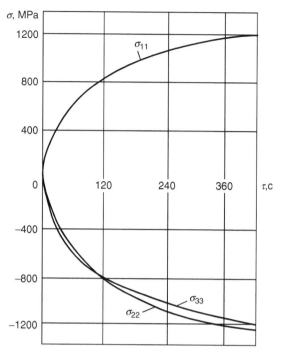

FIGURE 2.9 Stresses on the surface of a cylindrical sample (diameter of 300 mm) vs. time. The sample is made out of 45 steel; the quenching is not thorough (Bi > 100).

residual stresses at parts made out of the mentioned steel grades. The depth of the hard layer can be regulated in two ways. One can either enlarge the part or shift the CCT diagram by time. In both first and second cases, a different depth of hard layer is observed, which is regulated, in practice, for the account of changes in the chemical composition of the steel. At modeling, one must consider the relative depth of hard layer. With regard to these facts, the calculation of current and residual stresses was made for cylinders of different sizes with appropriate time shifts in CCT diagrams. As a result, it has been established that at the initial moment of cooling on the surface of parts, tensile stresses appear, and in inner areas, compressive stresses appear. As the part cools and martensite begins to form in the surface layer, stresses begin to reduce and become negative. Continued martenite formation moves to the core, compressive stresses gradually increase, and if the Biot number is sufficiently large, they may be as large as −1400 MPa. It has been noted that while the Biot number increases, the compressive stresses increase. However, there is an optimal depth of hard layer for which compressive stresses on the surface of part reach maximum values. It is well known that compressive stresses on the surface of parts to be quenched have greater strength and resistance to wearing, and as a result, they have longer service life in the case of cyclical loads. For this reason, machine parts, for example, car axles, should be made out of such steel grade that provides optimal depth of hard layer. The calculations given above show that compressive stresses in parts can be reached while using simple carbon steels without using steels of controlled hardenability like 47GT, melting of which is a very complicated process. In particular, at plant "GAZ" (Gorkogo, Russia), half-axles for GAZ cars are made out of 40G steel. The quenching process of those axles is made in intensive water flows under not big superfluous pressure.

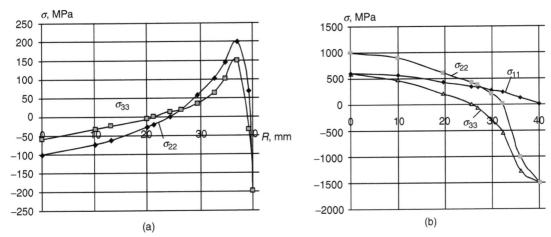

FIGURE 2.10 Stress distribution on the cross section of cylindrical sample (diameter of 80 mm) made out of 45 steel in the case of nonthorough hardening when Bi = 50: (a) after 2 s; (b) after quenching.

The results obtained can be also used for making machine parts, in which it is necessary to create high compressive stresses in surface layers of the steel, which essentially prolongs the service life.

In addition to creating high compressive stresses on the surface of parts to be quenched, it is important to know the character of stress distribution on cross sections of parts (Figure 2.10). As one can see from this figure, in inner layers, quite high compressive stresses are observed, which change to compressive stresses while moving to the surface. This is a distinctive feature for axial σ_{22} and hoop σ_{33} stresses. Radial stresses on the surface of quenched samples are zero; on the axis of cylinder, they are equal to hoop stresses.

The investigations made show that when quenching parts such that the Biot number is large (this is usually done for large-size machine parts), quench cracks can form inside parts.

To avoid quench cracks, for example, while quenching large-size turbine rotors, a hollow is formed at the axis of rotor, through which the quenchant penetrates in some time. In this case, in inner layers, compressive stresses appear, which are the factor preventing the quench cracking. The shortcoming of this method is essential costs for drilling hollows in large-size parts. For this reason, it is of great practical interest to develop the technology of hardening large-size parts preventing quench cracking in solid parts without axial hollows.

Joint investigations have been started between various companies for the introduction of intensive quenching methods.

2.5 SOME ADVICE ON HEAT TREATMENT OF MACHINE PARTS

It is of practical interest of investigate the effect of the heat transfer intensification process while austenite is transformed into martensite. Such investigation can be used in two-step quenching, when at the first step, the transformation of austenite into martensite is delayed because parts are quenched in hot oils or liquid medium under pressure, or in high-concentration aqueous salt solutions having high temperature of boiling.

The second step of cooling is made in the area of convective heat transfer, where the transformation of austenite into martensite mainly takes place. The process of the transformation during the convective heat transfer can occur in cases of high (large Biot numbers) and low cooling rates. During the transformation of austenite into martensite, the cooling process is usually slowed down in practice because of the fear that in the case of high cooling rates in the martensite range, there will be the fracture of material and big distortions. The investigations made earlier have shown that the heat transfer intensification in the martensite transformation range has a good effect upon the distribution of residual stresses and even results in the reduction of distortions. In particular, the example of two-step cooling of bearing rings made out of ShKh15 steel has shown that during the intensive cooling in the martensite range, the cone distortion of rings is significantly reduced [26]. The result obtained is practically important not only because the cooling intensification results in increase in cost and time savings, but also because in the case of intensive cooling in the martensite range, one can get high mechanical properties of material, the strength becomes higher, and plastic properties of material are improved [13].

For further proving the truth of the established fact, the simulation of carburized pin hardening has been made for a pin made out of 14KhGSN2MA steel. The carburized pin was firstly cooled at slow cooling rate in hot oil MS-20 to temperature of 220°C, and then after allowing temperature to become even on the cross section, it was quenched in quenchants having different heat transfer coefficients.

In accordance with the described technology, the simulation has been made for hardening process with setting different heat transfer coefficient at the second stage of cooling. By results of investigations, it has been established that while cooling is intensified, compressive stresses in the carburized layer grow and conical distortion of the pin is reduced. In this case, tensile stresses are changed insignificantly. Thus it has been established that cooling intensification in the martensite range has a good effect upon the distribution of residual stresses and reduces the distortion of parts quenched. This can be used for the development of advanced technology of bearing ring hardening allowing to reduce the oval distortion of rings and conical distortions while the heat transfer process is intensified; it can be also used for the development of advanced technology of heat treatment for tools and machine parts, in particular, cooling treatment. Thus, for example, carburized pins of gasturbine aviation engine after their oil quenching were treated by cooling at temperature of 70°C. This treatment is usually made by air cooling in special refrigerators. It is advisable in the case of cooling treatment to cool carburized parts as intensively as possible, which will have a positive effect for the reasons mentioned above.

2.6 SHELL HARDENING OF BEARING RINGS (THROUGH-SURFACE HARDENING)

The material given above proves the existence of the optimal depth of hard layer for bodies of simple shape. It has been emphasized that there is a similarity in the distribution of hardness on the cross section for bodies of simple shape with regard to the size, i.e., $\Delta r/R = $ const, where Δr is the depth of the martensite layer (shell) and R is the radius of cylinder or ball (or half-width of the plate). It should be noted that the optimal hard layer corresponds to the best stress distribution on the surface. In this case, compressive stresses are much higher than at hardening for martensite on all the cross section. In this paragraph, we will show that the same regularities are true for the bodies of complicated shape. In particular, we will consider the process of quenching for three different bearing rings presented in Figure 2.11 and Figure 2.12a through Figure 2.12c.

Let us investigate the stress distribution of these rings made of steel 52100, using mechanical properties steel as initial data. At the same time, the hardenability controlling we will make for the account of the shift on the diagram. In our case, we shifted the CCT diagram by 22 s to the left, which is shown in Figure 2.13.

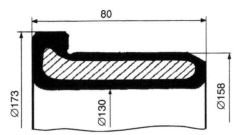

FIGURE 2.11 Dimensions and the configuration of hardened layer after TSH in inner ring of bearing for boxes of railway cars.

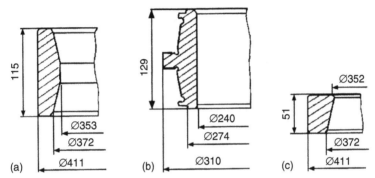

FIGURE 2.12 Dimensions of large bearing rings hardened by TSH at furnace or induction heating.

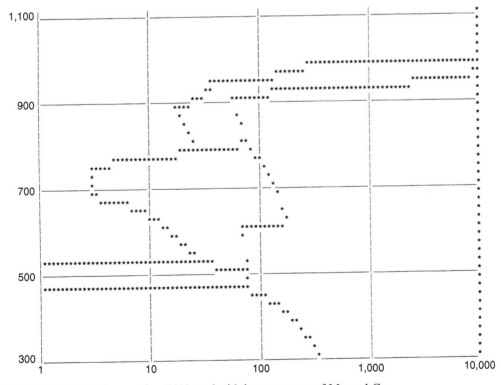

FIGURE 2.13 CCT diagram for 52100 steel with lesser content of Mn and Cr.

The results of calculations of the stress state in the case of through-surface hardening and shell quenching for the bearing ring shown in Figure 2.11 are presented below.

So Figure 2.14a shows the character of the phase distribution in the case of through-surface hardening and shell quenching. Besides, the through-surface hardening was made in oil where heat transfer coefficient $\alpha = 1800$ W/m^2 K. In the case of shell quenching (Figure 2.14b), the quenching was made by intensive cooling with $\alpha = 60,000$ W/m^2 K.

Now consider the character of hoop stress distribution for these two cases, namely, though-surface hardening in oil (Figure 2.14a) and shell hardening (Figure 2.14b). In the case of through-surface hardening in oil at the surface of bearing rings, residual hoop stresses are formed reaching the following values: 180 (on the left) and 260 (on the right) (see the middle of the cross section in Figure 2.15).

In the case of shell quenching with $\alpha = 60,000$ W/m^2 K, the stress distribution at the corresponding time is as follows: -800 (left) and -900 (right), so they are compressive, and in the core, $+500$ (tensile) (Figure 2.16).

Now consider the residual hoop stresses for bodies of complicated shapes for two cases: intensive through-surface hardening and intensive shell quenching. In both cases, $\alpha = 40,000$ W/m^2 K. Let us study the difference between the stresses at the surface.

Phase distribution fields for through-hardening and shell quenching for part shown in Figure 2.12b are presented in Figure 2.17.

In the case of through-surface hardening of bearing ring (Figure 2.17a) at $\alpha = 40,000$ W/m^2 K at the surface (in the middle of the cross section at the left and at the right), not big compressive stresses are formed, which are about -120 at the left and -160 at the right (Figure 2.18).

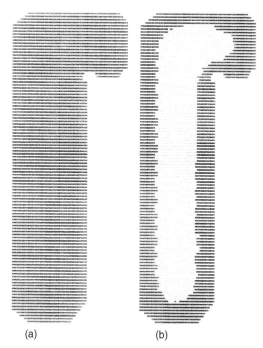

(a)   (b)

FIGURE 2.14 Phase distribution fields: (a) through-hardened, $\alpha = 1800$ W/m^2 K; (b) shell-hardened, $\alpha = 60,000$ W/m^2 K.

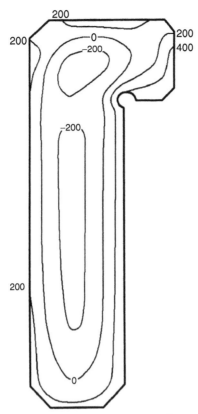

FIGURE 2.15 Residual hoop stresses S33 (MPa), $\alpha = 1800$ W/m^2 K.

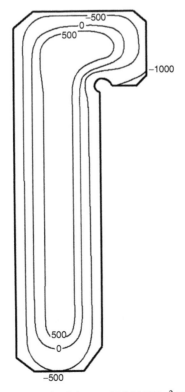

FIGURE 2.16 Residual hoop stresses S33 (MPa), $\alpha = 60{,}000$ W/m^2 K.

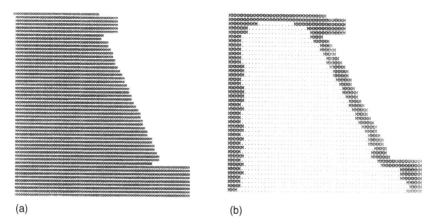

(a)   (b)

FIGURE 2.17 Phase distribution fields: (a) through-hardened, $\alpha = 40,000$ W/m^2 K; (b) shell-hardened, $\alpha = 40,000$ W/m^2 K.

At the same time, in the case of shell quenching, huge compressive stresses are-formed at both sides in the middle of the cross section (Figure 2.19). These compressive stresses reach the value -1050 at the left and -1450 at the right.

Thus through-surface hardening results in not big compressive stresses at the surface, while shell quenching results in huge compressive stresses at the surface. The calculations show that when the hard layer is closer to the optimum, higher compressive stresses are at the surface.

Consider one more example illustrating this very important regularity. Let us show this fact with the rings shown in Figure 2.20.

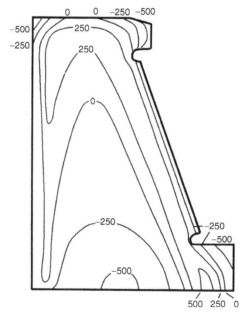

FIGURE 2.18 Residual hoop stresses S33 (MPa) through-hardened, $\alpha = 40,000$ W/m^2 K.

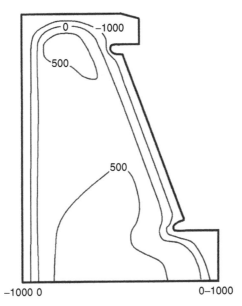

FIGURE 2.19 Residual hoop stresses S33 (MPa) shell-hardened, $\alpha = 40,000$ W/m^2 K.

The cooling is in the same conditions: $\alpha = 40,000$ W/m^2 K, but in the first case, it is through-surface hardening, and in the second case, it is shell quenching. In the case of through-surface hardening, hoop stresses in the middle are not big tensile: $+80$ (at the left) and $+480$ (at the right) (Figure 2.21). At the same time, in the case of shell quenching, these stresses are big compressive: from -1200 to -1500 MPa (Figure 2.22).

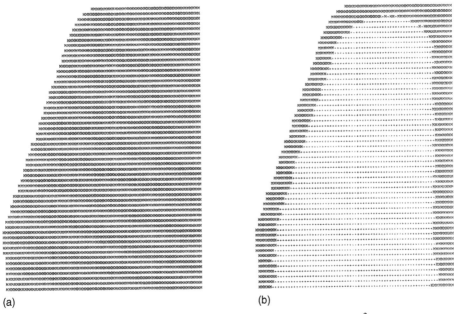

(a)   (b)

FIGURE 2.20 Phase distribution fields: (a) through-hardened, (40,000 W/m^2 K); (b) shell-hardened (40,000 W/m^2 K).

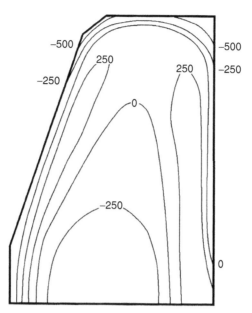

FIGURE 2.21 Residual hoop stresses S33 (MPa), through-hardened (40,000 W/m² K).

It should be noted that in the case of intensive through-hardening for bodies of big size and not complicated shape, it is possible to reach uniform compressive stresses at all the surfaces (Figure 2.23).

As one can see from calculations presented above, in the case of intensive cooling, if the steel is hardened through to lesser degree, higher compressive stresses are formed at the

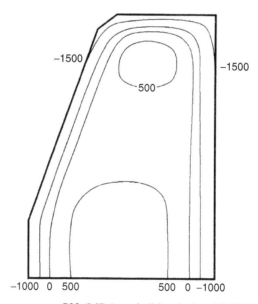

FIGURE 2.22 Residual hoop stresses S33 (MPa) at shell hardening (40,000 W/m² K).

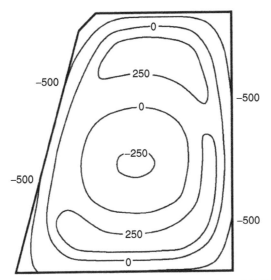

FIGURE 2.23 Residual hoop stresses S33 (MPa), through-hardened (40,000 W/m^2 K).

surface. This factor must have effect upon the increase in the service life of the part. To prove the above-mentioned, inner rings of railway bearing were made of different steel grades, namely, ShKh15SG, 18KhGT, and ShKh4:

1. Steel ShKh4 (GOST 801-78) contains 0.95–1.05% C, 0.15–0.30% Si, 0.15–0.30% Mn, 0.35–0.50% Cr, 0.015–0.050% Al, <0.020% S, <0.027% P, <0.25% Cu, <0.30% Ni, and (Ni + Cu) <0.50%. The steel is intended for bearing rings with wall thickness of 18 mm and more and rollers with diameter of more than 30 mm.
2. Steel ShKh4 (TU 14-19-33-87) is distinguished from the previous version by the regulation of alloying chromium addition depending upon the real content of nickel which is not added especially but comes to the steel from scrap at smelting. The regulation proceeds by the use of the following ratios of nickel and chromium contents:

| Ni (%) | <0.10 | 0.08–0.20 | 0.18–0.30 |
| Cr (%) | 0.35–0.50 | 0.27–0.42 | 0.20–0.35 |

These ratios were obtained on the basis of the following equation inferred by statistical treatment of data on hardenability for 80 industrial heats of ShKh4 steel:

$$D = 13.213[1 + (Ni)]^{0.764}[1 + (Cr)]^{0.895}$$

where D = ideal critical diameter (mm) and Ni and Cr = percentage of nickel and chromium, respectively.

The steel has a little lower and more stable hardenability level and is used for rings of railway bearings with wall thickness of 14.5 mm; steel ShKh2 has been elaborated and tested in laboratory conditions. The steel contains 1.15–1.25% C, 0.15–0.30% Si, 0.15–0.30% Mn, 0.015–0.030% Al, <0.020% S, <0.027% P, <0.15% Cr, <0.10% Ni, <0.12% Cu, and <0.03% Mo.

2.6.1 Heating during Through-Surface Hardening

For TSH, comparatively slow induction heating with isothermal holding at the hardening temperature was used. The required specific high-frequency generator power is usually about 0.05–0.2 kW per 1 cm^2 of surface area of heated parts. For example, for rings of railway car bearings, total heating time is about 3 min including isothermal holding at hardening temperature (820–850°C) for 50–60 s.

The heating procedure provides nearly uniform heating of rings of complex shape, required degree of carbides dissolution, and saturation of austenite by carbon (0.55–0.65%), the fine austenitic grain being the same (not worse than No.10, GOST 5639-82—the diameter of grains is an average of 0.01 mm).

2.6.2 Quenching during Through-Surface Hardening

The quenching is carried out in special devices by intense water stream or shower from pumps with pressure 1–4 atm. Total water consumption is not large because the closed-circuit water cycle is used (water tank–water pump-quenching device–water tank). A small amount of cold water is added into the tank to prevent water from heating above 50°C.

The design of quenching devices should guarantee velocity of water with respect to the surface of quenched parts in the order of 10–15 m/s. Water inside the devices should be under pressure (1.5–3 atm.). Time of intense water quenching should be limited to allow self-tempering of parts at 150–200°C. One of typical designs of the device for quenching of bearing rings is shown in Figure 2.24.

2.6.2.1 Through-Surface Hardening of Inner Rings of Bearings for the Boxes of Railway Cars

In the past, the bearing rings were made of traditional bearing steel ShKh15SG (1.0%C, 1.5%Cr, 0.6%Si, and 1.0%Mn) and hardened by a standard method to obtain HRC 58–62. After a short period in service, some rings failed because of brittle cracking and fatigue.

Since 1976, TSH has been used for bearings. The inner rings are made of steel ShKh4. Dimensions of the rings and the typical configuration of hardened layer after TSH are given in Figure 2.11.

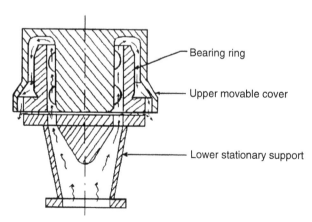

FIGURE 2.24 Schematic design of a quenching device for inner rings of railway bearings (water stream is shown by arrows).

At present, several millions of such bearings are successfully used in operation on car boxes in the rail-roads of Russia and other states of CIS. Automatic machine for TSH of the inner rings is shown in Figure 2.26.

In order to show advantages of TSH for bearings, the results of mechanical tests are given below for the inner rings of railway bearings made of different steels and strengthened by various heat treatments including:

1. Standard steel ShKh15SG (52100 Grade 4) with electroslag refining, traditional through-hardening with furnace heating and quenching in oil; low tempering.
2. Standard carburized steel 18KhGT (GOST 4543–71) containing on the average 0.2%C, 1.1%Cr, 1.0%Mn, and 0.05%Ti, carburizing, quenching, low tempering.
3. Steel ShKh4, TSH, low tempering.

These numbers of steels and heat treatment variants are referred in Table 2.1 which contains some structural characteristics and hardness of the rings. Characteristic curves of residual stresses in the cross section of hardened rings and their strength under static, cyclic, and impact loads are given in Figure 2.25 and Figure 2.26.

The technology developed has been implemented at GPZ-1 plant (Moscow).

2.6.2.2 Through-Surface Hardening of Rollers

Through-Surface hardening is very suitable and profitable for bearing rollers. It allows to obtain roller properties similar to those of carburized rollers made of expensive chromium–nickel steels, but at lower steel cost and production expenses.

TABLE 2.1
Structure and Hardness of Inner Rings of Railway Bearings Made of Various Steels

Parameters	Values for Rings Made of the Following Steels			
	ShKh15SG	18KhGT	ShKh4	ShKh4
1. Number of steel and heat treatment variant	1	2	3	4
2. Surface hardness (HRC)	60–61	59–61	62–64	62–64
3. Depth of hardened layer, mm:	Through-hardening			
HRC > 58		0.7–0.9	1.5–2	2.5–4
HRC > 55		1.8–2.1	2.4–2.7	3–4.5
4. Core hardness (HRC)	60–61	32–35	36–40	40–45
5. Microstructure in the core	Martensite, carbides	Low-carbon martensite	Sorbite, troostite	Troostite
6. Percentage of retained austenite in surface	14–16	8–10	6–8	6–8
7. Number of austenitic grain size[a]	9–10	9	10–11	10–11

[a] GOST 5639–82.

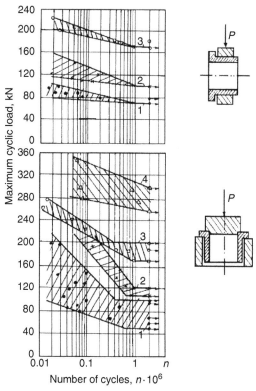

FIGURE 2.25 Fatigue testing results for inner rings of railway car bearings made of various steels (the numbers of the curves correspond to numbers of steel and heat treatment variants in Table 2.1).

FIGURE 2.26 Automatic machines for TSH of inner rings for railway car box bearings designed and put into operation at GPZ-1 (Moscow) in 1995.

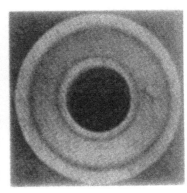

FIGURE 2.27 Macrostructure of through-the-wall sections of hollow rollers for railway car box bearings.

Through-surface hardening was successfully used at GPZ-9 for tapered rollers of 40–50-mm diameter made of ShKh4 steel.

Recently, new technology of TSH was developed for hollow rollers of railway car box bearings. Application of a design of hollow rollers with a central hole is a promising direction in bearing production. In this case, contact area between the ring and the roller increases because of a small elastic deformation of the rollers, contact stresses decrease, and durability of the bearing increases by several times.

According to service tests, durability of railway bearings with hollow rollers increases by 2.8 times. The rollers had the following dimensions: external diameter of 32 mm, length of 52 mm, and diameter of the central hole of 12 mm. Macrostructure and hardness of the rollers hardened by TSH are shown in Figure 2.27. While zones on the surface correspond to the martensitic structure, the dark core has a troostite and sorbite structure.

The primary condition for heat treatment of hollow rollers consists in providing surface hardening on the external and internal surfaces. This leads to high compressive internal stresses in the hardened layers which prevents formation of fatigue cracks in service. It should be noted that through-hardening is dangerous for hollow rollers. This is demonstrated by fatigue tests of the rollers under maximum load of 200 kN. Through-hardened rollers made of 52100 steel were destroyed after 15,000–30,000 cycles, while TSH-hardened rollers made of ShKh4 steel survived after 10 million cycles (Figure 2.28).

More detailed information about results of experimental studies can be found in Refs. [2, 27–33].

2.7 DESIGNING OF STEEL-INTENSIVE QUENCH PROCESSES IN MACHINE CONSTRUCTION

2.7.1 THROUGH-SURFACE HARDENING OF SMALL-SIZE DRIVING WHEELS MADE OUT OF 58 (55PP) STEEL

The original technology of through-surface quenching has been developed for driving wheels with module of 4–6 mm, which are typical for construction and road machines, metal-cutting machines, transport devices, and other machines. They are produced in big varieties but in small lots. The peculiarities of the technology are the use of carbon steel with the low content of admixtures, and thus, low hardenability; untraditional choice of current frequency and

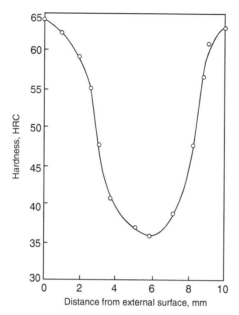

FIGURE 2.28 Hardness of through-the-wall sections of hollow rollers for railway car box bearings.

conditions of induction heating; performance of high-quality quenching of driving wheels of various kinds and sizes with the use of a single inductor and cooling device; minimum quenching distortion; and ecological purity of the quenching process. The design of universal easily readjustable quenching installment has been developed.

In modern engineering, a great number of driving wheels are used that have diameter of 80–120 mm and module of 4–6 mm. A part of them, satellite driving wheels, require hardening for only teeth crown, while the other body and the hole cannot be hardened. The greatest number of driving wheels having a hole, according to operation conditions, must be hardened to high hardness on all surfaces, including teeth crown, flange surface, and surface of hole.

At present, such driving wheels are hardened by either surface induction under a layer of water, when they are produced out of medium-carbon construction steels, or carburizing with the further quenching and low tempering with the use of carburized steels.

Quenching under a layer of water very often results in the appearance of quench cracks on the surface, with which one can do nothing but allow; however, it is very difficult to determine and control the limits of such allowance. The use of carburizing and nitrocarburizing on many plants is prevented by the absence and high cost of modern thermal equipment and automatic regulation devices, especially as for the regulation of components of gaseous saturating atmosphere. Making carburizing and nitrocarburizing at out-of-date universal equipment does not guarantee the required level and stability of the quality of heat treatment.

Due to the stated above, one of the processes having high prospects for hardening driving wheels is through-surface quenching with induction heating. The advantages of this method are as follows:

1. Opportunity of high-quality and reliable hardening of steel parts
2. Opportunity to produce equipment for through-surface quenching directly at machine-construction plants

3. Electric power saving, absence of quench oil in the production process, and gaseous-controlled atmospheres, which cause ecological problems
4. Lesser production area in comparison with chemical heat treatment, better labor conditions, and opportunity to place induction installments in shops of mechanical treatment

In a laboratory of Moscow State Evening Metallurgic University, engineers developed the technology of through-surface quenching for driving wheels with module of 4–6 mm. The wheels have been made out of 58 (55PP) steel having the following chemical content: 0.56% C, 0.30% Si, 0.09% Mn, 0.10% Cr, 0.08% Cu, 0.04% Ni, 0.004% S, 0.006% P, and 0.008% Al. The low content of admixtures in the steel is due to production at Oskolskyi Steel Works, which is the main producer of this steel.

For production of driving wheels, rolling 130-mm diameter circle was used. The specimens were obtained through forging, further cutting, and preparation for induction. Five kinds of driving wheels presented in Figure 2.29 were used in tests, which are used at excavator plants and having a module of 4.6 mm.

Heating of all wheels has been made at one-spiral cylindrical inductor with the inner diameter of 150 mm and height of 110 mm; cooling has been performed by quick water flows provided at outer, inner, and flange surfaces of driving wheels.

In the experiment, heating induction was at current frequencies of 2.5 and 8 kHz from machine generators of 100-kW power.

On the basis of traditional principles of current frequency selection for teeth crowns of driving wheels with module of 4–6 mm, with regard to uniform heating of teeth and holes while heating in ring inductors, the optimal current frequency must be as follows:

$$f_{opt} = 600/M^2 = 600/(4-6)^2 = 40\ldots20.$$

where f_{opt} is optimal current frequency (kHz) and M is a module of a driving wheel (mm).

For this reason, making a choice out of available industrial induction devices for such driving wheels, more preferable are devices for 8 kHz than 2.5 kHz.

FIGURE 2.29 Macrostructure of driving wheels with module of 4–6 mm after through-surface quenching: light surfaces are the surface hard layer having martensite structure and hardness greater than HRC 60; dark surfaces are areas with the structures of troostite or sorbite of quenching and hardness of HRC 45–30.

This was proved to be true for satellite driving wheels for which surface quenching of hole is not necessary, in connection with which the through-surface quenching is fulfilled with not thorough heating but quite deep induction heating of teeth crown only.

The induction heating of such cylindrical driving wheels of 115-mm diameter on the circle of teeth tips and 48-mm height has been performed at the current frequency of 8 kHz for 30 s, and the teeth have been heated thoroughly, and cavities at the depth of 3.4 mm. The cooling time, cooling being performed by intensive water flows, was 4 s. After quenching on the surface of teeth and cavities, the uniform hardness within the range of HRC 63–66 has been received. The macrostructure of driving wheels after through-surface quenching with detected hard-layer by itching with "triple" agent is given in Figure 2.29 (lower line of wheels). Tempering of hard driving wheels was made in electric furnace equipped with ventilator, at temperature of 165°C, for 3.5 h.

The results of measuring dimensions of driving wheels on the mean line of the common normal and diameter of the hole in the original state before quenching, after quenching, and after quenching with tempering are given in Figure 2.30. The reduction of mean length of common normal is in average as follows: 0.014 mm in quenching, 0.021 mm in tempering, and the total change in quenching and tempering is 0.035 mm in comparison with the original state after mechanical treatment. The diameter of the hole reduced in average by 0.019 mm in quenching, 0.026 mm additionally in tempering, and the total reduction in quenching and tempering was 0.045 mm. The values of distortion obtained are quite small and completely meet the requirements for the heat treatment of satellite driving wheels.

The majority of standard driving wheels are driving wheels with inner-gear hole, where it is necessary to harden at high hardness not only teeth crown, but also surface of inner-gear hole, and in this case, the allowance of distortion on the diameter of inner-gear hole is quite severe and usually it is not more than 0.12 mm.

It has been noticed earlier that proceeding from conditions of uniformity of heating for cavities and teeth of outer teeth crown for driving wheels with module of 4–6 mm, it is more preferable to use induction heating with the current frequency of 8 kHz. However, experiments fulfilled showed that for such driving wheels with regard to their geometry and peculiarities of the work conditions, the use of 2.5-kHz frequency gives a number of advantages and allows to expand the range of effective application of through-surface quenching and 58 (55PP) steel for the production towards the reduction of module of driving wheels to 4 mm while providing the clearly expressed effect of outline quenching of teeth crown.

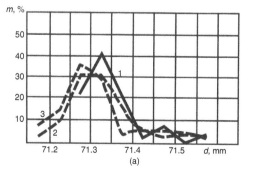 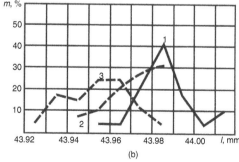

FIGURE 2.30 Chang in the diameter (d) of hole for placement (a) and length (l) for common normal (b) of satellite wheels in the case of through-surface quenching (m is a group frequency); 1—sizes of driving wheels in the original state after mechanical treatment; 2—after through-surface quenching; 3—after through-surface quenching and tempering.

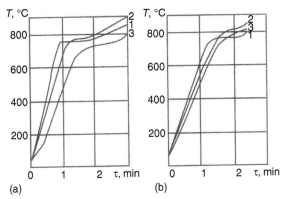

FIGURE 2.31 Thermal curves of induction heating for driving wheels with module of 4 mm at current frequency of 8 kHz (a) and 2.5 kHz (b). Specifications of the driving wheel: outer diameter is 91.1 mm, inner diameter is 45 mm: 1 is the tip of a tooth; 2 is the cavity of teeth crown; 3 is the surface of inner-gear hole.

This is explained by the following circumstances:

1. Using the current frequency of 8 kHz, even in the case of the computer step-by-step control of the voltage generator and time of heating of 3 min, since it is necessary to reach the minimum quenching temperature of 790–800°C on the surface of inner-gear hole, the result is the considerable overheating of teeth crown to 850–880°C (Figure 2.31a), due to which the teeth are hardened to high depth or even thoroughly (Figure 2.32). Although such quenching can be done in many cases, outline quenching is more preferable.
2. Using current frequency of 2.5 kHz, the heating is more uniform with respect to thickness of driving wheels, and by time of finishing heating, the temperature in cavities of teeth crown is 850°C (Figure 2.31b), which provide the 1.7-mm thickness of hard layer in cavity with up to HV600 hardness. At the same time, the temperature on the tip of teeth is essentially lower, 790–800°C, which is allowed for quenching with regard to constructive hardness of driving wheels since the tips of teeth are areas with the lowest load. The specified difference of temperatures on the outline of teeth crown gives

FIGURE 2.32 Macrostructure of cross section of a driving wheel with module of 4 mm after through-surface quenching with induction heating at the current frequency of 8 kHz.

advantages for receiving the optimal configuration of hard layer as for the height of the teeth (see Figure 2.29, upper line of driving wheels).

With heating at current frequency of 8 kHz, through-surface quenching has been performed for batches of three types of driving wheels, macrostructure of which after quenching is presented in Figure 2.29 (upper line of driving wheels). The time of heating for different driving wheels was within the range of 2–2.5 min; cooling was performed directly in inductor by quick water flows within 4–8 s, providing self-tempering at temperature of 150–200°C. The driving wheels after quenching were tempered at 165°C for 3.5 h.

Figure 2.33 presents results of hardness measurement in various zones on the cross section of a driving wheel with module of 4 mm after through-surface quenching and tempering in comparison with driving wheels made out of 18KhGT steel after carburizing and heat treatment by serial technology of TEQ. As one can see, through-surface quenching through one session of thorough heating and cooling has provided efficient hardening of both teeth crown and zone of inner-gear hole with the hardness of HRC 58–60 and thickness of hard layer within 2–5 mm in different cross sections and, at the same time, hardening of the wheel core to HRC 30–40. This combination of properties is advantageous for high service life and operational reliability.

Table 2.2 gives results of measuring the distortion of wheels in the case of through-surface quenching with low tempering. As one can see, the technology of through-surface quenching of driving wheels provided quite low values of distortion not exceeding 0.05 mm on the mean length of common normal and 0.075 mm on the diameter of inner-gear hole, which meets the highest requirements for the distortion.

Batches of driving wheels hardened by this through-surface quenching technology were used in the serial production of excavators. The design of industrial induction heating installment for through-surface quenching of driving wheels is outlined in Figure 2.34 and

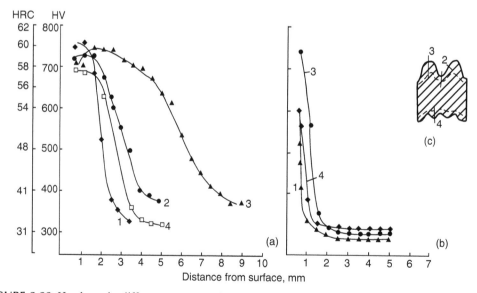

FIGURE 2.33 Hardness in different zones on the cross section of a teeth crown and inner-gear hole: (a) is for a driving wheel made out of 58 (55PP) steel after through-surface quenching and tempering; (b) is for a driving wheel made out of 18KhGT steel after carburizing and heat treatment; (c) is for cross section of a teeth crown and inner-gear hole; 1–4 are zones of the cross section.

TABLE 2.2
Distortion Characteristics for Driving Wheels with Inner Inner-Gear Hole after Through-Surface Quenching

Dimensions of Driving Wheels (mm)	Mean Length of Common Normal (mm)		Diameter of Inner Hole (Size of Inner Gear for Rolls) (mm)	
OD = 91.1, ID = 45, H = 37, M = 4	31.77–31.84	31.805	31.27–31.30	31.285
	31.78–31.84	31.81	31.14–31.28	31.28
OD = 91.1, ID = 51, H = 52, M = 4	31.80–31.84	31.82	40.13–40.17	40.15
	31.80–31.85	31.825	40.07–40.11	40.09
OD = 89.6, ID = 45, H = 50, M = 5	39.78–39.80	39.79	40.12–40.17	40.145
	39.80–39.81	39.805	40.06–40.14	40.10

ID is the inner diameter; OD is the outer diameter; H is height of the wheel; M is the module of teeth crown. The data above line are sizes of wheels in the original state after mechanical treatment; and data below line are sizes of those wheels after through-surface quenching with low tempering.

Figure 2.35. The installment is universal for a wide range of driving wheels of different sizes and easily readjusted for different kinds of driving wheels through changing upper and lower mandrels (Table 2.3).

Thus the typical technology of through-surface quenching of driving wheels with diameter of 80–120 mm and module of 4–6 mm includes as follows:

1. Production of driving wheels out of 58 (55PP) steel at Oskolskyi Steel Works.
2. Deep (for satellite wheels) or thorough (for wheels with inner-gear hole) induction heating for 0.5–3 min in cylindrical one-spiral inductor. The optimal current frequency is 8 kHz for satellite wheels and 2.5 kHz for wheels with an inner-gear hole, which are to be hardened.
3. All-side intensive cooling at a quenching device linked directly to the inductor by quick water flows fed to outer, inner, and flange surfaces of driving wheels.
4. Low tempering at 160–170°C for 3–4 h.

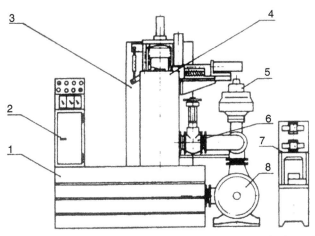

FIGURE 2.34 Installation for through-surface quenching of driving wheels through induction heating: 1—water tank; 2—control panel; 3—quenching block; 4—case of quenching device; 5—pneumatic-hydraulic valve; 6—valve; 7—pump; and 8—hydraulic station.

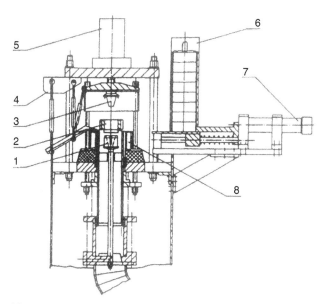

FIGURE 2.35 Quenching device: 1—changeable mandrel; 2—driving wheel to be quenched; 3—cone; 4—removing device; 5—hydraulic cylinder for vertical movement and wheel fixing; 6—receiver of parts, changeable; 7—hydraulic cylinder for wheel feeding; and 8—inductor.

The technology developed provided the optimal configuration of hard layers on all work surfaces (Figure 2.29), providing high constructive durability of driving wheels with the minimum quench distortion of no more than 0.1 mm along the common normal and diameter of inner-gear hole.

TABLE 2.3
Installation Specifications

Dimensions of driving wheels to be quenched:	
Maximum outer diameter	145
Maximum height	90
Parameters of frequency transformation:	
Power (kW)	100
Current frequency (kHz)	2.5–10
Specific consumption of electric power (kWh/Mt)	300–500
Quenching water-supply pump:	
Output (m³/hr)	300
Pressure (MPa)	0.1...0.4
Water consumption (m³/h):	
For cooling electrical elements	3–4
For cooling wheels during quenching	1–1.5
Type of work drives	Hydraulic
Dimensions (mm):	
Length	3000
Width	2800
Height	2600

TABLE 2.4
Mechanical Properties at the Core of Conventional (Oil) and Intensive (Chloride Medium) Quenched Steels (Cylindrical 50-mm Diameter Specimen Used)

Steel	Quenchant	Tensile Strength, R_m (MPa)	Yield Strength, $R_{p0.2}$ (MPa)	Elongation, A (%)	Reduction in Area, Z (%)	U-Notch Impact Energy, KSU (J/cm^2)	Hardness (HRB)
40×	Oil	780	575	21	64	113	217
	Chloride	860	695	17	65	168	269
35 × M	Oil	960	775	14	53	54	285
	Chloride	970	820	17	65	150	285
25 × 1 M	Oil	755	630	18	74	70	229
	Chloride	920	820	15	68	170	285

Source: Mukhina, M.P. Kobasko, N.I.; Gondejeva, L.V. *Metalloved. Term. Obrab. Metall.* 1989, (9), 32–36.

The cooling rate of the core of the specimen also is high under these conditions, but will drop when the compressive stress reaches its maximum value. The core's microstructure consists of bainite or tempered martensite, and its mechanical properties are higher than those of oil-quenched steel (Table 2.4).

The work of Shepelyakovskii and Ushakov [35] supports these observations. They developed steels: 58 (55PP), 45C, 47GT, and ShX4—having limited hardenability (LH) and controlled hardenability (CH). These steels contain small amounts of aluminum, titanium, and vanadium. They are less expensive than conventional alloy grades because they use two to three times less total alloying elements. The researchers emphasize that the depth of the hard layer is determined by steel hardenability.

One of the new steels is first induction through-surface hardened, and then slowly heated (2–10°C/s) either through the section or to a depth beyond that required for a fully martensitic layer. The slow-heating time is usually 30–300 s. Parts are then intensively quenched in a water shower or rapidly flowing stream of water. Production parts heat-treated using this method are listed in Table 2.5. Note that the durability of intensive-quenched, noncarburized low-alloy steel parts is at least twice that of carburized and conventionally oil-quenched alloy steel parts.

TABLE 2.5
Production Applications of Intensive-Quenched Limited-Hardenability Steels

Applications	Former Steel and Process	New Steel and Process	Advantages
Gears, modulus, $m = 5$–8 mm	18KhGT	58 (55PP)	No carburizing; steel and part costs decrease; durability increases.
Large-modulus gears, $m = 10$–14 mm	12KhN3A	ShKh4	No carburizing; durability increases 2 times; steel cost decreases 1.5 times.
Truck leaf springs	60C2KhG	45S	Weight decreases 15–20%; durability increases 3 times.
Rings and races of bearings thicker than 12 mm	ShKh15SG and 20Kh2N4A	ShKh4	No sudden brittle fracture in service; durability increases 2 times; high production rate.

Source: From Shepelyakovski, K.Z.; Ushakov, B.K Induction surface hardening—Progressive technology of XX and XXI Centuries. Proceedings of the 7th International Congress on Heat Treatment and Technology of Surface Coatings, Moscow, Russia 2(11–14), 33–40, 11–14 Dec., 1990.

Other production applications for these new low-alloy steels include [35]:

- Gear, 380-mm diameter and 6-mm modulus, steel 58 (GOST 1050-74). Note: modulus = pitch diameter/number of teeth
- Cross-head of truck cardan shaft, steel 58 (GOST 1050–74)
- Bearing for railway carriage, 14-mm ring thickness, steel ShKh4 (GOST 801–78)
- Truck axle, 48-mm bar diameter, steel 47GT

The optimum depth of hard layer is a function of part dimensions. Therefore it is recommended that the steel (and its hardenability) be tailored to the part to ensure that it will be capable of hardening to this depth [36], making reaching the optimum depth a certainty.

With the right steel, this method can provide an optimum combination of high surface compressive stress, a high-strength, wear-resistant quenched layer of optimum depth, and a relatively soft but properly strengthened core. The combination is ideal for applications requiring high strength and resistance to static, dynamic, or cyclic loads.

2.8 NEW METHODS OF QUENCHING

There are two methods of reaching the optimal depth of hard layer at the surface:

1. For each specific steel part, a special steel grade is selected, which provides the optimal hard layer and maximum compressive stresses at the surface. When the sizes of the part are changed, different steel grades are selected providing meeting the condition $\Delta r/R = \text{const}$.
2. Steel part quenching is made so that $0.8 \le \text{Kn} \le 1$ and the process of intensive cooling is interrupted at the time of reaching maximum compressive stresses at the surface. In this case, the optimal depth of hard layer is reached automatically. This method was protected by inventor's certificate in 1983, which became a patent of Ukraine in 1994 [36].

In order to confirm the above statements, one can refer to Ref. [37] where a method of IQ-quenching steel parts by using intensive jet cooling is described. The author points out that the method of intensive cooling is applied to superficial hardening of small parts (shafts, axes, pinions, etc.) made of alloy steels. Here a very high intensity of cooling is achieved that allows to obtain a 100% martensite structure in the outer layer and high residual compression stresses. It should be noted that under the conditions of very intensive cooling, the strain decrease is observed. While treating the parts of complex configuration, it is necessary to use several combined jets in order to prevent the steam jacket formation. A disadvantage of this method is a high cost of the equipment.

Let us consider in more detail the character of changes in the current stress at the surface of the parts being quenched depending on various intensities of cooling. It has been stated that in the course of time, small tensile stresses occur at first at the surface of the specimen to be quenched, and then during the martensite layer formation, these stresses transform into the compressive ones that reach their maximum at a certain moment of time and then decrease (Figure 2.36). The current stresses become the residual ones that can be either tensile or compressive depending on the cooling intensity. The maximum tensile stresses correspond to the maximum compressive stresses in the control layer of the part being quenched (Figure 2.37).

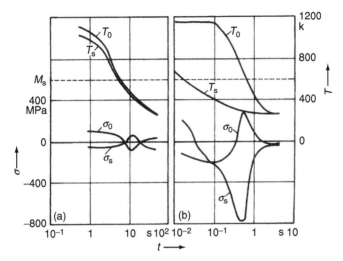

FIGURE 2.36 (a-b) Change of current hoop stresses at the surface (σ_s) and in the center (σ_c) of the cylindrical specimen being cooled under various heat transfer conditions. (a) Slow cooling; (b) intensive cooling. (From Replace ref. 24 from 764.)

The mechanism of the current stress formation is as follows. When the part is completely in the austenite state, there arise tensile stresses that transfer into the compressive ones in the process of the martensite phase formation and due to increase of its specific volume. The larger part of austenite is transformed to martensite and the larger the martensite layer is, the higher the compression stress is. The situation goes on until a sufficiently thick martensite crust is formed resembling a rigid vessel that still contains the supercooled austenite in the supercooled phase. The further advance of martensite inside the part causes the effect

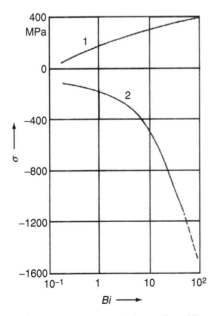

FIGURE 2.37 Maximum compressive hoop stresses at the surface (2) and tensile stresses in the center (1) of the cylindrical specimen of 6 mm in diameter vs. Biot number.

of water freezing in a glass vessel [23]. Due to the core volume increase at martensite transformations, either decrease of compressive stresses in the surface layers or destruction of the external layer will take place if the phase specific change is large enough and the external layer is insufficiently stretched and strong. Under such conditions, compression stresses in the surface layer change over to stress state that cause destruction in the surface layer.

Reduction of compressive stresses at the further advance of martensite into the part to be quenched is caused by the parting action that is attributed to variation in the phase specific volume in the core.

If the process of intensive cooling is stopped at the moment of achieving the maximum compression stresses, and isothermal holding is realized at the temperature of the martensite start (M_s), then the martensite phase advance will cease and sufficiently high compression stresses can be fixed. They will slightly decrease due to the isothermal holding at which stress relaxation takes place.

An optimal depth of the quenched layer that depends on part dimensions corresponds to maximum compression stresses.

Using the calculation methods developed and the potentialities of the software package TANDEM- HART [1,8], it is easy to find with the help of computer the time of achieving the maximum for bodies of arbitrary axisymmetric form being quenched under various heat transfer conditions.

The degree of intensive cooling can be characterized by Bi_v number or by Kondratjev number Kn. There is a universal interconnection between these numbers

$$Kn = \Psi Bi_v = \frac{Bi_v}{\sqrt{Bi_v^2 + 1.437 Bi_v + 1}}, \tag{2.11}$$

which is valid for bodies of various configurations.

The author of the well-known handbook, *Theory of Heat Conduction*, Lykov [38] has called Equation 2.1 an important relation of the theory of regular conditions. Criterion $Kn = \Psi Bi_v$ is the main value determining the heat transfer mechanism of the body. It was named Kondratjev number (criterion) in honor of the outstanding thermal scientist G.M. Kondratjev.

It appeared that the curves $Kn = f(Bi_v)$ for geometrically different bodies (sphere, parallelepiped, cylinder, etc.) were located so close to each other that practically all the family could be replaced by a single averaged curve [38].

The parameter criterion Ψ characterizing the temperature field nonuniformity is equal to the ratio of the body surface excess temperature to the mean excess temperature over the body volume. If the temperature distribution across the body is uniform $(Bi_v \to 0)$, then $\Psi = 1$. The higher is the temperature nonuniformity, the less is Ψ. At $\Psi = 0$, the temperature distribution non-uniformity is the highest $(Bi \to \infty$, while $T \to T_\infty)$.

Thus Kondratjev number characterizes not only the temperature field nonuniformity, but also the intensity of interaction between the body surface and the environment.

Kondratjev number is the most generalizing and the most universal value, which may serve to describe the cooling conditions under which compression stresses occur at the surface of various bodies.

For rather high compression stresses to occur at the surface of the part being quenched, it is sufficient to meet the following condition:

$$0.8 \leq Kn \leq 1.$$

On the basis of regularities mentioned above, a new method of quenching was elaborated. The essence of the new method is that alloy and high-alloy steel parts are cooled under the conditions of high-intensive heat transfer (Kn ≥ 0.8) up to the moment of reaching maximum compression stresses at the surface with the following isothermal holding under temperature M_s [39]. A year later, similar quenching method was proposed in Japan [40].

In accordance with the method mentioned, alloyed steel parts are quenched in such a way that a very hard surface layer of the given depth and an arbitrarily hard matrix are obtained. An example of such method realization is given below. An alloy steel specimen containing 0.65–0.85% C, 0.23–0.32% Si, 0.4–0.9% Mn, 2% Ni, 0.5–1.5% Cr, and 0.1–0.2% Mo is heated up to 800–850°C and spray-quenched with water fed under pressure of 0.4–0.6 MPa during 0.2–0.8 s. The specimen is subject further to isothermal heating at 150–250°C for 10–50 min [40].

It is obvious that the spray quenching under high pressures provides intensive cooling (Kn > 0.8) that is completed when a certain depth of the composition-quenched layer is achieved.

For the steel composition cited, the temperature of the martensite start is within the range of 150–200°C.

The isothermal holding time at this temperature (about 10–15 min) is chosen from CCT diagrams of supercooled austenite dissociation in such a way to provide this dissociation into intermediate components in the part central layers.

The analysis of the methods described shows that various authors have come independently to an identical conclusion that is a rather pleasant coincidence because it testifies to urgency and authenticity of the technology being studied.

Structural steel transformations during quenching are accounted through dependencies of thermal–physical and mechanical properties of the material on the temperature and time of cooling in accordance with CCT diagram for the transformation of overcooled austenite. The method has been proved by a number of test problems [41]. The error of calculations was $\leq 3\%$ for temperature and $\leq 12\%$ for stresses, which provides grounds for using this method for the study of regularities of changes in thermal and stress–strain state of parts to be quenched with regard to cooling conditions and character of structural steel transformations.

The calculation of current and residual stresses for cylindrical sample of 6-mm diameter made out of 45 steel was made for different heat transfer coefficients, so that Bi changed from 0.2 to 100. The temperature of sample heating is 1300 K.

The investigations have shown that as far as the process of cooling is intensified, the residual stresses on the surface of cylindrical sample firstly increase reaching the maximum value at Bi $= 4$, and then when Bi $= 18$, they become negative, and as far as Bi grows, they become compressive. At Bi $= 100$, the hoop stresses σ_{33} reach the value of 600 MPa.

The facts observed can be grounded as follows: when Bi is small, there is insignificant temperature gradient in the body. As far as austenite is transformed into martensite, due to the great specific volume of martensite, the stresses appearing first on the surface are not big and compressive. However, when the martensite forms at the center of the sample, big forces moving aside appear, which result in the tensile stresses on the surface. In the case of intensive cooling (Bi > 20), martensite transformations start in thin surface layer of the sample, while the temperature at its other points is high. The greater the Bi number is, the greater the gradient in the surface layer is, and the further from the axis the layer of freshly formed martensite is. As far as inner layers become cooler, two processes fight against each other: process of shrinking for the account of the temperature reduction and process of expansion of the material for the account of the formation of martensite having big specific volume in comparison with austenite. In the case Bi > 20, the process of shrinking prevails in inner points of the sample. Thus in the cooled sample, the surface layer appears to be shrunk

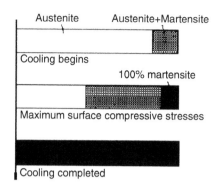

FIGURE 2.38 Relative amounts of microstructural phases present at the optimal hard depth in a steel specimen at the beginning and end of intensive quenching, and at the time when the surface compressive stress reaches its maximum value.

because shrunk inner layers of the sample try to move initially formed layer of martensite closer to the axis. In the case of not big temperature gradient (Bi ≪ 18), the outer layer of freshly formed martensite, in comparison with cold state, is lesser shifted from the axis; for this reason, in this case, tensile residual stresses will appear for the account of increase in the specific volume of the material during martensite transformations in inner layers. It is obvious that there exist such values of Bi that forces connected with material shrinking compensate each other. In this case, on the surface of quenched sample, residual stresses are zero (Bi = 18–20).

Let us note that the quenching method [36,42] provides the optimal depth of the hard layer for any alloy steel.

The optimum depth of hard layer is that which corresponds to maximum surface compressive stresses.

To obtain very high surface compressive stresses, it is sufficient to meet the condition $0.8 \leq Kn \leq 1$, where Kn is the Kondratjev number. This condition can be satisfied by intensive quenching using water jets or rapidly flowing water. Additional strengthening (superstrengthening) of the surface layer will also result. The high compressive stresses and superstrengthening both help enhance the durability and prolong the service life of machine parts (Figure 2.38).

2.9 DISCUSSION

The results of investigations made evidence that shell hardening is one of the advantageous directions since it brings big economical savings for the account of the following.

Reaching the optimal depth of hard surface layer can be done with lesser amount of alloying elements, and, in this case, the cost of the steel is reduced. For shell hardening, simple water is used. Since in the case of optimal hard layer, optimal compressive stresses are reached at the surface, the cyclical durability of ready steel products increases, which is proved in Ref. [2].

Despite all of these facts and that idea of the existence of the optimal hard layer has been published in 1983 [1], by the present time, this issue has not been studied carefully yet. Probably this is because of the lack of software and databases for these calculations. Experimental way of studying this is almost impossible since it is a huge load of work and requires a lot of finances. Only on the basis of our software TANDEM-HART and generalization of big experimental data of many authors it became possible to ground the existence of the optimal hard layer, which can be widely used in practice for designing of steel-intensive quench processes.

In our opinion, designing of steel-intensive quench processes should be done as follows:

1. Depending on the shape and size of the part, conditions of intensive quenching are selected by calculating the water flow speed or intensity of the shower cooling at the surface of the part to be quenched.
2. The proper steel grade is selected which provides the optimal depth of hard layer and maximum compressive stresses at the surface or the intensive cooling process is interrupted at the time of reaching this optimal depth.
3. On the basis of software TANDEM-HART, calculations of parameters of intensive cooling and thermal and stress–strain state are made.
4. Special equipment and devices are designed for the implementation of intensive cooling.

The final steps after designing are as follows:

1. The equipment and devices designed are produced.
2. The quenching conditions and equipment are adjusted in industrial conditions on site to prevent possible failure with the full automation of the whole industrial hardening process. The automation is very important because the work of the line must be stable. Cracks can be formed in the case of failure to meet all the conditions designed.

A steel quenching method where the depth of the quenched surface layer is controlled, which increases the service life, is described in publication [43]. The above-mentioned steels have the following characteristics: low depth of hardened layer and fine grain with arrested growth of austenite grains at high temperatures. Due to limited hardenability, on the surface of parts, residual compressive stresses appear and the fine grain has the effect of high strength of-material. In addition to the increase in the service life, there is an opportunity to replace expensive materials with cheaper materials and fire-dangerous quench oils with simple water. Steel quenching where the depth of the hardened surface layer is controlled is made in intensive water jets. The service life of parts made of steels where the depth of the hardened surface layer is regulated increases by some times [43] in comparison with oil quenching. The weak aspect of this quenching method where the depth of the hardened surface layer is regulated is that steel having the effect of optimal depth of hardened surface layer must be chosen each time for definite shape and dimensions. In the case of the change in the shape or dimensions, it is necessary to change the steel grade to have the effect of high compressive stresses on the surface. As known, the changes in the dimensions of the part and the optimum depth have the following correlation [13]:

$$\frac{\Delta\delta}{D} = \text{idem},$$

where idem means constant. When the depth of the hardened layer is greater or less, the compressive stresses are lower. If the layer is thin, in the transition zone, high extensive stresses can appear, which results in cracking. In this quenching method, there are no criteria allowing to calculate the rate of water flow for each concrete part. High water flow rate is chosen for all kinds of parts which is not always justified and results in unnecessary energy spending and makes the industrial process more complicated. The high service life of parts where the depth of hardened surface layer is regulated is considered as an advantage of these steel grades; however, for various steel grades, the effect of superstrengthening and high residual compressive surface stresses can be reached. In this method, the induction heating is mainly used and there are no data regarding oven heating, including such data for carburized parts. The industrial regimes are not optimized. In the method mentioned above, it is advised to use only those steels where the depth of the hardened surface layer is regulated; however, there are some engineering problems related to melting of such steels where the depth of the hardened surface layer is regulated.

There is a steel quenching method [5] dealing with shell hardening, which means uniform quenching of all the surface to the insignificant depth until reaching high hardness on the basis of using intensive jet cooling. In this method, the examples of the application of meanalloy steel 45 are given. The main advantage of this method is the opportunity to increase the service life of steel parts while using usual carbon steels, not using steels where the depth of the hardened surface layer is regulated. This method also has weaknesses the same as in the method above, namely, no consideration is given to conditions optimizing the depth of the hardened surface layer, and the following correlation is ignored:

$$\frac{\Delta\delta}{D} = \text{idem},$$

(that is, in the case of the change in dimensions of parts, the depth of the hardened surface layer should be changed correspondingly). This method does not have criteria allowing to calculate the rate of cooling quench flow which would prevent the development of self-regulated thermal process. The technological process is not optimized.

An intensive steel quenching method has been also developed in Japan [40]. In accordance with this method, alloyed steel parts are quenched in such a way that a very hard surface layer of the given depth and an arbitrarily hard matrix are obtained. For given steel grades, according to this method, ranges for hardening regimes are found by experiments to increase the service life of such parts. An example of such method realization is given below. An alloy steel specimen containing 0.65–0.85% C, 0.23–0.32% Si, 0.4–0.9% Mn, 2% Ni, 0.5–1.5% Cr, and 0.1–0.2% Mo is heated up to 800–850°C and spray-quenched with water fed under pressure of 0.4–0.6 MPa during 0.2–0.8 s. The specimen is subject further to isothermal heating at 150–250°C during 10–50 min [40]. The weakness of the method above is that it considers just high-carbon alloyed steels. The depth of the hard surface layer is not optimal for various dimensions of parts. Because of this steel, superstrengthening is not reached. It does not consider conditions of the optimization of the quenchant circulation rate.

The analysis of the existing methods of steel quenching used in various countries (in Russia, United States, Japan, and Ukraine) shows that the intensive steel quenching with the formation of hard surface layer of the given depth has greater advantages than for thorough quenching. The common weakness of all these methods is that there is no change in the optimum depth of the hard surface layer in cases of the changes in dimensions of parts, and that the quenchant circulation rate is not optimized for the prevention of the development of self-regulated thermal process and reaching material superstrengthening.

The intensive steel quenching method [36] has been chosen as a prototype of this invention, including heating, intensive cooling until the appearance of maximum compressive surface stresses, isothermal heating, and tempering. The method is based on the following: intensive cooling is formed in the range of $0.8 \leq \text{Kn} \leq 1$, where Kn is the Kondratjev number, until reaching maximum compressive surface stresses, and then it is isothermally heated at martensite start temperature M_s until the complete transformation of the overcooled austenite of the matrix, then it is tempered.

The basic weakness of this method is that it deals just with alloyed steels. To reach the maximum compressive stresses on the surface, the intensive cooling is stopped and due to it, the effect of superstrengthening does not show itself in full. There is no concrete method of the calculation of optimal rate of quenchant flow to ensure reaching material superstrengthening.

The proposed quenching method is new since in the part quenched for any steel grades, alloyed and not alloyed, high, mean, and low carbon, the depth of the hard surface layer reached is optimal that the maximum compressive stresses are formed. If the depth of the hard surface layer is greater or less, the compressive stresses are lower.

The creation of conditions to reach maximum compressive stresses is reached due to greater effect of additional strengthening (superstrengthening) of material. In the compressed surface layer during the quench process, martensite transformations take place. Due to greater specific volume of martensite plates (than for the resulting phases), the plastic deformation of austenite occurs which is located between the martensite plates. The higher the compressive surface stresses in the layer being hardened and the higher the cooling rate is in the martensite area, the greater the deformations in the austenite which is between martensite plates. In this case, martensite plates function like "microhammers" due to which high density dislocations are reached under high pressure. While cooling is fast, these dislocations are "frozen" in the material.

In such conditions, the effect of low-temperature mechanical heat treatment is present. After this treatment, the material hardened can have higher mechanical and plastic properties in comparison with usual hardening. Thus the optimal depth of the quenched layer is necessary for not only reaching maximum compressive surface stresses, but also for the formation of optimal conditions under which the effect of additional material strengthening (superstrengthening) is present in full. The additional strengthening (superstrengthening) of material and high compressive stresses in the surface layer quenched result in the increase in their service life.

In conclusion, it should be noted that the main idea of this chapter is the existence of the optimal depth of hard layer for which the maximum compressive stresses are reached at the surface. In our opinion, it is a principally new approach which can be used for designing of steel-intensive quench processes. By the present time, the progress has been made in two ways. One of them is through-surface hardening, at which special steel grades are used that provide shell hardening [2–4,46]. In this case, it does not deal with the consideration of the optimal depth of hard layer, and when changing to other sizes with the same steel grade, the optimal depth of hard layer is never reached.

The other way lies in the interruption of intensive cooling when maximum compressive stresses are formed at the surface, as mentioned above [36,39]. In this case, the optimal depth of the hard layer is reached automatically.

The first method was used with induction heating and steel grades of controlled hardenability. In found wide application in Russia [2–4,6,46]. The second method was used with furnace heating and interrupted cooling at the time of reaching maximum compressive stresses at the surface. It is widely used in Ukraine, other countries of the former Soviet Union, and the United States [36,39,49–51]. The purpose of this chapter is to unite efforts of these two methods for the optimization of intensive quenching processes. For the optimization of these processes, TANDEM-HART software has been developed, which belongs to Intensive Technologies Ltd. and distributed by it. The software package consists of two parts: TANDEM-HART package and TANDEM-HART ANALYSIS. TANDEM-HART package makes numerical calculations of fields of temperature, stress, deformation, and safety factor. TANDEM-HART ANALYSIS finds the time of reaching optimal stress distribution and makes quick calculations of quenching conditions. For this software to produce good results, it is necessary to acquire data of mechanical properties, CCT and TTT diagrams for new steel grade of controlled hardenability, and boundary conditions. We believe that uniting efforts of many specialists will help in solving these problems.

Thus the application of the proposed steel quenching method allows as follows:

1. To reach the effect of material superstrengthening and high compressive surface stresses when using arbitrary steel grade.
2. Alloyed and high-alloyed steels can be replaced with simple carbon steels that has the effect that the depth of the hard layer is optimal. In this case, the effect of material superstrengthening is greater. Due to it, the service life of such parts increases.
3. Expensive and fire-dangerous oils can be replaced by water and water solutions.
4. The labor efficiency increases.
5. The ecological state of the environment is improved.

2.10 CONCLUSIONS

As a result of investigating the kinetics of phase transformations in bodies of complicated configuration, the following regularities have been found:

1. When the Biot number increases, the axial and hoop residual stresses on the surface of a part to be quenched firstly grow reaching the maximum at $Bi = 4$ and then reduced and become negative at $Bi \geq 20$.
2. The distribution of residual stresses in parts with full and partially controlled hardenability while the heat transfer is highly forced ($Bi > 20$) has different character. In parts of controlled hard-enability on the surface, high compressive stresses appear, which gradually change to tensile stresses at the center of the part. In the case of thorough hardening, while $Bi > 20$ in the surface layer, there are high tensile stresses changing to compressive stresses on the surface. With the elapse of the time, parts made out of steel of controlled hardenability have compressive residual stresses on the surface growing all the time, while parts hardened thoroughly have stresses that are compressive and grow until a certain moment of time at which they reach the maximum, then they are reduced.
3. Methods of numerical investigation of the kinetics of phase transformation in bodies of arbitrary shape have been developed. Having CCT diagrams with physical and mechanical properties of structural components for these diagrams, one can determine the structure, hardness, and hardenability of parts having a complicated configuration and forecast the mechanical properties of the material [26,44–48]. For this purpose, TANDEM-HART software has been developed.
4. It has been established that there is optimal depth of hard layer for a part at which compressive stresses on the surface reach the maximum.
5. There is similarity of the distribution of current and residual stresses in bodies having different sizes and the conditions when this similarity is observed are given above.
6. The practical application of shell hardening for car box rollers and wheels is described.
7. For the wide application of these methods, it is necessary to improve further the software and to develop databases of initial data for solving the problem of calculating the optimal depth of hard layer for various steel grades, which would provide the optimal distribution of compressive stresses at the surface and in the core.
8. The optimal depth of the hard layer can be reached for the account of either proper selection of steel grade or interrupting the cooling at the time of reaching the optimal maximum compressive stresses at the surface.

ACKNOWLEDGMENTS

The authors would like to present their thanks to V.V. Dobrivecher for the help in the preparation of the material and translation of this chapter into English.

REFERENCES

1. Kobasko, N.I.; Morhuniuk, W.S. *Investigation of thermal and stress–strain state at heat treatment of power machine parts* (*Issledovanie teplovogo i napryagenno-deformirovannogo sostoyaniya pri termicheskoy obrabotke izdeliy energomashinostroyeniya*); Znanie: Kiev, 1983; 16 pp.
2. Ouchakov, B.K.; Shepelyakovskii, K.Z. *New Steels and Methods for Induction Hardening of Bearing Rings and Rollers, Bearing Steels: Into the 21st Century, ASTM STP 1327*; Hoo, J.J.C., Ed.; American Society for Testing and Materials: 1998.
3. Ushakov, B.K.; Lyubovtsov, D.V.; Putimtsev, N.B. Volume-surface hardening of small-module wheels made of 58 (55PP) steel produced at OEMK. *Mater. Sci. Trans.* Mashinostroenie: Moscow, 1998, (4), 33–35.
4. Beskrovny, G.G.; Ushakov, B.K.; Devyatkin, N.B. Raising of longevity of car box bearings with use of hollow rollers hardened by volume-surface hardening. *Vestnik*; VNIIZhT: Moscow, 1998, (1), 40–44.
5. Kern, R.F. Intense quenching. *Heat Treat.* 1986, (1), 19–23.
6. Shepelyakovskii, K.Z. *Hardening Machine Parts by Surface Quenching Through Induction Heating*; Mashinostroenie: Moscow, 1972; 288 pp.
7. Kobasko, N.I.; Morhuniuk, W.S.; Gnuchiy, Yu.B. *Investigation of technological machine part treatment* (*Issledovanie tekhnologicheskikh protsesov obrabotki izdeliy mashinostroeniya*); Znanie: Kiev, 1979; 24 pp.
8. Kobasko, N.I.; Morhuniuk, W.S. *Investigation of thermal stress state in the case of heat treatment of power machine parts* (*Issledovanie teplovogo i napryagennogo sostoyaniya izdeliy energomashinostroyeniya pri termicheskoy obrabotke*); Znanie: Kiev, 1981; 16 pp.
9. Inoue, T.; Arimoto, K.; Ju, D.Y. Proc. First Int. Conf. Quenching and Control of Distortion; ASM International: 1992; 205–212.
10. Inoue, T.; Arimoto, K. Development and implementation of CAE system "HEARTS" for heat treatment simulation based on metallo- thermo-mechanics. *JMEP* 1997, 6 (1), 51–60.
11. Narazaki, M.; Ju, D.Y. Simulation of distortion during quenching of steel—effect of heat transfer in quenching. Proc. of the 18th ASM Heat Treating Society Conference & Exposition, Rosemont, Illinois, USA, October 12–15, 1998.
12. Reti, T.; Horvath, L.; Felde, I.A comparative study of methods used for the prediction of non-isothermal austenite decomposition. *JMEP* 1997, 6 (4), 433–442.
13. Kobasko, N.I. *Steel quenching in liquid media under pressure* (*Zakalka stali v zhidkikh sredakh pod davleniem*); Naukova Dumka: Kiev, 1980; 206 pp.
14. Kobasko, N.I.; Morhuniuk, W.S.; Lushchik, L.V. Investigation of thermal stress state of steel parts in the case of intensive cooling at quenching. *Thermal and Thermomechanical Steel Treatment* (*Termicheskaya i termo-mekhanicheskaya obrabotka stali*), Metallurgy: Moscow, 1984; 26–31.
15. Kobasko, N.I.; Morhuniuk, W.S.; Dobrivecher, V.V. Calculations of cooling conditions of steel parts during quenching. Proc. of the 18th ASM Heat Treating Society Conference & Exposition, Rosemont, Illinois, USA, October 12–15, 1998.
16. Totten, G.E., Howes, A.H., Eds.; *Steel Heat Treatment Handbook*; Marcel Dekker, Inc.: New York, 1997; 1192 pp.
17. Kobasko, N.I.; Prokhorenko, N.I. Cooling rate effect of quenching on crack formation in 45 steel. *Metalloved. Term. Obrab. Metall.* 1964, (2), 53–54.
18. Kobasko, N.I. Crack formation at steel quenching. *MiTOM* 1970, (11), 5–6.
19. Bogatyrev, JuM.; Shepelyakovskii, K.Z.; Shklyarov, I.N. Cooling rate effect on crack formation at steel quenching. *MiTOM* 1967, (4), 15–22.
20. Ganiev, R.F.; Kobasko, N.I.; Frolov, K.V. On principally new ways of increasing metal part service life. *Dokl. Akad. Nauk USSR* 1987, *194* (6), 1364–1473.

21. Kobasko, N.I. Increase of service life of machine parts and tools by means of cooling intensification at quenching. *MiTOM* 1986, (10), 47–52.

22. Kobasko, N.I.; Nikolin, B.I.; Drachinskaya, A.G. Increase of service life of machine parts and tools by creating high compression stresses in them. *Izvestija VUZ (Machinostrojenie)*, 1987, (10), 157.

23. Kobasko, N.I. Increase of steel part service life and reliability by using new methods of quenching. Metal-loved. *Term. Obrab. Metall.* 1989, (9), 7–14.

24. Kobasko, N.I.; Morhuniuk, W.S. Numerical study of phase changes, current and residual stresses at quenching parts of complex configuration. Proc. of 4th Int. Congr. Heat Treatment Mater, Berlin, 1985; 466–486.

25. Kobasko, N.I. On the possibility of controlling residual stresses by changing the cooling properties of quench media. *Metody povyshenija konstruktivnoi prochnosti metallicheskikh materialov*; Znanije RSFSR: Moscow, 1988; 79–85.

26. Kobasko, N.I.; Morhuniuk, W.S. *Investigation of Thermal and Stress State for Steel Parts of Machines at Heat Treatment*; Znanie: Kyiv, 1981; 24 pp.

27. Shepelyakovskii, K.Z.; Devjatkin, V.P.; Ouchakov, B.K. Induction surface hardening of rolling bearing parts. *Metalloved. Term. Obrab. Metall.* January 1974, (1), 17–21. *in Russian.*

28. Shepelyakovskii, K.Z. Surface and deep and surface hardening of steel as a means of strengthening of critical machine parts and economy in material resources. *Metal Science and Heat Treatment (A translation of Metallovedenie i Termicheskaya Obrabotka Metallov)*; Consultants' Bureau: New York, November–December, 1993, (11,12), 614–622.

29. Ouchakov, B.K.; Efremov, V.N.; Kolodjagny, V.V. New compositions of bearing steels of controlled hardenability. *Steel.* October 1991, (10), 62–65. *in Russian.*

30. Shepelyakovskii, K.Z.; Devyatkin, V.P.; Ushakov, B.K.; Devyatkin, V.F.; Shakhov, V.I.; Bernshtein, B.O. Induction surface hardening of swinging bearing parts. *Metalloved. Term. Obrab. Metall.* 1974, (1), 17–21.

31. Devyatkin, V.P.; Shakhov, V.I.; Devin, R.M.; Mirza, A.N. Application of hollow rollers for the prolongation of service life of cylindrical rolling bearings. *Vestn. VNIIZhT* 1974, (3), 20–22.

32. Polyakova, A.I. Comparative tests of car bearings 42726 and 232726 with solid and hollow rollers. *Bearing Industry, Issue 8*; NIINAvtoprom: Moscow, 1974; 1–10.

33. Rauzin, Ya.P. *Heat Treatment of Chrome-Containing Steel*; Mashinostroenie: Moscow, 1978; 277 pp.

34. Mukhina, M.P.; Kobasko, N.I.; Gordejeva, L.V. Hardening of structural steels in chloride quenching media. *Metalloved. Term. Obrab. Metall.* 1989, (9), 32–36.

35. Shepelyakovskii, K.Z.; Ushakov, B.K. Induction surface hardening—Progressive technology of XX and XXI centuries. Proceedings of the 7th International Congress on Heat Treatment and Technology of Surface Coatings, Moscow, Russia 11–14 December 1990, 2 (11–14), 33–40.

36. Kobasko, N.I. Patent of Ukraine: UA 4448, Bulletin No. 6–1, 1994.

37. Sigeo, O. Intensive cooling. *Kinzoku Metals Technol.* 1987, *57* (3), 48–49.

38. Lykov, A.V. *Theory of Heat Conduction*; Vysshaya Shkola: Moscow, 1967; 560 pp.

39. Kobasko, N.I. Method of part quenching made of high-alloyed steels, Inventor's certificate 1215361 (USSR), Bulletin of Inventions No. 12., Applied 13.04.1983., No. 3579858 (02-22), 1988.

40. Naito, T. Method of steel quenching. Application 61-48514 (Japan), 16.08.1984, No. 59-170039.

41. Loshkarev, V.E. Thermal and stress state of large-size pokovok at cooling in heat treatment. Dissertation abstract. Sverdlovsk, 1981; 24 pp.

42. Kobasko, N.I. *Intensive Steel Quenching Methods, Theory and Technology of Quenching*; Liscis, B. Tensi, H.M., Luty, W., Eds.; Springer-Verlag: New York, NY, 1992; 367–389.

43. Shepelyakovskii, K.Z.; Bezmenov, F.V. New induction hardening technology. *Adv. Mater. Process.* October 1998; 225–227.

44. Morhuniuk, W.S. Thermal and stress–strain state of steel parts with complicated configuration at quenching. Dissertation abstract, Kyiv, 1982; 24 pp.

45. Morhuniuk, W.S.; Kobasko, N.I.; Kharchenko, V.K. On possibility to forecast quench cracks. *Probl. Procn.* 1982, (9), 63–68.

46. Shepelyakovskii, K.Z. Through-surface quenching as a method of improving durability, reliability and service life of machine parts. *MiTOM* 1995, (11), 2–9.

47. Bashnin, Yu.A.; Ushakov, B.K.; Sekey, A.G. *Technology of Steel Heat Treatment*; Metallurgiya: Moscow, 1986; 424 pp.
48. Kobasko, N.I. Self-regulated thermal process at steel quenching. *Prom. Teploteh.* 1998, *20* (5), 10–14.
49. Kobasko, N.I. Generalization of results of computations and natural experiments at steel parts quenching. *J. Shanghai Jiaotong Univ.* June 2000, *E-5* (1), 128–134.
50. Kobasko, N.I. Thermal and physical basics of the creation of high-strength materials. *Prom. Teploteh.* 2000, *22* (4), 20–26.
51. Aronov, M.A.; Kobasko, N.I.; Powell, J.A. Practical application of intensive quenching process for steel parts. Proc. of the 12th Int. Federation of Heat Treatment and Surface Engineering Congress, (Melbourne, Australia), 29 October–2 November 2000; 51 pp.

3 Vacuum Heat Processing

Bernd Edenhofer, Jan W. Bouwman, and Daniel H. Herring

CONTENTS

3.1 INTRODUCTION

The term vacuum heat processing refers to heat treatment processes in which ferrous and nonferrous components are subjected to the application of thermal heat energy in a vacuum environment [1].

What is a vacuum? The word vacuum originates from Latin and means empty or empty space. By empty space one thinks of a space entirely devoid of matter. Such a space does not exist nor can it be produced. In technical terms a vacuum can be considered a space with highly reduced gas density. Such a space is produced by removing the air (and other gases) from a gastight container using, for example, a vacuum pumping system.

The quality of a vacuum is described by the degree of reduction of gas density, i.e., gas pressure. Gas pressure is commonly measured in torr, where 1 torr = 1.33 millibar (mbar). Atmospheric pressure at standard temperature and sea level is 760 torr (1013 mbar). The conversions between common vacuum units such as torr, bar, and pascal are shown in Table 3.1.

One distinguishes four different vacuum levels or qualities as shown in Table 3.2. The heat treatment of steel is carried out in three qualities of vacuum—rough, fine, and high. The majority of applications are processed in the fine vacuum range.

What is the purpose of utilizing vacuum for heat processing? The heat treatment of metal components such as steel in air leads to surface oxidation. The type and thickness of oxide layers produced are dependent on the temperature of the heat treatment, the duration of exposure, and, also on the type of steel.

To avoid surface oxidation there are several options available. The air in the heat treatment furnace must be replaced by an atmosphere that does not contain oxygen. The use of an inert atmosphere, such as nitrogen (N_2), is one method. Another is the use of a protective atmosphere that utilizes nitrogen in combination with reducing elements such as hydrogen (H_2) and carbon monoxide (CO). Either method requires the furnace to be purged completely of air, especially when using protective atmospheres, as they are typically combustible. Figure 3.1 shows the reduction of the oxygen level, through purging, in a gastight furnace with increasing numbers of volumetric changes.

Another way to avoid surface oxidation is to reduce the amount of air surrounding the workpieces during the thermal processing by evacuation to such a low level that the remaining oxygen is below the oxidation level of the material. From the known temperature dependence of formation energies of oxides [2], the oxidation boundaries of iron and typical alloying elements have been calculated and are shown in Figure 3.2 [3]. These curves, for example,

TABLE 3.1
Conversion of Pressure Units

Pressure Unit	torr	bar	Pa
1 Pa = 1 N/m²	7.5006×10^{-3}	10^{-5}	1
1 bar	750.06	1	10^5
1 torr = 1 mm Hg	1	1.3332×10^{-3}	133.32
1 m/H₂O	73.56	9.807×10^{-2}	9807
1 atm	760.0	1.0133	1.0133×10^5
1 μm	10^{-3}	1.3332×10^{-6}	0.13332
1 in. Hg	25.4	3.3864×10^{-2}	3386.4
1 lb/ft²	0.3591	4.788×10^{-4}	47.88

TABLE 3.2
Classification of Vacuum

Quality of Vacuum	Pressure Range (torr)
Rough	1–760
Fine	10^{-3}–1
High	10^{-7}–10^{-3}
Ultrahigh	$<10^{-7}$

---

show that to avoid the oxidation of iron at 1832°F (1000°C), the partial pressure of oxygen in air has to be reduced from 1.5×10^2 torr (0.2 bar) to below 1×10^{-15} torr (1.3×10^{-15} mbar). This is not only impossible, but it is also not required.

Practical experience shows that fine vacuum of 10^{-2} to 10^{-3} torr (1.3×10^{-2} to 1.3×10^{-3} mbar) suffices to produce bright surfaces on most steels. Thus, even though the oxygen in fine vacuum oxidizes the steel surface, the degree is insufficient to produce a visible effect. Also, certain furnace materials getter oxygen (by producing oxides themselves) and thus reduce the partial pressure of oxygen within the furnace even further. All materials that have lower oxide formation energies than iron such as carbon, chromium, and manganese contribute to this effect if they are sufficiently hot.

3.2 COMPARISON TO ATMOSPHERIC PROCESSES

Almost all heat treatment processes carried out at normal pressure in protective atmospheres have an equivalent counterpart in vacuum processing. Table 3.3 gives an overview.

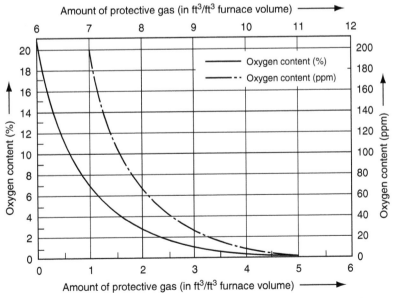

FIGURE 3.1 Reduction of oxygen level in a furnace through purging with a protective atmosphere.

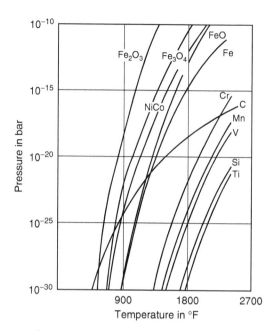

FIGURE 3.2 Partial pressure of oxygen for the formation of iron oxides and oxides of other elements.

If atmosphere furnaces have no separate purge chambers, an inert atmosphere is not sufficient to produce bright surfaces. Therefore, protective atmospheres usually contain certain amounts of reducing gases, like H_2 and CO, to counter the oxidizing and decarburizing effects of constituents such as water vapor (H_2O) and oxygen (O_2) that enter the furnace during door openings.

A general difference between atmospheric and vacuum processes is given by the cleanliness and the substantially reduced environmental impact of vacuum heat treatments. Vacuum furnaces produce no fumes or exhaust gases, possess no flames, and are usually cold. They behave more like machine tools than like furnaces. In addition, the visual appearance of the treated steel workpieces differs. Pieces treated in atmospheric furnaces usually exhibit a light gray color even in reducing atmospheres, whereas the surfaces of vacuum heat-treated components remain bright and shiny. This is due to the evaporation effect of residues and the reduction of oxides on the surfaces at the low-pressure levels achieved in vacuum. In some cases it may also be advisable to employ multiple pump down cycles or feed hydrogen at low partial pressures into the vacuum furnace to counter the effect of water vapor adsorbed on the inside of the cold-furnace walls or insulation during door openings.

Low pressures, i.e., high vacuum, will not only evaporate adsorbed material from the surface, they also cause the volatilization of those alloyed elements in the steels that have a high vapor pressure. Manganese and chromium are particularly susceptible.

Carrying out thermochemical processes like carburizing and nitriding in vacuum furnaces at low pressures can be done using gases that produce high mass transfer rates where the thermal dissociation of the process gas yields large amounts of reactive elements. In vacuum carburizing, for example, this is done only by the use of acetylene [4]. Other gases such as propane, ethylene, and methane dissociate on first contact, resulting in both soot formation and a higher mass transfer on the outside of a dense load than in the load center. An alternative is activating and ionizing these process gas molecules by applying an electric field of high voltage that will overcome their weakness by substantially increasing the number

TABLE 3.3
Comparison of Atmospheric^a and Vacuum Heat Treatment Processes

Atmospheric Processes		Vacuum Processes		
Process	Common Gas Constituents	Process	Pressure (torr)	Gas Type
Annealing	N_2 $N_2 + CH_4$ (C_3H_8) $N_2 + CH_3OH$ Exothermic gas, endothermic gas	Annealing	10^{-3}–10^{-1}	None N_2 $N_2 + H_2$
Hardening	See annealing	Hardening	10^{-3}–10^{-1}	See annealing
Tempering	See annealing	Tempering	10^{-2}–10^3	See annealing
Carburizing	$N_2 + CH_3OH$ $+ CH_4$(C_3H_8) Endothermic gas $+ CH_4$ (C_3H_8)	Low-pressure carburizing	3–300	C_2H_2, C_3H_8, CH_4, C_2H_4
		Plasma carburizing	1.5–7.5	CH_4(+H_2), $C_3H_8 + H_2$(+Ar)
Carbonitriding	Carburizing + NH_3	Low-pressure carbonitriding	1.5–10^2	C_2H_2, C_3H_8, CH_4, C_2H_4, +NH_3
		Plasma carbonitriding	1.5–7.5	Plasma carburizing + N_2
Nitriding	NH_3 or $NH_3 + N_2$(H_2) or $NH_3 + O_2$(+N_2)	Plasma nitriding	0.4–7.5	$N_2 + H_2$
Nitrocarburizing	NH_3 + endothermic gas $N_2 + H_2 + CO_2$	Plasma nitrocarburizing	0.4–7.5	$N_2 + H_2 + CH_4$(CO_2)
Brazing	H_2, exothermic or endothermic gas	Brazing	10^{-4}–10^{-2}	None or H_2
Sintering	H_2 or $N_2 + H_2$ or endothermic gas	Sintering	10^{-3}–10^{-2}	None or N_2 (+H_2)

^{a}At normal (760 torr) pressure.

of reactive gas species and distributing them uniformly throughout the load. Processes that include electric activation of the gases are called plasma processes, and are shown in Table 3.3.

3.3 VOLATILIZATION, DISSOCIATION, AND DEGASSING

During vacuum heat treatment of steel, it is always necessary to consider the evaporation or sublimation of the alloying constituents. For example, manganese and chromium have relatively high vapor pressures (Table 3.4). It is known that operating a vacuum furnace at higher pressure causes more frequent collisions of gas molecules, resulting in fewer metal atoms escaping from the metal surface. Hence, the vaporization rate is directly related to the furnace pressure.

Elements with very high vapor pressures, such as zinc and cadmium, will evaporate very rapidly when heated even at low vacuum levels. Alloys with high concentrations of these volatile elements, such as brass and certain brazing alloys, are seldom heat treated in vacuum because of the risk of dealloying the brass and contaminating everything inside the furnace with condensed zinc.

TABLE 3.4
Temperatures at Which Common Elements Exhibit Specific Vapor Pressures

	Temperature (°C) at Which Vapor Pressure Is				
Element	1×10^{-4} torr	1×10^{-3} torr	1×10^{-2} torr	1×10^{-1} torr	750 torr
Aluminum	808	890	997	1124	2058
Antimony	525	595	678	780	1441
Arsenic	—	220	—	310	610
Barium	544	626	717	830	1404
Beryllium	1030	1131	1247	1396	—
Bismuth	537	609	699	721	1421
Boron	1141	1240	1356	1490	—
Cadmium	180	220	264	321	766
Calcium	463	528	605	701	1488
Carbon	2290	2473	2683	2928	4831
Cerium	1092	1191	1306	1440	—
Cesium	74	110	153	207	691
Chromium	993	1091	1206	1343	2484
Cobalt	1363	1495	1650	1834	—
Copper	1036	1142	1274	1433	2764
Gallium	860	966	1094	1249	—
Germanium	997	1113	1252	1422	—
Gold	1191	1317	1466	1647	2999
Indium	747	841	953	1089	—
Iridium	2156	2342	2558	2813	—
Iron	1196	1311	1448	1604	2737
Lanthanum	1126	1243	1382	1550	—
Lead	548	620	718	821	1745
Lithium	378	439	514	608	1373
Magnesium	331	380	443	515	1108
Manganese	792	878	981	1021	2153
Molybdenum	2097	2297	2535	3011	5573
Nickel	1258	1372	1511	1680	2734
Niobium	2357	2541	—	—	—
Osmium	2266	2453	2669	2922	—
Palladium	1272	1406	1567	1760	—
Platinum	1745	1905	2092	2295	4411
Potassium	123	161	207	265	643
Rhodium	1816	1973	2151	2359	—
Rubidium	88	123	165	217	679
Ruthenium	2040	2232	2433	2668	—
Scandium	1162	1283	1424	1596	—
Silicon	1117	1224	1344	1486	2289
Silver	848	921	1048	1161	2214
Sodium	195	238	291	356	893
Strontium	413	475	549	639	1385
Tantalum	2601	2822	—	—	—
Thallium	461	500	607	661	1485
Thorium	1833	2000	2198	2433	—
Tin	923	1011	1190	1271	2272
Titanium	1250	1385	1547	1725	—

TABLE 3.4 (continued)
Temperatures at Which Common Elements Exhibit Specific Vapor Pressures

	Temperature (°C) at Which Vapor Pressure Is				
Element	1×10^{-4} torr	1×10^{-3} torr	1×10^{-2} torr	1×10^{-1} torr	750 torr
Tungsten	2769	3019	3312	—	5932
Uranium	1586	1731	1899	2099	—
Vanadium	1587	1726	1889	2081	—
Yttrium	1363	1495	1630	1834	—
Zinc	248	290	343	405	908
Zirconium	1661	1818	2003	2214	—

Consider another example: a stainless steel with 14% chromium shows no discernible chromium loss at 1800°F (990°C) and a vacuum level of 10^{-2} torr (1.3×10^{-2} mbar) after 2.5 h. However, at 10^{-4} torr (1.3×10^{-4} mbar), with identical time and temperature conditions, a surface chromium loss of 0.5% was measured.

Most ferrous and nonferrous materials can be processed in vacuum by using a partial pressure, which exceeds the vapor pressure of the volatile alloying elements at the temperature involved. It is common practice to use a partial pressure of inert gas sometimes blended with a small percentage of hydrogen between 10^{-1} torr (1.3×10^{-1} mbar) and 7.5 torr (10 mbar) at temperatures above 1475°F (880°C) to preclude the vaporization of elements such as chromium, copper, and manganese from steels in their normal heat treatments.

Like the pure metals, metal–hydrogen, metal–oxygen, and metal–nitrogen compounds will also decompose when heated to sufficiently high temperatures at correspondingly low pressures. For this reason, vacuum treating can be used both to dissociate these compounds and to evacuate the evolved gases, leaving an undisturbed and clean base metal behind.

Gas incorporated in steel during manufacturing or posttreatment such as pickling and welding can also be removed in vacuum. To degas a metal, the pressure in the furnace vessel must be lower than the pressure of the gas in the metal. Under these conditions, the gas will diffuse out of the metal into the vacuum. As the outer layer of the metal degasses, a gradient is set up effecting gas desorption from the interior to the surface. The desorption rate is accelerated by higher temperatures. In the degassing of thicker parts, the gas diffusion rate through the mass becomes a limiting factor. Gases are removed from the metal surface into the vacuum surrounding the workload and are finally evacuated by the vacuum pumps and exhausted from the system.

The degassing of oxygen, nitrogen, and hydrogen from refractory metals is particularly important in view of the improvements in properties, including ductility and fatigue, that can be achieved as a result of reducing the interstitial alloy content in these materials.

3.4   VACUUM FURNACE EQUIPMENT

Heat treatment in vacuum furnaces is characterized by special conditions with regard to the design of the furnaces as well as the control of temperature and vacuum level during the heat treatment. The design of the furnaces generally depends on the size of the load, pressure, and temperature to be attained, and the medium to be used in cooling the load.

The main parts of a vacuum furnace are the vessel, pumping system, hot zone, control system, and cooling system. Each of these will be discussed below.

3.4.1 VESSEL

Vacuum furnace vessels can be grouped into so-called hot-wall and cold-wall designs. A typical hot-wall furnace has a retort that is commonly metallic or ceramic depending on the temperature (Figure 3.3a). The heating system is usually located outside of the retort and consists of resistive heating elements or an induction coil. Limitations of this retort-type furnace are the restricted dimensions of the heating zones and the restricted temperature range of the metallic retort, usually limited to 2000°F (1100°C) maximum arising from low mechanical strength, gas permeability, evaporation of alloying elements, and slow cooling rates. Other designs employ ceramic fiber insulation attached directly to the interior vessel walls. These avoid the temperature limitation concerns but the insulation tends to be hydroscopic (water absorbing), causing extended pumping times.

The demands of the heat treatment industry are for higher temperatures, lower pressures, rapid heating and cooling capabilities, and higher production rates. The cold-wall vacuum furnace has become the dominant design for high-temperature furnaces since the late 1960s.

With cold-wall furnaces, the vacuum vessel is cooled with a cooling medium (usually water) and is kept near ambient temperature during high-temperature operations (Figure 3.3b). In comparison to the hot-wall furnace, the features of the cold-wall furnace are:

- Higher operating temperature ranges between 2400°F (1315°C) standard and 3000°F (1650°C)
- Lower heat losses and less heat load released to the surroundings
- Faster heating and cooling performance
- Greater temperature uniformity control

A disadvantage over the retort design is the greater adsorption of gases and water vapors on the cooled furnace walls and in the insulation after opening of the furnace.

3.4.2 PUMPING SYSTEM

The construction of the pumping system depends on the following factors:

- The volume of the vessel
- The surface area of the vessel and the type of furnace internals
- The outgassing of the workload and related fixturing
- The time required for evacuation down to the final pressure

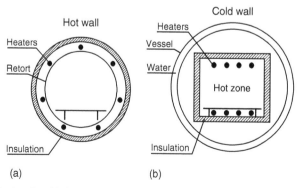

(a)   (b)

FIGURE 3.3 Vessel design for (a) hot-wall and (b) cold-wall vacuum furnaces.

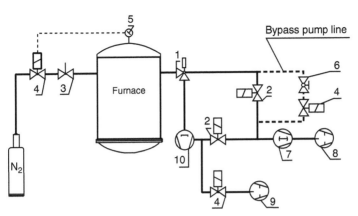

FIGURE 3.4 Typical vacuum system with mechanical and diffusion pumps, connection valves and lines. 1, Main valve; 2, butterfly valve; 3, needle valve; 4, solenoid valve; 5, Pirani gauge; 6, ball valve; 7, roots pump; 8, rotary vane pump; 9, holding pump; 10, diffusion pump.

It is important to note that the pumping system must maintain the process vacuum level without being overwhelmed by the outgassing of the workload. Pumping systems are usually divided into two subsystems, pumps for rough vacuum (micron range) and pumps for high vacuum (submicron range). For certain applications a single pumping system can handle the entire range and cycle. A typical system is illustrated in Figure 3.4. Pumps are usually classified into two general categories, mechanical pumps and diffusion pumps. There are other specialized types of vacuum pumps for use in achieving higher vacuum ranges such as ejectors, ion pumps, cryo pumps, turbomolecular pumps, and chemical getter pumps.

Table 3.5 gives a survey of the types of pumps, or pump combinations, for achieving various final vacuum levels. The various vacuum ranges are grouped in Table 3.6 together with their specific application in the heat treatment of steel in vacuum.

3.4.3 HEATING CHAMBER

For the insulation of the hot zone the following designs and materials are in common use (see Figure 3.5):

TABLE 3.5
Final Vacuum Levels for Certain Pump Combinations

	Pressure (torr)			
Pump	$1–10^{-3}$	$10^{-3}–10^{-4}$	$10^{-4}–10^{-7}$	$<10^{-7}$
Rotary vane piston pump	Required	Required	Required	Required
Roots pump	Recommended	Required	Recommended	Required
Oil diffusion pump	Unsuitable	Recommended	Required	Unsuitable
Turbomolecular pump	Unsuitable	Unsuitable	Unsuitable	Required

TABLE 3.6
Pressure Ranges Used in Vacuum Heat Treatment of Steel

Pressure Range (torr)	Applications
Rough vacuum, 760–1	Unalloyed steel
Medium vacuum, $1-10^{-3}$	Alloyed steels; stainless steel (non-Ti-alloyed)
High vacuum, $10^{-3}-10^{-7}$	Stainless steel (Ti-alloyed); CVD-coated tools; critical brazing treatments

- All metallic (radiation shields)
- Combination of radiation shields and other (ceramic) insulating material
- Multiple-layer (sandwich) insulation
- All graphite (board, fiber, carbon–carbon composite)

Radiation shields are manufactured from:

- Tungsten or tantalum having a maximum operating temperature of 4350°F (2400°C)
- Molybdenum having a maximum operating temperature of 3100°F (1700°C)
- Stainless steel or nickel alloys having a maximum operating temperature of 2100°F (1150°C)

Mostly all metallic designs consist of a combination of materials, for example, three molyb-denum shields backed by two stainless steel shields would be typical for 2400°F (1150°C) operation. Radiation shields adsorb only small amounts of gases and water vapors during

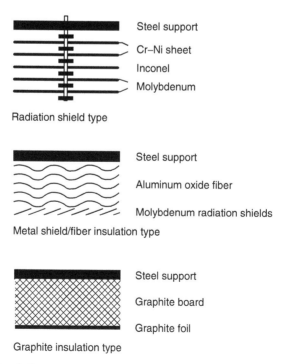

Radiation shield type

Metal shield/fiber insulation type

Graphite insulation type

FIGURE 3.5 Vacuum insulation.

opening of the furnace. They are, however, expensive to purchase and maintain, and often require greater pumping capacity to remove any moisture trapped between the shields. Compared with other types of insulation, their heat losses are high and become higher with loss of emissivity (reflectivity) due to the gradual contamination of the shields.

Sandwich insulation is composed of one or more radiation shields typically with ceramic insulation wool between them. Combinations of graphite fiber sheets and ceramic insulation wool are also used. These versions are cheaper to buy and maintain but adsorb higher levels of water vapor and gases (due to the very large surface area of the insulation wool). Their heat losses are considerably lower than those of radiation shields.

Graphite fiber insulation designs cost a little more than sandwich insulation. However, as their heat losses are lower, a smaller thickness is sufficient. In these designs, the adsorption of gases and water vapor is considerably reduced. Furthermore, the heating costs are lower, and the lifetime of this type of insulation is much longer. The maximum operating temperature is around 3630°F (2000°C). The lifetime depends strongly on the purity of the graphite. In some applications such as brazing, a sacrificial layer is used to protect the insulation beneath.

For most heat treatments in vacuum furnaces, graphite insulation is used.

3.4.4 HEATING SYSTEM

In general, the heating elements for heating systems in vacuum furnaces are made from one of the following materials:

- Nickel–chromium alloys that can be used up to 2100°F (1150°C). Above 1475°F (800°C) there is a risk of evaporation of chromium.
- Silicon carbide with a maximum operating temperature of 2200°F (1200°C). There is a risk of evaporation of silicon at high temperatures and low vacuum levels.
- Molybdenum with a maximum operating temperature of 3100°F (1700°C). Molybdenum becomes brittle at high temperature and is sensitive to changes in emissivity brought about by exposure to oxygen or water vapor.
- Graphite can be used up to 3630°F (2000°C). Graphite is sensitive to exposure to oxygen or water vapor, resulting in reduction in material thickness due to the formation of carbon monoxide (CO) that will be evacuated by the pumps. The strength of graphite increases with temperature.
- Tantalum has a maximum operating temperature of 4350°F (2400°C). Tantalum, like molybdenum, becomes brittle at high temperatures and is sensitive to changes in emissivity brought about by exposure to oxygen or water vapor.

Uniformity of temperature is of great importance to heat treatment results. The construction of the heating system should be such that temperature uniformity in the load during heating is optimal; it should be better than ±10°F (5°C) after temperature equalization. This is realized with single or multiple temperature control zones and a continuously adjustable supply of heating power (using thyristor or transductor control) for each zone.

In the lower temperature range, below 1550°F (850°C), the radiant heat transfer is low and can be increased by convection-assisted heating. For this purpose, after evacuation the furnace is backfilled with an inert gas up to an operating pressure of 1–2 bar, and a built-in convection fan circulates the gas around the heating elements and the load. In this way, the time to heat different loads, especially those with large cross-section parts to moderate temperatures, for example 1000°F (550°C), can be reduced by as much as 30–40%. At the same time the temperature uniformity during convection-assisted heating is much better, resulting in less distortion of the heat-treated part.

3.4.5 Control System

One of the requirements of industrial furnaces is to achieve certain metallurgical, mechanical, or physical properties in the part such as hardness or a hardness profile with minimal energy consumption and by a method that is exactly reproducible from cycle to cycle. Microprocessor-controlled programming systems have become indispensable for satisfying this requirement.

A temperature sensor attached to the workpiece is an additional safeguard against unacceptable process deviations between the workpiece and the furnace temperature. The fundamental advantage of a process control system with a load thermocouple is that the user, when preparing the cycle recipe, does not need to consider any temperature–time lags between furnace and load and need only to input the duration of soaking times for the load at the various stages. Also, the varying mass of different loads is rendered unimportant even if it significantly influences the actual heating up rate.

Thermocouple breakage can also be detected by computer program control followed by an automatic thermocouple changeover so that the continuity of the automatic cycle is not interrupted. Hence, microprocessor-controlled programming systems make it possible to prevent unacceptable temperature differences between the furnace and the workpieces and also within the workpiece and related areas during heat-up phase. This feature, coupled with uniform gas quenching, enables reduced distortion in the heat treatment process.

3.4.6 Thermocouples

The thermocouples usually used in vacuum applications are manufactured from:

- Nickel–nickel and chromium (type K) used up to 2150°F (1175°C)
- Nickel–chromium–silicon and nickel–chromium (type N) used up to 2280°F (1250°C)
- Nickel–nickel and molybdenum used up to 2450°F (1350°C)
- Platinum–platinum and rhodium (type S) used up to 2900°F (1600°C)
- Tungsten–tungsten and rhenium (type W) used up to 4900°F (2700°C)

At higher temperatures the platinum–platinum and rhodium can only be used in the form of a compacted ceramic thermocouple because rhodium evaporates at these temperatures. These elements are also sensitive to the adsorption of contaminants, resulting in deviations.

A growing number of temperature control systems, either stand-alone or integrated into a combination equipment and process control scheme are used today to achieve precise and uniform heating of the workload.

3.4.7 Cooling System

The following media (listed in order of increasing intensity of heat transfer) are used for the cooling of components in vacuum furnaces:

- Vacuum
- Subatmospheric cooling with static or agitated inert gas (typically Ar or N_2)
- Pressurization (up to 20 bar or more) cooling with a highly agitated, recirculated gas (Ar, N_2, He, H_2, or mixtures of these gases)
- Oil, still or agitated

After heating in vacuum, the bright surface of the components must be maintained during the cooling. Today, sufficiently clean gases are available for cooling in gas. Permissible levels of impurities amount to approximately 2 ppm of oxygen and 5–10 ppm of water by volume. Normally nitrogen is used as a cooling medium because it is inexpensive and relatively safe.

With multichamber furnaces such as a vacuum furnace with an integral oil quench, an additional cooling medium, namely oil, is also available. These oils are specially formulated (evaporation-resistant) for vacuum operation.

3.4.8 WORKLOAD SUPPORT

Materials such as stainless steel, nickel-based alloys (such as Inconel), molybdenum, and graphite are very often used for baskets, trays, and fixtures. Under vacuum conditions and at high temperatures, these materials might react with the workload; for example, graphite and stainless steel form a eutectic that melts at 2057°F (1125°C).

Table 3.7 shows the maximum permissible temperature for selected materials in mutual contact under vacuum. These materials must be separated, typically by using ceramic materials, if the working temperature in the furnace exceeds these maximum permissible temperatures.

3.4.9 TYPES OF VACUUM FURNACES

Vacuum furnaces can be classified, according to the mode of loading, into horizontal and vertical furnaces. A large number of configurations exist that are described in detail in the literature [5]. Here the description of these furnaces is restricted to some of the latest developments in cold-wall furnaces.

3.4.9.1 Horizontal Batch Furnaces

Figure 3.6a and Figure 3.6b show typical horizontal vacuum furnaces having internal and external cooling systems.

As vacuum furnaces have evoked a great deal of importance emphasis was originally placed on increasing the speed of cooling and this is achieved by the use of highly reliable high-gas pressure quench systems. Today, vacuum furnaces with gas-quenching capabilities of up to 20 bar gas pressure and more exist. With high-pressure gas cooling, the influencing factors are well known, enabling the cooling rate to be controlled.

TABLE 3.7
Maximum Temperature (°F) of Materials in Mutual Contact under Vacuum

	W	Mo	Al_2O_3	BeO	MgO	SiO_2	ThO_2	ZrO_2	Ta	Ti	Ni[a]	Fe[a]	C
W	4600	3500	3310	3200	2500	2500	4000	2900			2300	2200	2700
Mo		3500	3310	3200	2500	2500	3450	3450	3500	2300	2300	2200	2700
Al_2O_3			3310						3310				
BeO				3200	2500		3200	3200	2900				3200
MgO					2500		2500	2500	2500				2500
SiO_2						2500							2500
ThO_2							3600	4000	3450				3600
ZrO_2								3700	2900				2900
Ta									4250	2300	2300	2200	3500
Ti										2300	1700	1900	2300
Ni											2300	2200	2300
Fe												2200	2000
C													4000

[a] Also for Ni, Fe, Cr alloys.

(a)

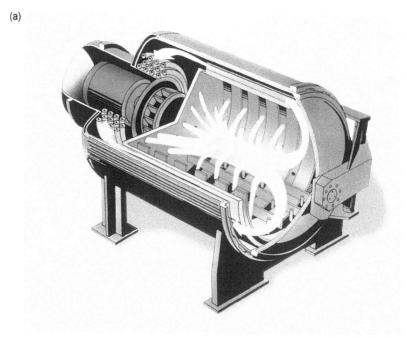

(b)

FIGURE 3.6 Horizontal vacuum furnace with (a) internal and (b) external gas cooling system. (From Segerberg, S. and Truell, E., High pressure gas quenching using a cold chamber to increase cooling capacity, *HTM*, 1997, pp. 21–24.)

However, it has become equally if not more important to recognize that a balance is required between achieving the required hardness by employing a sufficiently high rate of cooling, and minimizing the distortion of the quenched component. This means that high pressure is not always the answer. In the first place, distortion is not strictly a function of the speed of cooling itself, but by the uniformity of cooling. The uniformity of cooling in every section of the components depends on the design of the furnace, i.e., the design of the cooling system and the way the components are positioned in the charge. Figure 3.7 shows several cooling systems in use with today's vacuum-hardening furnaces.

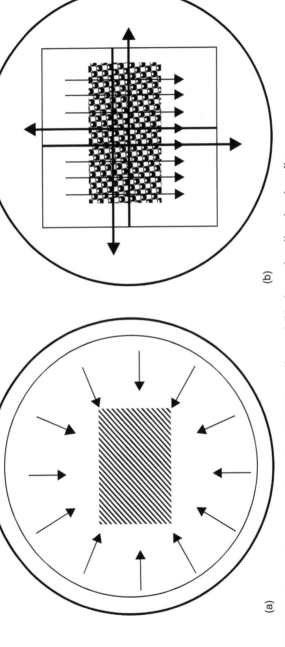

FIGURE 3.7 Gas cooling system alternatives (a) 360° cooling and (b) alternating directional cooling.

The all-around jet system (Figure 3.7a) directs streams of gas onto the workload from all sides. With large, heavy workpieces, this leads to uniform cooling. However, with a dense load of many small components, such a jet system creates a zone of less movement of the cooling gas in the center of the load, consequently with a reduced cooling rate in this area.

Cooling systems with straight-through gas flow (Figure 3.7b) realized by the opening of two opposite dampers in the heating chamber provide a good cooling rate in the center of a load of many small components. The disadvantage is the distortion of large components, which receive the flow from one side only. However, by alternating the direction of flow (e.g., 10 s from above, 10 s from below) and oscillating the jet stream horizontally, this disadvantage is reduced to a minimum. Similarly, the ability to use horizontal gas flow, in certain workload configurations, dramatically improves uniformity.

Figure 3.8 shows a schematic of a horizontal vacuum furnace with a dual dynamic gas cooling system. This design also illustrates the aforementioned convection-assisted heating system designed to improve the heating rate and temperature uniformity in the lower temperature ranges up to 1550°F (850°C).

Another variation is the plasma or ion furnaces. The basic differences, as shown in Figure 3.9, are the electrical isolation of the load from the furnace vessel via load support isolators; the plasma current feed-through; the high-voltage generator, which creates the plasma; and the gas dosage and distribution system. Plasma furnaces also utilize conventional vacuum furnace chamber and pumping systems.

Plasma furnaces exist in all variations, horizontal in single or multiple chamber configurations, as well as vertical designs such as bell furnaces and bottom loaders. Depending on the specific application, they are either low-temperature furnaces up to 1400°F (750°C) for plasma (ion) nitriding or high-temperature furnaces up to 2400°F (1100°C) for plasma (ion) carburizing. Low-temperature furnaces for plasma nitriding are constructed as cold-wall or hot-wall furnaces. High-temperature furnaces are usually cold-wall furnaces with water-cooled double walls. They can be equipped either with a high-pressure gas-quench system or an integrated oil quench tank.

The generator needed to create a plasma glow discharge inside a plasma furnace has to be a high-voltage DC generator (up to 1000 V). Currently there are two types of generators in use; one type has continuous-current outputs and the other has pulsed-current outputs.

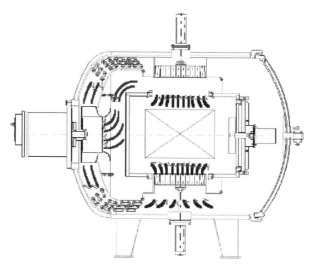

FIGURE 3.8 Horizontal vacuum furnace with directional vertical gas cooling sytstem and convection heating fan.

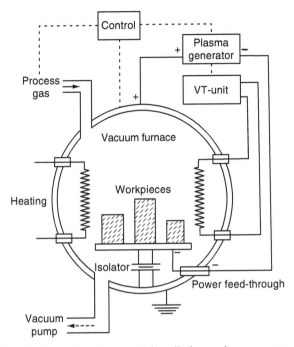

FIGURE 3.9 Schematic of a plasma heat treatment installation and components.

3.4.9.2 Horizontal Multichamber Furnaces

To increase the throughput, multichamber furnaces can be used. Figure 3.10 shows a three-chamber furnace. The heating chamber module is in the middle and is separated from the loading chamber and gas–oil quench chamber by vacuum-sealed doors. The heating chamber remains constantly evacuated and normally maintained at operating temperature.

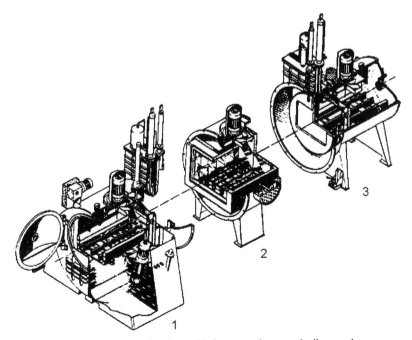

FIGURE 3.10 Three-chamber vacuum furnace with integrated gas and oil quench.

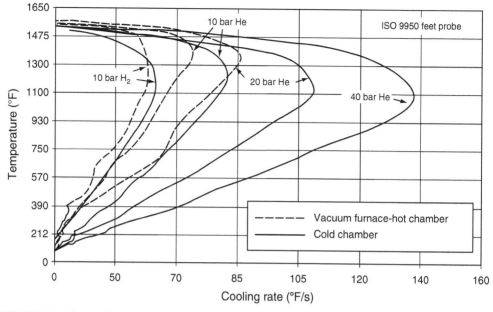

FIGURE 3.11 Comparison of cooling rate in hot- and cold-chamber designs.

Furnaces like this can also manage to successfully quench components with low hardenability or whose cross-sectional thickness is such that they cannot be fully hardened in single-chamber vacuum furnaces by gas quenching. Oil or gas can be chosen as a cooling medium. In this way, various furnace modules can be combined to achieve greater versatility in vacuum heat processing for a wide variety of applications. Figure 3.11 [6] shows data that compare results achieved in a hot chamber (single-chamber batch furnace) and cold chamber (multichamber) quench arrangement.

3.4.9.3 Vertical Furnaces

The previously described high-pressure gas-quench technology of horizontal furnaces has also been applied to vertical vacuum furnaces. Figure 3.12 shows schematically a vertical vacuum furnace for the hardening of long and large tools (e.g., broaches) with weights up to 2200 lb (1000 kg) and lengths up to 10 ft (3 m). Again, these furnaces are equipped with a convection-assisted heating system to reduce heating time and increase temperature uniformity during heating, thereby minimizing distortion.

3.5 HEAT TREATMENT PROCESSES

3.5.1 ANNEALING

Annealing treatments are undertaken primarily to soften a material, to relieve internal stresses, and to modify the grain structure. These operations are carried out by heating to the required temperature and soaking at this temperature for sufficient time to allow the required changes to take place, usually followed by a slow cooling at a predetermined rate. The choice of vacuum annealing is primarily influenced by the cleanliness and high quality of

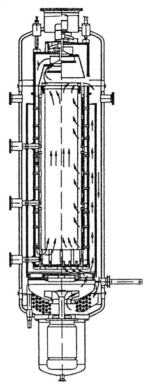

FIGURE 3.12 Cross section of a vertical vacuum-hardening furnace with convection heating and over-pressure gas quenching.

surface finish that can be obtained relatively easily compared to controlled atmosphere heat-treatment operations.

3.5.1.1 Stainless Steel

Although many grades of stainless steel have been successfully annealed in low-dew-point hydrogen atmospheres for many years, there is now a considerable quantity of all grades of stainless steel annealed in vacuum furnaces. Processing of stainless steel components in vacuum furnaces is often specified not only because of the cleanliness of the finished product, but also because the fast inert gas-quench capability supports a high production rate. Austenitic and ferritic stainless steels are usually gas quenched in nitrogen for general commercial applications. However, austenitic steel grades stabilized with titanium and niobium are argon quenched, particularly for nuclear energy and aerospace applications so as to avoid nitrogen pickup in the surface of the material, which degrades the corrosion resistance.

Some chromium evaporation can take place during the annealing of stainless steels, but normally the amount lost is not significant because of the short time at heat and the slow diffusion rates of chromium in steel. The annealing parameters for a range of stainless steels are presented in Table 3.8.

3.5.1.2 Carbon and Low-Alloy Steels

Increasing use is made of vacuum annealing of carbon and low-alloy steels where it is economically justifiable because of the cleanliness of the products and the prevention of carburization or decarburization of the part surface.

TABLE 3.8
Annealing Parameters for Stainless Steels

Type	Typical Analysis (%)	Annealing Temperature Range (°F)	Vacuum Range (torr)
Ferritic	12–27% Cr, 0.08% C maximum	1150–1600	10^{-3}–10^{-4}
Martensitic	12–14% Cr, 0.2% C and 16–18% Cr, 0.9% C	1525–1650	10^{-3}–10^{-4}
Austenitic unstabilized	18% Cr, 8% Ni	1850–2050	10^{-3}–10^{-4}
Stabilized	18% Cr, 8% Ni, 1% Nb or Ti	1750–2050	10^{-5}–10^{-6}

3.5.1.3 Tool Steels

The vacuum environment that surrounds a part typically does not react with the materials in process, so it is possible to anneal tools that have already been hardened, modify their design to meet required changes, and reharden them in vacuum. This is impractical with other types of furnace equipment, as all working surfaces of the tools could be affected to the extent that they might have to be reground, thus losing the dimensional precision required.

3.5.2 Hardening by Oil Quenching

Liquid quenching in vacuum furnaces is usually done in an integral oil quench design (Figure 3.13). Occasionally parts such as large dies are transferred, through air, into a salt quench tank sitting next to a vacuum furnace. The economics of high-pressure gas quenching is a stumbling block yet to be overcome. Changes in material chemistry as well as component design and fixturing will help to achieve this goal.

The design of an oil quench tank in a vacuum furnace is very similar to its atmosphere counterpart. There are oil circulation agitators or pumps on one or both sides of the tank and

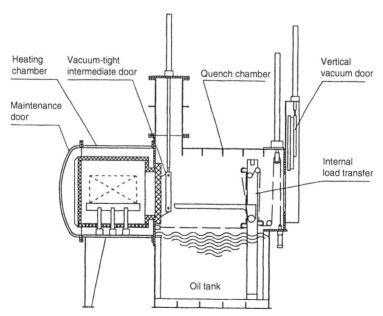

FIGURE 3.13 Schematic of a horizontal chamber plasma carburizer with integrated oil quench.

baffles to guide the respective oil flow below the load (Figure 3.14). The oil is heated and its temperature is controlled. It is cooled via an external oil cooler, usually employing air as cooling medium (for safety reasons).

The peculiarity of quenching in vacuum furnaces is the low pressure above the oil, which causes standard quench oils to degas violently. The duration of this degassing process depends on the amount of air or nitrogen absorbed by the oil during the loading and unloading of the furnace. Vacuum oils have been created to minimize these problems. Oils that are not degassed properly have a worse quenching severity and produce discolored components. Vacuum quench oils are distilled and fractionated to a higher purity than normal oils, which is important in producing the better surface appearance of quenched parts. In practice, the quenching in vacuum furnaces is frequently done with a partial pressure of nitrogen above the oil. Pressures between 35 torr (50 mbar) and 150 torr (200 mbar) are very common. This pressure increase just before initiating the quench serves mainly to reduce the evaporation of the oil. It is well known, however, that such a pressure increase also changes the oil cooling characteristics of the quench oils. As Figure 3.15 [7] demonstrates, the pressure increase shortens the vapor blanket phase, thus increasing the quench severity at high temperatures (in the pearlite–ferrite transformation). On the other hand, it lowers the quench rate in the convective cooling phase, i.e., in the lower temperature range of the bainite or martensite transformation. Thus, high partial pressures above the oil can be advantageous in producing full hardness on unalloyed or very low alloy materials, whereas low nitrogen pressures above the oil produce higher hardness and lower distortions on components of alloyed steels.

Very low pressures above the oil <35 torr (<50 mbar) and very high quenching temperatures in the area of 2200°F (1200°C) can lead to carbon deposition and pick-up on the

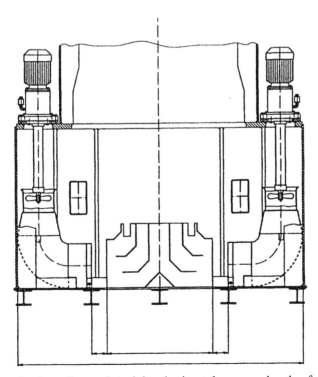

FIGURE 3.14 Schematic of an oil quench tank in a horizontal vacuum chamber furnace.

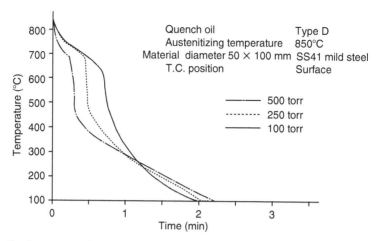

FIGURE 3.15 Cooling curves of a quench oil with three different nitrogen pressures (100, 250, 500 torr) above the oil. (From U.S. Patent 5,702,540, 1997.)

surface of the workpieces. This is due to the thermal decomposition of the quench oil as has been experienced in hardening high-speed steels [8]. The carbon originates from the oil vapor in contact with the hot surface of the load in the initial phase of the quench process. High nitrogen pressures >150 torr (>200 mbar) tend to reduce or eliminate this effect.

3.5.3 HARDENING BY GAS QUENCHING

The emphasis today is on finding alternatives to the use of oil quenching by substituting for high-pressure gas quenching. The capabilities of vacuum furnaces are continually examined to improve their productivity by reducing cycle time and to improve the metallurgical applications by improving the quenching capability. In recent years, great attention has been paid to obtaining more rapid gas quenches [9].

3.5.3.1 Cooling Properties of Gases

With the advent of rapid gas-quenching systems in vacuum furnaces, it was recognized that the flow rate and density of the cooling gas blown onto the surface of the load were the decisive factors for reaching high heat transfer, i.e., high cooling rates. In addition to high-gas velocities, high-gas pressures are needed to through-harden a wide variety of steel parts with appreciable dimensions. The first vacuum-quenching furnace to operate at a pressure of 2 bar was developed in 1975, and furnace pressures have been increasing ever since. Today, commercial designs up to 20 bar are available.

Calculations of heat transfer show that the heat transfer coefficient alpha (α) is proportional to the product of gas velocity and gas pressure as given in the equation below:

$$\alpha \approx (vP)^n \tag{3.1}$$

where v is the gas velocity and P the pressure of the gas. The exponent n depends on the furnace design, the load, and the properties of gas. It lies typically in the range of 0.6–0.8. The exponential behavior of the heat transfer makes it clear that the difference in the increase of heat transfer is considerable with the first few bars of pressure but decreases with increasing pressure.

TABLE 3.9
Correlation of Material Requirements and Cooling Rates or α-Values

Steel	Core Hardness (HRC)	Cooling Rate 1475–930°F (°F/s)	λ	Heat Transfer Coefficient α Diameters (in.) 1.00	1.25	1.50
4118	≥30	72.0	0.13	1250	1800	>2000
4140	≥54	140.0	0.05	>2000	>2000	>2000
4140 Modified	≥54	7.0	0.80	350	460	600
4320	≥30	7.2	0.75	200	320	400
5120	≥30	27.0	0.20	800	1150	1500
D3	≥64	8.9	0.63	260	370	480

The critical transformation range for most steels is between 1475°F (800°C) and 930°F (500°C). The lambda (λ) value, a number that represents the time required to pass through this temperature range divided by 100 s, is a relative measure of the cooling rate achievable and is shown in Table 3.9 [10].

The increase of cooling speed with higher gas velocity (higher flow rate) is shown in Figure 3.16.

Figure 3.17 shows, for different types of cooling gas, that argon gives the slowest cooling, followed by nitrogen, helium, and hydrogen, in the order of increasing cooling rate.

Theoretically, there is no limit to the improvement in cooling rate that can be obtained by increasing gas velocity and pressure. Practically, however, very high pressure and very high velocity systems are difficult and costly to construct. In particular, the power required for gas

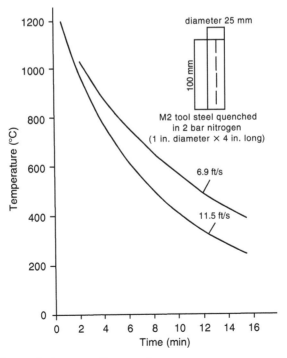

FIGURE 3.16 Effect of gas velocity on cooling speed.

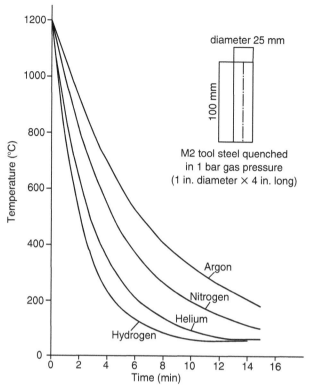

FIGURE 3.17 Effect of type of gas on cooling speed during gas quenching.

recirculation increases faster than benefits accrue. On the other hand, there are pressure–gas combinations that provide heat transfer coefficients within the range of those produced by agitated oil quenchants (Table 3.10). Naturally, the power requirements for gas circulation

TABLE 3.10
Effects of Quenching Parameters on Heat Transfer Coefficient

Medium and Quenching Parameters	Heat Transfer Coefficient [W/(m² K)]
Air, no forced flow	50–80
Nitrogen, 6 bar, fast[a]	300–400
Nitrogen, 10 bar, fast[a]	400–500
Helium, 6 bar, fast[a]	400–500
Helium, 10 bar, fast[a]	550–650
Helium, 20 bar, fast[a]	900–1000
Hydrogen, 6 bar, fast[a]	450–600
Hydrogen, 10 bar, fast	750[b]
Hydrogen, 20 bar, fast	1300[b]
Hydrogen, 40 bar, fast	2200[b]
Oil, 70–175°F, no flow	1000–1500
Oil, 70–175°F, agitated	1800–2200
Water, 60–80°F, agitated	3000–3500

[a]Fan speed of 3000 rpm.
[b]Calculated values.

TABLE 3.11
Guidelines for α-Values of Different Gas-Quenching Systems

Furnace	Loading	6 bar N_2	10 bar N_2	10 bar He
Single-chamber furnace[a]	Maximum	300–400	450–600	600–700
	Half	350–450	500–650	700–800
Cold chamber: high-gas volume flow	Maximum	600–750	750–900	1000–1200
	Half	600–750	750–900	1000–1200
Cold chamber: nozzle field	Single layer	Approx. 2000	—	—

[a]*Note*: Above data apply to a furnace having a workload size of 24 in. wide × 24 in. high × 36 in. long having a 1200 lb gross load capacity.

decrease dramatically when less dense gases such as helium or hydrogen are used. Thus, gas quenching can be used to produce full hardness in many oil-hardening steels.

Gas quenching has certain advantages over liquid (oil and salt bath) quenching. Because the cooling rate can be easily changed by altering gas velocity or pressure, the same vacuum furnace and gaseous cooling medium can be used to quench a wide variety of materials. A furnace using nitrogen at 6 bar, for example, can quench a variety of tool steels, stainless steels, and high-alloy steels in production-sized load configurations. With an increase in pressure and a change of cooling gas to one less dense, medium, and even low-alloy steels can be properly transformed as shown in Table 3.11.

The effect of load weight on the resultant cooling speed during gas quenching is more pronounced than, for example, in liquid or salt bath quenching. Figure 3.18 shows the cooling speed for similar load weights obtained with similar part dimensions during gas quenching with 6 bar nitrogen. This type of data will vary from furnace to furnace depending on design and loading capacity. However, the data provide an indication of the magnitude of the parameters involved. The determination of such data for any particular furnace is of considerable use to a heat treater who needs to obtain the maximum performance from a furnace. The heat treater should also consider the largest section size that can be through-hardened and compare this to the appropriate continuous cooling diagram for the material to be processed, with respect to the data obtained above.

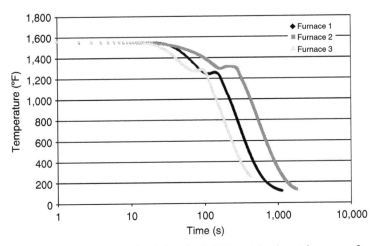

FIGURE 3.18 Cooling curves of similar load sizes in 3 diffferent horizontal vacuum furnaces.

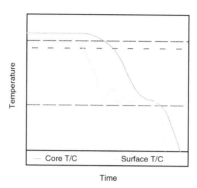

FIGURE 3.19 Gas cooling with isothermal hold.

It is also possible in gas quenching to change the cooling rate during a single cycle, whereas multiple baths are required in liquid-quenching systems. Figure 3.19 shows the controlled cooling with an isothermal hold in vacuum furnaces. The isothermal hold requires two load thermocouples. The cooling speed of the workpiece is controlled by variation in gas speed and pressure coupled with reheating of the furnace if necessary. This ability to perform interrupted quenching is particularly important in applications where parts can distort or crack if cooling occurs too rapidly over the entire quenching range.

3.5.3.2 Tool Steels

The advantages of vacuum furnaces for the heat treatment of tool steels are well known today. Vacuum furnaces provide nonreactive conditions for the materials under treatment that ensure development of the optimum properties desired. Parts processed in this way have surfaces that are neither carburized nor decarburized and consequently exhibit superior performance. Such a condition is practically impossible to obtain when using protective atmospheres generated by cracking some types of hydrocarbons, because their carbon potential cannot be kept in balance with the carbon content of the material under processing over the total range of temperatures experienced in the heat treatment cycle. Even dry hydrogen atmospheres present problems in maintaining the dew points necessary to prevent some degree of decarburization in the furnace.

The bright hardening of tool steels from the air-hardening category has proven in many situations to be economically viable relative to salt bath treatments and is the most important application of gas-quench hardening in vacuum furnaces. The ability to minimize shape distortion and in many situations to virtually eliminate finishing costs on complex shapes through the use of vacuum furnaces is particularly advantageous.

Table 3.12 presents data on tool steels that have been successfully treated in vacuum for many years now. In general, the air-hardening tool steel parts are hardened in much the same way as in atmosphere furnaces. They are preheated, heated to a high austenitizing temperature, and cooled at a moderate rate. The medium-alloy air-hardening steels in the A series and the high-carbon, high-chromium steels in the D series are regularly hardened in gas-quench furnaces (nitrogen up to 6 bar).

Today many grades of hot-work tool steels such as H13 are high-pressure gas quenched. Specifications that once called for cooling rates of 50°F/min are revised upward since vacuum furnace designs today can achieve cooling rates of greater than 100°F/min. Oil quench rates of 300°F/min can be reached but distortion issues negate any positive benefits achieved.

Also tungsten and molybdenum high-speed steels can be hardened by a high-pressure nitrogen quenching without any loss of essential properties in comparison to salt bath heat

TABLE 3.12
Types of Tool Steels Successfully Treated in Vacuum
Furnaces with Gas Quenching

A series	Air-hardening medium-alloy cold-work tool steels
D series	High carbon, high chromium tool steels
H series	Hot work tool steels (Cr, Mo, W)
T series	Tungsten high speed steels
M series	Molybdenum high speed steels.

treatment. Figure 3.20 shows no difference with relation to carbide precipitation on grain boundaries between milling cutters (M2 steel), which were hardened in salt bath and in vacuum with high-gas pressure.

It should be noted that only a rough vacuum of 10^{-1} to 1 torr (1.3×10^{-1} to 1.3 mbar) should be used during the heat treatment of tool steels. This level of vacuum is required mainly because of the relatively high vapor pressures of chromium, manganese, and other easily vaporized elements in the tool steel. The values of the vapor pressures of these alloying elements in pure form can be seen as a function of temperature in Table 3.4. The vapor pressures of the elements in solid solution should be somewhat lower than these values, but the actual values will depend on the alloy concentration in the steel.

3.5.3.3 Martensitic Stainless Steels

The martensitic stainless steels can be adequately hardened in vacuum furnaces. Vacuum furnace processing has eliminated many heat treatment variables, particularly with atmosphere control, which previously often resulted in high rejection rates with these heat treatment-sensitive materials.

All grades of martensitic stainless steels have been processed in vacuum furnaces using the same austenitizing temperatures and considerations as those used in atmosphere furnaces. Since the austenitizing temperatures are usually below 2000°F (1100°C), vacuum levels in the range of 10^{-3} torr (1.3×10^{-3} mbar) are very often used, which result in clean and bright parts upon unloading from the vacuum furnace. To avoid evaporation of alloying

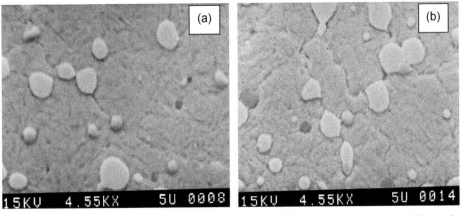

FIGURE 3.20 Electron microscope image (magnification 4550×) of milling cutters (M2 steel) after (a) salt bath hardening and (b) vacuum hardening with high-pressure gas quenching.

TABLE 3.13

Gas Quenching of Some Oil-Hardening Steels in Vacuum Furnaces with 6 Bar Nitrogen

SAE/AISI	Hardness HRC		
	Maximum Value Attained	20 mm Diameter	40 mm Diameter
4130	50	28	
4135	55	39	38
300M	59	59	59
4340	58	57	56
431	51	51	51

elements, processing is also done at vacuum levels ranging from 10^{-3} to 10^{-1} torr (1.3×10^{-3} to 1.3×10^{-1} mbar), but the lower levels resulting in cleaner and brighter workpieces. Due to the differences in the hardenability of the various martensitic stainless alloys, there is a limitation on the section sizes that can be fully hardened by recirculated nitrogen gas quenching. Other types of cooling gases can be used but the economic benefits must be carefully considered. The actual values of section size limits depend on the type of cooling system and the capability of the specific furnace employed.

3.5.3.4   Heat-Treatable Construction Steels

Parts for the automotive and aerospace industry are very often made of heat-treatable steels that are normally oil hardened. Table 3.13 shows that at least some of these steels can also be successfully hardened with gas pressure quenching. Further applications of gas quenching involve an increase in quenching speed obtained by using a higher gas pressure (up to 20 bar) and gases other than nitrogen (e.g., helium). This enables some low-to-medium alloy steels and case-hardening steels to be hardened.

3.5.4   CASE HARDENING BY CARBURIZING AND CARBONITRIDING

Over 180 chemical equations are necessary to describe the reactions occurring during atmosphere carburizing. One of the most important is the water gas reaction, which can be written as

$$CO + H_2O \leftrightarrow CO_2 + H_2 \tag{3.2}$$

Control of the atmosphere carburizing process is done by looking at the CO/CO_2 and H_2O/H_2 ratios of this equation using instruments such as dew point analyzers, infrared analyzers, and oxygen (carbon) probes. In atmospheres containing CO and H_2, the carbon transfer is dominated by the CO adsorption reaction

$$CO \rightarrow CO_{ad} \rightarrow [C] + O_{ad} \tag{3.3}$$

and the oxygen desorption reaction

$$O_{ad} + H_2 \rightarrow H_2O \tag{3.4}$$

that together yield an alternate form of the water gas reaction, namely

$$CO + H_2 \leftrightarrow [C] + H_2O \tag{3.5}$$

Thus the transfer of carbon in atmospheres containing CO and H_2 is connected with a transfer of oxygen, giving rise to an oxidation effect in steel with alloying elements such as silicon, chromium, and manganese. This phenomenon is known as internal or intergranular oxidation of steel.

A popular alternative carburizing method used today is low-pressure vacuum carburizing using various hydrocarbon gases as the source of carbon to be absorbed by the steel. In this case the carburizing gas contains no oxidation-producing constituents resulting in surfaces free of internal oxidation (Figure 3.21a and Figure 3.21b).

Low-pressure carburizing can be defined as carburizing under 7.5 torr (10 mbar) typically at temperatures between 1475°F (790°C) and 1900°F (1040°C).

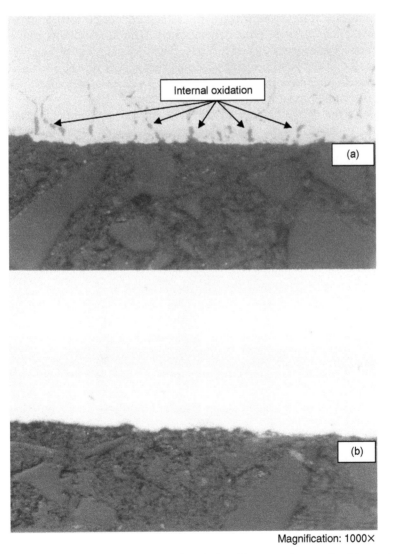

Magnification: 1000×

FIGURE 3.21 Internal oxidation of steel occurring during (a) atmosphere carburizing but not during (b) low-pressure carburizing.

Hydrocarbon gases currently used for vacuum carburizing are methane (CH_4), ethylene (C_2H_4), propane (C_3H_8), and acetylene (C_2H_2). At normal atmospheric pressures they deliver carbon via reactions such as

$$CH_4 \rightarrow [C] + 2H_2 \tag{3.6}$$

$$C_3H_8 \rightarrow 3[C] + 4H_2 \tag{3.7}$$

These reactions come into equilibrium only at very high hydrogen levels. Thus undiluted hydrocarbons will produce excess carbon and will therefore quickly soot up the furnaces. To control their carbon availability, one either dilutes the hydrocarbons with nitrogen or hydrogen, thus reducing their partial pressure, or lowers the total pressure in the furnace. The latter technique is known as vacuum carburizing.

Production of vacuum carburizing began in the 1970s and over the years has changed from high-pressure vacuum carburizing, done in the range of 50 torr (65 mbar) to 600 torr (800 mbar) to the low-pressure process done today. The need for this change was due to excessive sooting of the furnaces and the associated low uptime productivity and higher maintenance costs. In the past, propane was the primary hydrocarbon gas used for vacuum carburizing. Developmental work done in the last few years has demonstrated that the best performing gas for low-pressure carburizing is acetylene.

This is due to the vast difference in the chemistry of acetylene and propane. Propane dissociation can occur before the gas has come in contact with the surface of the steel to be carburized producing free carbon, or soot. This uncontrolled soot formation results in poor carbon transfer to the part.

The idea of injecting propane into the vacuum furnace at such a high rate, i.e., high-gas velocity, that the thermal dissociation of propane into methane and ethylene and further into methane, carbon, and hydrogen,

$$C_3H_8 \rightarrow CH_4 + C_2H_4 \rightarrow CH_4 + 2C + 2H_2 \tag{3.8}$$

would not occur before the propane gas reached the workpieces of the load, does not work for large and dense loads. The limitation in load size and density is necessary for the propane gas to reach the central part of the load in sufficient time and quantity. Soot formation, however, still occurs.

Hydrocarbon gas dissociation is given in Table 3.14 [11]. An example of the efficiency of acetylene low-pressure vacuum carburizing can be seen in Figure 3.22a, which illustrates the limited penetrating power of propane or ethylene into holes and gaps.

TABLE 3.14
Dissociation of Hydrogen Gases Used in Low-Pressure Carburizing[a]

Gas	Symbol	Dissociation Equations
Acetylene	C_2H_2	$C_2H_2 \rightarrow 2C + H_2$
Ethylene	C_2H_4	$C_2H_4 \rightarrow C + CH_4$
Methane	CH_4	$CH_4 \rightarrow CH_4$ (no reaction)
Propane	C_3H_8	$C_3H_8 \rightarrow C + 2CH_4$
		$C_3H_8 \rightarrow C_2H_4 + CH_4 \rightarrow C + 2CH_4$
		$C_3H_8 \rightarrow C_2H_2 + H_2 + CH_4 \rightarrow C + 2CH_4$

[a] $T = 1650°F$ (900°C)–1900°F (1040°C).
$P \leq 15$ torr (≤ 20 mbar).

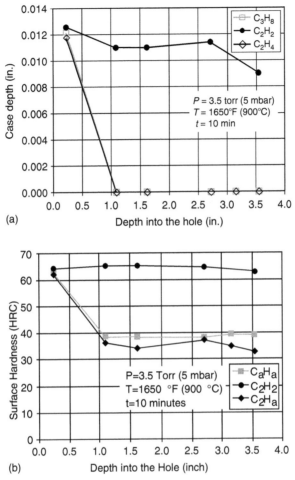

(a)

(b)

FIGURE 3.22 Carburizing capacity of propane (C$_3$H$_8$), ethylene (C$_2$H$_4$), and acetylene (C$_2$H$_2$) after carburizing (10 min/1650°F/3.5 torr) a blind hole of 0.12 in. diameter and 3.6 in. length.

The control of the low-pressure vacuum carburizing process can be realized on a time basis to control the carbon transfer and the diffusion phases. The carbon transfer rates are usually found empirically as a function of temperature, gas pressure, and flow rate and applied to the respective kinetic and diffusion models to determine the boost and diffuse times of the cycle. Simulation programs have been created for this purpose.

Ammonia can be added in addition to the selected hydrocarbon gas or in the diffusion phases in a process known as carbonitriding. Plasma (ion) carburizing is also a low-pressure technique that can be used. The main advantage is the capability of using simple mechanical masking techniques for selective carburizing, while at the same time applying methane (or propane) as carburizing gas.

3.5.5 CASE HARDENING BY NITRIDING

In vacuum, plasma (ion) nitriding is a widely utilized process. Improvements to range of application and economics of operation of the process have been achieved by the use of pulsed power generators and insulated furnace workload designs.

Plasma nitriding uses nitrogen gas at low pressures in the range of 0.75–7.5 torr (1–10 mbar) as the source for nitrogen transfer. Nitrogen is a neutral gas that does not react with steel surfaces at temperatures below 1832°F (1000°C), and becomes reactive when an electric field of a few hundred volts (300–1200 V) is applied. The electric field is established in such a way that the workload is at the negative potential (cathode) and the furnace wall is at ground potential (anode). The nitrogen transfer is caused by the attraction of the positively charged nitrogen ions to the cathode (workpieces), with the ionization and excitation processes taking place in the glow discharge near the cathode's surface. The rate of nitrogen transfer can be adjusted by diluting the nitrogen gas with hydrogen. The addition of hydrogen to the gas mixture has to be above 75% in order for a reduction of the nitrogen transfer rate to take place. This influences the thickness of the compound layer (white layer) as Table 3.15 shows. The higher the nitrogen concentration, the thicker the compound layer.

The compound layer consists of iron and alloy nitrides that develop in the outer region of the diffusion layer after saturation with nitrogen. According to the iron–nitrogen phase diagram basically two iron nitrides are possible, the nitrogen-poor γ' phase (Fe_4N) and the nitrogen-rich ε phase ($Fe_{2-3}N$).

The temperature of the workpiece is another important control variable. The depth of the diffusion layer also depends strongly on the nitriding temperature, part uniformity, and time. For a given temperature the case depth increases with the square root of time.

A third process variable is the plasma power or current density, which has a certain influence on the thickness of the compound layer [12]. Adding small amounts (1–3%) of methane or carbon dioxide gas to the nitrogen–hydrogen gas mixture will produce a carbon-containing ε compound layer ($Fe_{2-3}C_XN_Y$). This process is called plasma nitrocarburizing. It is normally used only for unalloyed steels and cast irons.

The different possible layer structures and large number of process variables make it difficult to make the correct choice for each application. Table 3.16 gives some indications of which layer structure to choose for certain applications and steels [13]. Guidelines for case depths, expected surface hardness, white layer thickness, and applicable nitriding temperatures for some typical steels and cast irons are presented in Table 3.17 [13]. Table 3.18 presents typical treatment cycles and results for a variety of steels.

Plasma nitriding increases the wear resistance and sliding properties of components, as well as their load-bearing capacity, fatigue strength, and corrosion resistance. Corrosion

---

TABLE 3.15

Thickness and Structure of the Compound Layer after Plasma Nitriding of AISI 4140 Steel at 1050°F in N_2–H_2 Gas Mixtures with Varying N_2 Contents

Amount of Nitrogen in the N_2–H_2 Gas (vol%)	Thickness of Compound Layer (μm) after Plasma Nitriding at 1050°F for		Structure of Compound Layer
	2 h	5 h	
2	0	0	None
5	2.5	3	γ
25	6.5	8	γ
50	7	9	$\gamma + \varepsilon$
100	7	3.5	$\gamma + \varepsilon$

TABLE 3.16
Choice of Surface Layers for Different Steels and Applied Stresses

	Unalloyed Steels, Cast Irons, Annealed Alloy Steels	Alloy Steels (Hardened), Nitriding Steels, Hot-Work Steels	Stainless, Heat-Resistant, Precipitation Hardening	High-Chromium, High-Speed Tool Steels
Adhesive wear				
Low load	ε	γ	γ	γ
High load	N/A	γ	γ or diffusion	Diffusion
Abrasive wear				
Low load	ε	γ	diffusion	Diffusion
High load	N/A	γ	Diffusion	Diffusion
Pitting or impact				
Low load	N/A	Diffusion	Diffusion	Diffusion
High load	N/A	Diffusion	N/A	Diffusion

N/A, not applicable.

resistance can be especially enhanced by a plasma postoxidation treatment. Dimensional changes are minimal; the masking method is simple and effective.

The process is nonpolluting and creates excellent working conditions. It is used in many applications in the automotive industry, gearing, machinery (plastic extrusion, agricultural, food, etc.), chemical apparatus, and tooling (dies, molds, drills, punches, etc.). Also, the plasma nitriding of sintered iron materials is very effective. The nitriding effect is limited to the surface area, keeping the growth of the components very small.

TABLE 3.17
Typical Property Ranges for Plasma-Nitrided Materials

Steel	Nitriding Temperature Range (°F)	Surface Hardness (HRC)	Case Depth (in.)	Compound Layer Thickness (in. × 10⁻⁴)
Alloy				
1045	1025–1075	35–49	0.012–0.031	1.6–6.0
4150	925–1025	52.5–58	0.008–0.024	1.6–3.1
6150	925–1025	52.5–58	0.008–0.024	1.6–3.1
9310	975–1025	49–58	0.012–0.028	1.6–3.1
Nitralloy 135	950–1025	67 to >70	0.008–0.020	0.8–3.9
Tool steel				
H13	840–1060	67 to >70	0.004–0.012	0.8–2.4
A2	840–1025	64–63.5	0.004–0.012	0.8–2.7
D2	900–1025	67 to >70	0.004–0.008	0.0–1.2
M2	900–950	69 to >70	0.001–0.002	None
Stainless steel				
304	1025–1075	69 to >70	0.002–0.008	None
430	1025–1060	69 to >70	0.004–0.008	None

The Compound Layer Thickness uses $\times 10^{-4}$ units.

TABLE 3.18
Representative Treatment Cycles and Results for Plasma Nitriding

Steel	Temperature/ Time (°F/h)	Gas	Core Hardness (HRC)	Surface Hardness (HV)	Case Depth (in.)	Compound Layer (in. × 10^{-4})
Ductile iron	1000/6	N_2–H_2–CH_4	29	370	0.006	2.4–4.3 (ε)
1045	1040/6	N_2–H_2–CH_4	<20	350	—	3.1–3.9 (ε)
4140	1025/15	N_2–H_2	30	600	0.016	1.6–3.2 (Y)
9310	990/14	N_2–H_2	—	640	—	—
Nitralloy 135	950/10	N_2–H_2	30	>1100	0.013	1.2–2.0 (Y)
H13	990/10	N_2–H_2	48	970	0.009	0.8 (Y)
D2	1025/10	N_2–H_2	—	930	0.003	0.4–1.2 (Y)
M2	925/0.33	N_2–H_2	60	840	—	None
304	1025/16	N_2–H_2	<20	>1180	0.006	None

3.5.6 VACUUM BRAZING

The dollar volume of work produced by vacuum brazing probably far exceeds that of any other process in which vacuum furnaces are utilized. The development of heat-resisting alloys containing aluminum and titanium, which are difficult if not impossible to braze in very dry hydrogen gas atmospheres, promoted vacuum brazing as the first large commercial application of vacuum furnaces. The transportation industry has provided the impetus for the increasing use of vacuum furnaces for brazing, as the use of lightweight, high-strength materials such as titanium has expanded. It also has inspired the development of the multichamber vacuum furnaces for high-production, fluxless brazing of aluminum heat exchangers.

Vacuum furnaces have many advantages, some of which are as follows:

- Process permits brazing of complex, dense assemblies with blind passages that would be almost impossible to braze and adequately clean using atmospheric flux brazing techniques.
- Vacuum furnace processing at 10^{-4} to 10^{-5} torr (1.3×10^{-5} to 1.3×10^{-4} mbar) removes essentially all gases that could inhibit the flow of brazing alloy, prevents the development of tenacious oxide films, and promotes the wetting and flow of the braze alloy over the vacuum conditioned surfaces.
- Properly processed parts are unloaded in a clean and bright condition that often requires no additional processing.

Today a wide variety of steels, cast irons, stainless steel, aluminum, titanium, nickel alloys, and cobalt-base heat-resisting alloys are brazed successfully in vacuum furnaces without the use of any flux. In fact, flux would contaminate the vacuum furnace and degrade the work results.

Many different types of nickel-base, copper, copper-base, gold-base, palladium-base, aluminum-base, and some silver-base brazing alloys are used for the filler metal. Generally, alloys that contain easily vaporized elements for lowering the melting points are avoided. With respect to the heat treatment of steel, the copper and the nickel-base brazing alloys are the most widely used filler metals.

3.5.6.1 Brazing with Copper

Copper filler metal applied either as paste, foil, or clad copper to the base metal, can be vacuum brazed recognizing that the high vapor pressure of copper at its melting point causes some evaporation and undesirable contamination of the furnace internals. To prevent this action, the furnace is first evacuated to a low pressure 10^{-2} to 10^{-4} torr (1.3×10^{-2} to 1.3×10^{-4} mbar) to remove residual air. The temperature is then raised to approximately 1750°F (955°C) to allow outgassing and to remove any surface contamination. Finally, the furnace is heated, usually between 2000°F (1100°C) and 2050°F (1120°C) brazing temperature, with a partial pressure of up to 7.5×10^{-1} torr (1 mbar) to inhibit evaporation of copper. When brazing is completed, usually within minutes once the setpoint temperature has been reached, the work is allowed to slow cool to approximately 1800°F (980°C) so that the filler metal will solidify, then the parts can be rapidly cooled by gas quenching.

3.5.6.2 Brazing with Nickel-Base Alloys

Brazing with nickel-base alloys is usually done without any partial pressure at the vacuum levels attainable by the furnace 10^{-3} to 10^{-4} torr (1.3×10^{-3} to 1.3×10^{-4} mbar). Normally a preheat soak at 1700°F to 1800°F (920°C to 980°C) is used to ensure that large workloads are uniformly heated. After brazing, the furnace temperature can be lowered for additional solution or hardening heat treatments before gas cooling and unloading.

3.5.6.3 Brazing with Aluminum and Aluminum Alloys

In the brazing of aluminum components it is important that vacuum levels be maintained in the 10^{-5} torr (1.3×10^{-5} mbar) range or more. Parts are heated to 1070°F (575°C) to 1100°F (590°C) part temperature depending on the alloy. Temperature uniformity is critical and multiple zone temperature-controlled furnaces are common. Cycle times are dependent on furnace type, part configuration, and part fixturing. Longer cycles are required for large parts and very dense loads.

3.5.7 Sintering

Vacuum furnaces are used for an ever-expanding range of powder metal applications and materials. Vacuum sintering and secondary heat treatment operations are used on both conventional P/M materials as well as metal injection molded (MIM) components. This transition away from controlled atmosphere sintering arises not only from the increased volume of production of sintered parts, but also from factors such as:

- The purity of the vacuum environment and its effect on part microstructure
- The use of subatmospheric pressure to improve the efficiency of the sintering reactions especially with highly alloyed materials that require elevated sintering temperatures
- The ability of the vacuum process to reduce pore size and improve pore size distribution
- The higher furnace temperature capabilities that permit faster sintering reactions carried out much closer to the melting point and with alloys of higher melting point interstitial elements

The limitation on the application of sintering in vacuum furnaces is the vapor pressure of the metals processed at the sintering temperature. If the vapor pressure is comparable with the working pressure in the vacuum furnace, there will be considerable loss of metal by vaporization unless a sufficiently high partial pressure of a pure inert gas is used. In certain instances the partial pressure gas will react with the surface of the part creating a surface layer that may need to be removed.

3.5.7.1 Stainless Steel

Vacuum sintering of stainless steel powder metal parts is a common process, employed for the AISI 410, 420, 303, 316, and 17-4PH grades of stainless steel. These products are very often superior to those sintered in hydrogen or dissociated ammonia atmospheres with respect to their corrosion resistance and mechanical properties.

3.5.7.2 High-Speed Steel

Powder metallurgy manufacturing methods have been developed for producing finished and full-density cutting tools of high-speed tool steel. Applications include such items as complex geometry hobs, pipe taps, and reamers. Special isostatic compacting techniques have been developed that use neither lubricants nor binders for these types of components. The pressed compacts are sintered in vacuum furnaces under precise control of heating rate, sintering time, temperature, and vacuum pressure in order to eliminate porosity. The result is predictable densification of the pressed compact with final size tolerances of ± 0.5–1.0%. Full-density, sintered high-speed steel tools have been shown to be at least equivalent to conventional wrought material in cutting properties. Grindability is dramatically improved, in particular for the high-alloy grades such as M4 and T15. This is attributed to the finer and more uniform carbide distribution.

3.5.8 Tempering and Stress Relieving

Where surface finish is very critical and clean parts are desired to avoid any additional processing, heat treaters now employ vacuum furnaces for tempering and stress relief. At temperatures below which radiant energy is an efficient method for heating, the furnace is evacuated and backfilled with an inert gas such as dry nitrogen or argon to a pressure slightly above atmospheric but below 2 bar. A fan in the furnace recirculates this atmosphere, and parts are heated by both convection and conduction. This process is used when tempering high-speed steel components and a variety of other items made from many types of steels. Also, special furnaces have been designed to combine the tempering treatment with other hardening or case hardening treatments in the same equipment.

REFERENCES

1. Edenhofer, B. and Bouwman, J.W., *Vacuum Heat Treatment*, Ipsen International, Report 91e Kleve, GmbH.
2. Richardson, F.D. and Jeffes, J.H., The thermodynamics of substances of interest in iron and steel making 111, *J. Iron Steel Inst.*, *171*: 165–175 (1952).
3. Hoffmann, R., Oxidation events during heat treatments, *Härterei-Techn. Mitt.*, *39* (2): 61–70 (1984) (in German).
4. U.S. Patent 5,702,540, 1997.
5. ASM, Heat Treating, in *Metals Handbook*, 10th ed., Vol. 4, ASM International, Materials Park, Cleveland, OH, 1991, pp. 463–503.
6. Segerberg, S. and Truell, E., High pressure gas quenching using a cold chamber to increase cooling capacity, *HTM*, 1997, pp. 21–24.
7. Herring, D.H., Sugiyama, M., and Uchigaito M., Vacuum furnace oil quenching influence of oil surface pressure on steel hardness and distortion, *Industrial Heating Magazine*, June 1986, pp. 14–17.

8. Reynoldson, R.W. and Harris, K.C., The vacuum heat treatment of tool steels, *Metal Treating*, 1970, pp. 3–24.
9. Edenhofer, B. and Bouwman, J.W., Progress in design and use of vacuum furnaces with high pressure gas quenching systems, *Industrial Heating Magazine*, February 1988, pp. 12–16.
10. Lohrmann, M. and Gräfen, W., *Advanced Process and Furnace Technology for Case Hardening Using Low-Pressure Carburisation (LPC) with or without Plasma Assistance in Combination with Gas Quenching*, ASM International, Heat Treating Conference Proceedings, 2000.
11. Herring, D.H., *An Update on Low Pressure Carburizing Techniques and Experiences*, ASM International, Heat Treating Conference Proceedings, 2000.
12. Conybear, J.G. and Edenhofer, B., Progress in the control of plasma nitriding and carburizing for better layer consistency and reproducibility, in *Sixth International Congress on Heat Treatment of Materials*, Chicago, September 28–30, 1988, ASM International, Materials Park, OH.
13. Conybear, J.G., Guidelines for choosing the ion nitriding process, *Heat Treating Magazine*, *19* (3): 33–37 (1987).

4 Induction Heat Treatment: Basic Principles, Computation, Coil Construction, and Design Considerations

Valery I. Rudnev, Raymond L. Cook, Don L. Loveless, and Micah R. Black

CONTENTS

4.1 INTRODUCTION

As modern technology has changed from the abacus to the computer, a similar change has occurred in the field of induction heat treatment (IHT). In modern industry the requirements for the induction heating process have become quite strict. Some of these requirements could not possibly have been satisfied 15 or even 10 years ago. In order to provide a successful design for modern induction heating it is now necessary to take into account many details of the process. Many years ago a basic knowledge of electromagnetic fields, a calculator, and an engineering intuition were all that were available to create an induction heating system. Now they are not enough. The designer of induction heating systems today must have special software tools that allow effective simulation of the process. The designer must also be aware of new developments in theory and practice in the area of IHT.

Several useful books and lessons have been published on the basic principles of induction heating design [1–12]. This material is very useful for the first-time designer with limited experience in induction heating. A significant portion of the reference material describes various aspects of induction heating, such as mass heating for forging, extrusion, and rolling. IHT has many features that make it a unique process from the standpoint of coil design as well as computational methods, process operation, and equipment maintenance.

Traditional methods of calculation for the induction heating process (e.g., Baker's method) were based on the equations for an infinitely long coil and workpiece. Unfortunately, this assumption is not always valid in induction heat treating, where induction coils typically have no more than a few turns and cannot be considered infinitely long.

During the last decade a considerable amount of experience has been accumulated on the computation of induction heating and heat treatment problems by using numerical techniques. Unfortunately, the descriptions of particular methods are contained in a variety of internal reports, scientific journals, or special literature concerned with numerical methods. They are usually presented in a form almost inaccessible to engineers who have limited experience in numerical analysis. Also, the textbooks on numerical analysis usually emphasize the mathematical methods. They do not get into the details of the physical aspects of the problem that are often crucial to the success of a simulation. It is not our aim to describe all of the available numerical methods. However, in order to make the right choice, one should have some knowledge of the possibilities as well as the limitations of the methods that are most suitable for the heat treatment process.

In most previous publications, induction heating was introduced as a so-called uncoupled process without emphasis on the interrelated features of electromagnetic and heat transfer phenomena. The studies often consider the two phenomena separately. Such an approach could lead to substantial errors in predicting the required power, frequency, and heat treatment pattern. The nature of IHT is a combination of the two phenomena, which are tightly interrelated because physical properties depend strongly on both magnetic field intensity and temperature; therefore, this effect should be taken into account in contemporary design. Besides that, today's heat treater should consider the nonconstant behavior of the thermal and electromagnetic properties of the workpiece such as specific heat, thermal conductivity, magnetic permeability, and electrical resistivity during the heating cycle.

In addition, some induction heating coils can work very effectively with certain types of power supplies and ineffectively with others. Therefore, the optimal design of a modern IHT system should consider the features of IHT not as a physical stand-alone process but as a combination of the inductor, load-matching station, and power supply (i.e., solid-state inverter).

The major goal of this chapter is to embark upon the next step in the study and design of the modern IHT process and equipment. We concentrate on the features of modern IHT design and introduce advanced methods used in the study and evaluation of different types of heat treatment processes. The study includes the systematization of new knowledge accumulated in recent years at INDUCTOHEAT Inc., and by our colleagues around the world. Some materials presented here are new and have never been published before. Others have existed only in the articles or internal reports of the INDUCTOHEAT Group.

Many useful and practical recommendations are presented in this chapter as well as a description of some subtle aspects of the electromagnetic and heat transfer phenomena that are imperative for modern heat treatment practitioners and engineers to know.

Upon completion of this chapter, readers will have an orientation in modern computational methods for electroheat problems encountered in IHT. They will be able to evaluate the important features of the induction process and have knowledge that will help to avoid many of the unpleasant surprises one might encounter in the design of IHT systems. A basic knowledge of electrical engineering and understanding of basic induction heating principles should be essential to grasping some of the modern computation introduced in this chapter. Current information will be presented on the use of magnetic field plots, temperature profiles as well as sketches, drawings and photographs of practical applications. An attempt will be made to bridge the gap between purely theoretical information and that which is of practical use to the heat treatment specialists.

A reader with great experience in the field of induction heating and heat treatment will also find this chapter useful because he or she will discover the reasons for intuitive engineering decisions made in the past. College students preparing for a career involving induction heating or metal heat treating will see how the theory and mathematical methods they are studying are used to solve the problems encountered in everyday practice. This chapter consists of information that will be useful to a manager as well, because he or she will better understand the complexity of the process and the attention to detail required to obtain cost-effective, high-quality IHT results.

This chapter is not intended to describe in detail the specific mathematical methods or deep theoretical aspects of electrodynamics, thermodynamics, and optimal control involved in the process of designing modern induction heating equipment. For that the reader would need to be well versed in many advanced theoretical subjects. As we mentioned above, a basic knowledge of electrical engineering principles and common sense will be enough to understand the materials presented here. However, if someone finds this material too advanced and too difficult to understand or desires only a very basic or intuitive introduction to the subject, we would recommend the book *Basics of Induction Heating* by Chester Tudbury [2], which is available from INDUCTOHEAT Inc.

Finally, it is a most comforting discovery for most engineers that someone has actually built a production system to accomplish the same task they are trying to perform. Plots of electromagnetic fields and photographs of a variety of production installations are provided to show not only that the task has been previously accomplished but also how it has been done.

For those who would like to acquire more information about certain subjects, we have provided several references. Readers are also welcome to contact the authors of this chapter directly at INDUCTOHEAT Inc., in Madison Heights, Michigan.

4.2 PRELUDE TO DISCUSSION OF INDUCTION HEAT TREATMENT

Induction heating is often one of the most effective heat treatment processes available for a variety of applications including:

- Surface hardening
- Through hardening
- Tempering and stress relief (low-temperature)
- Annealing and normalizing (high-temperature)
- Weld seam annealing
- Sintering of powdered metals

In most of these applications, induction heating is used to selectively heat only the portion of the part that requires treatment. This usually means that the process can be accomplished in a relatively short time and with high efficiency because energy is applied to the part only where it is needed. The ability to heat treat in-line, as opposed to batch processing, with high productivity, less distortion, and a clean environment is an obvious benefit of IHT.

IHT is a segment of the much larger technical field of induction heating (Figure 4.1), which combines many other industrial processes using the phenomenon of heating by induction. Examples of these processes are:

- Heat treating
- Melting
- Levitation of molten metal

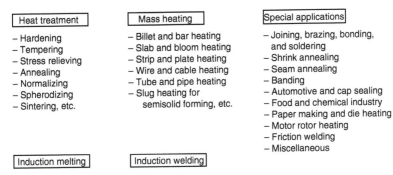

FIGURE 4.1 Industrial applications of induction heating..

- Through heating for forging, rolling, extrusion, etc.
- Brazing, soldering, and welding
- Shrink fitting
- Mold core meltout
- Adhesive curing
- Sealing
- Induction plasma

This handbook is devoted to heat treatment. Therefore, this chapter focuses specifically on the features of modern heat treatment by induction.

Any electrically conductive material can be induction heated. The ability to couple energy into a specific part of a workpiece by induction and obtain the desired heat treatment depends on many factors that are explored in detail in this chapter. Some of the more important factors are:

- Thermal and electromagnetic properties of the material
- Power applied
- Frequency selected
- Rate of heating
- Cooling process
- Ability of the material to respond to heat treatment
- Workpiece shape
- Inductor and quench design

Process steps used to accomplish the heat treatment of a part (workpiece) by induction are usually considerably fewer than are required by alternative processes [13–15]. For example, the induction contour-hardening process is accomplished in a much shorter time with less energy and also results in far less part distortion and grain size growth than carburizing. Figure 4.2 shows the pattern obtained by induction contour hardening of gear teeth.

IHT is well suited to in-line processing and flexible manufacturing cells. In many cases, the IHT machine receives the part by conveyor from an automatic grinding machine and includes part washing, a go and a no-go quality check of critical dimensions, induction hardening and tempering, eddy current check of the heat treatment, part marking, and data logging. The part is then transferred automatically by a conveyor or a robot to the next cell. Due to its unique features, IHT is very flexible and has been successfully used in a variety of heat-treatment processes. Figure 4.3 shows inductor for processing automative front wheel drive parts.

Let us briefly examine the most popular applications of heat treatment by induction.

FIGURE 4.2 Pattern obtained by induction contour hardening of gear teeth.

4.2.1 SURFACE HARDENING

Surface hardening of carbon steel and iron is the most common form of IHT because the heating can be localized to the areas where the metallurgical changes are desired. This process is a complex combination of electromagnetic, heat transfer, and metallurgical phenomena that occur when a workpiece is heated rapidly to a temperature above that required for a phase transformation to austenite and then rapidly quenched. As shown in Chapter 3 through Chapter 5, during the heating and quenching of the workpiece the metal actually undergoes a change in crystalline structure. The goal in surface hardening is to provide a martensitic layer on specific areas of the workpiece to increase hardness and wear resistance while allowing the remainder of the part to be unaffected by the process [6,11,12,18–20].

FIGURE 4.3 Induction heat treatment machine for automotive front wheel drive part.

Figure 4.4 shows a constant-velocity automotive front wheel drive component that has been cut and etched to show the pattern obtained by induction heat treating [93]. This component requires two areas of hardness with different strengths, loads, and wear requirements. The stem needs torsional strength as well as a hard outer surface, whereas the soft core must be ductile and therefore able to handle the mechanical shock from constant pulsing. The inner surface of the bell needs hardness for wear purposes, as ball bearings ride in the track or raceways. The thread of this component holds the wheel on. For heavy-duty applications the thread is also hardened and then annealed to produce a very tough thread. The annealing is almost always done with induction, using a separate inductor.

Surface hardening to increase wear resistance requires a shallow-hardened case [6,18–24]. Case depth (hardness depth) is typically defined as the surface area where the microstructure is at least 50% martensite. Below the case depth the hardness begins to decrease drastically. This depth is measured in a cut section of the hardened workpiece using a hardness tester to determine the hardness at various distances from the surface. The distance below the part surface where the hardness drops to 10 HRC lower than the surface hardness is often called the effective case depth. Hardness distribution along the workpiece radius depends primarily on the temperature distribution, the microstructure of the metal, its carbon content, quenching conditions, and the hardenability of the steel.

Bearings, rocker arms, pump shafts, and skid plates are examples of parts that require a shallow-hardened case for wear resistance. Case depths for these parts that will enable them to handle light loads are usually in the range of 0.25–1 mm (0.010–0.040 in.). In the following sections of this chapter, control of the depth of induction heating by selection of frequency,

FIGURE 4.4 Constant-velocity automotive front wheel drive component.

power density, and workpiece and coil geometry is discussed in detail. Generally speaking, heating for shallow case depths requires high frequency, low energy, and high power density. In some cases of shallow case depth surface hardening (less than 1 mm), it is possible to use self-quenching or mass quench. Because the heated surface layer is very fine and its mass is negligibly small compared to the mass of the cold core, it is possible to have rapid surface cooling due to heat conducted toward the cold core. In this case, the mass of the cold core acts as a large heat sink. Therefore, self-quenching excludes the use of fluid quenchants except as a cool down to allow handling of parts.

The parts that require both wear resistance and moderate loading such as camshafts and crankshafts are usually induction-hardened to case depths of 1–2.5 mm (0.040–0.1 in.). Deeper case depths strengthen the part dramatically because load stresses drop exponentially from the surface. The induction heating frequency that is required to obtain this case depth is usually between 8 and 50 kHz.

Parts that must withstand a heavy load require greater case depths; these include axle shafts, wheel spindles, and large heavy-duty gears. The cross-sectional areas of the part and the magnitude of the load they must handle determine the appropriate case depth.

In many of these applications the induction heating pattern can encompass a significant part of the cross-sectional area. As a result, some distortions may appear. Where distortion is present, it may be necessary to provide additional stock and case depth to allow for a final grinding after hardening. These heavy workpiece applications usually require a case depth in the range from 3.5 mm to as much as 12 mm (0.125–0.47 in.). Here a higher energy at 500 Hz to 3 kHz is usually called for.

Induction surface hardening also causes residual compressive stresses induced by the transformation hardening of the surface. In the great majority of heat-treating applications it is desirable to have compressive stresses at the surface. However, when it is necessary the effects of these stresses may be eliminated by a subsequent stress-relief process for which induction is also quite a popular choice.

Induction surface hardening is characterized by high dimensional stability of the heat-treated parts. The ability to keep part distortion as low as possible after the heat treatment process is due to the fact that induction heating is a very fast process and concentrates the heat sources in a thin surface layer called the current penetration depth. An analysis of this feature is given in Section 4.4.

Surface hardening is accomplished by raising the required depth of material above the transformation temperature (A_{c3} from Fe–C on the phase diagram) to the point where it is transformed to austenite and then cooling the workpiece rapidly to produce martensite. This rapid cooling or quenching is covered in detail in Chapter 9 of *Steel Heat Treatment: Metallurgy and Technologies*. The heating time to complete this process normally ranges from 1 to 8 s per component. The end use of the part to be hardened, as discussed briefly above, will determine the depth of hardening and may also indicate the specific type of material to be selected based on hardenability. The hardenability of metals is covered in detail in Chapter 5 of this handbook.

The general classification of steels surface-hardened by induction is as follows:

High-carbon steels are used primarily for tools such as drill bits and other cutting tools due to their ability to achieve high hardness. Commonly used steels are in the AISI 1050–1080 range.

Medium-carbon steels are often used in the automotive industry; they are used, for example, for front wheel drive components and drive shafts. These steels may be in the range of AISI 1035–1050.

Low-carbon steels are used where toughness rather than high hardness is required such as in clutch plates or pins for farm equipment. These may be AISI 1020–1035.

Alloy steels are used in bearings and automotive transmissions. These components may have several functions and therefore require a combination of materials. More information about alloy steels is presented in Chapter 4.

When discussing surface hardening by induction it is necessary to mention the phenomenon that is usually called superhardening. Due to this phenomenon, for a given carbon steel the surface hardness of the part that has been heat treated by induction is typically 3–4 HRC higher than that for a furnace-hardened steel. Superhardening can be attributed to the finer grain size of steel that has been induction-hardened. This occurs because the steel is at the austenizing temperature for a very short time, which results in finer grain sizes [6,12]. Another factor that is involved in superhardening is the higher lattice strain from residual compressive stresses at the surface of the part when its internal regions and the core remain underheated or even cold [12]. This phenomenon is particularly true for steel with a carbon content of 0.35–0.6%.

4.2.2 Through Hardening

Through hardening may be needed for parts requiring high strength such as snowplow blades, springs, chain links, truck bed frames, and certain fasteners such as nails or screws. In these cases the entire workpiece is raised above the transformation temperature and then quenched. Selection of the correct induction heating frequency is very important to achieve uniform temperature surface-to-core in the shortest time with the highest efficiency. The factors related to frequency selection are discussed in detail later in this chapter. Depending on the workpiece, the inductors may be in any of three stages: preheat, midheat, and final heat. The different stages, sometimes with different heating frequencies, allow the heat to soak into the workpiece. Induction heating for through hardening and for forging or extrusion are very similar; they all are accomplished by heating the entire workpiece uniformly to the desired temperature and at the required process time.

It is necessary to mention that the cooling intensity of the surface layers during quenching of a surface-hardened workpiece is greater than the intensity of cooling of a through-hardened workpiece. This is so because in surface hardening, additional cooling of the surface layers takes place due to the cold core, which plays the role of a heat sink. This greater cooling rate results in the ability to obtain a deeper martensitic structure after induction surface hardening. Consequently, the surface compressive stresses and hardness of through-hardened parts are typically lower than the hardness of a surface-hardened workpiece.

4.2.3 Tempering

The tempering process takes place after steel is hardened, but is no less important in metal heat treatment. Tempering temperatures are usually below the lower transformation temperature. The main purpose of tempering is to increase the steel's toughness, yield strength and ductility, to relieve internal stresses, to improve homogenization, and to eliminate brittleness [6,11,18,19,21,149–151]. The transformation to martensite through quenching creates a very hard and brittle structure. Untempered martensite is typically too brittle for commercial use and retains a lot of stresses. As discussed above, in surface hardening, only a thin surface layer of the workpiece is heated. The surface is raised to a relatively high temperature in a short period of time. A significant surface-to-core temperature difference and steel transformation phenomena results in the buildup of internal stresses. Reheating the steel for tempering after hardening and quenching leads to a decrease or relaxation of these internal stresses. In other words, because of tempering it is possible to improve the mechanical properties of the workpiece and to reduce the stresses caused by the previous heat treatment stage without losing too much of the achieved hardness.

Tempering temperatures are usually in the range of 120–600°C (248–1112°F). If the steel is heated to less than 120°C (248°F), there is no change in the metal structure, and tempering will not take place. Low-temperature tempering is typically performed at temperatures of 120–300°C (248–572°F). The main purpose of low-temperature tempering is stress relieving. Hardness reduction typically does not exceed 1–2 points HRC. If the tempering temperature is higher than 600°C (1112°F), essential changes in the structure of the steel may result that can lead to a significant loss in hardness. Hardness reduction exceeds 15 points HRC and maximum hardness is typically in a range of 36–44 HRC. Therefore, tempering is always a reasonable compromise between maintaining the required hardness and obtaining a low-stress and ductile microstructure in the metal.

There is a common misconception that tempering removes all the internal stresses. It does not, but tempering significantly decreases stresses. Tempering makes the steel softer and reduces the chance of distortion and the possibility of cracking. It is important that the time from quench to temper be held to a minimum. If this transient time is long enough, the internal stresses may have enough time to allow shape distortion or cracking to take place. Therefore, a long transient time between quenching and tempering will decrease or eliminate the tempering benefits. The soaking time of the tempering is much longer than both the heating times of hardening and quenching. It may take seconds to induction harden and minutes to induction temper.

If tempering has been done correctly, there will be only a slight loss in hardness. The benefits obtained, such as internal stress relief, the creation of the required ductility or toughness, shifting of the dangerous maximum of tensile stresses, which is located under the hardened surface layer further down toward the core of the workpiece, and improvement in the machinability of the steel, will offset the slight reduction in hardness.

4.2.4 NORMALIZING

Normalizing can be done by heating the steel above the upper transformation temperature and then allowing it to cool to room temperature. Normalizing usually takes a very long time (much longer than tempering), and can take as much as several hours. Normalizing is done to return the metal structure to a state in which the next heat treatment application can take place. Some parts are spheroidized to make them easier to form. It is often necessary to normalize them before hardening.

4.2.5 ANNEALING

Annealing is very similar to normalizing except that the soak time during cooling is even longer than in the normalizing process. Annealing temperatures are usually 750–900°C (1350–1650°F). The purpose of annealing is much like that of tempering in that the hardness is decreased and the ductility is increased [12,19,25]. This can be done by applying higher temperature and results in a more ductile structure. A good example is the annealing of the thread of a pinion. The entire pinion surface is hardened to approximately 58 HRC, and then the thread is annealed to 35–40 HRC to produce a thread that is harder than the core material and is very ductile. This results in production of a very strong part without the possibility of thread breakage when a heavy torque is applied.

Normalizing and annealing reduce hardness and strength while providing the following benefits:

1. Improved formability and machinability
2. Refined steel structure
3. Significant relief of internal stresses

4.2.6 SINTERING

Sintering is the bonding of the molecular structure in a powdered metal. It is done by heating the metal to a high temperature that is below its melting point to produce recrystallization. High-temperature sintering produces a more consistent structure than casting or forging and provides added hardenability with more consistent heat treatment results. Sintering powdered metal is like normalizing carbon steel and is recommended if a subsequent heat treatment is to be performed. The process of powder metallurgy is discussed in detail in Chapter 13 of *Steel Heat Treatment: Metallurgy and Technologies*.

There are many ways to heat treat metal parts including the use of gas-fired furnaces, fluidizing bed furnaces, salt baths, lasers, plasma systems, infrared heaters, electric- and fuel-fired furnaces, etc. Each method has its own advantages as well as disadvantages. There is obviously no universal method that is the best in all cases of metal heat treating.

In the past two decades heat treating by induction has become more and more popular. A major reason is the ability to create high heat intensity very quickly at well defined locations on the part. This leads to low process cycle time (high productivity) with constant quality. Induction heating is also more energy efficient and environmentally friendly than most other heat sources used for heat treatment such as carburizing systems or gas-fired furnaces. It usually requires far less start-up and shutdown time and a lower labor cost for machine operators. In many cases, IHT will require a minimum floor space and produce less distortion in the workpiece.

IHT is a complex combination of electromagnetic, heat transfer, and metallurgical phenomena involving many factors and components. The main components of an IHT system are an induction coil, power supply, load-matching station, quenching system, and the workpiece itself. The induction coil or inductor is usually designed for a specific application and is therefore found in widely varying shapes and sizes. The features involved in the design and operation of modern IHT processes are discussed later in this chapter.

4.3 THEORETICAL BACKGROUND AND MATHEMATICAL MODELING OF MODERN INDUCTION HEAT TREATMENT PROCESSES

Mathematical modeling is one of the major factors of successful IHT design. Theoretical models may vary from a simple hand-calculated formula to a very complicated numerical analysis, which can require several hours of computational work using modern supercomputers. The choice of a particular theoretical model depends on several factors, including the complexity of the engineering problems, required accuracy, time limitations, and cost.

Before an engineer starts to provide a mathematical simulation of any process he or she should have a sound understanding of the nature and physics of the process. Engineers should also be aware of the limitations of applied mathematical models, assumptions, and possible errors and should consider correctness and the sensitivity of the chosen model to some poorly defined parameters such as boundary conditions, material properties, or initial temperature conditions. One model can work very well in certain applications and give absolutely unrealistic results in another. An underestimation of the features of the process or overly simple assumptions can lead to an incorrect mathematical model (including chosen governing equations) that will not be able to provide the required accuracy. It is very important to remember that any computational analysis can at best produce only results that are derived from the governing equations. Therefore, the first and most important step in any mathematical simulation is to choose an appropriate theoretical model that will correctly describe the technological process or phenomena. Let us briefly look at the basic physical concepts related to the IHT process.

As mentioned above, an IHT is a complex combination of electromagnetic, heat transfer, and metallurgical phenomena. The metallurgical aspects have been discussed in the previous chapters. Therefore, in this chapter we will concentrate on the first two phenomena. Heat transfer and electromagnetics are closely interrelated because the physical properties of heat-treated materials depend strongly on both magnetic field intensity and temperature. In conventional induction heating, all three modes of heat transfer—conduction, convection, and radiation—are present [26–36]. Heat is transferred by conduction from the high-temperature region of the workpiece toward the low-temperature region. The basic law that describes heat transfer by conduction is Fourier's law,

$$q_{\text{cond}} = -\lambda \, \text{grad}(T)$$

where q_{cond} is the heat flux by conduction, λ the thermal conductivity, and T the temperature.

As one can see, according to Fourier's law, the rate of heat transfer in a workpiece is proportional to the temperature gradient (temperature difference) and the thermal conductivity of the workpiece. In other words, a large temperature difference between the surface and the core (which, for example, typically takes place during surface hardening), and a high value of thermal conductivity of the metal result in intensive heat transfer from the hot surface of the workpiece toward the cold core. Conversely, the rate of heat transfer by conduction is inversely proportional to the distance between regions with different temperatures.

In contrast to conduction, heat transfer by convection is carried out by fluid, gas or air (i.e., from the surface of the heated workpiece to the ambient area). Convection heat transfer can be described by the well-known Newton's law. This law states that the heat transfer rate is directly proportional to the temperature difference between the workpiece surface and the ambient area,

$$q_{\text{conv}} = \alpha(T_{\text{s}} - T_{\text{a}})$$

where q_{conv} is the heat flux density by convection, W/m^2 or W/in.^{2}; α is the convection surface heat transfer coefficient, W/(m^2°C) or W/(in.^{2}°F); T_{s} is the surface temperature, °C or °F; and T_{a} is the ambient temperature, °C or °F. The subscripts s and a denote surface and ambient, respectively.

The convection surface heat transfer coefficient is primarily a function of the thermal properties of the workpiece, the thermal properties of the surrounding fluids, gas or air, and their viscosity or the velocity of the heat-treated workpiece if the workpiece is moving at high speed (e.g., induction heat treating of a strip or spinning disk). It is particularly important to take this mode into account when designing low-temperature induction heating applications (i.e., tempering, stress relieving, paint curing or other processes with a maximum temperature <500°C (932°F)). This heat transfer mode plays a particularly important role in the quenching process where the surface heat transfer coefficient describes the cooling process during quenching.

In the third mode of heat transfer, which is heat radiation, the heat may be transferred from the hot workpiece into a nonmaterial region (vacuum). The effect of heat transfer by radiation can be introduced as a phenomenon of the electromagnetic energy propagating due to a temperature difference. This phenomenon is governed by the Stefan–Boltzmann law of thermal radiation, which states that the heat transfer rate by radiation is proportional to a radiation loss coefficient C_{s} and the value of $T_{\text{s}}^4 - T_{\text{a}}^4$. Due to the fact that radiation losses are proportional to the fourth power of temperature, these losses are a significant part of the total heat losses in high-temperature applications (for example, surface hardening, through hardening, induction heating before forging, rolling, etc.). The radiation heat loss coefficient

TABLE 4.1
Emissivity of Metals

	Aluminum	Carbon Steel	Copper	Brass and Zinc
Polished	0.042–0.053	0.062	0.026–0.042	0.03–0.039
Commercial	0.082–0.40	0.71–0.8	0.24–0.65	0.21–0.50

includes emissivity, radiation shape factor, and surface conditions. For example, polished metal will radiate less heat to the surroundings than a nonpolished metal. The radiation heat loss coefficient can be determined approximately as $C_s = \sigma_s \varepsilon_1$, where ε_1 is the emissivity of the metal and σ_s is the Stefan–Boltzmann constant ($\sigma_s = 5.67 \times 10^{-8}$ W/(m^2 K^4)). The values of ε_1 for metal that are typically used in the heat treatment are given in Table 4.1.

The above-described determination of radiation heat loss is a valid assumption for mathematical modeling of a great majority of induction heating and heat treatment problems. However, there are a few applications where the radiation heat transfer phenomenon can be complicated and such a simple approach would not be valid. Complete details of all the three modes of heat transfer can be found in several references [26–36]. In a typical induction heating and heat treatment, heat transfer by convection and radiation reflects the value of heat loss. A high value of heat loss reduces the total efficiency of the induction heater.

In general, the transient (time-dependent) heat transfer process in a metal workpiece can be described by the Fourier equation

$$C\gamma \frac{\partial T}{\partial t} + \text{div}(-\lambda \, \text{grad} \, T) = Q \qquad (4.1)$$

where T is the temperature, γ is the density of the metal, C is the specific heat, λ is the thermal conductivity of the metal, and Q is the heat source density induced by eddy currents per unit time in a unit volume (so-called heat generation). This heat source density is obtained by solving the electromagnetic problem.

The values of specific heat and thermal conductivity of some typical metals commonly used in heat treating are shown in Figure 4.5 [37,38]. As one may note, the specific heat and thermal conductivity are functions of temperature. Both specific heat and thermal conductivity have not only pure mathematical meaning but also a concrete engineering interpretation. The value of specific heat indicates the amount of energy that would have to be absorbed by the workpiece to achieve the required temperature change. A high value of specific heat corresponds to a higher required power.

Thermal conductivity λ designates the rate at which heat travels across the workpiece. A material with a high λ value will conduct the heat faster than a material with a low λ. In choosing a material for an inductor's refractory, a lower value of λ is required and will correspond to higher efficiency and lower heat loss. Conversely, when the thermal conductivity of the metal is high, then it is easier to obtain a uniform temperature distribution along the workpiece thickness, which is important in through-hardening processes. Therefore, from the point of view of obtaining temperature uniformity, a higher thermal conductivity of metal is preferable. However, in surface- or selective-hardening applications, a high value of λ is quite often a disadvantage because of its tendency to equalize the temperature distribution within the workpiece. As a result of temperature equalization the temperature rise will take place not only in the surface layer of the workpiece, which is to be hardened, but also in internal areas as well, which are not. The temperature increase in the internal areas of the

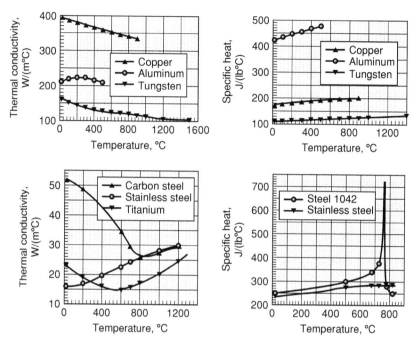

FIGURE 4.5 Thermal conductivity and specific heat of metals.

workpiece results in cooling of its surface and therefore requires more power to heat the surface layer to the desired final temperature. A large amount of heated mass in the workpiece can also cause a distortion of the workpiece geometry.

Our experience shows that in the great majority of tempering, stress relieving, annealing and normalizing applications, a rough approximation of thermal conductivity in simulations of the heat treatment process will not lead to significant errors in temperature distributions. However, in surface hardening and through hardening a rough approximation of λ can create unacceptable results. At the same time, regardless of application, a rough approximation of specific heat could create significant errors in obtaining the required coil power and temperature profile within the workpiece.

Equation 4.1, with suitable boundary and initial conditions, represents the three-dimensional temperature distribution at any time and at any point in the workpiece. The initial condition refers to the temperature profile within the workpiece at time $t = 0$; therefore, that condition is required only when dealing with a transient heat transfer problem where the temperature is a function not only of the space coordinates but of time also. The initial temperature distribution is usually uniform and corresponds to the ambient temperature. However, the initial temperature distribution is sometimes nonuniform due to the previous technological process (i.e., preheating, quenching, or continuous casting applications).

For most engineering induction heating problems, boundary conditions combine the convective and radiative losses. In this case the boundary condition can be expressed as

$$-\lambda \frac{\partial T}{\partial n} = \alpha(T_s - T_a) + C_s(T_s^4 - T_a^4) + Q_s \qquad (4.2)$$

where $\partial T/\partial n$ is the temperature gradient in a direction normal to the surface at the point under consideration, α is the convection surface heat transfer coefficient, C_s is the radiation heat loss coefficient, Q_s is the surface loss (i.e., during quenching), and n denotes the normal to the boundary surface.

As one may see from Equation 4.2, the heat losses at the workpiece surface are highly variable because of the nonconstant behavior of convection and the radiation losses. The analysis shows that convection losses are the major part of the heat loss in low-temperature induction heating applications (i.e., aluminum, lead, zinc, tin, magnesium, steel at a temperature lower than 350°C (or 662°F)). In high-temperature applications (such as induction hardening and heating of steels, titanium, tungsten, nickel, etc.), radiation losses are much more significant than convection losses (Figure 4.6).

If the heated body is geometrically symmetrical along the axis of symmetry, the Neumann boundary condition can be formulated as

$$\frac{\partial T}{\partial n} = 0 \tag{4.3}$$

The Neumann boundary condition implies that the temperature gradient in a direction normal to the axis of symmetry is zero. In other words, there is no heat exchange at the axis of symmetry. This boundary condition can also be applied in the case of a perfectly insulated workpiece.

In the case of heating a cylindrical workpiece, Equation 4.1 can be rewritten as

$$C\gamma \frac{\partial T}{\partial t} = \frac{\partial}{\partial Z}\left(\lambda \frac{\partial T}{\partial Z}\right) + \frac{1}{R}\frac{\partial}{\partial R}\left(\lambda R \frac{\partial T}{\partial R}\right) + Q(Z,R) \tag{4.4}$$

The same equation (4.1) can be shown in Cartesian coordinates (i.e., heat transfer in slab, strip, or plate) as

$$C\gamma \frac{\partial T}{\partial t} = \frac{\partial}{\partial X}\left(\lambda \frac{\partial T}{\partial X}\right) + \frac{\partial}{\partial Y}\left(\lambda \frac{\partial T}{\partial Y}\right) + \frac{\partial}{\partial Z}\left(\lambda \frac{\partial T}{\partial Z}\right) + Q(X,Y) \tag{4.5}$$

Equation 4.4 and Equation 4.5 with boundary conditions (4.2) and (4.3) are the most popular equations for mathematical modeling of the heat transfer processes in conventional heat treatment.

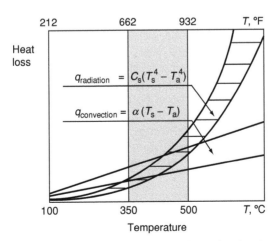

FIGURE 4.6 Convection and radiation heat losses in typical induction heating.

The technique of calculating the electromagnetic field depends on the ability to solve Maxwell's equations. For general time-variable electromagnetic fields, Maxwell's equations in differential form can be written as [39–52]

$$\text{curl}\,\mathbf{H} = \mathbf{J} + \frac{\partial \mathbf{D}}{\partial t} \text{ (from Ampere's law)} \tag{4.6}$$

$$\text{curl}\,\mathbf{E} = -\frac{\partial \mathbf{B}}{\partial t} \text{ (from Faraday's law)} \tag{4.7}$$

$$\text{div}\,\mathbf{B} = 0 \text{ (from Gauss' law)} \tag{4.8}$$

$$\text{div}\,\mathbf{D} = \rho \text{ (from Gauss' law, magnetic)} \tag{4.9}$$

where $\mathbf{E}$ is the electric field intensity, $\mathbf{D}$ is the electric flux density, $\mathbf{H}$ is the magnetic field intensity, $\mathbf{B}$ is magnetic flux density, $\mathbf{J}$ is the conduction current density, and ρ is the electric charge density.

Maxwell's equations not only have a purely mathematical meaning, but also have a concrete physical interpretation as well. For example, Equation 4.6 shows that the curl of $\mathbf{H}$ always has two sources: conductive ($\mathbf{J}$) and displacement ($\partial\mathbf{D}/\partial t$) currents. A magnetic field is produced whenever there are electric currents flowing in surrounding objects. From Equation 4.7 one can conclude that a time rate of change in the magnetic flux density always produces the curling electric field. In other words, a changing magnetic field always produces induced current in the surrounding area, and therefore it produces an electric field in the area where such changes take place. The minus sign in Equation 4.7 determines the direction of that induced electric field. This fundamental result can be applied to any region in space.

Let us consider how Equation 4.6 and Equation 4.7 can be used to explain the basic electromagnetic process that takes place in IHT. The application of alternating voltage to the induction coil will result in the appearance of an alternating current (AC) in the coil circuit. According to Equation 4.6, an alternating coil current will produce in its surrounding area an alternating (changing) magnetic field that will have the same frequency as the source current (coil current). That magnetic field's strength depends on the current flowing in the induction coil, the coil geometry, and the distance from the coil. The changing magnetic field induces eddy currents in the workpiece and in other objects that are located near that coil. By Equation 4.7, induced currents have the same frequency as the source coil current; however, their direction is opposite to that of the coil current. (This is determined by the minus sign in Equation 4.7.) According to Equation 4.6, alternating eddy currents induced in the workpiece produce their own magnetic fields, which have opposite directions to the direction of the main magnetic field of the coil. The total magnetic field of the induction coil is a result of the source magnetic field and induced magnetic fields.

As one would expect from an analysis of Equation 4.7, there can be an undesirable heating of tools or other electrically conductive structures that are located near the induction coil. From another perspective, Equation 4.7 can partially explain the undesirable existence of the magnetic field at the operator (the concern about the effects of electromagnetic field exposure on human beings). Both electric and magnetic fields induce weak electric currents in the operator's body.

In induction heating and heat treatment applications, an engineer should pay particular attention to such simple relations as (4.8) or (4.9). The popular saying "The best things come in the smallest packages" is particularly true here. The short notation of Equation 4.8 has real significance in induction heating and the heat treatment of metals. To say that the divergence of magnetic flux density is zero is equivalent to saying that **B** lines have no source points at which they originate or end, in other words **B** lines always form a continuous loop. A clear understanding of such a simple statement will allow one to explain and avoid many mistakes in dealing with IHT and the induction heating of workpieces with complicated geometry.

The above-described Maxwell's equations are in indefinite form because the number of equations is less than the number of unknowns. These equations become definite when the relations between the field quantities are additional and hold true for a linear, isotropic medium.

$$\mathbf{D} = \varepsilon\varepsilon_0\mathbf{E} \tag{4.10}$$

$$\mathbf{B} = \mu\mu_0\mathbf{H} \tag{4.11}$$

$$\mathbf{J} + \sigma\mathbf{E} \text{ (Ohm's law)} \tag{4.12}$$

where the parameters ε, μ, and σ denote, respectively, the relative permittivity, relative magnetic permeability, and electrical conductivity of the material; $\sigma = 1/\rho$, where ρ is electrical resistivity. The constant $\mu_0 = 4\pi \times 10^{-7}$ H/m (or Wb/(A m)) is called the permeability of free space (the vacuum), and similarly the constant $\varepsilon_0 = 8.854 \times 10^{-12}$ F/m is called the permittivity of free space. Both relative magnetic permeability μ and relative permittivity ε are nondimensional parameters and have very similar meanings. The relative magnetic permeability indicates the ability of a material to conduct the magnetic flux better than vacuum or air. This parameter is very important in choosing materials for magnetic flux concentrators. Relative permittivity (or dielectric constant) indicates the ability of a material to conduct the electric field better than vacuum or air.

By taking Equation 4.10 and Equation 4.12 into account, Equation 4.6 can be rewritten as

$$\text{curl }\mathbf{H} = \sigma\mathbf{E} + \frac{\partial(\varepsilon_0\varepsilon\mathbf{E})}{\partial t} \tag{4.13}$$

For most practical applications of the IHT of metals, where the frequency of currents is less than 10 MHz, the induced conduction current density **J** is much greater than the displacement current density $\partial\mathbf{D}/\partial t$, so the last term on the right-hand side of Equation 4.13 can be neglected. Therefore, Equation 4.13 can be rewritten as

$$\text{curl }\mathbf{H} = \sigma\mathbf{E} \tag{4.13a}$$

After some vector algebra and using Equation 4.6, Equation 4.7, and Equation 4.11, it is possible to show that

$$\text{curl}\left(\frac{1}{\sigma}\text{curl }\mathbf{H}\right) = -\mu\mu_0\frac{\partial\mathbf{H}}{\partial t} \tag{4.14}$$

$$\text{curl}\left(\frac{1}{\mu}\text{ curl }\mathbf{E}\right) = -\sigma\mu_0\frac{\partial\mathbf{E}}{\partial t} \tag{4.15}$$

Since the magnetic flux density **B** satisfied a zero divergence condition (Equation 4.8), it can be expressed in terms of a magnetic vector potential **A** as

$$\mathbf{B} = \text{curl }\mathbf{A} \tag{4.16}$$

and then, from (4.7) and (4.16), it follows that

$$\text{curl } \mathbf{E} = -\text{curl} \frac{\partial \mathbf{A}}{\partial t} \tag{4.17}$$

Therefore, after integration, one can obtain

$$\mathbf{E} = -\frac{\partial \mathbf{A}}{\partial t} - \text{grad } \varphi \tag{4.18}$$

where φ is the electric scalar potential. Equation 4.12 can be written as

$$\mathbf{J} = -\sigma \frac{\partial \mathbf{A}}{\partial t} + \mathbf{J}_s \tag{4.19}$$

where $\mathbf{J}_s = -\sigma \text{ grad}(\varphi)$ is the source (excitation) current density in the induction coil.

Taking the material properties as piecewise continuous and neglecting the hysteresis and magnetic saturation, it can be shown that

$$\frac{1}{\mu\mu_0} (\text{curl curl } \mathbf{A}) = \mathbf{J}_s - \sigma \frac{\partial \mathbf{A}}{\partial t}$$

It should be mentioned here that for the great majority of induction heat-treating applications (such as through hardening) a heat effect due to hysteresis losses does not exceed 7% compared to the heat effect due to eddy current losses. Therefore, an assumption neglecting the hysteresis loss is valid. However, in some cases, such as induction tempering, paint curing, heating before galvanizing and lacquer coating, hysteresis losses can be quite significant compared to eddy current losses. In these cases, hysteresis should be considered.

It can be shown that for the great majority of IHT applications it is possible to further simplify the mathematical model by assuming that the currents have a steady-state quality. Therefore, with this assumption we can conclude that the electromagnetic field quantities in Maxwell's equations are harmonically oscillating functions with a single frequency. Thus a time-harmonic electromagnetic field can be introduced. This field can be described by the following equations, which are derived after some vector algebra from Equation 4.14, Equation 4.15, and Equation 4.20, respectively:

$$\frac{1}{\sigma} \nabla^2 \mathbf{H} = \mathbf{j}\omega\mu\mu_0 \mathbf{H} \tag{4.21}$$

$$\frac{1}{\mu} \nabla^2 \mathbf{E} = \mathbf{j}\omega\sigma\mu_0 \mathbf{E} \tag{4.22}$$

$$\frac{1}{\mu\mu_0} \nabla^2 \mathbf{A} = -\mathbf{J}_s + \mathbf{j}\omega\sigma\mathbf{A} \tag{4.23}$$

where ∇^2 is the Laplacian, which has different forms in Cartesian and cylindrical coordinates. In Cartesian coordinates,

$$\nabla^2 \mathbf{A} = \frac{\partial^2 A}{\partial X^2} + \frac{\partial^2 A}{\partial Y^2} + \frac{\partial^2 A}{\partial Z^2} \tag{4.24}$$

In cylindrical coordinates (axisymmetric case),

$$\nabla^2 \mathbf{A} = \frac{1}{R}\frac{\partial}{\partial R}\left(R\frac{\partial A}{\partial R}\right) + \frac{\partial^2 A}{\partial Z^2} \tag{4.25}$$

In other words, it has been assumed that harmonics are absent in both the impressed and the induced currents and fields. The governing Equation 4.21 through Equation 4.23, for the time-harmonic field with the appropriate boundary condition can be solved with respect to $\mathbf{H}$, $\mathbf{E}$, or $\mathbf{A}$.

Equation 4.21 through Equation 4.23 are valid for general three-dimensional fields and allow one to find all of the required parameters of the induction system such as current, power, coil impedance, and heat source density induced by eddy currents. Although there is considerable practical interest in three-dimensional problems, a great majority of engineering problems in induction heating tend to be handled with a combination of two-dimensional assumptions. Several factors discourage three-dimensional field consideration:

1. Computing costs are much higher for three-dimensional cases (especially taking into account tightly interrelated features of electromagnetic and heat transfer phenomena in induction heating).
2. User must have special knowledge in many theoretical subjects and should have specific experience working with three-dimensional software.
3. Representation of both results and geometric input data could create a well-known problem of working with a three-dimensional image.

For many IHT applications the quantities of the magnetic field (such as magnetic vector potential, electric field intensity, and magnetic field intensity) may be assumed to be entirely directed. For example, in the longitudinal cross section of the induction coil, both $\mathbf{A}$ and $\mathbf{E}$ vectors have only one component, which is entirely Z-directed. In the case of a transverse section, $\mathbf{H}$ and $\mathbf{B}$ vectors have only one component. This allows one to reduce the three-dimensional field to a combination of two-dimensional forms. For example, in the case of magnetic vector potential, Equation 4.23 can be expressed as shown below:

For a two-dimensional Cartesian system,

$$\frac{1}{\mu\mu_0}\left(\frac{\partial^2 A}{\partial X^2} + \frac{\partial^2 A}{\partial Y^2}\right) = -\mathbf{J}_s + \mathbf{j}\omega\sigma\mathbf{A} \tag{4.26}$$

For the axisymmetric cylindrical system,

$$\frac{1}{\mu\mu_0}\left(\frac{\partial^2 A}{\partial R^2} + \frac{1}{R}\frac{\partial A}{\partial R} + \frac{\partial^2 A}{\partial Z^2} - \frac{A}{R^2}\right) = -\mathbf{J}_s + \mathbf{j}\omega\sigma\mathbf{A} \tag{4.27}$$

The boundary of the region is selected such that the magnetic vector potential $\mathbf{A}$ is zero along the boundary (Dirichlet condition) or its gradient is negligibly small along the boundary compared to its value elsewhere in the region (Neumann condition $\partial\mathbf{A}/\partial\mathbf{n} = 0$). Therefore, Equation 4.4 and Equation 4.27 with their initial and boundary conditions fully describe the electrothermal processes in a great majority of conventional cylindrical IHT systems.

By using analogous vector algebra manipulations it is possible to obtain governing equations similar to Equation 4.26 and Equation 4.27 that can be formulated with respect to $\mathbf{E}$, $\mathbf{B}$, or $\mathbf{H}$. Therefore, any given electromagnetic problem in IHT may be worked in terms of either $\mathbf{A}$, $\mathbf{E}$, $\mathbf{B}$, or $\mathbf{H}$. Part of the art of mathematical modeling of electromagnetic fields derives from the right choice of field representation, which could be different for different applications.

Partial differential equations that are formulated with respect to **A** or **E** are very convenient for describing the electromagnetic field in a longitudinal cross section of the induction heating system. However, the electromagnetic field distribution in a transverse cross section of the workpiece can be more conveniently described by the governing equations formulated with respect to **B** or **H** [49,50]. Field representations that are typically used by the IHT designer for describing electromagnetic processes in the conventional induction surface or through hardening of cylindrical workpieces are shown in Figure 4.7.

As mentioned above, the contemporary design of IHT equipment requires that several features of the process be taken into account, which makes its analysis quite complicated.

One of the major difficulties in electromagnetic filed and heat transfer computation is the nonlinearity of material properties. As mentioned earlier, this exists because the thermal conductivity, specific heat, and electrical conductivity of metals (i.e., steel) are functions of the temperature but their relative magnetic permeability is a function of two parameters: magnetic field intensity and temperature. Figure 4.8 shows electrical resistivities of typical metals as functions of temperature. The relative magnetic permeability as a function of temperature and magnetic field intensity is shown in Figure 4.9. In everyday engineering language, the induction heating specialists often call the relative magnetic permeability simply as magnetic permeability. In this chapter we will refer to it also as magnetic permeability.

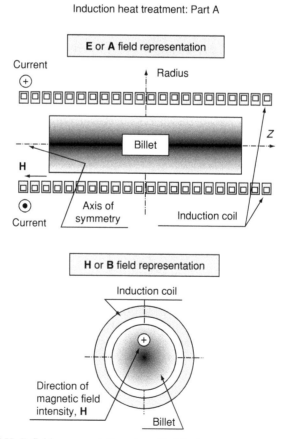

FIGURE 4.7 **E, A,** and **H, B** field representations in cylindrical system.

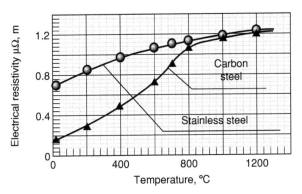

FIGURE 4.8 Electrical resistivity of steels.

Even a cursory look at the behavior of the material properties in Figure 4.5, Figure 4.8, and Figure 4.9 reveals the danger in using the methods (e.g., analytical or equivalent circuit coil design methods) that cannot take into account the total variation of properties. Owing to their nature, these methods can consider only average values, which can lead to unpredictable and often erroneous results, with possible financial loss to both the customer and the producer of the heat treatment equipment.

Another difficulty in the analysis and computation of the IHT process is related to the fact that the shapes of the workpiece and induction coil can be complicated. Therefore, to provide the required accuracy of evaluation it is necessary to reformulate the computational model for almost every problem.

Besides the considerations mentioned above, the induction heater configuration can consist of several coils, flux concentrators, flux robbers, shunts, and shells, and finally, there are tightly coupled (interrelated) effects of the electromagnetic and heat transfer phenomena. This requires the development of special computational algorithms that are able to deal with these coupled effects. The highly pronounced variation of the physical properties makes the IHT problem nonlinear. There are several ways to couple electromagnetic and heat transfer problems. One of the most common approaches is called the indirect coupling method. This method calls for the iteration process. An iteration process consists of an electromagnetic computation and then recalculation of heat sources in order to provide a heat transfer computation. This assumes that the temperature variations are not significant, which means that the material properties remain approximately the same, and during certain

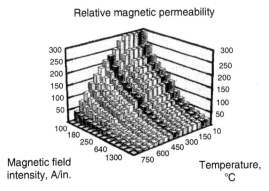

FIGURE 4.9 Magnetic permeability as a function of magnetic field intensity $\mathbf{H} = 100$–1500 A/in. and temperature $T = 10$–$750°C$ ($T = 50$–$1382°F$).

times the heat transfer process continues to be solved without correcting the heat sources. The temperature distribution within the workpiece obtained from the time-stepped heat transfer computation is used to update the values of specific heat and thermal conductivity at each time step. As soon as the heat source variations become significant (due to the variations of electrical conductivity and magnetic permeability), the convergence condition will no longer be satisfied, and recalculation of the electromagnetic field and heat sources will take place.

For most induction heating applications, an indirect coupling approach is valid and very effective. However, there are situations (i.e., intensive induction heating of carbon steel just above the Curie point, high power density induction contour-hardening applications, or low-power density strip annealing) where this approach could lead to significant errors. In these cases, only the direct coupling method should be applied [51,53].

The analytical methods and equivalent circuit coil design methods popular in the 1960s and 1970s no longer satisfy the modern designer because of the inherent restrictions outlined above. The designer must be aware that in many applications, very erroneous and inadequate results can be obtained when such methods are used. The development of modern computers and the increasing complexity in most modern induction heating applications have significantly restricted the solution to the application of simple formulas, analytical and seminumerical methods. These methods can be useful only in obtaining approximate results in simple cases. Rather than use simple computation techniques with many restrictions, modern IHT specialists are currently turning to highly effective numerical methods such as finite-difference method (FDM), finite-element method (FEM), impedance boundary element (IBE), and integral equation methods that are widely and successfully used in the computation of electromagnetic and heat transfer problems in heat treatment. Each of these methods has certain advantages and has been used alone or in combination with others.

Because of the extraordinarily large amount of information that is available in special scientific literature, even an experienced engineer can be easily confused if too many of the nuances of computer modeling of IHT problems are introduced. Therefore, we decided to discuss the state of the art of the subject of modern electroheat numerical computation while simplifying the materials so that they are understood by IHT specialists who have limited experience in numerical modeling. Thus our goal is to provide the reader with a general orientation in advanced numerical simulation methods.

Before we discuss the features and applications of the most popular numerical methods it is necessary to point out one of their important qualities: all numerical methods give an approximate solution to the modeled problem (i.e., heat transfer or electromagnetic). Therefore, there is always a danger of obtaining inappropriate results when those methods are used. The fact that the solution is approximate, and is not absolutely accurate, should not discourage engineers from using numerical methods. On the contrary, it should stimulate them to carefully study the features of these methods and transform them into a powerful computational tool that will allow heat treatment analysts to control accuracy of simulation and to produce information that cannot be measured or obtained by using analytical, semianalytical, or other kinds of methods, including physical experiments. It is wise to remember that the correct use of numerical methods will provide approximate, but acceptable, engineering solutions that will satisfy the requirements of modern technology from a practical standpoint.

Many mathematical modeling methods and programs exist or are under development. Work in this field is done in universities, research institutes, inside large companies such as INDUCTOHEAT Inc., and by specialized software companies such as MacNeal-Schwendler Corp., Integrated Engineering Software Inc., Infolytica Inc., Ansoft Corp., Structural Research and Analysis Inc., Cosmos Corp., Vector Fields Inc., and others.

For each particular problem or family of similar problems, certain numerical methods or software are preferred. It is obviously quite difficult if not impossible to find a single medicine that will be equally effective for both constipation and diarrhea. The medicine will most likely stop diarrhea or it will help constipation but not both. The same reasoning holds when searching for an efficient and universal computational tool. There is no cure-all or single universal computational method that fits all the cases and is optimum for solving all the induction heating problems. Our experience in the use of different numerical methods has shown that it is preferable to have a number of methods and programs rather than searching for one universal program for solving a wide variety of tasks. The right choice of computational method and software depends on the application and features of the specific problem or phenomenon. It is important for the IHT designer to know the advantages and limitations of a variety of computational techniques. This will allow the analyst to select an appropriate tool that will provide an accurate evaluation of the problem and allow the engineer to predetermine the temperature distribution within the workpiece in all stages of the process cycle, which includes heating, soaking, and quenching. This will result in obtaining the required heating conditions and the desired metallurgical properties of the work-piece.

Because of space limitations, this chapter does not give an exhaustive list of the methods available for electromagnetic field and heat transfer calculations. There are many publications that describe the features and applications of mathematical modeling methods. An interested reader can study the description of the most popular computational techniques used for simulation of heat transfer and electromagnetic processes in Refs. [5,30,47–92]. Here we briefly analyze and describe some of the methods that are successfully used at INDUC-TOHEAT. We hope that knowledge obtained will afford engineers and designers an understanding of how the numerical methods work and what kind of numerical method can be the most suitable for a particular IHT application.

4.3.1 FINITE-DIFFERENCE METHOD

The FDM was the earliest numerical technique [27,49,51–59] used for mathematical modeling of heat treatment processes. The FDM has been used extensively for solving both heat transfer and electromagnetic problems. It is particularly easy to apply when the modeling area has cylindrical or rectangular shapes. The area of study is represented by the mesh of lines that are parallel to the coordinate axes (so-called orthogonal grid of mesh nodes). The orthogonal mesh discretizes the area of modeling (i.e., induction coil, workpiece, flux concentrator, etc.) into a finite number of nodes (Figure 4.10).

Because of the orthogonal mesh, the discretization algorithm is quite simple. An approximate solution of the governing equation is found at the mesh points defined by intersections of the lines. The computation procedure consists of replacing each partial derivative of the governing Equation 4.4, Equation 4.5, Equation 4.26, or Equation 4.27 by a finite-difference stencil that couples the value of the unknown variable (i.e., temperature or magnetic vector potential) at a node of approximation with its value in the surrounding area. This method has universal application because of its generality and its relative simplicity.

By Taylor's theorem for two variables, the value of a variable at a node on the mesh can be expressed in terms of its neighboring values and separation distance (space step) h as in the following expressions (stencils):

$$\frac{\partial T}{\partial X} \Rightarrow \frac{T_{i+1} - T_i}{h} + O(h) \text{ (forward difference)} \tag{4.28}$$

$$\frac{\partial T}{\partial X} \Rightarrow \frac{T_i - T_{i-1}}{h} + O(h) \text{ (backward difference)} \tag{4.29}$$

$$\frac{\partial T}{\partial X} \Rightarrow \frac{T_{i+1} - T_{i-1}}{2h} + O(h) \text{ (central difference)} \tag{4.30}$$

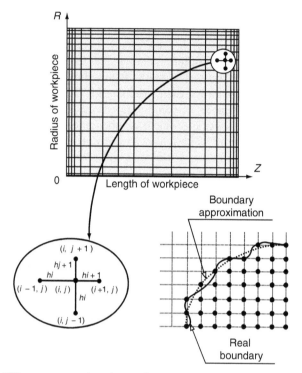

FIGURE 4.10 Finite-difference approximation and rectangular network of mesh.

$$\frac{\partial^2 T}{\partial X^2} \Rightarrow \frac{T_{i+1} - 2T_i + T_{i-1}}{h^2} + O(h^2) \qquad (4.31)$$

Here the notation $O(h)$ is used to show that the error involved in the approximation is on the order of h. Similarly, $O(h^2)$ is for the approximation error on the order of h^2, which is more accurate than one on the order of h.

Substituting the finite-difference stencils into the electromagnetic and heat transfer partial differential equations gives the local approximation. By assembling all local approximations and taking into account the proper initial and boundary conditions, one can obtain a set of linear algebraic equations that can be solved with respect to unknown variables (i.e., T, **A**, **E**, **H**, or **B**) at each node of the mesh, either by iterative techniques or by direct matrix inversion methods. Since the matrices are sparsely occupied (nonzero only in the neighborhood of the diagonal), some simplification in the computational procedure is used.

In FDM, it is important that the boundaries of the mesh region coincide with the boundaries of the appropriate regions of the IHT system. Experience in using FDM in IHT computations has shown that noncoincidence of the boundaries has a strong negative effect on the accuracy of the calculation.

4.3.2 FINITE-ELEMENT METHOD

As with FDM, the FEM is a numerical technique for obtaining an approximate solution for different technical problems, including those encountered in IHT. FEM has had a particularly great deal of success in engineering practice and has become one of the most popular numerical tools for a variety of scientific and engineering tasks.

In the last two decades, several variations of finite-element models have been developed. Here are some of them:

- Weighted residual method (weak form of the governing equations)
- Different kinds of the Ritz method
- Different types of the Galerkin method
- Pseudovariational methods
- Methods based on minimization of the energy functional, etc.

Those models are used in applications such as electric machines, motors, circuit breakers, transformers, magnetic recording systems, test equipment, electrical heating, and IHT. Even among the FEM only, there are situations where one type of FEM works just fine and in another case will not be able to do a good job at all.

As described above, the FDM provides pointwise approximations; however, FEM provides an elementwise approximation to the governing equations. The FEM does not require a direct calculation of the governing equation. Due to the general postulate of the variational principle, the solution of electromagnetic field computation is typically obtained by minimizing the energy functional that corresponds to the governing equation (e.g., Equation 4.26 or Equation 4.27) instead of solving that equation directly. The energy functional is minimized for the integral over the total area of modeling, which includes workpiece, coil, flux concentrators, and the surrounding area.

The well-known principle of minimum energy [55,60–64,66,68,71,75,80–82] requires that the vector potential distribution correspond to the minimum of the stored field energy per unit length. As a result of that assumption, it is necessary to solve the global set of resulting simultaneous algebraic equations with respect to the unknown, for example, magnetic vector potential at each node in the region of evaluation. The formulation of the energy functional, its minimization to obtain a set of finite-element equations, and the solution techniques (the solver) were created for both two-dimensional and axisymmetric heat treatment problems.

Many geometric arrangements and shapes of finite elements are possible. Their deformability allows them, in fact, to satisfy regions of any shape, i.e., any geometry of workpiece, inductors, flux concentrators, flux robbers, etc. The most common two-dimensional finite element is the first-order triangle (Figure 4.11). In the axisymmetric cylindrical case, such a finite-element mesh may be represented as a set of rings. Each ring revolves around the axis of symmetry and has a triangular cross section (so-called triangular torus element).

The following are some general remarks concerning the finite-element discretization (mesh generation):

- The area of study is subdivided into nonoverlapping finite elements (so-called finite-element mesh, Figure 4.11). The sides of finite elements intersect at nodes. The number and location of these elements are matters of personal judgment, but to obtain reasonable accuracy of the numerical solution, the finite-element mesh has to be relatively fine (sizes of finite elements must be smaller) in the regions where the rate of change of the unknown (i.e., the magnetic vector potential) is high. Experience in using finite-element software at INDUCTOHEAT Inc., has shown that special effort should be made to obtain a fine mesh within three penetration depths in the workpiece. A higher frequency requires a finer mesh.
- All the finite elements should have the same unit depth in the Z-direction.
- The current density, flux density, conductivity, magnetic permeability, and other material properties are postulated to be constant within each element. At the same time, they can be different from element to element.
- The designer should take advantage of the symmetry involved in the system geometry, for example by disappearing normal derivative values of $\mathbf{A}$ along the symmetry.

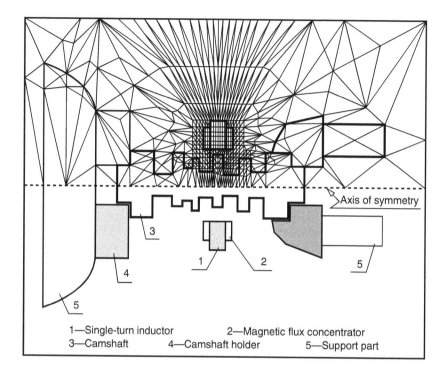

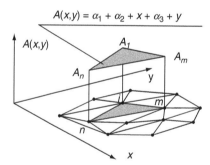

$$A(x,y) = \alpha_1 + \alpha_2 + x + \alpha_3 + y$$

FIGURE 4.11 Finite-element discretization.

In most typical cases, when it is necessary to obtain the electromagnetic field distribution and temperature profile along the length of a cylindrical workpiece, the FEM has been used in solving the governing equation with respect to magnetic vector potential for the electromagnetic problem (Equation 4.26 and Equation 4.27), and with respect to the temperature for the heat transfer problem (Equation 4.4 and Equation 4.5).

After solving the system of algebraic equations and obtaining the distribution of the magnetic vector potential in the region of modeling, it is possible to find all of the required output parameters of the electromagnetic field.

Induced current density in conductors:

$$\mathbf{J}_e = -j\omega\sigma\mathbf{A} \tag{4.32}$$

Total current density in the conductor:

$$\mathbf{J} = \mathbf{J}_s - j\omega\sigma\mathbf{A} \tag{4.33}$$

Magnetic flux density components $\mathbf{B}_x$ and $\mathbf{B}_y$ can be calculated from relationship (4.16) as follows [63,64]:

$$\frac{\partial \mathbf{A}}{\partial Y} = -\mathbf{B}_x; \quad \frac{\partial \mathbf{A}}{\partial X} = \mathbf{B}_y$$

Therefore, flux density can be obtained as

$$\mathbf{B} = \left[\mathbf{B}_x^2 + \mathbf{B}_y^2\right]^{1/2} \tag{4.34}$$

For the axisymmetric case of a cylindrical workpiece, the magnetic flux density components $\mathbf{B}_R$ and $\mathbf{B}_z$ can be calculated as

$$\mathbf{B}_R = -\frac{\partial \mathbf{A}}{\partial Z}; \qquad \mathbf{B}_z = \frac{\partial \mathbf{A}}{\partial R} + \frac{\mathbf{A}}{R}$$

Magnetic field intensity:

$$\mathbf{H} = \mathbf{B}/\mu\mu_0 \tag{4.35}$$

Electric field intensity:

$$\mathbf{E} = -j\omega\mathbf{A} \tag{4.36}$$

Electromagnetic force density in current-carrying conductors and the workpiece can be computed from the cross product of the vector of total current density and the vector of magnetic flux density:

$$F_x = \mathbf{J} \times \mathbf{B}_y; \quad F_y = -\mathbf{J} \times \mathbf{B}_x$$

From a vector potential solution it is possible to compute the other important quantities of the process such as stored energy, flux leakage, total power loss, and coil impedance.

Many worthwhile texts, conference proceedings, and articles have been written on the subject of finite-element modeling [55,60–83]. The large number of papers on the subject of FEM applications for electromagnetic field computations makes it impossible to mention all of the contributions. At the same time, some of the proposed finite-element models are similar in form. However, we would like to mention here that the first general nonlinear variational formulation of magnetic field analysis using FEM was presented by Chari [60] and Chari and Silvester [61]. Essential input into the development of FEM was provided by Lord, Udpa, Lavers, and many others [63–66,69,70,72,81,83]. The software that was developed and is in use at INDUCTOHEAT Inc., is based on the finite-element concept proposed by Lord [63,64] and Udpa and coworkers [72,73]. This concept was adapted to IHT needs and successfully used for several years.

4.3.3 A COMPARISON

Superficially, the FDM and FEM appear to be different; however, they are closely related. As outlined above, FDM starts with a differential statement of the problem of interest, and requires that the partial derivative of the governing equation be replaced by a finite-difference stencil to provide pointwise approximation. FEM starts with a variational statement and provides an elementwise approximation. Both methods discretize a continuous function (e.g., magnetic vector potential or temperature) and result in a set of simultaneous algebraic equations to be solved with respect to its nodal values. Therefore, the two methods are, in fact, quite similar.

Finite-difference stencils overlap one another, and in the case of complicated workpiece geometry they could have nodes outside the boundary of the workpiece, coil, or other components of the induction heating system. Finite elements do not overlap one another, do not have nodes outside the boundaries, and fit the complicated shape boundary perfectly. As shown above, in electromagnetic field computation finite elements are usually introduced as a way to minimize a functional. In fact, FDM can also be described as a form of functional minimization (so-called finite-difference energy method).

Therefore, FDM and FEM are different only in the choice of mesh generation and the way in which the global set of algebraic equations is obtained. They have approximately the same accuracy; however, required computer time and memory are often less when FDM is used. For example, the computer time needed to form global matrices is usually four to nine times greater with FEM than with FDM. As one would expect, a comparison of the efficiency of the two methods depends on the type of problem and program organization. Experience with both methods at INDUCTOHEAT shows that FDM is not well suited for an IHT system with complicated boundary shapes or in the case of a mixture of materials and forms (e.g., heat treating of camshafts, crankshafts, gears, and other critical components). In this case, FEM has a distinct advantage over FDM.

As shown above, both FDM and FEM require a network mesh of the area of modeling. That network includes induction coils, the workpiece, flux concentrators, etc. Unfortunately, to suit the condition of smoothness criteria and continuity of the governing differential equation it is also necessary to take into general consideration the airspace regions. Electromagnetic field distribution in the air, in most cases of coil designs, can be considered useless information, of interest only during the final design stage when evaluating electromagnetic field exposure of the induction heater. The need to always compute the electromagnetic field distribution in air can be considered a disadvantage of both the FDM and FEM. Another difficulty, which appears when using FDM or FEM for electromagnetic field computation, is how to treat an exterior region that extends to infinity. This deals with an infinite nature of electromagnetic wave propagation. Several methods have been used taking into account this phenomena of the infinite exterior region. Some of the methods are ballooning method, mapping technique, and combination of finite and infinite elements. However, each of the above-mentioned methods has certain shortcomings.

Therefore, one of the important improvements in using FDM or FEM in computation of IHT problems is the use of IBE. The theory of IBE is not easily explained. The mathematics required is more advanced than that needed for FDM or FEM. The interested reader will find several texts, conference proceedings, journal articles, and surveys [49,51,83–90] that describe various modifications of IBE. Here we will just mention that IBE will allow one to consider only conductive bodies in the computation. This significantly simplifies mesh generation, and decreases the computation time and required computer memory. Such improvements make this numerical software significantly more efficient.

4.3.4 Summary

We may summarize our introduction into numerical methods used for the simulation of IHT problems very simply. Each of the above-described methods has certain advantages. In many applications it is effective to use a combination of methods. The right choice of method depends on the specific application and features of the induction heat-treating process.

The use of modern software does not guarantee correct computational results. It must be used in conjunction with experience in numerical techniques and engineering knowledge to achieve the required accuracy of mathematical modeling. This is especially so because even in modern commercial software, regardless of the amount of testing and verification, a computation program may never have all of its possible errors detected. The engineer must consequently be on guard against various kinds of errors. The more powerful the software, the greater the probability of errors. Common sense and engineering gut feeling are always the analyst's helpful assistants.

Computer modeling provides the ability to predict how different factors may impact the transitional and final heat-treating conditions of the workpiece and what must be accomplished in the design of the induction heating system to improve the effectiveness of the process and guarantee the desired heat-treating results. As an example, Figure 4.12 shows the results of the magnetic field computation in the selective hardening of a workpiece with a complicated shape (e.g., a section of a camshaft or crankshaft). The proper choice of the coil and magnetic flux concentrator geometry, frequency, and power density allows the heating of the selected areas of the workpiece. Without using the flux concentrators, the electromagnetic

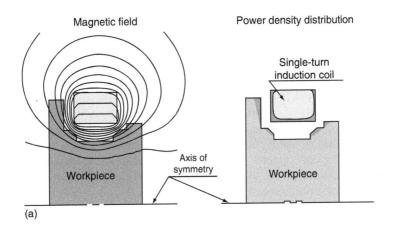

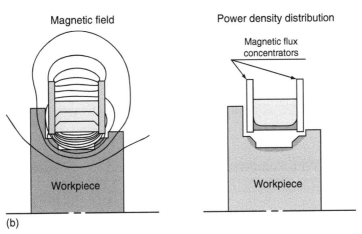

FIGURE 4.12 Results of numerical simulation of magnetic vector potential in selective hardening. Shown are field distributions (a) without and (b) with flux concentrator.

field distribution will cause the eddy current to be induced not only in the desired areas of the workpiece but also significantly in the adjacent areas. Heat produced in these areas as a result of the induced current could cause undesirable metallurgical changes.

The correct choice of flux concentrators, their location, and properties allows the designer to avoid undesirable heating of adjacent areas of the workpiece. Furthermore, as a result of using flux concentrators, the heat source intensity in the adjacent areas decreases because the concentrator provides the preferred path for the magnetic field. We show this magnetic field distribution (Figure 4.12) as an example of the use of numerical techniques for modeling the heat treatment process and obtaining a desired heating pattern. Later we discuss in detail the use of magnetic flux concentrators, their advantages and disadvantages. It will also be shown that in order to increase coil efficiency the flux concentrator should have a C shape. This will provide a preferred magnetic path in the sides and in the area external to the coil.

By combining the most advanced software with an outstanding computational and engineering background, modern IHT specialists possess a unique ability to analyze in a few hours complex induction heating problems that could take days or even weeks to solve using other methods including physical experiment. This leads to the saving of prototyping dollars and facilitates the building of reliable, competitive products in a short design cycle.

4.4 BASIC ELECTROMAGNETIC PHENOMENA IN INDUCTION HEATING

As shown earlier, an alternating voltage applied to an induction coil (e.g., solenoid coil) will result in an AC in the coil circuit. An alternating coil current will produce in its surrounding a time-variable magnetic field that has the same frequency as the coil current. This magnetic field induces eddy currents in the workpiece located inside the coil. Eddy currents will also be induced in other electrically conductive objects that are located near the coil. Induced currents have the same frequency as the coil current; however, their direction is opposite to the direction of the coil current. These currents produce heat by the Joule effect (I^2R). A conventional induction heating system that consists of a cylindrical load surrounded by a multiturn induction coil is diagrammed in Figure 4.13.

Due to several electromagnetic phenomena, the current distribution within an inductor and workpiece is not uniform. This heat source nonuniformity causes a nonuniform temperature profile in the workpiece. Nonuniform current distribution can be caused by several

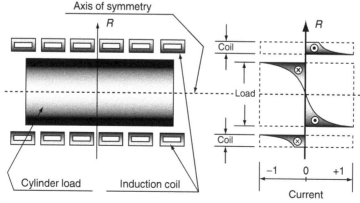

FIGURE 4.13 Current distribution in coil–workpiece induction heating system.

electromagnetic phenomena, including (1) the skin effect, (2) the proximity effect, and (3) the ring effect. These effects play an important role in understanding the induction heating phenomena [1–12,40,50–53,92,118–121].

4.4.1 Skin Effect

As one may know from the basics of electricity, when a direct current (DC) flows through a conductor that stands alone (bus bar or cable), the current distribution within the conductor's cross section is uniform. However, when an AC flows through the same conductor, the current distribution is not uniform. The maximum value of the current density will always be located on the surface of the conductor; current density will decrease from the surface of the conductor toward its center. This phenomenon of nonuniform current distribution within the conductor cross section is called the skin effect. The skin effect always occurs when there is an AC. Therefore, the skin effect will also be found in the workpiece located inside the induction coil (Figure 4.13). This is one of the major factors that cause the concentration of eddy current in the surface layer (skin) of the workpiece. Due to the circumferential nature of the eddy current induced in the workpiece, there is no current flow in its center.

Because of the skin effect, approximately 86% of the power will be concentrated in a surface layer of the conductor. This layer is called the reference (or penetration) depth (δ). The degree of skin effect depends on the frequency and material properties (electrical resistivity and relative magnetic permeability) of the conductor. There will be a pronounced skin effect when a high frequency is applied or when the radius of the workpiece is relatively large (Figure 4.14). The distribution of the current density along the workpiece thickness (radius) can be calculated by the equation

$$I_R = I_0 e^{-y/\delta} \tag{4.37}$$

where I_R is current density at distance R from the surface, A/m^2; I_0 is current density at the workpiece surface, A/m^2; y is the distance from the surface toward the core, m; and δ is penetration depth, m.

Penetration depth is described as

$$\delta = 503(\rho/\mu F)^{1/2} \text{ [m]} \tag{4.38}$$

where ρ is the electrical resistivity of the metal, Ω m; μ is the relative magnetic permeability; and F is the frequency, Hz (cycles per second), or

$$\delta = 3160(\rho/\mu F)^{1/2} \text{ [in.]} \tag{4.39}$$

where electrical resistivity is in Ω in.

Mathematically speaking, the penetration depth δ in Equation 4.37 is the distance from the surface of the conductor toward its core at which the current decreases exponentially to 1/exp its value at the surface. The power density at this distance will decrease to 1/exp^2 its value at the surface. Figure 4.15 shows the percentage reduction of current density and power density from the surface toward the core. As one can see from Figure 4.15, at one penetration depth from the surface ($y = \delta$), the current will equal 37% of its surface value. However, the power density will equal 14% of its surface value. From this we can conclude that 63% of the current and 86% of the power in the workpiece within a surface layer of thickness δ, will be concentrated.

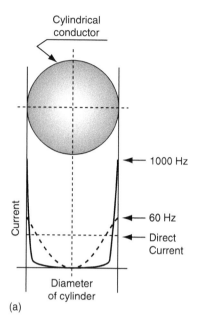

(a)

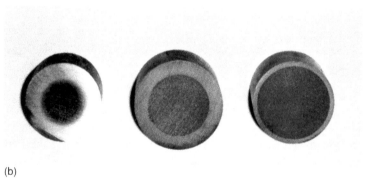

(b)

FIGURE 4.14 Skin effect. (a) Current distribution as function of frequency. (b) Variation in depth of skin effect results in different hardness patterns.

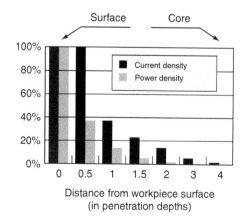

FIGURE 4.15 Current density and power density distribution due to the skin effect.

An analysis of Equation 4.38 and Equation 4.39 shows that the penetration depth has different values for different materials and is a function of frequency. The magnetic permeability μ of a nonmagnetic workpiece is equivalent to that of air and is assigned a value of 1. The electrical resistivity of metals ρ is a function of temperature (Figure 4.8). During the heating cycle ρ can increase to four to five times its initial value. Therefore, even for nonmagnetic metals, during the heating cycle the penetration depth can increase significantly. Table 4.2 shows some penetration depths of metals that are most commonly used with induction heating.

In contrast to nonmagnetic metals, the permeability of magnetic steels commonly used in heat treating can vary from 1 to more than 500, depending on the magnetic field intensity **H** and temperature. The magnetization curve describes the nonlinear relationship between magnetic flux density **B** and magnetic field intensity **H**. The nonlinear variation of $\mu = \mathbf{B}/(\mathbf{H}\mu_0)$ for a typical carbon steel is shown in Figure 4.16. Permeability is given by the ratio **B/H**. The maximum permeability occurs at the knee of the curve. The magnetic field intensity $\mathbf{H}_{cr}$ that corresponds to the maximum permeability is called a critical value of **H**. When $\mathbf{H} > \mathbf{H}_{cr}$, the magnetic permeability will decrease with increasing **H**. If $\mathbf{H} \to \infty$, then $\mu \to 1$. In conventional induction heat treating, the magnetic field intensity $\mathbf{H}_{surf}$ at the workpiece surface is much greater than $\mathbf{H}_{cr}$. However, there are some heat treatment applications where this is not true and this phenomenon will play an important role.

Because of the items discussed above, the same kind of carbon steel at the same temperature and frequency can have different penetration depths due to differences in the intensity of the magnetic field. Similar to the current distribution, the magnetic field intensity is at its maximum value at the surface of the homogeneous workpiece and falls off exponentially toward the core (Figure 4.17). As a result, the magnetic permeability varies within the magnetic body. At the surface, μ_{surf} corresponds to the surface magnetic field intensity $\mathbf{H}_{surf}$. In quick calculations $\mathbf{H}_{surf}$ can be considered as the field intensity in the air gap between the coil and the workpiece. With increasing distance from the surface, μ increases and after reaching its maximum value at $\mathbf{H} = \mathbf{H}_{cr}$ begins to fall off (Figure 4.17).

While discussing the behavior of μ within the ferromagnetic workpiece, it is necessary to mention that a definition of penetration depth of current into a magnetic body in its classical forms (Equation 4.38 and Equation 4.39), does not have a fully determined meaning because of the nonconstant distribution of μ within the workpiece. In engineering practice, the value of relative magnetic permeability at the surface of the workpiece is typically used to give a determination of those equations in definite form. Here we will also use the value of μ_{surf} to determine the penetration depth in the magnetic workpiece. Figure 4.18 and Table 4.3 show the value of the penetration depth in carbon steel (1045) at an ambient temperature (21°C or 70°F) as a function of frequency and magnetic field intensity **H** at the workpiece surface.

From another perspective, penetration depth is a function of temperature as well. At the beginning of the heating cycle, the current penetration into the carbon steel workpiece will increase slightly (Figure 4.19) because of the increase in electrical resistivity of the metal with temperature. With a further rise of temperature (at approximately 550°C or 1022°F), μ starts to decrease more and more. Near a critical temperature T_c known as the Curie temperature or Curie point, permeability drastically drops to unity because the metal becomes nonmagnetic.

As a result, the penetration depth will increase significantly (Table 4.4). After heating above the Curie temperature, the penetration depth will continue to increase due to the increase in electrical resistivity of the metal (Figure 4.19). However, the rate of growth will not be as significant as it was during the transition through the Curie temperature.

Typically, the variation of δ during the induction heating of a carbon steel workpiece drastically changes the degree of skin effect. It is especially important to take this phenomenon

TABLE 4.2
Penetration Depth of Nonmagnetic Metals (mm)

Metal	T °C	°F	ρ μΩ m	μΩ in.	0.06	0.5	1	2.5	4	8	10	30	70	200	500
Aluminum	20	68	0.027	1.06	10.7	3.70	2.61	1.65	1.30	0.92	0.83	0.48	0.31	0.18	0.12
	250	482	0.053	2.09	15.0	5.18	3.66	2.32	1.83	1.29	1.16	0.67	0.44	0.26	0.16
	500	932	0.087	3.43	19.2	6.64	4.69	2.97	2.35	1.66	1.48	0.86	0.56	0.33	0.21
Copper	20	68	0.018	0.71	8.81	3.05	2.16	1.36	1.08	0.76	0.68	0.39	0.26	0.15	0.10
	500	932	0.050	1.97	14.5	5.03	3.56	2.25	1.78	1.26	1.12	0.65	0.43	0.25	0.16
	900	1652	0.085	3.35	19.3	6.67	4.72	2.98	2.36	1.67	1.49	0.86	0.56	0.33	0.21
Brass	20	68	0.065	2.56	16.6	5.74	4.06	2.56	2.03	1.43	1.28	0.74	0.48	0.29	0.18
	400	752	0.114	4.49	21.9	7.60	5.37	3.40	2.69	1.90	1.70	0.98	0.64	0.38	0.24
	900	1632	0.203	7.99	29.3	10.1	7.17	4.53	3.58	2.53	2.27	1.31	0.86	0.51	0.32
Stainless steel	20	68	0.690	27.2	53.9	18.7	13.2	8.36	6.61	4.67	4.18	2.41	1.58	0.93	0.59
	800	1472	1.150	45.3	69.6	24.1	17.1	10.8	8.53	6.03	5.39	3.11	2.04	1.21	0.76
	1200	2192	1.240	48.8	72.3	25.1	17.7	11.2	8.86	6.26	5.60	3.23	2.12	1.25	0.79
Silver	20	68	0.017	0.67	8.34	2.89	2.04	1.29	1.02	0.72	0.65	0.37	0.24	0.14	0.09
	300	572	0.038	1.50	12.7	4.39	3.10	1.96	1.55	1.10	0.98	0.57	0.37	0.22	0.14
	800	1472	0.070	2.76	17.2	5.95	4.21	2.66	2.10	1.49	1.33	0.77	0.50	0.30	0.19
Tungsten	20	68	0.050	1.97	14.5	5.03	3.56	2.25	1.78	1.26	1.12	0.65	0.43	0.25	0.16
	1500	2732	0.550	21.7	48.2	16.7	11.8	7.46	5.90	4.17	3.73	2.15	1.41	0.83	0.53
	2800	5072	1.040	40.9	66.2	22.9	16.2	10.3	8.11	5.74	5.13	2.96	1.94	1.15	0.73
Titanium	20	68	0.500	19.7	45.9	15.9	11.3	7.11	5.62	3.98	3.56	2.05	1.34	0.80	0.50
	600	1112	1.400	55.1	76.8	26.6	18.8	11.9	9.41	6.65	5.95	3.44	2.25	1.33	0.84
	1200	2192	1.800	70.9	87.1	30.2	21.3	13.5	10.7	7.54	6.75	3.90	2.55	1.51	0.95

Frequency (kHz) spans columns 0.06 through 500.

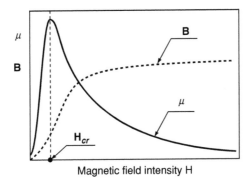

FIGURE 4.16 Magnetic field density (**B**) and relative magnetic permeability (μ).

into account when designing for induction through hardening of carbon steel where the core-to-surface temperature difference is primarily a result of the skin effect.

4.4.2 ELECTROMAGNETIC PROXIMITY EFFECT

When we discussed the skin effect in conductors or cables, we assumed that a conductor stands alone and that there were no other current-carrying conductors in the surrounding area. In most practical applications this is not the case. Most often there are other conductors in close proximity. These conductors have their own magnetic fields, which interact with nearby fields, and as a result the current and power density distribution will be distorted.

An analysis of the effect on current distribution in a conductor when another conductor is placed nearby is given below. Figure 4.20a shows the skin effect and magnetic field distribution in a conductor (i.e., cylindrical bar) that stands alone. When another conductor is placed near the first one, the currents in both conductors will redistribute. If the currents flowing in the bars have opposite directions, then both currents (Figure 4.20b) will be concentrated in the areas facing each other (internal areas). However, if the currents have the same direction, then these currents will be concentrated on opposite sides of the conductors (Figure 4.20c). The same will be true with bus bars (Figure 4.21).

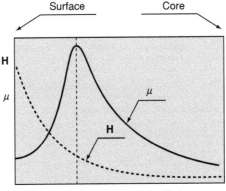

FIGURE 4.17 Distribution of magnetic field intensity (**H**) and relative magnetic permeability (μ) along the radius of a carbon steel homogeneous cylinder.

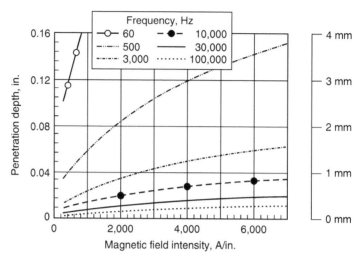

FIGURE 4.18 Current penetration depth in carbon steel (1045) at ambient temperature 21°C (70°F).

When the currents flow in opposite directions, a strong magnetic field forms in the area between the bus bars (Figure 4.21). This occurs because in this area the magnetic field lines that are produced by each bus bar have the same direction. Therefore, the resulting magnetic field between the bus bars will be very strong. However, because the current is concentrated in the internal areas, the external magnetic field will be weak. The external magnetic fields will have opposite directions and will tend to cancel each other. This phenomenon is used in coaxial cables. The opposite is true if the currents have the same direction, for then the magnetic field lines will have opposite directions in the area between bus bars, and therefore they will cancel each other in that area. Because of this cancellation, a weak magnetic field will exist between the bus bars. However, the external magnetic field will be quite strong because the magnetic lines produced by the two conductors will have the same direction in the external area.

TABLE 4.3
Penetration Depth in Carbon Steel 1040 at Ambient Temperature of 21°C (70°F)

		\multicolumn{12}{c}{Frequency (Hz)}										
		\multicolumn{2}{c}{60}	\multicolumn{2}{c}{500}	\multicolumn{2}{c}{3,000}	\multicolumn{2}{c}{10,000}	\multicolumn{2}{c}{30,000}	\multicolumn{2}{c}{100,000}
\multicolumn{2}{c}{Magnetic Field Intensity}	\multicolumn{12}{c}{Penetration Depth}											
A/mm	A/in.	mm	in.	mm	in.	mm	in.	mm	in.	mm	in.	mm	in.
10	250	2.5	0.100	0.88	0.034	0.36	0.014	0.2	0.008	0.11	0.004	0.06	0.002
40	1,000	4.7	0.185	1.63	0.064	0.67	0.026	0.36	0.014	0.21	0.008	0.12	0.005
80	2,000	6.3	0.249	2.2	0.086	0.9	0.035	0.49	0.019	0.28	0.011	0.16	0.006
120	3,050	7.76	0.306	2.69	0.106	1.1	0.043	0.6	0.024	0.35	0.014	0.19	0.007
160	4,050	8.76	0.345	3.03	0.119	1.24	0.049	0.68	0.027	0.39	0.015	0.21	0.008
200	5,100	9.63	0.379	3.33	0.131	1.36	0.054	0.75	0.029	0.43	0.017	0.24	0.009
280	7,100	11.2	0.442	3.89	0.153	1.59	0.062	0.87	0.034	0.50	0.02	0.27	0.011

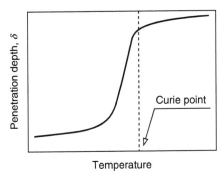

FIGURE 4.19 Typical variation of current penetration depth during induction heating of carbon steel workpiece.

If the distance between bars increases, then the strength of the proximity effect will decrease. Proximity effect in the case of nonsymmetrical systems is shown in Figure 4.22.

The phenomenon of proximity effect can be directly applied in induction heating. Induction systems consist of two conductors [41]. One of these conductors is an inductor that carries the source current (Figure 4.23), and the other is the workpiece that is located near the inductor. Eddy currents are induced in the workpiece by an external alternating magnetic field of the source current (Figure 4.23b). As shown above, due to Faraday's law (Equation 4.7), eddy currents induced within the workpieces have a direction opposite to that of the source current of the inductor. Therefore, due to the proximity effect, the coil current and workpiece eddy currents will concentrate in the areas facing each other (Figure 4.23a). This is the second factor that causes a current redistribution in an induction heating system as shown in Figure 4.13.

Figure 4.24 shows how the electromagnetic proximity effect produces different heating patterns. A carbon steel cylinder is located nonsymmetrically inside a single-turn inductor. If the cylinder is static (does not spin), then two different patterns will develop in its cross section. The appearance of these patterns is caused by a difference in the eddy current distribution in the cylinder. As shown in Figure 4.24, the eddy currents have a higher density in the workpiece area

TABLE 4.4
Penetration Depth in Carbon Steel 1040 at Ambient Temperature of $621^{\circ}C$ ($1,150^{\circ}F$)

		Frequency (Hz)										
		60		500		3,000		10,000		30,000		100,000	
Magnetic Field Intensity		Penetration Depth											
A/mm	A/in.	mm	in.	mm	in.	mm	in.	mm	in.	mm	in.	mm	in.
10	250	5.6	0.337	2.97	0.117	1.21	0.048	0.66	0.026	0.38	0.015	0.21	0.008
40	1,000	15.5	0.611	5.38	0.212	2.20	0.086	1.20	0.047	0.69	0.027	0.38	0.015
80	2,000	20.9	0.824	7.25	0.285	0.96	0.117	1.62	0.064	0.94	0.037	0.51	0.02
120	3,050	24.5	0.966	8.50	0.335	3.47	0.137	1.9	0.075	1.1	0.043	0.60	0.024
160	4,050	27.4	1.08	9.48	0.373	3.87	0.152	2.12	0.083	1.22	0.048	0.67	0.026
200	5,100	29.8	1.17	10.3	0.406	4.2	0.166	2.31	0.091	1.33	0.052	0.73	0.029
280	7,100	33.5	1.32	11.6	0.457	4.74	0.187	2.60	0.102	1.50	0.059	0.82	0.032

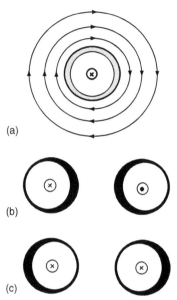

FIGURE 4.20 Proximity effect in cylindrical conductors: (a) single cable; (b) cables with opposite currents; (c) cables with similar currents.

where the coil–workpiece air gap is small (good coupling). Therefore, there will be intense heating due to the Joule effect. As a result, the heat pattern will be relatively narrow and deep. A lower frequency will result in a wider heat pattern and vice versa.

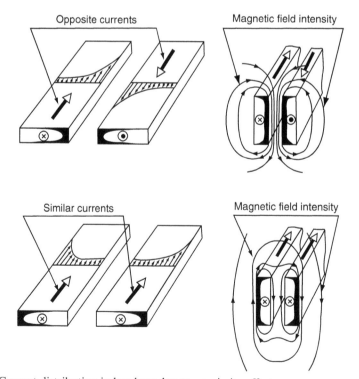

FIGURE 4.21 Current distribution in bus bars due to proximity effect.

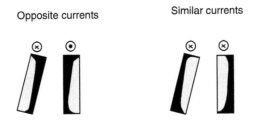

FIGURE 4.22 Proximity effect in nonsymmetrical systems.

In the area with the larger air gap (poor coupling) the temperature rise will not be as significant as in the case of good coupling. Also, the heat pattern will be much wider and more shallow.

In the case of an unequal coil–workpiece air gap, an almost identical heat pattern can be obtained by rotating the workpiece.

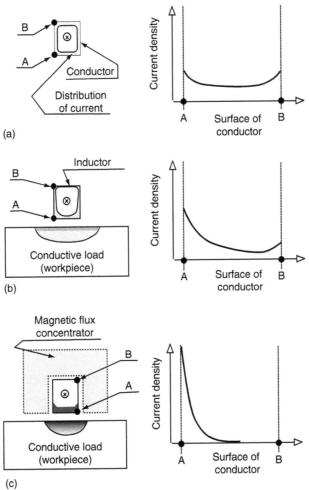

FIGURE 4.23 (a) Current distribution in a straight conductor. (b) Current redistribution due to the proximity effect. (c) The slot effect.

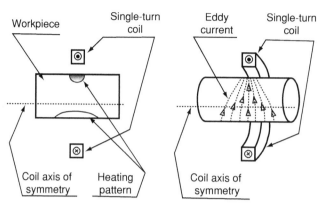

FIGURE 4.24 Proximity effect in nonsymmetrical single-turn inductors.

An understanding of the physics of the electromagnetic proximity effect and skin effect is important not only in induction heating but also in power supply and bus bar design. The proper design of a bus bar network will significantly decrease its impedance.

There is another electromagnetic effect that is related to the proximity effect. This is called the slot effect.

4.4.3 ELECTROMAGNETIC SLOT EFFECT

When we discussed the proximity effect, we first introduced the current distribution in a stand-alone conductor (Figure 4.23a) and then observed the current redistribution when a conductive load (workpiece) was located near this conductor (Figure 4.23b). As shown in Figure 4.23b, a significant part of the conductor's current will flow near the surface of the conductor that faces the load. The remainder of the current will be concentrated in the sides of the conductor [141].

Continuing our study, let us locate an external magnetic flux concentrator (e.g., C-shaped laminations) around this conductor as shown in Figure 4.23c. As a result, practically all of the conductor's current will be concentrated on the surface facing the workpiece. The magnetic concentrator will squeeze the current to the open surface of the conductor, in other words to the open area of the slot. It is necessary to mention here that the slot effect will also take place without the workpiece. In this case, the current will be slightly redistributed in the conductor, but most of it will still be concentrated in the open surface area. This effect always occurs when there is a conductor located within the magnetic slot. The actual current distribution in

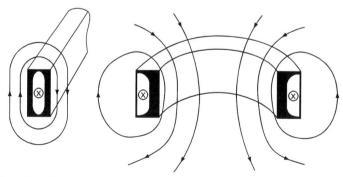

FIGURE 4.25 Ring effect in rectangular conductors.

the conductor depends on the frequency, magnetic field intensity, geometry, and electromagnetic properties of the conductor and the concentrator.

Slot and proximity effects play a particularly important role in the proper design of coils for selective induction hardening including channel, hairpin, odd-shaped, spiral–helical, and pancake types of inductors.

The slot effect is widely used not only in connection with induction heating but also in the design of other industrial machines such as motor generators and AC and DC machines.

4.4.4 ELECTROMAGNETIC RING EFFECT

Up to now we have discussed current density distribution in straight conductors. One such conductor, a rectangular bus bar, and its current distribution are shown in Figure 4.25. If that bar is bent to shape it into a ring, then its current will be redistributed also. Magnetic flux lines will be concentrated inside the ring, and therefore the density of the magnetic field will be higher inside the ring. Outside the ring, the magnetic flux lines will be disseminated. As a result, most of the current will flow within the thin inside surface layer of the ring [5]. As one can see, this ring effect is somewhat similar to the proximity effect. Figure 4.26 also shows the appearance of the electromagnetic ring effect in cylinders. As one can see, this effect leads to a concentration of current on the inside surface of the induction coil. The ring effect takes place not only in single-turn inductors but also in multiturn coils. Therefore, it is the third electromagnetic effect that is responsible for the current distribution in an induction system shown in Figure 4.13.

The appearance of the ring effect can have a positive or negative effect on the process. For example, in conventional induction heating of cylinders, when the workpiece is located inside the induction coil this effect plays a positive role because in combination with the skin and proximity effects it will lead to a concentration of the coil current on the inside diameter (I.D.) of the coil. As a result, there will be close coil–workpiece coupling, which leads to good coil efficiency.

The ring effect plays a negative role in the induction heating of internal surfaces (so-called I.D. heating), where the induction coil is located inside the workpiece. In this case, this effect leads to a coil current concentration on the I.D. of the coil. This makes the coil–workpiece coupling poor and therefore decreases coil efficiency. However, despite the ring effect, the proximity effect here tends to move the coil current to an outside surface of the coil. Therefore, the coil current distribution in such application is a result of two counteracting phenomena: the proximity and ring effects. It should be mentioned here that in order to help the proximity effect dominate the ring effect, in a great majority of I.D. heating applications, a magnetic flux concentrator is located inside of the coil. This allows a slot effect to appear

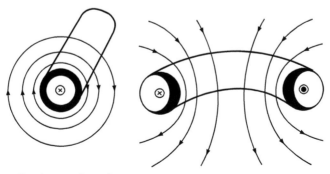

FIGURE 4.26 Ring effect in round conductors.

and assist the proximity effect to increase the coil efficiency and overcome the ring effect. The ring effect plays an important role in power supply design also. Because of this effect, the current is concentrated in areas where bus bars are bent, which leads to undesirable over-heating of certain areas of the bus bar. To avoid local overheating it is necessary to take this effect into account when designing the cooling circuit for the bus bar.

4.4.5 HEATING AND COOLING DURING INDUCTION HEAT TREATMENT

Up to this point we have been describing the electromagnetic process that takes place in induction heating. This must now be combined with the generation of heat that is produced by the Joule effect in the workpiece (I^2R). To obtain a complete picture of the induction heating process, we must also examine the heat transfer process. To simplify this, let us discuss the features of induction heating of a carbon steel cylinder located inside a solenoidal coil. The process of induction heating of a magnetic cylinder from an ambient temperature or a temperature below the Curie point to a temperature above it has several features that should be taken into account when designing a modern induction system.

Figure 4.27 shows the dynamics of the change of surface and core temperatures with constant current applied to the coil. As shown above, if the temperature of carbon steel is changed, some of the properties that affect the electroheating process also change. The

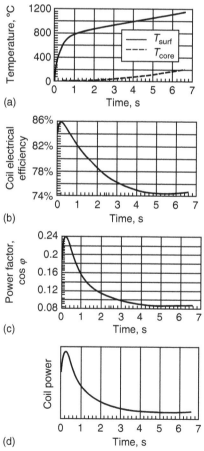

FIGURE 4.27 Change of coil parameters with time during surface hardening of carbon steel cylinder. (Coil current is constant.)

most significant of these are specific heat, electrical resistivity, and magnetic permeability (Figure 4.5, Figure 4.8, and Figure 4.9). At the first stage of the heating cycle the entire workpiece is magnetic, magnetic permeability is quite large, current penetration depth is very small (according to Equation 4.38 and Equation 4.39), and therefore the skin effect is pronounced. At the same time, because of the relatively low temperature, the heat losses from the cylinder surface at this stage are relatively low. Because of the pronounced skin effect, the induced power appears in the fine surface layer of the workpiece. This leads to a rapid increase in temperature at the surface with no change at the core. Figure 4.28a shows the temperature and heat source distribution along the radius of the workpiece at this stage. The maximum temperature is located at the surface. This stage is characterized by intensive heating and the existence of a large temperature differential within the workpiece. As one can see from Figure 4.28a, the temperature profile does not match the heat source profile because of thermal conductivity λ, which spreads the heat from the surface toward the core.

During this stage, the electrical efficiency (Figure 4.27b) increases due to an increase in electrical resistivity ρ of the metal with temperature (Figure 4.8). At the same time, the magnetic permeability μ remains high, and a slight reduction of μ does not affect the rise in electrical efficiency. After a short time, electrical efficiency reaches its maximum value and

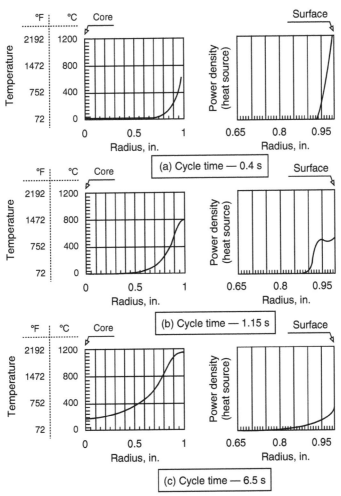

FIGURE 4.28 Temperature and power density (heat source) profiles at different stages of induction heating of 2 in. cylinder (frequency is 25 kHz).

then starts to decrease. The surface will reach the Curie temperature first, and after that the heat intensity at the surface will significantly decrease. This will take place primarily for two reasons:

1. Specific heat has its maximum value (a peak) near the Curie point (Figure 4.5). The value of the specific heat denotes the amount of energy that must be absorbed by the metal to achieve the required heat. Therefore, that peak leads to a decrease in heat intensity in the surface.
2. Steel in the surface area loses its magnetic properties and μ drops to 1. As a result, the surface power density and heat sources also will decrease.

Figure 4.28b shows the temperature profile and heat source distribution along the radius of the cylinder after the surface temperature passes the Curie point (second stage). At this stage, the electrical resistivity of the carbon steel increases approximately two- to threefold compared to its value in the initial stage. A decrease of μ and an increase of ρ cause a six- to tenfold increase in penetration depth from its value in the initial stage [93]. Most of the power is now induced in the surface and the internal layers of the workpiece. This stage can be characterized as the biproperties stage of steel. The workpiece surface becomes nonmagnetic; however, the internal layers of the bar remain magnetic. This stage takes place until the thickness of the nonmagnetic layer is less than the penetration depth in hot steel. Heat sources have a unique waveshape (waveform) that is different from the classical exponential distribution. Figure 4.28b shows that at the cylinder surface there is a maximum of the heat sources. Then the heat sources decrease toward the core. However, at a distance of 1.4 mm from the surface, the heat sources start to increase again. This takes place because of the remaining magnetic properties of the steel at this distance. It is necessary to mention here that in some applications, due to the biproperties phenomenon, the maximum value of heat sources can be located in an internal layer of the workpiece and not on its surface.

It is particularly important to take into account the existence of the biproperties structure in designing the contour-hardening processes.

Finally, the thickness of the surface layer with nonmagnetic properties exceeds the penetration depth in hot steel and the biproperties phenomenon is not pronounced and will finally disappear. The power density will then have its classical exponential distribution (Figure 4.28c).

Time and temperature are the two major factors responsible for establishing the required metal structure and providing efficient control of process parameters such as distortion. Rapid heating will tend to heat only the outer surface layer; in effect the hardened depth is approximately equal to the heated depth. This results in a very small transition area. As time is increased and power density is reduced, the temperature within the required depth of hardening will remain approximately the same; however, the temperature of internal areas of the workpiece will increase.

As seen in Figure 4.29a, a heating time of 1.5 s shows no rise in core temperature and a slight increase in temperature in the internal areas of the workpiece. A heating time of 3.0 s (Figure 4.29b) leads to a small rise in core temperature, while a heating time of 10.0 s (Figure 4.29c) leads to a significant rise in core temperature. In all the three cases, the required temperature at the hardened depth (2 mm from the surface of the workpiece) is approximately the same. Below the hardened depth is where the difference is seen as the temperature of internal layers begins to increase. Physics indicates that the more heat induced in the workpiece, the greater the mass heated and the greater the expansion, thus leading to more distortion.

From the point of view of decreasing part distortion, it is desirable to have the heating time as short as possible. However, there are more limitations. First, the material must reach the minimum required temperature at the depth to be hardened. If the frequency or surface

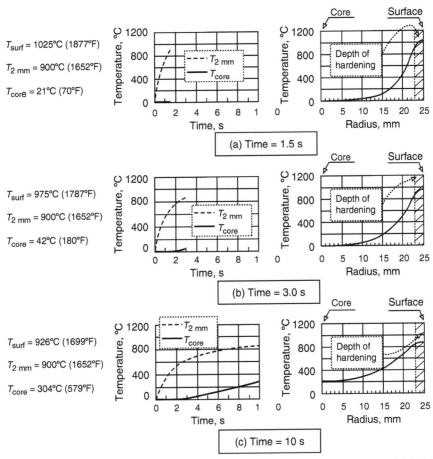

FIGURE 4.29 Temperature profiles during different stages of induction hardening (10 kHz). $T_{2\,mm}$ internal temperature (2 mm below surface).

power density is too high and the skin effect is pronounced when the temperature at this depth reaches the required value, then the surface could overheat or even melt. Second, as a result of the short cycle time, large temperature gradients will be present and thermal stresses can reach their critical value. Chapter 10 of *Steel Heat Treatment: Metallurgy and Technologies* of consists of a detailed analysis of the cracking, distortion, and residual stresses that appear during heat treatment of the metals.

As one can see, the choice of frequency is not as easy a task as it would seem to be at first glance and requires a detailed evaluation of the entire process. Practical recommendations in choosing the frequency and other design parameters are presented in the next section.

Up to now, we have discussed the heating process during IHT. However, as shown in Chapter 9 of *Steel Heat Treatment: Metallurgy and Technologies*, quenching is also one of the most important components of the heat treatment process. As an example, Figure 4.30 shows the dynamics of the induction heating of a carbon steel cylinder and its cooling during quenching [93]. As discussed later in this chapter, with IHT of carbon steel, the quenching must typically begin immediately after the required heating temperature is reached. As shown in Figure 4.30, after 4.1 s of induction heating the surface layer reached its required final temperature (approximately 1050°C or 1922°F). The core temperature did not rise significantly because of several factors such as a pronounced skin effect, the quite intensive heating process,

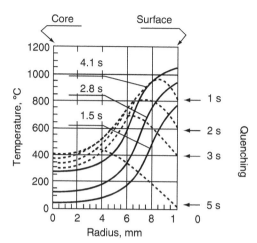

FIGURE 4.30 Dynamics of induction heating (—) of a carbon steel cylinder (20-mm outside diameter) and its cooling during quenching (---) ($F = 10$ kHz).

and short heating time. Because of these factors the heat soak from the surface of the workpiece toward its core was not sufficient.

After the heating stage is complete, the quenching process begins. In the first stage of quenching the high temperature of the workpiece surface layer begins to lessen. Figure 4.30 shows that after 2 s of quenching the surface temperature will be reduced by as much as 450°C. This results in a workpiece surface temperature of 600°C (1112°F). At this point, the maximum temperature will be located at a distance of 3 mm inward from the surface. It is necessary to note that at this stage two heat transfer phenomena will take place. First, the surface layers will cool as discussed above. The second phenomenon is the heating (soaking) effect of the core. After 5 s of quenching the surface temperature will decrease almost to the temperature of the quenchant. At the same time, the core is still quite warm. Core temperature is less than 400°C (752°F). In some cases, heat treaters do not cool the part completely. After the part is unloaded from the induction coil, it is kept for some time on the shop floor at an ambient temperature. During that time, the heat of the warm core soaks toward the surface. In time, the temperature distribution within the part will equalize. In this case, the remaining heat is used for slight temper-back, which gives the part a nonbrittle structure.

The rate of cooling during the quenching process depends on several factors, including the type of quenchant, temperatures, flow rate, geometry, and material properties of the metal. The design features of quenching are discussed in the following section.

4.5 COIL CONSTRUCTION AND DESIGN CONSIDERATIONS

The terms inductor, coil, or work coil refer to a current-carrying conductor near a part or workpiece to be heated. The AC in this inductor creates a time-variable magnetic field that links the inductor and the workpiece.

As shown earlier, by Faraday's law (Equation 4.7), this varying magnetic field causes a voltage to be induced or magnetically coupled from the inductor to the workpiece. The magnitude of this induced voltage is determined by the rate of change of the magnetic flux in the workpiece as well as the number of turns in the induction coil ($e = -N \, d\Phi/dt$). The presence of voltage or potential difference in the workpiece leads to the flow of current and subsequent I^2R loss or power dissipated in the workpiece. This power loss depends on the workpiece resistivity and manifests itself in the heating of the workpiece. This is the unique process referred to as induction heating.

The heat-treating coil is usually mounted to a transformer for impedance matching. Impedance (load) matching in IHT is discussed in Chapter 5. In this section we discuss the basics of inductor design and some of the thought processes it involves. The major intent of this section is to discuss some principles of coil design and to share experience and knowledge in building IHT equipment. We also discuss cases in which workpieces or parts can only be heat treated by induction and others in which induction can be considered an option along with other kinds of heat treatment. Practical advice and recommendations are presented here as well.

Inductors or coils are typically made of copper because of its high electrical conductivity and therefore low-power loss. Other important reasons for using copper in coil manufacture are that copper is relatively inexpensive and has good thermal and mechanical properties. Oxygen-free high-conductivity (OFHC) copper is commonly used for high-power or high-frequency applications. Coils made of OFHC copper are expected to have a longer life than coils constructed of commercial grade copper. Inductors are constructed of tubing or solid machined blocks and are almost always water-cooled. The tubing may be as small as 1/8 in. in diameter and is usually no longer than a rectangular size of 0.5×1.5 in. Tubing can be wrapped or brazed together to produce wider turns. It must be large enough to permit adequate flow of water for cooling. In cases where water passages will not allow adequate flow, a high-pressure booster pump is required. With machined coils, water passages must be smooth and free of burrs.

The formula that expresses adequate flow in the coil is

$$\text{gpm} = \frac{PK_1 K_2}{K_3 \Delta T} \qquad (4.40)$$

where gpm is gallons per minute, P is the total coil power, kW; K_1 is a tubing coefficient (for the great majority of high-frequency induction heat-treating applications $K_1 = 0.5$); $K_2 = 3415$ is a convention constant that is derived from Btu/kWh; K_3 is a conversion constant that represents the heat capacity of water (typically $K_3 = 500$); and ΔT is the permissible temperature rise in the cooling water, normally 40°F or less, to allow for proper cooling of the copper.

ΔT can be measured by installing an in-line thermometer as close to the water outlet as possible. However, it is necessary to keep in mind that some coil designs in heat treatment applications are dictated by the required heat treatment pattern. Therefore, there could be situations where adequate water flow is not obtained, thus the coil life will be limited.

In past years, IHT equipment was cooled by using process water from cooling towers or from a local source such as city water. Water quality would vary widely from facility to facility. It was found that electrolysis can occur in power equipment with high DC voltage potentials in the water paths, which can result in corroded water pipes and plugged circuits. This problem is compounded greatly by the presence of any ferrous metals in the water. Thus there was a need for a closed-loop water system, in which water quality can be controlled very closely. The entire loop can be constructed of nonferrous metals. Once the closed-loop system is charged with nonconductive, acid-free water it will stay relatively nonconductive with very little maintenance. Sometimes slightly alkaline water can be used.

In applications that require a water-recirculating system, that system can be bolted directly to the power supply or it can be a stand-alone unit. It may also cool more than one induction system, including coils and solid-state power supplies. Usually closed-loop water-recirculating systems have a plate heat exchanger, which allows heat to be removed quickly and efficiently. Typically, the heat exchanger is made with carefully selected, well-engineered components that resist corrosion and ion contribution, ensuring a long life and trouble-free operation. Plumbing requires a special consideration. Ferrous metals or aluminum should not be used for plumbing. Copper, brass, polyvinyl chloride (PVC), or certain types of stainless steel are good choices for pipes, tubes, hoses, etc. Because ionized water or any magnetic particles in the system can cause electrolysis or arcing, deionized water is used to protect the electronic devices within the power supply. However, in newly designed power supplies (i.e., IGBT and FET styles) the water does not come in contact with high AC voltage potentials, thus eliminating the electrolysis problem. The water system, if used, should be sized for the entire induction system (i.e., based on the total heat dissipation required). Two of the most popular compact closed-loop water-cooling and -recirculating systems are shown in Figure 4.31 along with some data on three models of each [95,96].

Coil-cooling design is a very important part of induction heater design. Because of electromagnetic phenomena discussed earlier (skin effect, proximity effect, and ring effect), the current distribution in the conductors can be significantly nonuniform. Therefore, in some applications the location of cooling is critical. It is obvious that the coil cooling should be as

Stand-alone water system

Model No.	Plant water flow (gpm)	Output water flow (gpm)	Heat dissipation (Btu/HR)	Floor plan dimensions
46P0210	37	100	200,000	29 in. × 53 in.
46P0410	73	100	400,000	29 in. × 53 in.
46P0815	146	150	800,000	29 in. × 53 in.

(a)

FIGURE 4.31 (a) INDUCTOHEAT's stand-alone closed-loop water-cooling and-recirculating systems.

Bolt on stlye

Model Desc.	Minimum flow (gpm)	Heat dissipation (Btu/HR)	Floor plan dimensions
UNICOOL	10	90,000	12 in. × 23.5 in.
UNICOOL	20	170,000	12 in. × 23.5 in.
UNICOOL	40	265,000	12 in. × 23.5 in.

(b)

FIGURE 4.31(Continued) (b) INDUCTOHEAT's compact closed-loop water-cooling and-recirculating systems.

close as possible to the heating face. The heating face wall thickness should increase as the frequency decreases. This fact is directly related to the penetration (reference) depth of current in the copper and holds for both solid machined and tubing coils. For quick estimation of the penetration depth in the copper, δ_1 can be calculated with the formulas

$$\delta_1 = \frac{70}{\sqrt{F}} \text{ [mm]} \quad \text{or} \quad \delta_1 = \frac{2.75}{\sqrt{F}} \text{ [in.]} \qquad (4.41)$$

where F is frequency, in Hz.

The most effective thickness of the conductive part of the tubing wall (d_1) can be calculated as $d_1 \cong 1.6\delta_1$. A tubing wall of the induction coil smaller than $1.6\delta_1$ results in a reduction in coil efficiency.

In some cases the tubing wall may be thicker than the calculated penetration depths. This is because it may be not mechanically practical to use a tubing wall thickness of 0.01 in. (e.g., in the case of high or radio frequency). Some guidelines for wall thickness are shown in the following table (where RF, radio frequency; AF, audiofrequency; and LF, low frequency).

Copper Wall Thickness (in.)	Frequency (kHz)
0.032–0.048	450–50 (RF)
0.065–0.090	25–8.3 (AF)
0.125–0.156	10–3 (AF)
0.156–0.250	3–1 (LF)

As shown above, the penetration depth is the actual depth at which the current flows on the copper surface. The rest of the coil serves other mechanical purposes such as cooling, quench pocket design, and support against mechanical flexing caused by electromagnetic forces. At higher frequencies, coil currents are typically lower. As frequency is lowered, more attention must be paid to coil support and brazed joints. There is also more vibration at lower frequencies, especially at the turns near both ends of the multiturn solenoidal coil or split-return inductor. Nonmagnetic metal studs held together with an insulator can be added for support. Brazed joints and copper may work harder and develop fatigue from the on–off cycling of power. The clearance in silver solder joints should be held to a minimum. The silver should flow from capillary action in joints that are critical. The areas that do not participate in the actual heating can be less precise. These include cooling tubes, covers for water pockets, studs for coil support, and any other items that are not expected to carry current. Commonly used silver solder contains 35–45% silver. Experience in using silver solder with this silver content has shown that it flows well and has relatively low electrical resistance.

4.5.1   FOREWORD TO INDUCTOR DESIGN

Before the inductor design can begin, some design parameters must be established. The first would be the style of inductor. Several things must be analyzed such as current path, workpiece geometry, cycle time, power, and frequency. Electric current, like water in a stream, takes the path of least impedance. This is true in the workpiece as well as the inductor. The workpiece must provide a complete path for current (Figure 4.32). Care must be taken to avoid current paths that cancel each other (Figure 4.33, left), because the current cancellation can lead to a significant decrease in electrical efficiency.

Typically, frequencies used in surface-hardening applications result in a pronounced skin effect, which prevents eddy currents induced within the workpiece from canceling each other. Therefore, in surface hardening, current cancellation typically does not take place.

In heat treatment applications such as through hardening, tempering, normalizing, and annealing, applied frequencies are much lower than in surface-hardening applications, and care should be taken to avoid current cancellation. Table 4.5 lists the recommended frequencies for induction heating of steel cylinders above the Curie temperature.

If the workpiece has an irregular shape (e.g., C-shaped tubes, odd-shaped parts, or slotted cylinders, Figure 4.32), the eddy current will flow on the inside area of the part in order to

 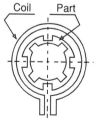

FIGURE 4.32 C-shaped and slotted parts.

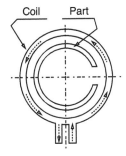
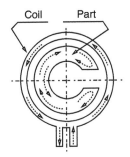

Workpiece (part) is too
thin compared to current
penetration depth (there
is current cancellation)

Workpiece (part) is thick
enough compared to current
penetration depth. There is
a distinguished current path
(no cancellation of current)

FIGURE 4.33 Current cancellation in induction heating of C-shaped parts.

provide an uninterrupted current loop and current cancellation can take place. The basic rule of thumb that will allow one to avoid current cancellation is that the current penetration depth should be no more than one third of the thickness of the current-conducting path.

In some applications, it is very effective to use a dual-frequency design. This requires the use of a low frequency during the stage when workpiece surface layers retain their magnetic properties (below the Curie point). In the second stage, when the workpiece becomes non-magnetic and penetration depth is increased up to two to five times, it is more efficient to use a higher frequency.

Workpiece geometry and hardness patterns are two of the major factors that can determine the inductor shape and style. For instance, some patterns may be obtained only with a single-shot inductor. Along the same line of thinking, the desired cycle time may be obtained only with a particular style of inductor. The coil geometry is affected by the frequency also.

In induction surface hardening the required frequency is primarily dictated by the hardened depth and hardness pattern and is directly related to the reference depth. As a basic rule of thumb, the required frequency can be found from the condition

$$\left(\frac{4}{X_{hd}}\right)^2 < \text{frequency} < \left(\frac{16}{X_{hd}}\right)^2$$

where X_{hd} is required hardened depth in mm (1 mm $< X_{hd} <$ 4 mm).

Required power can be approximated by using 5–15 kW/in.^{2} of workpiece surface area (Table 4.6). High power density will produce a shallow pattern; conversely, low power density will produce a deep pattern. When higher power density is used, the life of the inductor is limited.

The data in Table 4.5 and Table 4.6 are very useful for quick ballpark estimates of inductor parameters. The entire machine concept and material handling are based on the

TABLE 4.5
Frequency for Deep Heating Steel Cylinders above Curie

Frequency (kHz)	0.06	0.5	1	3	10	30
Outside diameter (in.)	> 12	3.5–7.5	2.5–5	1.5–3.5	1–2	0.5–1.5

TABLE 4.6
Heating Frequency and Power Density to Obtain Various Hardening Depths in Carbon Steel

Frequency (kHz)	Hardened Depth (in.)	Power Density (kW/in.^{2})	
		Low	High
450	0.015–0.045	7	12
	0.045–0.090	3	8
10	0.060–0.090	8	15
	0.090–0.160	5	13
3	0.090–0.120	10	17
	0.160–0.200	5	14
1	0.200–0.280	5	12
	0.280–0.350	5	12
Contour gear hardening			
450–200[a]	0.015–0.045	15	25

[a] A low power density preheat at 3 or 10 kHz is recommended for contour gear hardening.

coil style selected for the particular application. For this reason, it is important to establish the above features before the first line is drawn on paper.

4.5.2 TYPICAL PROCEDURE FOR DESIGNING INDUCTOR-TO-WORKPIECE COUPLING GAPS

The first step in designing an inductor is to draw an outline of the workpiece to be heated. This must be drawn accurately and must show the minimum and maximum case depths. Special attention should be given to features such as holes, fillets, sharp corners, snap-ring grooves, and keyways. These areas have a tendency to overheat or underheat due to heat conduction and electromagnetic edge effects. The features of induction heating a part with such elements are discussed later.

The inductor can then be drawn to conform to the workpiece, depending on the required heat pattern. If the inductor is machined, the copper wall may be made thicker to allow for some shaping of the inductor in the test or development stage. Once the coil is developed, the wall thickness should be sized according to the reference depths. The coupling gaps should be held to a minimum. However, they are sometimes dictated by material handling, workpiece tolerances, and thermal expansion of the workpiece during heating. Frequency is also a consideration in establishing coil-to-workpiece coupling gaps. With higher frequency the coupling gap is more critical and more sensitive. Any change in the coupling gap has much less effect at lower frequency than in higher frequency applications. This is because the flux lines stay closer to the coil at higher frequencies. In cases where high frequencies (10–450 kHz) are used, it is extremely important that the relationship between the workpiece and the inductor remain the same from cycle to cycle.

Figure 4.34 shows the influence of coil-to-workpiece air gap on the hardness pattern. Choosing a proper coupling gap and keeping the necessary tool tolerances will ensure consistent results. Otherwise, severe changes can occur in the heat treatment patterns. Coupling gaps may be as small as 1/16 in. (1.5 mm) to as large as 1 in. (25.4 mm), depending on the frequency and the type of heating. Table 4.7 gives the recommended coupling gaps for three frequency ranges. However, as mentioned above, these gaps may vary for different applications.

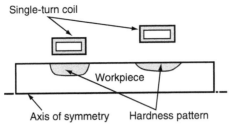

FIGURE 4.34 Heat patterns with a proper coupling gap compared to an improper one.

Inductors intended to heat fillet areas such as those found on axle shafts should have a larger gap on the shaft diameter and be coupled closely to the perpendicular flange surface. Scan inductors in general should have slightly more clearance because of the possibility of workpiece warping. This possibility increases as the part length increases because more heat is put into the part and more material is transformed into martensite. In the case of through heating, the coupling usually has to be larger than recommended in Table 4.7. There are several reasons for this. First of all, a low frequency is typically used for induction through hardening. Usually this type of inductor has a refractory cast liner to protect the coil from the high-temperature workpiece. The coil may also have a water-cooled stainless steel liner. These liners are required to protect the coil from wear and mechanical damage from moving workpieces. Because of these features, through hardening, annealing, or tempering inductors could have a relatively large coil coupling (i.e., more than 0.5 in.). These types of coils are typically multiturn, relatively long, and use lower power density than surface-hardening coils.

4.5.3 MOUNTING STYLES

The electrical connection is sometimes referred to as the mounting foot. There are several basic styles for this connection. A standard keyed foot has been developed to ensure good electrical contact. It is usually fastened with four 3/8-16 stainless steel bolts. The bolts should be tightened to approximately 35–40 ft lb. Overtightening of these bolts will pull the threads from the soft copper, and undertightening will not produce a good electrical contact. Stainless steel-threaded inserts should always be used where the electrical contact is made. When fast or frequent coil changeover is required, a quick-change type of mounting foot may be recommended. This can be in the form of a toggle clamp (Figure 4.35) [97], dovetail, or pneumatic. It can also be keyed for accurate location. One of the main advantages of this design is that no

TABLE 4.7
Typical Coil-to-Workpiece Gap with Various Heating Frequencies

	Frequency (kHz)		
	1–3	10–25	50–450
Coupling gap			
mm	6–3	3–2	1.5
in.	0.23–0.12	0.12–0.08	0.06

FIGURE 4.35 Quick-change coil design.

tools are required. The major shortcoming is that there are power and frequency limitations. For example, the toggle clamp and dovetail type should be limited to 300 kW with a frequency range of 3–450 kHz. All electrical contact areas must be clean and free of nicks or burrs, and they therefore require additional care and cleaning.

4.5.4 SPRAY QUENCH

The spray quench design can be just as important as the coil design. The quench must be designed for rapid heat removal to develop the desired hardness and metal structure. Nonintensive quenching results in soft pearlite-type and bainite-type metal structures. Uneven quenching makes the distortion problem more pronounced. The intensity of quenching depends upon the flow rate, the angle at which the quenchant strikes the workpiece surface, temperature, purity, and type of quenchant. There are three main considerations for quench design: heat treatment pattern requirements, style of the inductor, and workpiece geometry. Spray quenching works best if the workpiece is rotated during the quenching operation. This will ensure uniformity in quenching. The point of impingement will have a faster rate of cooling than an area that is quenched from a flood of quenching fluid. Rotating the workpiece simulates a constant impingement rather than many small impingements. Small orifices are required to agitate quench, to prevent steam pockets from developing.

Various fluids, such as aqueous polymer solutions, salt water, oils, and even plain water, are used in quenching. The features of the quenchants were discussed in detail in Chapter 9 of *Steel Heat Treatment: Metallurgy and Technologies*. Here we will just mention that plain water is seldom the best choice for quenching. Typically, steels with very low hardenability require the use of plain water. However, the use of water results in rust, and the workpiece distortion and cracking also become more pronounced. Polymer fluids are usually used as quenchants in IHT. Using a small percentage of polymer can make quenching even faster than plain water. The polymer helps to eliminate a vapor barrier that can occur with plain water.

The quench pattern is generally conformed to the part, just as the inductor is. When using a single-shot inductor it is preferable to quench from two sides of the workpiece. The quench holes should be placed facing the part at 3/16–1/4 in. intervals and have a staggered pattern. The orifice size is related to the shaft or workpiece diameter.

Shaft Diameter (in.)	Orifice Size (in.)
0.25–0.50	0.046–0.063
0.50–1.50	0.063–0.094
>1.50	0.125–0.156

For inductors where quench holes are drilled through the heating face, the holes should not occupy more than 10% of the heating face area. To control the direction of the stream, the wall thickness must be at least 1.5 times the orifice diameter. The inlet–outlet ratio should be at least 1:1. Baffles or deflectors should be added in the area of the inlet. These will help to obtain a more uniform pressure distribution. The baffle hole area should also have at least a 1:1 ratio. The total quench flow may need to be estimated for pump sizing and flowmeter sizing. This can be approximated by finding the flow of one quench hole from the chart shown in Figure 4.36. The flow found from this chart should be multiplied by the number of holes in the inductor.

The temperature of the quenchant can be critical. A great majority of induction-hardening applications require quench temperatures of 24–35°C (75–95°F). Therefore, with start-up of the equipment, some warm-up time can be required. Otherwise, start-up with cold quenchant can cause cracking and part distortion. On the other hand, the use of hot quenchant results in a reduction in hardness.

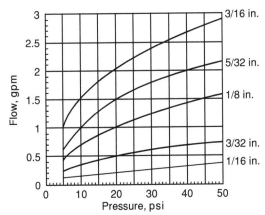

FIGURE 4.36 Flow of a single-quench orifice. Example: at 15 psi, with seventy-two 1/8 in. holes, gpm = 0.87 × 72 = 62.64.

4.5.5 INDUCTOR STYLES—SCAN INDUCTORS

There are many types of inductors. Three of the most common are scanning, static, and single-shot. Scanning is sometimes called progressive heating because either the workpiece moves through the inductor or the inductor moves and the workpiece is stationary [98–102]. With a static inductor, neither the workpiece nor the inductor move, except when the workpiece is moved in and out of the inductor. Thus during heating the workpiece remains in a static position. Single-shot inductors are also said to be static in the same way [103]. The main difference lies in the inductor design and construction. Single-shot inductors are discussed in Section 4.5.6.

The main advantage of the scanning type of inductor is its flexibility in running various lengths of parts. This type of inductor provides a repeatable, easily automated process that can quickly adapt to new heat treatment tasks and be easily integrated into the work cell.

Scan inductors are usually single turn (Figure 4.37) or have two or three turns. The number of turns is determined by the coil calculation. From another perspective, the number of coil turns plays an important role in the tuning of the heat station and coil with the power source. This problem is discussed in detail in Chapter 5.

The more turns there are, the faster the scan rate will be. Single-turn inductors are used where a sharp pattern runout is required, as is often the case, adjacent to a snap-ring groove. Single-turn inductors can be machined from a solid cooper bar, thus making them very rigid and durable.

Quench holes can be integrated into a single-turn scan inductor. This type of inductor is sometimes called a machined integral quench (MIQ) inductor (see Figure 4.38). The quench spray should hit the part approximately 3/4 in. from the heating face and be angled down to prevent the quench from washing back into the inductor. This dimension will vary with different types of steel and with different scan rates [98–102]. An additional quench follower must be added to inductors designed to run parts with varying diameters and grooves. This ensures a good quench in the fillet and radius areas that otherwise might be shadowed from the primary quench. A quench follower may also help to eliminate the barber pole effect. A narrow heating face is required for a sharp cutoff in pattern. A wider heating face or more coil turns can be used where a faster scan is desired. The main disadvantage to this is that it will produce a gradual pattern runout and may not meet some pattern specifications.

FIGURE 4.37 Scan inductor.

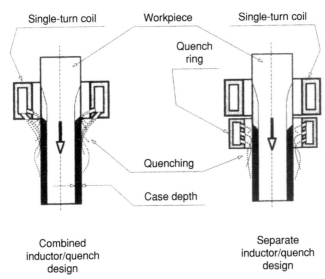

FIGURE 4.38 Combined and separate inductor/quench design.

Scan inductors that are intended to heat a fillet area as well as the shaft and the perpendicular flange must be focused into the fillet area. They typically have flux intensifiers (flux concentrators) to aid in focusing into the fillet. These are critical applications that require careful design to make the inductor work because the current will try to take the shortest path and stay in the shaft area. Therefore, all efforts must be made to focus it into the fillet. Scan inductors have a good chance of working without coil modifications. This is due to the flexibility that results from being able to vary power, scan rate, and delay of quench to achieve the desired hardness pattern. They are also able to run various lengths. A disadvantage of this inductor type is that it is generally limited to the shaft type of parts. Scanning is recommended for shafts where power is limited and shaft length is over 12 in.

Applications may call for horizontal or vertical scanning systems. As an example, Figure 4.39 shows a Statiscan vertical induction scanner with dual spindles. This system has been used for hardening and tempering a wide variety of parts such as shafts, bars, axles, and hubs. Parts are located and held between centers while they are scanned during the cycle. The part is rotated during the cycle by using a variable-speed motor. This particular model is provided with several power ratings: from 50 to 250 kW at frequencies from 3 to 200 kHz. A user-friendly keyboard on the Statiscan provides quick setup, changeover, and diagnostics capability. To simplify setup, upper tooling centers can be adjusted without tools. A selection of standard quick-change inductor mountings are also helpful in minimizing changeover time.

4.5.6   INDUCTOR STYLES—SINGLE-SHOT

Single-shot inductors [103] are made of tubing or machined from a solid, but unlike scan inductors they produce an axial rather than radial current path. Typically, they have two horseshoe-shaped loops that join the two legs. This type of inductor requires the most care in manufacturing because of the high power densities it must accommodate. Because of this, the brazed joints must accurately mate. Tongue-in-groove joints are preferred. The flux fields can usually be shaped to produce the exact hardness pattern desired; however, altering that shape is not always easy. Sometimes a completely new inductor head must be made.

Single-shot inductors are designed for a specific heat treatment pattern on a specific workpiece. The heat cycle for single-shot inductors is usually much shorter than for other

FIGURE 4.39 Statiscan unitized vertical dual spindles scanning system.

inductor styles. This leads to higher production rates. At high rates of production, a single-shot inductor typically has a shorter life than scan coils.

Electrically speaking, it is always important to keep in mind that the inductor is the weak link in an induction system and should be likened to a fuse. The inductor will fail if power is increased to the point at which water cannot adequately cool it. Additional cooling passages may be needed with single-shot coils, and a high-pressure booster pump is almost always required. Single-shot inductors should be used where the workpiece has varying diameters, radii, and fillets.

4.5.6.1 Striping Phenomenon

When discussing specifics of induction hardening, it is necessary to mention an effect that, because of its complexity, is not usually discussed in IHT publications: the striping phenomenon. Striping typically occurs during intensive induction hardening of carbon steels where high power densities are used. Because of this effect, the workpiece area under the coil may unexpectedly start to heat nonuniformly. The striping can be seen even in the case of a single-turn coil with a conventional cylindrical load (Figure 4.40). Shortly after the heating cycle begins, alternating hot bright areas (bright stripes) and cold areas (dark stripes) become visible. These bright and dark stripes encircle the cylinder and thus have the shape of rings.

The barber pole effect and the striping phenomenon have never been obtained by mathematical modeling. They have been viewed only in practical applications or during laboratory experiments on the induction heating of magnetic steels. In some applications striping suddenly occurs and then disappears. There is no single explanation of this phenomenon. The only attempt to explain it was made by Lozinskii [3] in the early 1940s. It was a very simple description based on the knowledge available to IHTs at that time. It should be

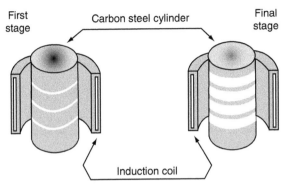

FIGURE 4.40 Striping phenomenon in induction heating of carbon steel cylinder.

mentioned that Lozinskii's hypothesis concerning striping has a certain logic and can be accepted as an introduction to this effect. However, from our point of view, that explanation oversimplifies the mechanism of the striping phenomenon. Since that time, there have been no further attempts to explain this effect. Therefore, here we will briefly introduce Lozinskii's hypothesis along with our own point of view, which is based on modern experience and new theoretical knowledge accumulated at INDUCTOHEAT during the development of various IHT processes.

Assume that a magnetic cylinder is located inside a single-turn inductor (Figure 4.40). As a result of the electromagnetic field produced by the induction coil, eddy currents will flow within the workpiece. Due to the skin effect, these eddy currents will appear primarily in the surface layer of the workpiece located inside the coil and cause the surface temperature of the workpiece to increase.

Realistically speaking, any workpiece has certain nonuniformities, microscopic defects, impurities, and nonhomogeneities. This includes structural and mechanical nonuniformities, metallurgical nonhomogeneities, etc. As a result, different surface regions of the workpiece will be heated slightly differently. Some will reach the Curie temperature first and lose their magnetic properties. The relative magnetic permeability of these areas will dramatically drop to unity ($\mu = 1$). This leads to a significant increase in the penetration depth in those areas. The resistance of these nonmagnetic regions will drastically decrease compared to neighboring surface areas that retain their magnetic properties. As a result, the density of induced currents in the low-resistance regions will increase. This leads to an increase in power density and an increase in heat sources in these areas. At the same time, there will be a redistribution of eddy currents in the workpiece surface. Eddy currents induced in areas that retain their magnetic properties (dark rings) will have a tendency to rush to complete their loops through the low-resistance paths (bright rings). This current redistribution leads to a further heat source reduction in the magnetic areas with low temperature (dark rings) and appears as additional heat sources in the nonmagnetic areas with high temperature (bright rings). Therefore, positive feedback will occur. As a result, one can view in the workpiece a mixture of ring-shaped stripes. Hot bright stripes will alternate with the relatively cold dark stripes. Experience shows that usually the thickness of the bright and dark stripes depends primarily on the frequency and power density and equals 1–3 current penetration depths in hot steel.

Besides the current redistribution, the striping phenomenon is a result of several other electromagnetic and heat transfer effects, including the electromagnetic edge effect of joining materials with different properties (i.e., magnetic and nonmagnetic metals). The electromagnetic edge effect of joining materials with different properties (EEJ effect) occurs when two different metals are located in a common magnetic field. To simplify the study of this effect,

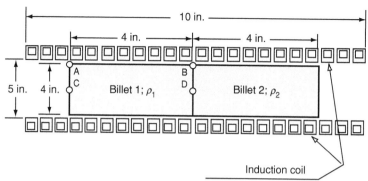

FIGURE 4.41 Sketch of the induction heating system.

let us consider the electromagnetic process in a conventional solenoidal induction coil with two workpieces, for example, two cylindrical billets (Figure 4.41). Assume that the billets have different material properties (e.g., different electrical resistivity ρ or magnetic permeability μ). When two billets with different material properties are joined together and are located inside an induction coil, the electromagnetic field in the joint area (so-called transition area) becomes distorted [10,50,51]. For example, if one billet has been heated above the Curie point (has become nonmagnetic) and the other still maintains its magnetic properties, then the distribution of the electromagnetic field will be as shown in Figure 4.42. If the billets are long enough, then the magnetic field intensity at their central areas will be approximately the same and correspond to the coil current. At the same time, the power densities at the surface of the magnetic and nonmagnetic billets will be rather different (Figure 4.42).

At the left tail end of the nonmagnetic billet (billet 1) and at the right tail end of the magnetic billet (billet 2), there will be a nonuniform power density distribution due to the end effects of the nonmagnetic and magnetic workpieces. At the transition area of the billets, the field distribution is quite complicated. At the right end of the nonmagnetic cylinder (billet 1), the magnetic field intensity and power density sharply increase. At the left end of the magnetic cylinder (billet 2), those parameters sharply decrease. This phenomenon is called the EEJ effect. Obviously, this effect plays an important role in the appearance of the striping phenomenon and leads to a significant redistribution of the electromagnetic field in the area of the dark rings (which retain their magnetic properties) and in the bright high-temperature rings (which have become nonmagnetic). As one might expect, this electromagnetic effect will cause the heat source distribution to differ from the classical form that is traditionally assumed in the study and design of IHT processes.

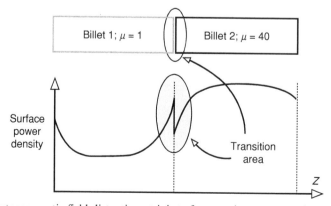

FIGURE 4.42 Electromagnetic field distortion at joint of magnetic–nonmagnetic steels.

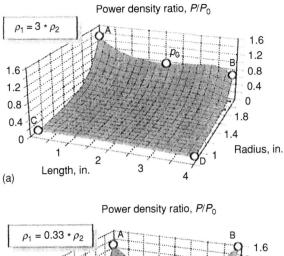

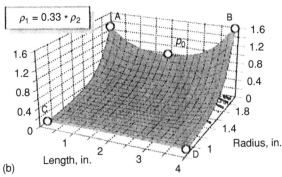

FIGURE 4.43 Power density distribution along the length of billet 1 (frequency $= 60$ Hz; $\rho_1 = 1.1$ μΩ in.).

When discussing the EEJ effect, it is necessary to mention that it also occurs when both workpieces are nonmagnetic but have different electrical resistivities (ρ). Figure 4.43 shows the power density distribution within billet 1 for the induction system shown in Figure 4.41. Both billets are nonmagnetic, but they have different electrical resistivities (ρ_1 and ρ_2). When the electrical resistivity of billet 1 (ρ_1) is three times that of billet 2 (Figure 4.43a), then power density is reduced in the joint area of billet 1. When $\rho_1 = 0.33\rho_2$ (Figure 4.43b), there is an increase in power density (heat source).

The effect of joining materials with different properties should be taken into account when designing other IHT and induction heating systems also. For example, it may have a significant influence not only on the striping phenomenon but also on the final temperature distribution, which should be considered when designing the billet heater, especially when the induction equipment will operate below or just above the Curie point [50,51,108].

Experience shows that striping can appear in several different ways. However, in the great majority of cases, very narrow bright stripes (rings) will appear at the beginning of the heating cycle (Figure 4.40). With time, the narrow stripes will widen. At this stage, the maximum temperatures will move from the center of each ring toward the edges of each bright hot ring. During the heating process, the stripes sometimes move back and forth along the workpiece surface area under the coil. When the length of the heating cycle is increased, typically the striping effect will not be pronounced and temperatures will equalize in the workpiece surface.

The appearance of the stripes depends on a complex function of the frequency, magnetic field intensity, and thermal, electrical, and magnetic properties of the steel. It can occur only when high power density is applied. If the power density is relatively low, then the temperature will equalize between the neighboring bright (high-temperature) and dark (low-temperature) rings because of the thermal conductivity of the steel.

4.5.7 Gear Hardening

In recent years, gear manufacturers have increased their technological knowledge of the production of quality gears. This knowledge has led to many improvements including lower noise, lighter weight, and lower cost as well as increased load-carrying capacity to handle higher speeds and torque with a minimum amount of generated heat [1].

Not all gears and pinions are well suited for induction hardening. External spur and helical gears, worm gears, and internal gears, racks, and sprockets are among the parts that are typically induction-hardened. Conversely, bevel gears, hypoid gears, and noncircular gears are rarely heat treated by induction.

In contrast to carburizing and nitriding, induction hardening does not require heating the whole gear. With induction the heating can be localized only to the areas where metallurgical changes are desired (e.g., flank, root, and gear tip can be selectively hardened) and the heating effect on adjacent areas is minimum. Depending upon the application a tooth hardness typically ranges from 42 to 60 HRC.

One of the goals of induction gear hardening is to provide a fine-grained fully martensitic layer on specific areas of the gear to increase hardness and wear resistance while allowing the remainder of the part to be unaffected by the process. The increase in hardness improves contact fatigue strength as well. A combination of increased hardness, wear resistance, and the ability to provide a fine martensite structure, often allows the substitution of inexpensive medium- or high-carbon steel or low-alloyed steel for more expensive highly alloyed steels.

It is not always possible to obtain a fully martensitic case depth. Depending upon the kind of steel and prior microstructure, the presence of some amount of retained austenite within the case depth is unavoidable (unless cryogenic treatment is used). This is particularly true for steels with high-carbon content and cast irons.

Up to a certain point, some amount of retained austenite do not noticeably reduce the surface hardness. However, it brings some ductility and provides better absorption of impact energy, which is imperative for heavily loaded gears. In addition, having an unstable nature, retained austenite has a tendency with time to transform into martensite. It introduces additional compressive residual stresses and increases the surface hardness. From this perspective, a small amount of retained austenite is not only harmless but may even be considered beneficial in some cases. However, an excessive amount of retained austenite can be detrimental because it may noticeably reduce the surface hardness, weaken bending fatigue properties, and can result in the appearance of a crucial amount of brittle untempered martensite during gear service life.

Another goal of induction gear hardening deals with the ability to provide significant compressive residual stresses at the surface. This is an important feature, since it helps to inhibit crack development as well as resist tensile bending fatigue.

The kind of steel used, its prior microstructure, and gear performance characteristics (including load condition and operating environment) dictate the required surface hardness, core hardness, hardness profile, gear strength, and residual stress distribution.

4.5.7.1 Material Selection and Required Gear Conditions before Heat Treatment

Steel selection depends upon features of the gear working conditions, required hardness, and cost. Low-alloy and medium-carbon steels with 0.4–0.55% carbon content (i.e., AISI 4140, 4340, 1045, 4150, 1552, 5150) are quite commonly used for gear heat treatment by induction.

When discussing induction hardening it is imperative to mention the importance of having favorable metal conditions before gear hardening. Hardness repeatability and the stability of the hardness pattern are grossly affected by the consistency of the microstructure before heat treatment (referred to as microstructure of a green gear) and the steel's chemical composition [1].

Favorable initial microstructure, including a homogeneous fine-grained quenched and tempered martensitic structure with hardness of 30–34 HRC leads to fast and consistent metal response to heat treating with the smallest shape and size distortion and a minimum amount of grain growth. This type of initial microstructure results in higher hardness and deeper hardened case depth compared to the ferritic–pearlitic initial microstructure.

If the initial microstructure of the gear has a significant amount of coarse pearlite and most importantly coarse ferrites or clusters of ferrites, then these microstructures cannot be considered favorable because gears with such structures will require longer austenization time and higher austenizing temperatures to make sure that diffusion-type processes are completed and homogeneous austenite is obtained. Ferrite is practically a pure iron and does not contain the carbon required for martensitic transformation. Pure ferrite consists of less than 0.025% carbon. Large areas (clusters) of free ferrite require a long time for carbon to be able to travel and diffuse into the poor carbon area of the ferrite. Otherwise, clusters of ferrites will act as one huge grain of ferrite that will be retained in the austenite and upon quenching a complex ferritic–martensitic microstructure with scattered soft and hard spots will take place.

In opposition to quenched and tempered prior microstructure, steels with large carbides (i.e., spheroidized microstructures) have poor response to induction hardening and also require in the need for prolonged heating and higher temperatures for austenization. Longer heat time leads to grain growth, the appearance of coarse martensite, data scatter, an extended transition zone, and essential gear shape distortion. Coarse martensite has a negative effect on tooth toughness and creates favorable conditions for crack development.

As opposed to other heat-treating techniques, heat treatment by induction is appreciably affected by variation in the metal chemical composition. Therefore, favorable initial metal condition also includes tight control of the specified chemical composition of steels and cast irons. Wide compositional limits cause surface hardness and case depth variation. However, tight control of the composition eliminates possible variation of the heat treat pattern, resulting from multiple steel and iron sources. Segregated and banded initial microstructures of green gears should be avoided.

The surface condition of the gear is another factor that can have a distinct effect on gear heat-treating practice. Voids, microcracks, notches, and other surface and subsurface discontinuities as well as other stress concentrators can initiate crack development during induction hardening when the metal goes through the expansion–contraction cycle; thermal gradients and stresses can reach critical values and open notches and microcracks. Conversely, a homogeneous metal structure with a smooth surface free of voids, cracks, notches, and the like improves the heat-treating conditions, and positively affects important gear characteristics such as bending fatigue strength, endurance limit, gear durability, and gear life.

Medium and high frequencies have a tendency to overheat sharp corners; therefore, gear teeth should be generously chamfered if possible for optimum results in the heating process.

Because gears provide transmission of motion and force, they belong to a group of the most geometrically accurate power transmission components. A gear's geometrical accuracy and ability to provide a required fit to its mate greatly affect gear performance characteristics. Typical required gear tolerances are measured in microns, therefore the ability to control such undesirable phenomena as gear warpage, ovality, conicality, out-of-flatness, tooth crowning, bending, growth, shrinkage, and the like plays a dominant role in providing quality gears.

   This is why hardness pattern consistency, minimum shape and size distortion, and distortion repeatability are among the most critical parameters that should be satisfied when heat-treating gears.

4.5.7.2   Overview of Hardness Patterns

The first step in designing an induction gear heat treatment machine is to specify the required surface hardness and hardness profile. Figure 4.44 shows typical locations of gear profile measurements. In some cases, depending upon the type of gear and its application, some customers create specific procedures for gear profile measurements.

   Insufficient hardness as well as an interrupted (broken) hardness profile at tooth contact areas will shorten gear life due to poor load-carrying capacity, premature wear, tooth bending fatigue, rolling contact fatigue, pitting, and spalling, and can even result in some plastic deformation of the teeth.

   A through-hardened gear tooth with a hardness reading exceeding 60 HRC is too brittle and will often experience a premature brittle fracture. Hardened case depth should be adequate (not too large and not too small) to provide the required gear tooth properties.

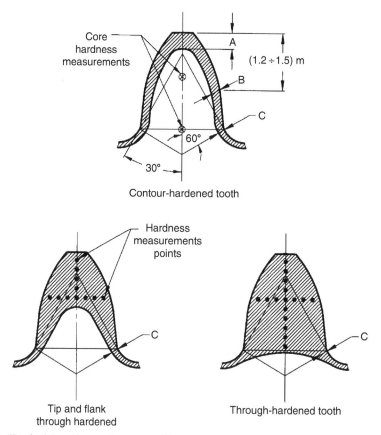

FIGURE 4.44 Typical locations of gear profile measurements. Depending on type of gear and its application, specific procedures are created for these measurements. (From V. Rudnev, D. Loveless, R. Cook, and M. Black, *Handbook of Induction Heating*, Marcel Dekker, New York, 2003. With permission.)

There is a common misconception that a uniform contour profile is always the best pattern for all gear-hardening applications. It is not. In many cases, a certain hardness gradient profile can provide a gear with better performance.

Operating load condition (whether there are occasional, intermittent, or continuous loads) has a pronounced effect on the type of gear, tooth geometry, and hardness profile.

Let us briefly evaluate a variety of hardening patterns (Figure 4.45) and their effect on gear load-carrying capacity and life [1,16]. Pattern A is a flank-hardening pattern that has been used since the late 1940s for hardening large gears (with teeth modulus of eight and larger). The hardened pattern occupies the tooth flank area and ends before the tooth fillet. This pattern provides the required wear resistance, but the typical failure mode of gears with this type of pattern is a bending fatigue caused by repeated loading. A crack typically initiates in the tooth root and fillet area. In the harden–nonharden transition region, the residual stresses change from compressive in the hardened area to tensile in the nonhardened area. The maximum tensile residual stresses are located just below the end of the hardened pattern. A combination of applied tensile stresses with tensile residual stresses creates a favorable condition for early crack development in the root and fillet area particularly for heavily loaded gears. Therefore, when pattern A is obtained, a mechanical hardening (i.e., roll or ball hardening) is typically required to harden the fillet area and develop within that area useful compressive residual stresses that will resist bending fatigue. When mechanical hardening is not used, it is typically recommended that one uses a pattern that hardens the root areas as well, such as that pictured in pattern I.

Pattern B is a flank and tooth-hardening pattern. This pattern has a shortcoming similar to the previous one, featuring a poor load-carrying capacity yet can be used in cases where wear resistance is of prime concern. Patterns E, F, and G provide better results when a combination of wear, tear, and bending fatigue resistances is required.

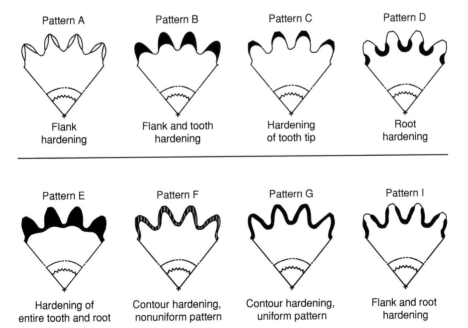

FIGURE 4.45 Induction-hardening patterns of gears. (From V. Rudnev, D. Loveless, R. Cook, and M. Black, *Handbook of Induction Heating*, Marcel Dekker, New York, 2003. With permission.)

Pattern C is a tooth tip-hardening pattern. In this case, the gear has minimum shape distortion. In addition, the application of gears with this pattern is extremely limited because the two most important tooth areas (flank and root) are not hardened. As a matter of fact, due to unfavorable residual stress distribution, the bending fatigue strength of a gear with this pattern, as well as patterns A and B, can even be 25% lower compared to gear strength before hardening (green gear) [17]. In most cases, patterns F and G would be better choices.

Pattern D is a root-hardening pattern. The maximum bending stresses are located in the tooth fillet area, therefore this pattern provides good fillet and root strengthening, meaning a combination of hard surface, sufficient case depth, and compressive stresses. The root is essentially reinforced, thus the maximum tensile residual stresses are shifted far away from the root and fillet surface to a depth where tensile residual stresses will not complement tensile applied stresses during service and create bending fatigue fracture. However, the application of this pattern is quite limited. Since the tooth flank is not hardened, this pattern provides poor wear resistance that may result in the removal or displacement of metal particles from the gear surface. Theoretically, it is possible to imagine the necessity of using this pattern as well as the previous one; however, it is more practical to use another pattern, such as pattern I.

Pattern E is one of the most popular induction-hardening patterns, particularly for small-size gears and sprockets. Because the body of the tooth is through hardened, some quench cracking may occur. In addition, there is a danger of having a brittle fracture in gears with through-hardened teeth, particularly those subjected to shock loads. The core should be able to withstand impact loads and prevent plastic deformation of the gear teeth. It should also have some ductility. This is why a low-temperature tempering is often applied. The core strength is measured by its hardness. Low-temperature tempering lowers the final hardness down to 52–58 HRC and provides some ductility and toughness to the gear teeth. This pattern offers good resistance to wear and pitting.

Patterns F and G are popular patterns for medium-size gears in many applications. According to pattern F, a case depth in the tooth root area is typically 30–40% of the depth in the tooth tip. Slightly larger hardness depth at the tooth pitch line compared to the root is beneficial as a preventive action against spalling and pitting. It is very important to harden the entire gear perimeter, including the flank and root areas. An uninterrupted hardened pattern of all contact areas of the tooth indicates good wear properties of the gear, and in addition typically ensures the existence of an uninterrupted distribution of desirable compressive stresses at the gear surface. Because gear teeth are not hardened through, a relatively ductile tooth core (30–44 HRC) and a hard surface (56–62 HRC) provide a good combination of such important gear properties as exceptional wear resistance, toughness, and bending strength, and provide superior gear durability.

Pattern I is one of the most popular choices for induction hardening of large gears and pinions (i.e., gears with 300 mm and even larger outside diameter, O.D.) with coarse teeth (modules greater than eight). This pattern provides an exceptional combination of fatigue and wear strengths as well as resistance to shock loading and scuffing (severe adhesion wear where metal particles transfer from one tooth to another tooth), which is very important for heavily loaded gears and pinions experiencing severe shock loads. It is recommended that for these applications surface hardness should not be very high, typically in the range of 55–59 HRC. If surface hardness exceeds 61–62 HRC, the gear might be too brittle and could experience some tooth bending failures.

4.5.7.3 Gear Coil Design and Heat Mode

The variety of required hardness profiles requires different coil designs and heat modes. Development, including coil design, is largely based on induction principles, the result

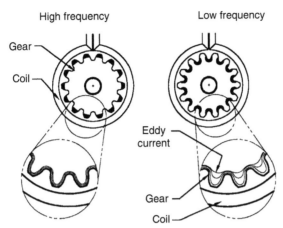

FIGURE 4.46 Frequency influence on hardness profile with encircling induction coil. (From V. Rudnev, D. Loveless, R. Cook, and M. Black, *Handbook of Induction Heating*, Marcel Dekker, New York, 2003. With permission.)

of mathematical evaluation, and experience with previous jobs. The development establishes not only process parameters, including cycle times and power levels, but also coil geometry.

4.5.7.3.1   Tooth-by-Tooth and Gap-by-Gap Inductors

Generally speaking, gears are induction heat-treated by either encircling the part with a coil (so-called spin hardening, Figure 4.46) or, in larger gears and pinions, heating them tooth-by-tooth or gap-by-gap (Figure 4.47) [1].

Tooth-by-tooth and gap-by-gap techniques require a high level of skill, knowledge, and experience in order to obtain the required hardness pattern. These techniques can be realized by applying a single-shot or scanning mode. Scanning rates can be quite high, reaching 15 in./min and even higher. Both tooth-by-tooth and gap-by-gap techniques are typically not very suitable for small- and fine-pitch gears (modules smaller than six).

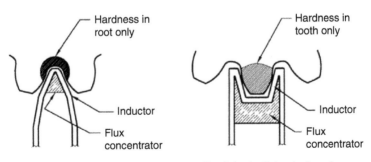

FIGURE 4.47 Gap-by-gap and tooth-by-tooth induction hardening. (From V. Rudnev, D. Loveless, R. Cook, and M. Black, *Handbook of Induction Heating*, Marcel Dekker, New York, 2003. With permission.)

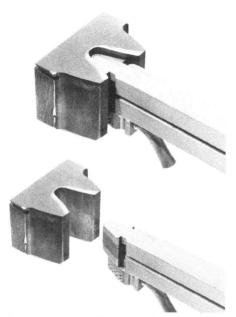

FIGURE 4.48 Gap-by-gap inductor and gear. (Courtesy of INDUCTOHEAT Inc.) (From V. Rudnev, D. Loveless, R. Cook, and M. Black, *Handbook of Induction Heating*, Marcel Dekker, New York, 2003. With permission.)

Coil geometry depends upon the shape of the teeth and the required hardness pattern. According to the tooth-by-tooth technique, an inductor encircles a single tooth or is located around it. Such an inductor design provides patterns B and C.

Gap-by-gap techniques require the coil to be symmetrically located between two flanks of two adjacent teeth. This inductor can be designed to heat only the root and flange of the tooth, leaving the tip and tooth core soft, tough, and ductile. There are many variations of coil designs applying these principles. Probably one of the most popular is the inductor shape shown in Figure 4.48. This type of inductor was originally developed in the 1950s by the British firm, Delapena. Figure 4.49 shows the pattern profile for gap-by-gap induction hardening.

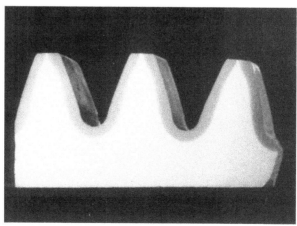

FIGURE 4.49 Gap-by-gap pattern profile. (From V. Rudnev, D. Loveless, R. Cook, and M. Black, *Handbook of Induction Heating*, Marcel Dekker, New York, 2003. With permission.)

As one can see from Figure 4.50, the path of the induced eddy current has a butterfly-shaped loop. The maximum current density is located in the tooth root area (the center part of the butterfly). In order to further increase the power density induced in the root, a magnetic flux concentrator is applied. A stack of laminations or powder-based magnetic materials are typically used as flux concentrators here. Laminations are oriented across the gap.

Although the eddy current path has a butterfly shape, when applied with a scanning mode, the temperature is distributed within gear roots and flanks quite uniformly. At the same time, since the eddy current makes a return path through the flank and, particularly through the tooth tip, proper care should be taken to prevent overheating the tooth tip. Overheating of the tip can substantially weaken the tooth.

Gears heat treated by using the tooth-by-tooth or gap-by-gap techniques can be fairly large, having outside diameters of 100 in. or more, and can weigh several tons. These techniques can be applied for external and internal gears and pinions. There is a limitation to applying these methods for hardening internal gears. Typically, it is required that the I.D. of the internal gear exceeds 8 in. and in some cases 10 in. or more.

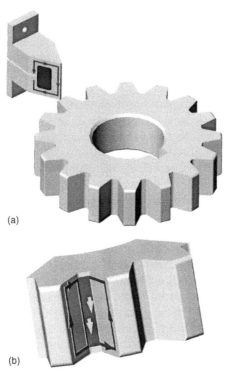

(a)

(b)

FIGURE 4.50 (a) Current flow on gap-by-gap inductor; (b) path of induced eddy currents has butterfly-shaped loop. (From V. Rudnev, D. Loveless, R. Cook, and M. Black, *Handbook of Induction Heating*, Marcel Dekker, New York, 2003. With permission.)

Both tooth-by-tooth and gap-by-gap hardening are time-consuming processes with low production rates. Power requirements for these techniques are relatively low and depend upon the production rate, type of steel, case depth, and tooth geometry. Modest power requirements can be considered an advantage, because if spin hardening is used a large gear would require an enormous amount of power that could diminish the cost-effectiveness of the heat-treating process.

Applied frequencies are usually in the range of 1–30 kHz. At the same time, there are some cases when a frequency of 70 kHz and even higher frequencies have been used. For example, the NATCO submerged technique [18] applied a radio frequency of 450 kHz.

Pattern uniformity is quite sensitive to coil positioning. The coil should be symmetrically located in the gap between the two teeth. Asymmetrical coil positioning results in a nonuniform hardness pattern. For example, an increase in the air gap between the coil copper and the fillet surface on one side will result in a reduction of hardness and shallower case depth there. Shallow case depth can diminish the bending fatigue strength of the gear. Excessive wear of the working (contacting) side of the gear tooth can also occur.

A decrease of the air gap can result in local overheating or even melting of the gear surface. Some arcing can occur between the coil and the gear surface. Precise coil fabrication techniques, rigidity of the inductor, and careful alignment are required. Special locators should be used to ensure proper inductor positioning in the tooth space. Thermal expansion of metal during heating should be taken into consideration when determining the proper coil-to-gear-tooth air gap.

There can be an appreciable shape and size distortion when applying tooth-by-tooth or gap-by-gap techniques for hardening large gears and pinions. Shape distortion is particularly noticeable in the last heating position. The last tooth can be pushed out 0.1–0.3 mm. In some cases, distortion can also be minimized by hardening every second tooth or tooth gap. Obviously this will require two revolutions to harden the entire gear. Therefore, final grinding is often required. There is a linear relationship between the volume of required metal removal and the grinding time. Thus excessive distortion leads to a prolonged grinding operation and increases the cost. Heat treat distortion can also be compensated for previous stages of gear design and manufacturing.

Even though there might be appreciable distortion when applying induction hardening to large gears and pinions (e.g., mill, marine, and large transportation gears, etc.), this distortion is not as significant compared to carburizing or nitriding. Both carburizing and nitriding operations require soaking of gears for many hours (in some cases up to 30 h or longer) at temperatures of 850°C (1562°F) to 950°C (1742°F). At these temperatures the large masses of metal expand to a much greater extent compared to a case when only the gear surface layer is heated. The expansion of a large mass of metal during heating and soaking and its contraction during cooling and quenching after carburizing results in much greater gear shape distortion compared to the distortion after induction hardening.

In addition, large gears held at temperatures of 850°C (1562°F) to 950°C (1742°F) for many hours have little rigidity; therefore, they can sag and have a tendency to follow the movement of their supporting structures during soaking and handling. During induction hardening, areas unaffected by heat as well as areas with temperatures corresponding to the elastic deformation range serve as shape stabilizers and lead to lower, more predictable distortion.

It is necessary to mention here that due to small coil–workpiece air gaps (0.5–1.5 mm) and harsh working conditions, these coils require intensive maintenance and have a relatively short life compared to inductors that encircle the gear.

When designing this type of induction heating process, particular attention should be paid to electromagnetic end or edge effects and the ability to provide the required pattern in the gear face areas (gear ends) as well as along the tooth perimeter.

When a single-shot mode is used, an active coil length has approximately the same length as the gear width. A single-shot mode is more limited in providing a uniform face-to-face hardness pattern compared to the scanning mode.

For the scanning mode, the coil length is typically at least twice as short as the gear thickness. In order to obtain the required face-to-face temperature uniformity, it is necessary to use a complex control algorithm, power, and scan rate versus coil position. A short dwell at the initial and final stages of coil travel is often used. Thanks to preheating due to thermal conductivity, the dwell at the end of coil travel is usually shorter compared to the dwell at the beginning of travel.

When applying the scanning mode for hardening gears with wide teeth, two techniques can be used: a design concept where the inductor is stationary and the gear is movable, and a concept that assumes the gear is stationary and the inductor is movable.

Tooth-by-tooth and gap-by-gap techniques can harden gears submerged in a temperature-controlled tank of the quenchant. This technique was applied in the original Delapena induction-hardening process. In this case, quenching is practically instantaneous and both controllability and repeatability of the hardness pattern are improved, although additional power is required. The fact that the gear is submerged in a quenchant also helps to prevent the tempering back problem as well as crack development in the tooth root. In addition, the quenchant serves as a coolant to the inductor. Therefore, in submerged hardening an induction coil does not have to be water-cooled.

4.5.7.3.2 Gear Spin Hardening (Encircling Inductors)
Spin hardening of gears utilizes a single- or multiturn inductor that encircles the part and usually requires gear rotation. It is typically used for gears with fine- and medium-sized teeth and is considered less time consuming and more cost-effective than the previously discussed processes. Therefore, it is strongly recommended to use spin hardening whenever it is possible. Unfortunately, spin hardening is not a cure-all and sometimes cannot be easily used for medium-sized helical and bevel gears and large module gears due to an enormously large amount of required power and difficulties in obtaining the desired hardness pattern.

Gears are rotated during heating to ensure an even distribution of energy across their perimeter. Rotation rates are chosen to suit process requirements. When applying encircling coils, there are five parameters that play a dominant role in obtaining the required hardness pattern: frequency, power, cycle time, coil geometry, and quenching conditions. A proper control of these parameters can result in totally different hardened profiles.

Figure 4.51 illustrates a diversity of induction-hardening patterns that were obtained on the same carbon steel shaft with variations in time, frequency, and power. As a basic rule, when it is necessary to harden the tooth tips only, a higher frequency and high power density should be applied (Figure 4.46, left). When hardening the tooth root, a lower frequency and lower power density should be used (Figure 4.46, right). A high power density generally gives a shallow pattern; conversely, a low power density will produce a deep pattern.

In addition to the process parameters mentioned above, hardness pattern uniformity and repeatability depend strongly upon the relative positioning of the gear and coil and the ability to maintain gear concentricity to the induction coil.

There are several ways to accomplish quenching for spin hardening of gears. One technique is to submerge the gear in a quenching tank. This technique is applicable for large-size

FIGURE 4.51 Diversity of induction-hardening patterns, with variations in time, frequency, and power. (Courtesy of J. LaMonte, INDUCTOHEAT Inc., Madison Heights, MI.) (From V. Rudnev, D. Loveless, R. Cook, and M. Black, *Handbook of Induction Heating*, Marcel Dekker, New York, 2003. With permission.)

gears. Small and medium gear sizes are usually quenched in place, using an integrated quench. The third technique requires the use of a separate concentric quench block (quench ring) located below the inductor.

As a general remark, it has been reported that more favorable compressive stresses within the tooth root were achieved with the gear spin-hardening technique than with the tooth-by-tooth or gap-by-gap approaches.

Figure 4.52 shows three of the most popular design concepts of the induction gear heat-treating processes that employ encircling-type coils: conventional single-frequency concept (CSFC), pulsing single-frequency concept (PSFC), and pulsing dual-frequency concept (PDFC). All three concepts can be used in either a single-shot or scanning mode.

4.5.7.3.2.1 Conventional Single-Frequency Concept

The CSFC is used for hardening gears with medium and small teeth. As one can see in Figure 4.45 (Patterns B and E), the teeth are often through hardened. Quite frequently CSFC can also be successfully used for medium-size gears. As an example, Figure 4.53 shows an induction gear-hardening machine that applies this concept. The gear heat treated in this application is an automotive transmission component with helical teeth on the I.D. and large teeth on the O.D. for a parking brake. Both the I.D. and O.D. require hardening (Figure 4.54). The hardening of the I.D. gear teeth requires a higher frequency than the O.D. Therefore, a frequency of 10 kHz was chosen for O.D. heating and 200 kHz was chosen for I.D. heating. After the heat is off, quenchant is applied to the hot gear in place;

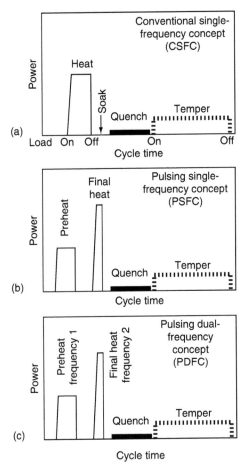

FIGURE 4.52 Concepts of gear hardening by induction. (From V. Rudnev, D. Loveless, R. Cook, and M. Black, *Handbook of Induction Heating*, Marcel Dekker, New York, 2003. With permission.)

that is, no repositioning is required. This practically instantaneous quench provides a consistent metallurgical response. During quenching there is minimal or no rotation to ensure that the quenchant penetrates all the areas of the gear evenly. Quenching reduces the gear temperature to the quenchant temperature or temperature suitable for gear handling.

On the machine referred to above (Figure 4.53) gears are conveyed to the machine, where they are then transferred by a cam-operated robot to the heat-treating station. Parts are monitored at each station and accepted or rejected based on all the major factors that affect gear quality. This includes energy input into the part, quench flow rate, temperature, quench pressure, and heat time. An advanced control and monitoring system verifies all machine settings to provide confidence in the quality of processing for each individual gear. The precise control of the hardening operations and careful attention to the coil design minimize part distortion and provide the desirable residual stresses in the finished gear. The hardened gear is inspected and moved to the next operation.

Although CSFC is the most suitable for small- and medium-size gears, there are cases when this concept can be successfully used for large gears as well. As an example, Figure 4.55

FIGURE 4.53 Equipment used to harden O.D. and I.D. of gear. (From V. Rudnev, D. Loveless, R. Cook, and M. Black, *Handbook of Induction Heating*, Marcel Dekker, New York, 2003. With permission.)

shows an induction machine for hardening large gears. A multiturn encircling inductor is used for hardening gears with a major diameter of 27.6 in. (701 mm), root diameter 24.3 in. (617 mm), and thickness 3.125 in. (79 mm). In this particular case, it was in the customer's best interest to harden and temper in the same coil using the same power supply. In other cases it might not be the best solution.

Quite often, in order to prevent problems such as pitting, spalling, tooth fatigue, and endurance, hardening the contour of the gear (contour hardening) is required. In some cases, this can be a difficult task due to the difference in current density (heat source) distribution and heat transfer conditions within a gear tooth.

FIGURE 4.54 Some gears require hardening both O.D. and I.D. (From V. Rudnev, D. Loveless, R. Cook, and M. Black, *Handbook of Induction Heating*, Marcel Dekker, New York, 2003. With permission.)

FIGURE 4.55 Induction equipment for hardening large gears. (Courtesy of INDUCTOHEAT Inc.) (From V. Rudnev, D. Loveless, R. Cook, and M. Black, *Handbook of Induction Heating*, Marcel Dekker, New York, 2003. With permission.)

Two main factors complicate the task of obtaining the required contour hardness profile. With encircling-type coils, the root area does not have good coupling with the inductor compared to the coupling at the gear tip. Therefore, it is more difficult to induce energy into the gear root. In addition, there is a significant heat sink located under the gear root (below the base circle, Figure 4.46).

4.5.7.3.2.2 Pulsing Single-Frequency Concept

In order to overcome these difficulties and to be able to meet customer specifications, the PSFC was developed (Figure 4.52b). In many cases, PSFC allows the user to avoid the shortcomings of CSFC and obtain a contour-hardening profile. Pulsing provides the desirable heat flow toward the root of the gear without noticeable overheating of the tooth tip. A well-defined, crisp, hardened profile that follows the gear contour (Patterns F and G) can be obtained using high power density at the final heating stage.

A typical dual pulse contour-hardening system, which applies a PSFC has been discussed in Ref. [1]. This machine is designed to provide gear contour heat treatment (including preheating, final heating, quenching, and tempering) with the same coil using one high-frequency power supply. Figure 4.52 illustrates the process cycle with moderate-power preheat, soaking stage, short high-power final heat, and quench followed by low-power heat for temper.

Preheating ensures a reasonable heated depth at the roots of the gear, enabling the attainment of the desired metallurgical result and decreasing the distortion in some materials.

Preheat times are typically from several seconds to a minute, depending on the size and shape of the gear. Obviously, preheating reduces the amount of energy required in the final heat.

After preheating, there might be a soak time ranging from 2 to 10 s to achieve a more uniform temperature distribution across the teeth of the gear. Final heat times can range from less than 1 s to several seconds.

As a general rule, for both CSFC and PSFC techniques the higher frequency is called for by finer pitch gears, which typically require a smaller case depth.

4.5.7.3.2.3 Pulsing Dual-Frequency Concept

A third concept, the PDFC, is not new. The idea of using two different frequencies to produce the desired contour pattern has been around since the late 1950s. This concept was primarily developed to obtain a contour-hardening profile for helical and straight spur gears. Several companies, have pursued this idea.

According to PDFC (Figure 4.52c), the gear is preheated within an induction coil to a temperature determined by the process features. This temperature is usually 350–100°C below the critical temperature A_{c1}. Preheat temperature depends upon the type and size of the gear, tooth shape, prior microstructure, required hardness pattern, acceptable distortion, and the available power source. It should be mentioned that the higher the preheat temperature, the lower the power required for the final heat. However, high preheat temperatures can result in an increased distortion.

As in previous gear spin-hardening concepts, PDFC can be accomplished using a single-shot mode or scanning mode. The scanning mode is typically applied for longer gears.

Preheating is usually accomplished by using a medium frequency (3–10 kHz). Depending upon the type of gear, its size, and material, a high frequency (30–450 kHz) and high power density are applied during the final heat stage. For the final heating stage, the selected frequency allows the current to penetrate only to the desired depth. This process gives excellent repeatability.

Depending upon the application, two coil design arrangements can be used when applying the scanning or single-shot modes. In the first arrangement (Figure 4.56), one coil and two power supplies are utilized to harden the gear. The sequence of operations is as follows:

1. Gar is located within the induction coil.
2. Gear rotation is started.
3. Low-frequency voltage is applied to the induction coil.
4. Coil begins to move along the gear length and preheats the full length of the gear.

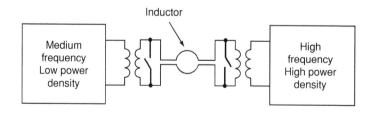

FIGURE 4.56 Using one coil and two inverters for gear hardening. (From V. Rudnev, D. Loveless, R. Cook, and M. Black, *Handbook of Induction Heating*, Marcel Dekker, New York, 2003. With permission.)

5. After completion of the preheating stage, the coil is disconnected from the low-frequency source.
6. Upon returning to the initial position, a high-frequency voltage is applied to the coil and a second scanning cycle begins.
7. Gear is heated to the hardening temperature and quenching is applied simultaneously or the gear is quenched after completion of the heating stage.

The first approach has many limitations including low reliability and complexity, therefore in a great majority of cases, the PDFC process utilizes another type of coil arrangement.

For the second coil arrangement (Figure 4.57), two coils and two power supplies are utilized. One coil provides preheating and another coil, final heating. Both coils work simultaneously if the scanning mode is applied. In the case of a single-shot mode, a two-step index-type approach is used.

Quenching completes the hardening process and brings the gear down to ambient temperature. In some cases, dual-frequency machines produce parts with lower distortion and a more favorable distribution of residual stresses compared to other techniques.

As mentioned above, when applying high frequency (i.e., 70 kHz and higher) it is important to pay special attention to gears with sharp corners. Due to the electromagnetic edge effect, high frequency has a tendency to overheat sharp edges and corners. This results in weakened teeth due to decarburization, oxidation, grain growth, and sometimes even local melting of sharp edges. Therefore, in order to improve the life of a gear, the sharp edges and corners should be broken and generously chamfered.

The main drawbacks of the PDFC process are its complexity and high cost, as it is necessary to have two different power supplies. In some cases, it is possible to use one dual-frequency power supply instead of two single-frequency inverters; however, the cost of these variable frequency devices is high.

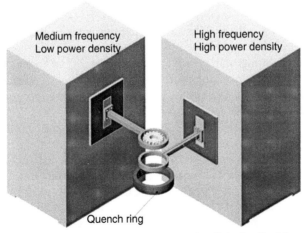

FIGURE 4.57 One coil is used for preheat and the second coil is for final heat. (From V. Rudnev, D. Loveless, R. Cook, and M. Black, *Handbook of Induction Heating*, Marcel Dekker, New York, 2003. With permission.)

4.5.7.4 Lightening Holes

Gears are often produced with holes to reduce the weight of the gear. In induction hardening of gears with internal lightening holes, including hubless spur gears and sprockets, cracks can develop below the case depth in the interhole areas. This crack development results from an unfavorable stress distribution during or after quenching. Proper material selection, improved quenching technique, and modification in gear design and required hardness pattern can prevent crack development in the lightening hole areas.

4.5.8 POWDER METALLURGY GEARS

Special attention should be paid when designing induction-hardening machines for powdered steel gears. These gears are affected to a much larger extent by variations in the material properties of powdered metals as compared to gears manufactuered by casting or forming. This is because the electrical resistivity, thermal conductivity, and magnetic permeability strongly depend on the density of the powdered metal. Variations in the porosity of the powdered steel lead to scattered hardness, case depth, and residual stresses data [1]. A detailed discussion on powder metallurgy parts can be found in Refs. [1,20].

4.5.9 THROUGH AND SURFACE HARDENING TECHNOLOGY FOR GEAR HARDENING

Impressive results can be achieved not only by developing a sophisticated process, but also by using existing processes with a combination of advanced steels. Through and surface hardening (TSH technology) is a synergistic combination of advanced steels and special induction-hardening techniques [21]. These steels were invented by Dr. Shepeljakovskii [21] and distributed by ERS Engineering [22]. The new low-alloyed carbon steels are characterized by very little grain growth during heating into the hardening temperature range. They can be substituted for more expensive steels that are typically hardened by conventional induction or carburizing.

The main features of TSH technology include the following:

- TSH steels are relatively inexpensive, incorporating significantly smaller amounts (three to eight times less) of alloying elements such as manganese, molybdenum, chromium, and nickel.
- They require a lower induction-hardening frequency (1–10 kHz), which reduces power supply cost.
- They exhibit high surface compressive residual stresses (up to 600 MPa/85 ksi).
- The hardened depth is primarily controlled by the steel's chemical composition and initial microstructure. This makes the heat-treating process repeatable and robust.
- They exhibit fine grain size (see Figure 4.58).
- They show a reduced chance of overheating part edges and sharp corners due to end effect.

One of the unique features of TSH gears is that instead of using a two-step approach (first O.D. heat and then I.D. heat or vice versa) that gear has been heated and quenched in a single step using only one inductor. O.D. and I.D. teeth have fine-grained martensite with a hardness of 62 HRC. The microstructure of the core is a combination of very fine pearlite and bainite having a hardness of 30–40 HRC.

TSH technology gears are stronger and more durable than some made of conventionally heat-treated standard steels. Typical applications include gears, bushings, shafts, bearings, and coil and leaf springs.

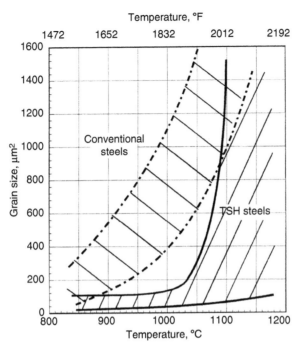

FIGURE 4.58 Grain growth for TSH steels versus that of conventional grades. (From V. Rudnev, D. Loveless, R. Cook, and M. Black, *Handbook of Induction Heating*, Marcel Dekker, New York, 2003. With permission.)

4.5.10 Inductor Styles—Specialty Inductors

Some special types of inductors include I.D. coils, pancakes, split returns, clamshells, and hairpin inductors. Generally speaking, internal heating inductors are not very efficient, mainly because the current travels on the I.D. of the inductor. This phenomenon automatically leads to the fact that the real coil coupling is larger than the actual air gap between the internal surface of the workpiece and the coil O.D. The physics of behind phenomenon has been previously explained in our discussion of electromagnetic ring effect.

I.D. coils almost always require the use of flux concentrators that are located inside the coil. If the frequency is less than 15 kHz, efficient heating can be provided for a workpiece I.D. greater than 1 in. (25.4 mm). Slightly smaller diameters may work with radio frequencies (Figure 4.59).

Pancake inductors have been used for selective surface heating of flats, disks, or plates. This type of inductor has the appearance of an electric stove burner. In fact, induction cooktop stoves are commercially available, all using the pancake coil (Figure 4.60).

Split-return inductors offer a unique distribution of current. The center leg of the inductor carries twice as much current as the side legs. This type of inductor produces a narrow band of heating.

Clamshell inductors are generally used for crankshafts, where part geometry will not allow close coupling or for unusually shaped cams (i.e., cams for racing engines). This type of inductor has a hinge on one side and opens and closes when the part is in position.

Hairpin inductors are usually in the form of bent tubing and conform to a workpiece to heat a selected area. This type is usually used with radio frequency. Hairpin inductors are often shaped during manufacturing using the workpiece as a pattern.

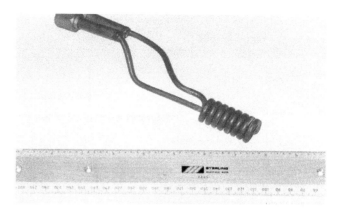

FIGURE 4.59 Inner diameter type of inductor.

4.5.11 TEMPER INDUCTOR STYLES

The decision to induction temper should be carefully considered. It may not be optimal in all the cases. Surprisingly, in many cases the parts can be successfully induction tempered. The main purposes of tempering have been discussed in the introduction of this chapter.

Before we discuss the features of induction tempering, let us examine how stresses appear during induction hardening. Detailed analyses of the nature of residual stresses are presented in Refs. [6,12,19,146–151]. Because of certain aspects of induction hardening, the mechanism of formation of residual stresses here is slightly different than in other heat treatment processes such as carburizing. Substantial work in developing a commercial induction-tempering system and study of the mechanism of residual stresses has been done by HWG (Germany) [149–151], VNIITVCh (Russia) [12], and Colorado School of Mines. Here we give only a short description of the features of induction tempering.

Generally speaking, there are two different types of stresses: thermal stresses and transformation stresses. Thermal stresses are caused by different magnitudes of temperature and

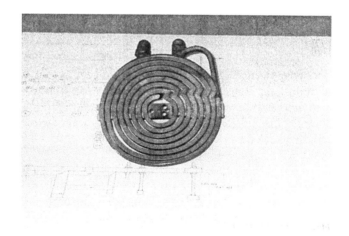

FIGURE 4.60 Pancake inductor.

temperature gradients. Transformation stresses occur due to microstructural changes taking place as a result of the formation of austenite or martensite. The total stress is a combination of both components. At different stages of heat treating the impact of both components into total stresses is different.

Figure 4.61 is a sketch of an induction-heated part (i.e., a steel cylinder). As a result of heating, the section of the part located under the coil will try to expand. At the first stage of the heating cycle, the temperature of the workpiece is relatively low (less than 500°C or 932°F). At this stage, carbon steels have a nonplastic condition and cannot easily expand. As a result of that, stresses build up within the workpiece. The temperature rise will result in the appearance of high compressive stresses at the surface (Figure 4.62). In the temperature range 580–750°C (1076–1382°F) the steels become more elastic and start to expand. As a result, the stresses start to decrease, and since the temperature exceeds 850°C (1562°F) the steel attains a plastic condition, the diameter of the heated area becomes slightly greater than its initial diameter, and stresses at the surface significantly decrease.

After the quenching fluid is sprayed onto the heated surface of the part, an intensive surface cooling will take place. Quite quickly, the surface layer loses its elastic properties

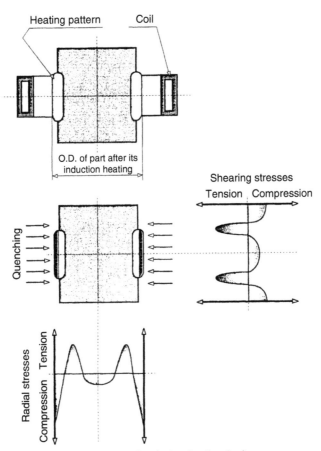

FIGURE 4.61 Formation of residual stresses after induction hardening.

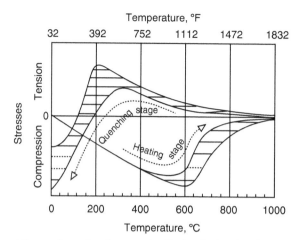

FIGURE 4.62 Stresses at the surface of carbon steel cylinder during heating–quenching cycle.

and tension stresses appear at the surface of the workpiece. There is a pronounced maximum of tensile stresses at the surface of the workpiece. This maximum typically takes place at a surface temperature from 380°C (716°F) to 200°C (392°F). In time, the austenite transforms to martensite or pearlite depending on the cooling conditions. This results in the appearance of compressive stresses at the surface (Figure 4.62).

Figure 4.61 shows the appearance of the thin surface layer where the austenite-to-martensite transformation takes place. Two mutually opposite stresses act in this layer. Because of the drastic temperature decrease in that surface layer, it tries to reduce its volume during the austenite-to-martensite transformation. As the process of quenching continues and martensite forms in the internal layers below the surface of the cylinder, stresses occur due to the fact that the outer layers are already hardened and in nonplastic condition. Finally, when the part is cooled, a combination of compression and tension stresses exists in the workpiece (Figure 4.61).

The compression stresses in the surface may show higher hardness values than are normally achieved with the given steel and also afford some protection against cracks caused by microscopic scratches. The overall stress condition, however, increases brittleness and notch sensitivity, which reduces part reliability. Therefore, it is necessary to relieve stress on the part.

A conventional way of relieving stress is to run the parts through a tempering furnace, which is typically located in a separate production area and therefore requires extra space, labor, and time for parts transportation. In addition, tempering in the furnaces is a time-consuming process that may take up 2–3 h. To overcome these disadvantages, in-line induction tempering was developed.

Basically, there are two ways to perform induction tempering: self-tempering (or tempering by residual heat) and induction tempering.

4.5.11.1 Self-Tempering

The principles of self-tempering after induction hardening are illustrated in Figure 4.63. As described in Section 4.4, during the initial stage of induction heating of the steel cylinder, an intensive heating of the surface layers takes place (Figure 4.63a). On completion of the heating cycle, the temperature profile as shown in Figure 4.63b is obtained. In this case, the temperature within the surface layer that is required for hardening should be within the limits of the hardening temperatures for the given steel. At the same time, the core temperature of the cylinder will be a little different from its initial temperature. After the heating stage is complete, the quenching begins. A hardened layer appears on the surface as a result of the intense cooling. The maximum temperature of the workpiece will be in the internal layer below the surface. At the same time, because of the thermal conductivity of the steel there will be an increase in the core temperature (the heat soaks from the surface toward the core). When the surface layer has been hardened and the temperature at the surface is reduced below the martensitic temperature, a considerable amount of heat is still retained inside the cylinder (Figure 4.63d). If at this moment the supply of quenching fluid is cut off, the part will begin to be heated through due to the accumulation of internal heat. After a certain time, the surface temperature will be increased to a value higher than it had when the quench was cut off. With proper selection of the quenching condition, the heat that is retained inside the part can be used for carrying out the tempering. Typically, self-tempering temperatures do not exceed 260–290°C (500–554°F). At the same time, in the case of surface hardening of selected areas of the workpiece, the self-tempering temperatures usually do not exceed 210–240°C (410–464°F). Figure 4.64 shows a typical temperature–time diagram of this process for carbon steel shaft hardening (shaft O.D. is 1 in.).

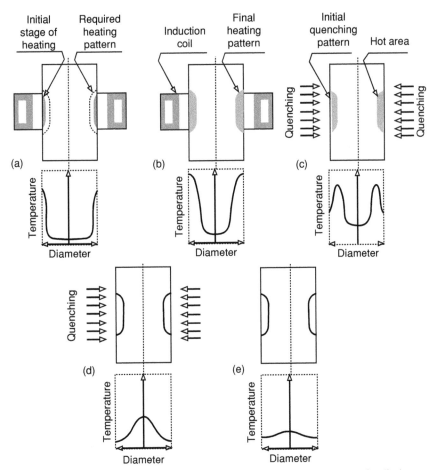

FIGURE 4.63 Induction heating, quenching, and self-tempering of a carbon steel cylinder.

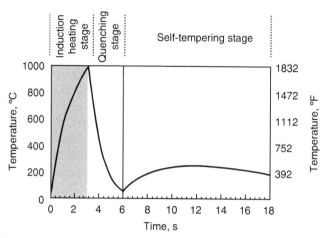

FIGURE 4.64 Surface temperature change during three stages of induction surface hardening: heating, quenching, and self-tempering.

To ensure that the process is always performed correctly, several precautions must be taken. The energy introduced to the part as well as the heating time has to be monitored to ensure a constant amount of residual heat. The quench flow, quench time, and quench temperature should be monitored and held within close tolerances to ensure that the same surface temperature after reheating is always achieved. Moreover, in many cases an infrared pyrometer is used to monitor the surface-tempering temperature. More details of process monitoring are discussed in Section 5.1 (see Chapter 5).

Typically, self-tempering is used when the part has a large mass and single-shot induction hardening is applied. In the case of scan hardening, it is more complicated to use this type of tempering because the core temperatures are different over the scanning length. The amount of heat stored as well as the heat sink underneath each section of the hardened case must be the same; otherwise, the temperatures achieved after heating will be different and the tempering result will be unacceptable.

Sometimes a quench–soak cycle is used. This includes the heating stage, first quenching stage, first self-tempering, second quenching stage, and final self-tempering. Such a method allows one to obtain unique properties of the workpiece.

4.5.11.2 Induction-Tempering Method

For those parts that cannot be self-tempered, the induction-tempering method can be applied. Typically, it is not recommended to use the same inductor for hardening and tempering. There are three reasons for this. The first reason has to do with the fact that in order to obtain the required hardness pattern of the workpiece, because of the workpiece shape, it is necessary to introduce more energy within certain areas. Second, the power densities during hardening are much higher than with tempering. With tempering it is necessary to heat the surface at a much slower rate to achieve a low-temperature gradient from the surface to case depth for otherwise the surface could exceed the required tempering temperature, which would result in an unacceptable soft surface. The third reason has to do with the fact that quite often it is preferable use of lower frequency for tempering because the tempering temperatures are below the Curie point. As a result, the heated part retains its magnetic properties and the skin effect is pronounced. Therefore, in induction tempering, in

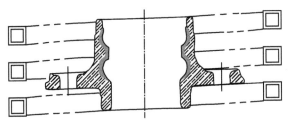

FIGURE 4.65 Induction tempering of a part that has been hardened on the inside surface.

order to increase the current penetration depth in the part, it is more effective to use a lower frequency compared to induction hardening. The temper inductor should heat the entire workpiece, not just the selected hardened area. A loosely coupled multiturn inductor can be used for this purpose. Because of the physics of the tempering process, the total tempering cycle is usually much longer than the hardening cycle.

In the case of induction tempering of complex-shaped parts, such as gears or other critical components, the choice of frequency, power density, and coil geometry is dictated by a need to apply enough energy into certain areas of the part. For example, in gear-tempering applications, it is necessary to induce enough energy into the root area of the tooth without overheating its tip. The root of the gear is the most critical area because the maximum concentration of residual stress is located there. As a result, cracks and distortion occur primarily in the root area. Therefore, this area needs to be stress relieved in the first order. However, there are three factors that make this task quite complicated. First of all, the root area does not typically have a good coupling with the induction coil compared to the gear tip. Because of that, it is more difficult to induce energy there. Secondary, the tempering temperatures are below the Curie point, therefore the gear is magnetic and the skin effect is pronounced. The use of high frequency will result in power surplus in the tip of the tooth compared to its root. The third factor deals with a fact that there is a significant heat sink located under the gear root (under the base circle). In order to overcome the above-mentioned difficulties, IHT manufacturers have developed several new design concepts, which have resulted in the development of advanced induction gear-tempering machines.

Figure 4.65 and Figure 4.66 show two of the most typical examples of induction tempering. Figure 4.65 shows a part that has been hardened on the inside surface. This is the most effective use of induction tempering. The tempering coil is located around the part, so

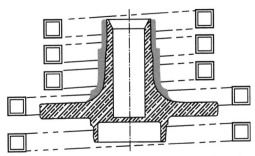

FIGURE 4.66 Induction tempering of a part that has been hardened on the outside surface.

the temperature can slowly increase from the outside surface toward the hardened layer on the inside surface. By using the power energy control, this is a very effective method to carry out the tempering without the danger of overheating the hardened surface. The same idea can be applied for a hollow workpiece that has been hardened on the outside surface. In this case, it is very effective to locate the tempering coil inside the hollow workpiece.

Figure 4.66 shows a hub. This kind of workpiece requires a specially developed induction coil that will allow the predetermined amount of energy to be induced into each section. This allows one to bring each hardened section to the required tempering temperature. It will not be possible to design an inductor that heats only the hardened case like the induction-hardening inductor does because of the wide variation of mass distribution in the hardened area, which results in a different heat sink effect. This inductor for such a workpiece should also induce the heat into areas that are not heat treated such as the flange. These areas will then act as a heat buffer and may be at a temperature slightly above the tempering temperature. This will allow for inconvenient areas to also be heated. With these types of inductors it is possible to induction temper workpieces with complicated shapes and mass distributions such as constant-velocity (CV) joints, which are hardened inside the bell and on the outside of the shaft.

Typically, the tempering inductor is loosely coupled to the workpiece. This allows the tempering coils the ability to heat lightly even the edges, grooves, and other critical regions.

Often a heat–soak cycle is used. This allows the heat to soak all the way through the workpiece. The number of heat cycles can be determined by taking hardness readings in the sectioned workpiece.

As a rule of thumb, the heating time for tempering is at least twice the hardening cycle time (heating and quenching). This means that an IHT machine can have one station for hardening (heat and quench) and two stations for tempering. In this case, the indexing time between the two tempering stations acts as soak time. As mentioned above, the frequency selected for induction tempering is quite often lower than the frequency for hardening, thus resulting in a deeper penetration of the electromagnetic field in the part.

A maximum of tension stresses is typically located below the hardened area, somewhere within the transition zone. This requires expanding the heated area not only to the case depth (hardened depth) but into the transition zone as well.

A cooling cycle may follow completion of the tempering cycle. A cooling station may be located just after tempering. Another practice is to have a cooling station separate from the tempering machine. In this case, the cooling station is located on an exit conveyor. This reduces the number of stations involved in the machine design.

The correct process parameters for induction tempering should be found by hardness measurements. Table 4.8 shows residual stresses in steel 1045 after surface hardening and tempering [151]. As the transformation from the tetragonal martensite to a tempered martensite is a function of time and temperature, one can see from the data that for the same drop in hardness it is necessary to have a higher temperature when using induction tempering. However, at the same time it will have better reduction in residual stresses.

Induction tempering can be commercially successful primarily because it offers the following advantages:

1. Less tempering time than furnace tempering
2. Time and labor savings
3. Low floor space requirements
4. Savings in investment cost
5. Can be incorporated in-line or in a work cell
6. Environmentally friendly

TABLE 4.8
Effect of Induction Tempering on 1045 Steel

Heat Treatment after Hardening	Surface Hardness (Rc)	Maximum Residual Stresses (kg/mm × mm)			
		Shearing		Axial	
		Compressive	Tensile	Compressive	Tensile
Ordinary tempering					
At 100°C (212°F)	60	70	45	48	28
At 200°C (392°F)	55	48	35	35	23.5
Induction tempering					
At 200°C (392°F)	60	63	38	32	23
At 300°C (572°F)	55	40	26	25	28

Source: From K. Weiss, *Industrial Heating*, December, 1995.

Fatigue and failure testing for induction tempering should be compared to that in furnace tempering of individual workpieces. It is important to remember that the entire workpiece, not just the hardened area, usually needs to achieve the desired temperature for proper tempering. The surface temperature alone is not a valid indication of a proper temper.

4.5.12 END AND EDGE EFFECTS, LONGITUDINAL AND TRANSVERSE HOLES, KEYWAYS

4.5.12.1 End and Edge Effects

The coil–workpiece geometry has a significant influence on the heat treatment pattern. Nonuniformity of the heating pattern at the coil or workpiece ends is related to the distortion of the electromagnetic field in those areas. This distortion is called the electromagnetic end effect. Electromagnetic end effect can result in either overheating or underheating of the workpiece end. Basically, it is a function of frequency, coil–workpiece geometry, and material properties. This effect will be discussed in the following subsection. Here we briefly discuss its appearance in the case of the single-turn coil.

Figure 4.67 shows the distribution of the coil current and induced eddy current for different coil locations relative to the end of the part. When a low frequency is applied, the eddy current distribution and heat pattern will be different compared to high frequency.

Generally speaking, electromagnetic end effects are considered one of the most complicated problems in IHT. Electromagnetic end effects can be studied by using advanced computational techniques such as finite-element analysis or IBEs.

As mentioned earlier, IHT is a very flexible process and has been used successfully in a variety of applications. However, in applications where it is required to heat-treat parts (camshaft, crankshaft, axle, transmission shafts, etc.) that contain longitudinal and transverse holes, sharp corners, and keyways (Figure 4.68), it presents certain difficulties. The existence of these features can result in an undesirable appearance of hot and cold spots, cracks, and shape distortion. In these cases, it is necessary to make a careful evaluation of the eddy current and temperature fields in order to achieve the required heat uniformities and meet process specifications.

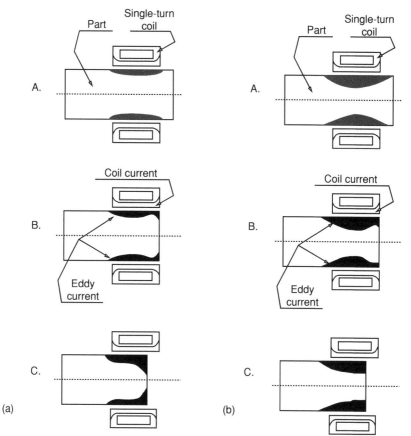

FIGURE 4.67 Electromagnetic end effect in induction heat treatment. (a) High frequency applied. (b) Low frequency applied.

4.5.12.2 Longitudinal Holes

The existence of longitudinal holes (Figure 4.69) or longitudinally oriented hollow areas within the part can cause a redistribution of the eddy current flow that can result in over-heating of certain regions (hot spots). The left-hand part of Figure 4.70 shows a segment of a cylindrical part and the normal current flow within it. If a longitudinal hole is located within the current penetration depth, then it blocks the normal eddy current path and leads to a current redistribution (Figure 4.70, right). Because of that redistribution, areas between the part's surface and the hole can become overheated or even melt.

Actually, there are two factors that cause the overheating effect. The first is an increase in the power density in that area due to the redistribution of induced current. The second is a lack of adjacent mass in that area. As a result, less heat will soak from the surface of the part toward its core. The decrease in this heat soaking is due to the fact that the thermal conductivity of the air is much smaller than the thermal conductivity of the metal. As one can see, these two factors may coincide and result in the overheating of certain areas of the part. In different applications the influence of the two may be different. If a longitudinal hole is located near the part surface within one penetration depth, then the first factor typically prevails and is primarily responsible for the heat surplus in that area (Figure 4.71, hole A).

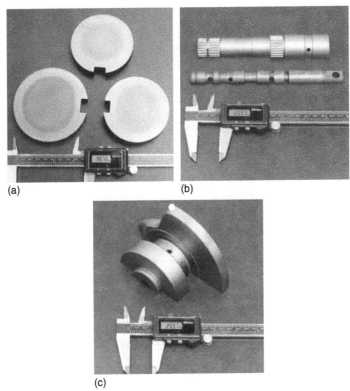

(a)   (b)

(c)

FIGURE 4.68 Induction heat treatment parts with holes, keyways, and sharp corners.

Intense heating with higher power density makes this overheating more pronounced. When the hole is located within one or two current penetration depths (hole B), then both factors have approximately the same influence on the heat surplus. If the longitudinal hole is located within two or three penetration depths and the heat cycle time is relatively long (8–12 s or more), the second factor makes a major contribution to the overheating of the area. When the hole is located within three to five current penetration depths or more under the surface of the part (Figure 4.71, hole D), the heat surplus due to the existence of a longitudinal hole is minor and the probability of overheating is very small.

4.5.12.3   Transverse Holes, Keyways, Various Oriented Hollow Areas

Transverse holes can also cause a redistribution of eddy current flow (Figure 4.69). Unlike the case of longitudinal holes, eddy current redistribution due to transverse holes can result in both an underheating and an overheating of the hole edges (Figure 4.72a). Because of the current concentration, overheating can occur at the hole edges, which are parallel to the eddy current flow [12,107]. On the other hand, the hole edges that are perpendicular to the eddy current flow will be underheated. With an increase of both the I.D. of the transverse hole and the frequency, the nonuniformity of the temperature distribution along the perimeter of the hole will be more pronounced. These heat nonuniformities can cause cracks and distortion in the vicinity of the hole edges.

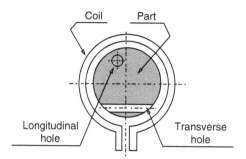

FIGURE 4.69 Longitudinal and transverse holes in induction heat treatment.

It is possible to obtain a relatively uniform temperature distribution along the hole perimeter by putting a plug in the hole. If the plug is made of the same metal as the part, the heat nonuniformities will be negligible and the temperature distribution can be considered uniform (Figure 4.72b). Despite their use in avoiding overheating and underheating at the hole edges, steel plugs present problems because it is often difficult to insert them and remove them after heat treatment because they may become welded in the hole.

An alternative to the use of steel plugs is the use of copper plugs. In this case, a nonuniform current distribution still occurs, which can lead to local overheating and underheating. However, the overheating and underheating phenomenon is opposite to that seen with steel plugs. Figure 4.72c shows that the eddy current will gather in the copper plug from the neighboring carbon steel regions. This takes place because the electrical resistivity of the copper is much less (approximately a factor of 10) than the resistivity of any steel and it is much easier for the eddy currents to flow through the low-resistance copper than through the high-resistance steel. Therefore, when copper plugs are used, the hole edges that are parallel to the eddy current flow are underheated and the hole edges that are perpendicular to the current flow are overheated. An actual distortion of the hardness patterns due to the use of copper plugs is shown in Ref. [107]. It is necessary to mention that when using copper plugs (Figure 4.72c), overheating of the hole edges is much less pronounced compared to the case when plugs are not used (Figure 4.72a). At the same time, the copper plugs eliminate the appearance of cracks in the hole areas during induction heating and quenching.

In some applications, instead of metal plugs, water-soaked wooden plugs have been used. Because wood is not electrically conductive, the wooden plugs do not change the eddy current distribution in the hole edge areas. Therefore, there will still be regions with high

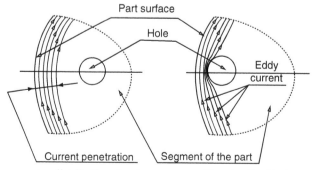

FIGURE 4.70 Eddy current redistribution due to presence of longitudinal hole.

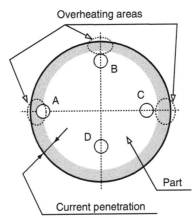

FIGURE 4.71 Overheating areas due to presence of longitudinal holes.

and low current densities as shown in Figure 4.72a. However, the use of water-soaked wooden plugs allows one to decrease the overheating of the hole edges up to a certain point due to the heat transfer from the high-temperature regions into the water-soaked wood by thermal conductivity. There is a benefit in using any plug that limits exposure of the hole to quenching.

Special care should be taken in the case when the heat-treated part contains angled holes or when there is a combination of closely located transverse holes. The temperature field in angled hole areas combines the features of nonuniform temperature distribution along the perimeter of the hole and an overheating effect due to the increase of current density and the lack of mass on one side of the hole. The last two effects are typical for longitudinal holes.

Certain difficulties can appear when the part consists of several closely located holes (Figure 4.72d). Applying IHT in such cases could cause a sequence of cold spots (poorly heated areas) and very hot spots (almost melted areas). In such cases, quite often, alternative heat treatment processes can be recommended.

Keyways can be considered as extreme cases of longitudinal holes. The size, shape, and orientation of the keyways have a substantial effect on the ability to obtain the required temperature profiles within the keyway area and to avoid undesirable hot and cold spots in these areas (Figure 4.72e).

Inserting plugs into holes and taking them out are very delicate and time-consuming processes. A popular saying among heat treaters who deal with parts containing holes is that if there is a possibility of avoiding nonuniform heating of the hole edges without using plugs (e.g., by using a special heating regime or coil design), then the extra development effort required is justified. Therefore, plugs should be used only as a last resort.

When discussing heat nonuniformities due to electromagnetic end and edge effects, longitudinal and transverse holes, and temperature field nonuniformities due to keyways, sharp edges, and corners, it is necessary to mention that smooth chamfers on edges and rounding of sharp corners can be a great help in decreasing the possibility of overheating or cracking.

Experience at INDUCTOHEAT Inc., during heat treating of different parts with the above-mentioned features shows that, surprisingly, the proper choice of design parameters (applied frequency, power density, coil geometry, etc.) allows the heat treater to obtain the required heat-treating pattern even in cases that seem unsuitable for heat treating by induction. For example, in some cases, even such complex-shaped parts as a ball bearing cage, which

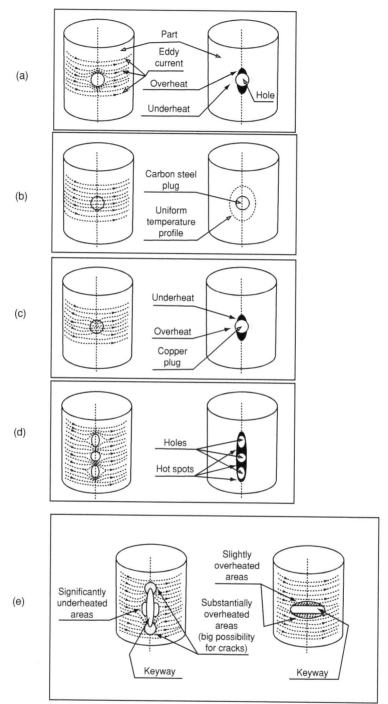

FIGURE 4.72 Eddy current distribution and heat nonuniformities due to presence of transverse holes. (a) Transverse hole, no plug; (b) carbon steel part and carbon steel plug; (c) carbon steel part and copper plug; (d) multiholed part, no plugs; (e) keyways in part.

consists of a number of closely located large holes, can be surface-hardened, or contour-hardened, by induction instead of using time- and space-consuming carburizing processes.

4.5.13 Induction Bar End Heater

The majority of the induction heating applications discussed so far in this chapter have dealt with complex shapes. However, a significant portion of the heat treatment applications deals with the classical cylindrical workpieces such as bars and tubes. Although many of the cylindrical workpieces manufactured today lend themselves to automatic processes in which a bar or tube can be fed into a roll former or other types of forming systems, certain parts require the forming of only the end of the bar. Some examples of these types of parts are sucker rods for oil country goods or various structural linkages in which an eye or a thread may be added to one or both ends of the bar [104,106,115–117].

Induction bar end heating is generally accomplished by placing the end of the bar into a multiturn coil and heating it for a specified amount of time. Multiple bar ends can be heated in a channel type of coil, a single multiturn oval coil, or a two-, three-, or four-coil arrangement that is configured out of individual conventional solenoidal coils [1].

In the case of the oval coil design shown in Figure 4.73, bars are loaded into a magazine from which they are pushed, one at a time, into the upper end of an inclined oval coil. They roll down as previous bars are removed. By the time they reach the bottom position they have reached the desired temperature and are pushed out for delivery to the forming machine. In the case of a channel coil (Figure 4.74), typically the magazine-loaded bars are removed by carriers on a belt, and the end of the bar to be heated passes through the channel coil. Upon leaving the coil, the end of bar is at the required temperature, and the bar moves to the next operation.

The choice of a particular type of induction bar end heater arrangement depends on the customer's requirements. Besides the variety of coil arrangements, the loading and unloading operation can be fully automated or semiautomated, the choice depending on the required production rate and the cost of equipment. The least expensive is the semiautomated design where the operator simply removes the hottest bar and replaces it with a cold bar before it is moved to the forming hammer, press, or upsetter.

Although there are a variety of bar end heating coil designs, the basic principles and rules of thumb in obtaining the required temperature profile within the bar end are quite similar. Consequently, we will discuss the features of the design and operation of the induction bar end heater by analyzing a conventional multiturn solenoidal coil design (Figure 4.75) [104,115]. However, the reader should again bear in mind that the principles and recommendations that

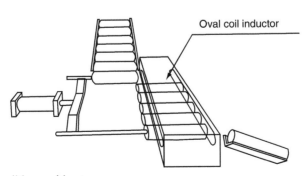

FIGURE 4.73 Oval coil bar end heater.

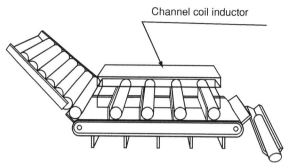

FIGURE 4.74 Channel coil bar end heater.

will be discussed for the conventional solenoidal heater can be extended to other types of induction bar end heater designs.

In progressive induction heating applications (e.g., in-line inductors), it is easier to predict the expected temperature distribution and obtain the required temperature profile in the bar than in the case of bar end heating applications. Basically, this is so because each part of the bar in a progressive induction heater [50,104,105,115,116] experiences the same magnetic flux with respect to time. On a bar end heater it is a much different story, and more variables are involved in obtaining the required temperature profiles within the bar. The large number of design parameters makes the design process quite complicated. Let us have a closer look at the features of the induction bar end heating process.

Typically, in the induction bar end heater the bar may be only partially inserted in the heating coil (Figure 4.75). This shows the difficulty of using analytical and equivalent circuit coil design methods [1–4,6–11] to calculate this process, because those methods are based

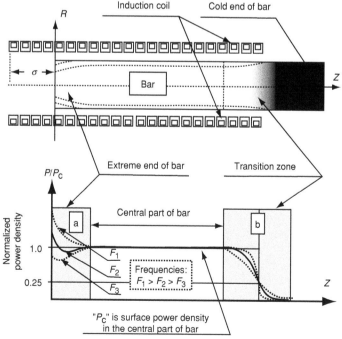

FIGURE 4.75 Sketch of induction heater and power distribution along the length of bar.

on the assumptions of an infinitely long coil and symmetrically located workpiece within the coil. Unfortunately, such assumptions are not valid for the induction bar end heater, which cannot be considered an infinitely long system. Nor, in most cases, is the bar located symmetrically within the coil. As a result of such limitations, those computation techniques cannot be used to accurately predict the power density and temperature distribution within the bar end.

From another perspective, a difference in the effective coil–workpiece resistance and reactance may affect the tuned frequency of the load and the available power. It also causes a much different situation for an analysis of the workpieces thermal condition and temperature distribution because the extreme end of the workpiece (Figure 4.75) offers no path for the conduction of heat, while its cold end provides a ready heat sink and easy conduction path.

For the designer, all of the above-mentioned variables usually cause unpleasant surprises and results in them trying a method of coil design rather than a true ability to predict what will actually happen.

The required temperature distribution within the bar end depends on the frequency, the coil and bar geometry (including the part-to-coil air gap and coil overhang), the material properties of the bar, emissivity, refractory, power density, and cycle time. Obviously, only modern numerical methods can be applied as a computational technique that allows accurate calculations of the results, saves prototyping time and dollars, and facilitates the manufacture of reliable, competitive products with short design cycles.

In the case of the induction bar end heater, the temperature distribution within the bar is affected, among other factors, by a distortion of the electromagnetic field in the extreme end of the bar and induction coil end zone (Figure 4.75). This field distortion and corresponding distribution of induced current and power are referred to as electromagnetic end effects. These end effects may be illustrated by the curves in the lower diagram of Figure 4.75, zone a [5,49–53,104,115,118]. Basically, the electromagnetic end effect in the extreme end of the cylindrical bar is defined by three variables, R/δ, the skin effect; σ, the coil overhang, and the ratio R_i/R, where R is the radius of the bar and R_i is the inside radius of the coil; δ is the current penetration depth which can be defined by Equation 4.38 or Equation 4.39.

An incorrect combination of these factors can lead to underheating or overheating the extreme end of the bar. Studies show that typically the electromagnetic end effect area extends toward the central region of the bar (Figure 4.75) no further than 1.5 times the bar diameter. Higher frequency and large coil overhang will lead to a power surplus in the extreme end of the bar. As a result, significant overheating may take place in that area. A low frequency and small coil overhang will cause a power deficit at the extreme end of the bar, which will therefore be underheated.

It should be pointed out here that a uniform power distribution along the extreme end of the bar will not correspond to its uniform temperature profile because of the additional heat losses (radiation and convection losses) at the bar end area compared to its central part. By the proper choice of design parameters, it is possible to obtain a situation where the additional heat losses at the end of the bar are compensated by the additional power (power surplus) due to the electromagnetic end effect. This will allow the designer to obtain a reasonably uniform temperature distribution within the bar at the end as well as at the center.

The magnetic field distribution in the bar near the right end of the coil in Figure 4.75 (zone b) depends primarily on the radii ratio R_i/R and the skin effect, which is most prominent. Due to the physics of the electromagnetic end effect in that area, there is always a power deficit under the coil tail end at any frequency. It is possible to show that in the case of a long multiturn induction coil with a long homogeneous bar (zone b), the density of the induced eddy current in the bar area under the coil tail end is only half that in the central part [1]. Therefore, the power density in that area is only one fourth of the power in its central part. Typically, the length of zone b is equal to 0.5–3.5 times the coil radius. High frequency and pronounced skin effect lead to a shorter end effect zone. An external magnetic flux concentrator will also give a shorter length of zone b.

It should be emphasized that in order to compensate for the power deficit within the bar area located near the cold end of the bar (zone b), the extreme end of the coil should be extended further toward the bar cold end. In other words, the coil length should be longer than the required uniformly heated area of the bar.

Another important feature that defines the coil length is the fact that in zone b, which is often called the transition zone, there is a significant temperature gradient along the length of the bar. As a result, heat is conducted from the high-temperature region of the bar toward its cold area, which is the heat sink phenomenon. This in turn results in a cooling effect at the other end of the required heated area (zone b). The proper choice of coil geometry (primarily coil length), power density, and frequency will allow one to compensate for this cooling effect and obtain the required uniform temperature profile for the bar.

Typically, there is a limitation on the permissible length of the transition zone. Therefore, the choice of coil length is always a reasonable compromise between obtaining the required heating profile and the minimum length of the transition zone. A short cycle time, high power density, small coil–workpiece air gap, and high frequency help to make this zone shorter.

It is important to understand that the choice of frequency is defined not only by the required temperature profile in the bar, but also by the requirement for high electrical efficiency.

Figure 4.76 shows a sketch of an induction bar end heater for a carbon steel bar (AISI-SAE 1035, bar O.D. is 1 in.) that has been heated at a frequency of 9000 Hz. Frequently, the need for a bar end heater involves equipment that can be used for a wide range of part sizes. This particular coil arrangement can provide satisfactory results in induction end heating for bar diameters as small as 3/8 in. The dynamics of an induction carbon steel bar end heating process (bar O.D. 1 in.) are shown in Figure 4.77.

The analysis shows that in the first heating stage the whole bar is magnetic (cycle time <3 s); therefore, the value of relative magnetic permeability is high. As a result, the skin

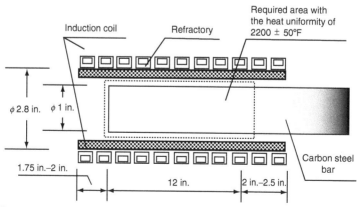

FIGURE 4.76 Geometry of induction bar end heater (frequency, 9 kHz; heating cycle, 36; 9 coil turns; length of bar, 18 in.).

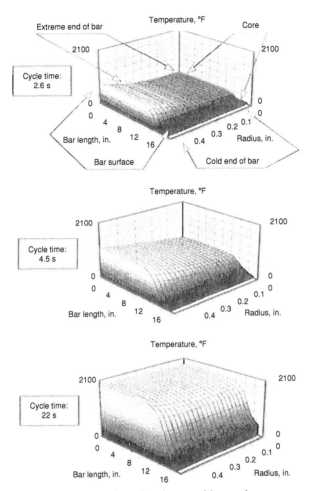

FIGURE 4.77 Dynamics of induction heating of carbon steel bar end.

effect is pronounced ($R/\delta > 20$). At the same time, because of the relatively low temperature, the heat losses from the bar surface are relatively low. Because of the pronounced skin effect, the induced power appears in the fine surface layer of the bar. No power is induced into its inner layers. The core of the bar is heated from the surface only by thermal conductivity. Consequently, the surface of the bar is heated much faster than the core. This stage is characterized by a high-temperature gradient along the radius of the bar, which can cause cracks in steels with 0.5% or more carbon.

With time (cycle time <5 s), as a result of surface temperature rise, the surface of the bar starts to lose its magnetic properties and the relative magnetic permeability in the surface layer drops to 1. Furthermore, because of the temperature rise, the electrical resistivity of the carbon steel increases to approximately 1.5–2.5 times its initial stage value. The decrease of relative magnetic permeability and increase of electrical resistivity cause an increase in the penetration depth. As a result, the essential portion of power is now induced in the internal layers of the bar. This stage can be characterized as the biproperties stage of the bar [5,51,93,104,152]. In this stage the bar surface becomes nonmagnetic; however, the internal area of the bar and its core remain magnetic. The surface-to-core temperature difference here is not as significant as it was in the first stage.

Finally (cycle time $\cong$ 15 s), the total bar area under the coil, including the core of the bar, becomes nonmagnetic. The penetration depth becomes quite large. Skin effect becomes less pronounced ($R/\delta < 2.5$). As a result, a significant amount of energy is induced within the internal area of the bar, and the core starts to heat much more intensely than in the previous two stages. Furthermore, the heat losses from the bar surface become relatively high (primarily due to radiation losses). Because of the large penetration depth, increased heat losses from the surface of the bar, and thermal conductivity within the bar, the temperature profile within the bar thickness will equalize.

The correct choice of induction coil geometry, power density, and frequency allows the induction heating designer to obtain the required temperature uniformity, high efficiency, short transition zone, small equipment space requirements, less scale, and high production rate for a variety of stock sizes.

One of the most successful developments of an induction bar end heater is shown in Figure 4.78 [106,115,116]. This portable induction system (Unipower UPF) is produced by INDUCTOHEAT and can be used in forging applications as well as for bright annealing stainless steel tubing, stress-relieving pipe and wire, and other induction heating applications. Machines with output frequencies of 1–30 kHz providing output power of 25–400 kW are available. Unipower UPF systems have a built-in control, output transformer (depending on the application), and heat station for stand-alone operation. Unipower and Uniforge types of heaters (both produced by INDUCTOHEAT Inc.) are actually a *de facto* standard for use in a large variety of induction heating applications including bar end heating. They have a broad power-matching capability to deliver full power throughout the heating cycle. Precise power settings are ensured within $\pm 2\%$ accuracy with $\pm 10\%$ line variance. All the controls necessary for independent operation are built into machines or located on an optional remote

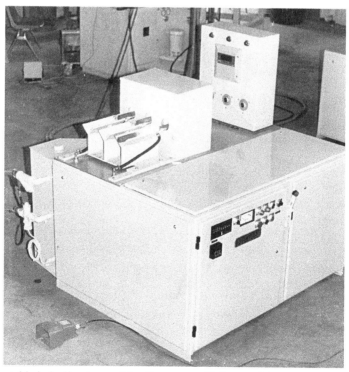

FIGURE 4.78 Portable induction bar end heating system Unipower UPF.

operator's panel. Controls include a single meter with a selector switch for monitoring kilowatts, volts, or frequency; light-emitting diodes to indicate heat on, kilowatts, capacitor volts, output volts, and output current limits; a manual–timed–auto selector switch; and built-in digital timer. Optional equipment includes diagnostic indicator lights for door open, water temperature, water pressure, capacity pressure, and side-mounted water-recirculating system. Additional heat levels and special coil assemblies can be added. Input power factors of such systems are 0.94 under all operating conditions.

Utilizing the Unipower and Uniforge design concept, the power supply, matching components, and heating coil are all grouped together in a single package that can easily be moved from place to place. The secret of the success of both machines is that their design combines outstanding practical experience, modern theoretical knowledge, and a full understanding of the intricacies of the process. Unipower and Uniforge machines offer the ultimate induction heating technology in the most economical package.

4.5.14 INDUCTION STRIP HEATING

As shown above, the induction phenomenon applies to heat treatment of a variety of metalworking processes. Induction strip heating is another area in which induction heating has been successfully used. This includes full and partial annealing, galvanizing and galvannealing, preheating before furnace heating, tempering, paint curing, lacquer coating, and drying. A particular application requires determination of the optimal coil design. There are three basic induction strip heating coil designs [119–121]:

1. Longitudinal flux inductor
2. Transverse flux inductor
3. Traveling wave inductor

Generally speaking, these designs are distinguished by the orientation of the main magnetic flux. Each design has its advantages, and all have been used either alone or in combination with others.

Due to several electromagnetic phenomena (end and edge effects), the distribution of heat sources (power density) induced in a rectangular workpiece (i.e., slab, plate, strip) is not uniform [51,53,73,92,104,119–121]. As a result, there is a problem obtaining the required temperature profile. However, unlike the case of the induction heating of slabs or thick plates [92,104,119–121], the heat source nonuniform distribution along the strip thickness will not typically cause a surface-to-core nonuniform temperature profile. This is because the thermal conductivity of the metal is able to quickly equalize the temperature nonuniformities within the strip thickness.

From another perspective, unlike the induction heating of round bars, strip heating will lead to temperature nonuniformity along the strip surface (width). The temperature distribution along the strip width is primarily affected by a distortion of the electromagnetic field in its edge areas. This is another instance of the electromagnetic edge effect, which can cause a major problem in developing effective induction strip heating equipment.

4.5.14.1 Longitudinal Flux Inductor

A longitudinal flux inductor can be described as a solenoidal heater (Figure 4.79) similar to that used in the induction heating of cylinders. The strip is surrounded by the induction coil. An AC flows through the coil turns and produces a longitudinally oriented time-variable magnetic field that induces eddy currents to circulate within the strip thickness. These currents produce heat by the Joule effect.

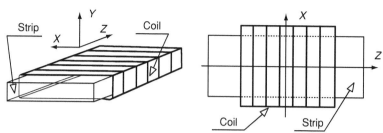

FIGURE 4.79 Longitudinal flux induction heater.

Traditionally, longitudinal flux induction heaters have high efficiency and uniform temperature distribution within the strip. There will be high coil efficiency when the ratio of strip thickness d_{st} to penetration depth δ is 2.5 or more. The optimal value of frequency that corresponds to the maximum coil efficiency can be determined as follows:

Nonmagnetic strip:

$$\frac{d_{st}}{\delta} \cong 3\text{--}3.3$$

Magnetic strip:

$$\frac{d_{st}}{\delta} \cong 2.7\text{--}2.9$$

The use of a frequency higher than the optimal will change the coil efficiency only slightly. At the same time, the use of very high frequencies tends to decrease the total electrical efficiency due to higher power losses in the coil and bus bars. If the chosen frequency is lower than the optimal value, the efficiency dramatically decrease. This is due to cancellation of the induced currents circulating in the opposite sides of the strip cross section.

This type of induction heating does not normally demand a very tight air gap, and it does not require big adjustments of the heater for strips with different widths and thicknesses. Longitudinal flux inductors are effectively used for low-temperature induction heating of thin magnetic strips when the final temperature of the strip is below the Curie point. However, in the case of nonmagnetic thin strip heating (i.e., aluminum, stainless steel, etc.) or for heating of magnetic strips above the Curie point, this type of inductor will require sufficient power at high frequencies of up to several hundred kilohertz or even megahertz and sufficient power. A new family of solid-state power supplies developed by INDUCTOHEAT can produce 1.5 MW at 800 kHz. In some applications this frequency is not high enough. The use of tube generators in the megawatt range of power can lead to well-known problems, such as concern for safety, low reliability, and low efficiency, yet they are the only choice until higher frequency solid-state power supplies are developed.

The process of induction heating of magnetic strips from an ambient temperature, or a temperature below the Curie temperature, to a temperature above the Curie point has several features that could call for a special design of the induction system. At the first stage, the whole strip is magnetic and the heating process is very efficient and intensive. After the strip is heated above the Curie point, the heat intensity will significantly decrease and the heating process will be inefficient. To avoid this, one could use a dual inductor. In the second stage when the load becomes nonmagnetic it is more efficient to use another type of inductor, such as the transverse flux inductor.

As mentioned earlier, one of the major difficulties in obtaining uniform temperature distribution within the strip is caused by the various geometries of the products (i.e., thickness or width) that are heated in the same coil. However, this problem is not as pronounced with the longitudinal flux inductor as with other types of induction coils (i.e., transverse flux or traveling wave inductors).

4.5.14.2 Transverse Flux Inductor

The transverse flux inductor (Figure 4.80) is one of the oldest induction heating techniques, having been developed for use in the induction aluminum alloy strip heating industry in the early 1940s. The principles of the process, simulation procedures, and experience of industrial utilization of transverse flux induction heaters were reported by Baker and Lamourdedieu as early as 1950 [122–125]. This process was established as a way to overcome the problems of induction heating of thin nonmagnetic strips in longitudinal flux inductors.

In the conventional transverse flux coil design, the strip passes through induction coil pairs that are located on both sides of the strip, and they create a common magnetic flux. This magnetic flux passes perpendicularly through the strip width. Unlike the longitudinal flux induction heater, in the transverse flux induction heater the induced eddy currents are circulated in the plane of the strip but not within the strip thickness. This allows induction heating of the thin strip to be carried out with high power densities using low and sometimes even line frequency. The transverse flux inductor often uses external magnetic flux concentrators (i.e., laminations or other high-permeability, low-reluctance materials). This type of inductor does not strictly require a small coil–strip air gap. However, electrically efficient heating can be provided only when strip thickness is 1.5–2 times the penetration depth or less. Without satisfying this condition, the transverse flux effect will disappear and typical proximity heating will take place.

The most difficult problem in using transverse flux inductors is in obtaining temperature uniformity along the width of the strip. The eddy current paths in the strip match the shape of the induction coils. When current reaches the edge area it must flow along the strip edge. As a result, the strip edges become overheated. Because of this natural phenomenon, the current's concentration will be higher in the strip edge area than in the central area of the strip. This undesirable effect can be corrected by the proper choice of inductor design parameters, such as pole step (pole pitch), coil opening, and frequency. Frequency that allows high electrical efficiency in the use of transverse flux induction heating of a nonmagnetic strip can be determined as

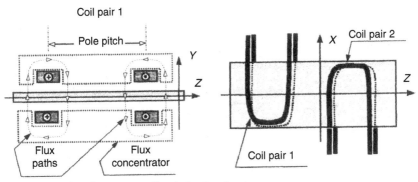

FIGURE 4.80 Conventional transverse flux induction heater.

$$F = 2 \times 10^6 \frac{\rho h}{\tau^2 t}$$

where ρ is the electrical resistivity, Ω m; h is the coil opening, m; τ is the pole step of the coil (pole pitch), m; t is the strip thickness, m. This formula is valid when

$$h/\tau = 0.2\text{--}0.55$$

One of the transverse flux inductor design parameter that has a significant influence on current and therefore on the heat source distribution in the plane of the strip is the shape of the induction coil. Over the last three decades, different three-dimensionally distributed transverse flux coils have appeared quite regularly [126–140]. In practice, most coil pairs have certain limitations in providing precise and effective induction heating when the strip width and thickness varies widely.

4.5.14.3 Traveling Wave Inductor

The traveling wave inductor is not as commonly used for strip heating applications as any of the heaters already discussed. The main reasons are probably the complexity of the process and the fact that there has not been enough experimental and research work done in developing the traveling wave induction heating process. The fundamental concept of that process is quite simple [10,119] and is similar to that of the conventional three-phase electric machine where the strip takes the place of the rotor and the induction coil can be considered a stator (Figure 4.81). The inductor turns are located quite close to each other and carry different phase currents. The inverse connections of the middle phases and the external magnetic flux concentrator have been used to eliminate current cancellation in the neighboring coils. The coils are located in the slots of the flux concentrator. The three-phase current flows through the coil turns, producing an electromagnetic field and heat sources in the strip. One of the main advantages of this system is its low level of vibration and industrial noise compared to other types of inductors. This is primarily important when induction heating nonmagnetic (aluminum, copper, etc.) thin strips. However, there are several difficulties that prevent common use of this type of heater.

First, the traveling wave heater strictly requires a small air gap between the coil and load. This can lead to difficult mechanical design problems. Second, even with the use of magnetic flux concentrators, because of the closeness of multiphase turns, there is a certain magnetic tie between the turns with different phases, and therefore magnetic fields still cancel in adjacent

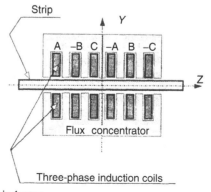

FIGURE 4.81 Traveling wave inductor.

sections. This leads to an overall decrease in the electrical efficiency of the inductor compared to the transverse flux or longitudinal induction heater. Finally, there is still the difficulty of producing a uniform temperature profile for strips of various widths.

No single type of coil will provide acceptable heating results in all strip heating applications. Therefore, the selection of the best type of coil and its specific design must rely on a detailed analysis of the application.

4.5.15 Magnetic Flux Concentrators (Flux Intensifiers, Flux Controllers)

Magnetic flux concentrators (flux intensifiers) have become an acknowledged standard in the IHT design. Modern, high-permeability, low-power-loss materials are now routinely used in a manner similar to that of magnetic flux diverters (cores) in power transformers or motors. A traditional function of flux concentrators in IHT has been to improve the magnetic coupling efficiency (by loss reduction) and to obtain effective selective heating in workpiece areas that are difficult to heat. The successful development of powdered metal concentrators based on Fe, Ni, Co, and other elements have dramatically increased the popularity of magnetic flux concentrators [1].

Let us examine what happens when a magnetic flux concentrator is applied. Without a concentrator, the magnetic flux would spread around the coil or current-carrying conductor and link with the electrically conductive surroundings (i.e., auxiliary equipment, metal support, tools, etc.). The flux concentrator forms a magnetic path to channel the main magnetic flux of the inductor in a well-determined area outside the coil. Figure 4.23a shows the current distribution in an isolated conductor. The current redistribution within this conductor after locating a conductive load (i.e., workpiece) near the conductor can be observed in Figure 4.23b. Due to the proximity effect, a significant part of the conductor's current will flow near the surface of the conductor that faces the load. The remainder of the current will be concentrated in the sides of the conductor.

When an external magnetic flux concentrator is placed around the conductor (Figure 4.23c), practically all of the conductor's current will be concentrated on the surface facing the workpiece. The magnetic concentrator will squeeze the current to the open surface of the conductor—in other words, to the open area of the slot (slot effect).

Current concentration within the coil surface facing the workpiece results in good coil–workpiece coupling and therefore improves the coil efficiency.

The actual current distribution in the conductor (Figure 4.23c) depends on the frequency, magnetic field intensity, geometry, and material properties of the conductor, the workpiece, and the concentrator.

There is a common misconception that the use of flux concentrators automatically leads to an increase in efficiency. Flux concentrators improve the efficiency of the process partly by reducing the coupling distance between the conductive part of the coil and the workpiece but also by reducing the stray losses (by reducing the reluctance of the air path). However, because the flux concentrator is an electrically conductive body and conducts high-density magnetic flux, there is some power loss generated as heat within it due to the Joule effect. This phenomenon could cause a reduction of electrical efficiency and the need to design a special water-cooling system to remove the heat from the concentrator. The first two factors will counteract the third, and the change in electrical efficiency will be a result of all three factors. In some applications, efficiency can actually be reduced. However, the appropriate use of a magnetic flux concentrator will typically allow an increase in process efficiency to be achieved. This can also result from the flux concentrator's ability to localize the magnetic field in a specific area. Because of this, the major portion of the magnetic field will not propagate behind the concentrator and the heated area will be localized. As a result, the heated mass of metal will be smaller, and therefore less power will be required to accomplish the desired heat treatment.

In most cases an application of magnetic flux concentrators does not require reengineering of the induction system. However, when a concentrator is used, higher power densities will be generated on the I.D. of the coil (Figure 4.23c). If the original coil design would be susceptible to stress cracking over time due to the copper work hardening, this condition will occur sooner. Therefore, consideration must be given to the coil I.D. wall thickness and the location of quench holes, which are frequently located near the coil surface edges. The propagation of stress fractures can almost always be minimized or eliminated by a well thought-out coil design. From another side, impedance of the straight coil can be much different compared to the coil with a flux concentrator. Therefore, after the flux concentrator has been installed in the straight coil it is necessary to check that the load properly matches to the power supply. The problem of load matching will be discussed in Chapter 5.

Using a magnetic flux concentrator leads to an increase in power density not only in the induction coil but also in the workpiece areas that must be heated selectively. Because of this, the slot effect plays a particularly important role in the proper design of coils for selective induction hardening, including channel, hairpin, odd-shaped, spiral–helical, and pancake inductors. Special care should be taken when applying flux concentrators to a multiturn coil. With this type of coil the voltage across the coil turns can be significant, and a short current path can develop through the concentrator. In this case, the reliability of the electrical insulation of concentrators plays an essential role in the coil design.

One of the major problems in the induction hardening of steel up to the austenite temperature range is that of undesirable heating of the adjacent areas that have previously been hardened (the so-called temper-back or annealing effect of adjacent parts). This is particularly important in the induction hardening of crankshafts, camshafts, gears, and other critical components [152]. The complexity of this problem arises from the fact that, due to electromagnetic field propagation, the eddy currents are induced not only in the workpiece that is located under the inductor but also in adjacent areas as well. A sketch of the induction carbon steel camshaft-hardening system is shown in Figure 4.82. Figure 4.83 shows the magnetic vector potential field distribution around a two-turn cylindrical induction coil. Without a concentrator, the magnetic flux would spread around the coil and link with electrically conductive surroundings, which include neighboring areas of the part (i.e., cam lobes, journals) and possibly areas of the machine or fixture. As a result of induced eddy currents, heat will be produced. This heat can cause undesirable metallurgical changes in these

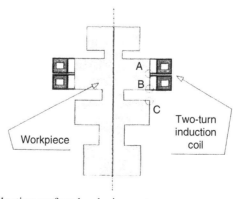

FIGURE 4.82 Sketch of induction surface-hardening system.

areas. At different stages of the heating cycle, the pronunciation of the heating rate of the adjacent areas can vary.

At the first stage of the heating cycle, the entire workpiece is magnetic, the inductor has good efficiency, and intensive heating of the areas located under the coil takes place. Because of better coupling, any surface areas of the workpiece located under the coil (Figure 4.82, regions A or B) will have much more intense heating than any other areas in the coil surroundings (for example, region C).

After a short time, the surface reaches the Curie temperature, the magnetic permeability μ drops to 1, the surface layer becomes nonmagnetic, and its heating intensity drastically decreases. At this stage, the coil will not have as good a coupling factor as it had during the first stage, when the whole workpiece was magnetic. Although the surface of the part has lost its magnetic properties, the adjacent areas retain theirs. Consequently, the coupling factor of these areas will not decrease, and a greater portion of the electromagnetic field will link with the adjacent areas.

In addition, in order to have a short cycle time and to keep the heat intensity of the surface areas located under the coil within the same range as during the first stage of heating, the system can automatically increase the coil power after the surface temperature passes the Curie point. This also will result in additional heating of magnetic parts located in the coil surroundings, which leads to the temper-back of these areas.

After a magnetic flux concentrator is located around the inductor (Figure 4.84), a small portion of the electromagnetic field of the coil will link with adjacent areas. For example, in the induction surface hardening of carbon steel camshafts, appropriate use of magnetic flux concentrators allows a 4- to 12-fold reduction in the power density induced in adjacent cams compared to using a bare coil.

Therefore, magnetic flux concentrators allow decoupling of the induction coil and the adjacent electrically conductive areas. This eliminates undesirable heating of these areas and as a result the temper-back (annealing) effect.

As one can see from Figure 4.12, Figure 4.83, and Figure 4.84, because of the magnetic field redistribution, a bare induction coil cannot provide the required heating pattern in the workpiece. The areas with high power density are observed in adjacent areas of the part but

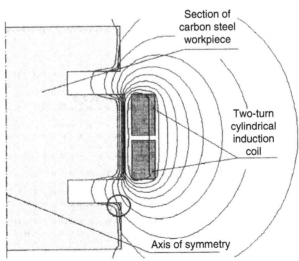

FIGURE 4.83 Electromagnetic field distribution in two-turn cylindrical coil without magnetic flux concentrator.

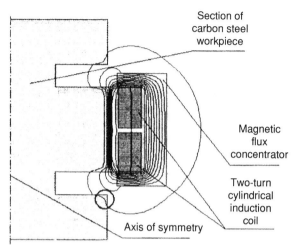

FIGURE 4.84 Electromagnetic field distribution in two-turn cylindrical coil with magnetic flux concentrator.

not in the required area. With the addition of a flux concentrator at the coil edges, practically all of the coil's current will be concentrated on the surface facing the workpiece. The magnetic flux concentrator will squeeze the current to the open internal face of the coil. Current concentration within the coil surface facing the workpiece results in good coil–workpiece coupling, with a consequent improvement in process efficiency. At the same time, the correct choice of flux concentrators and their location, geometry, material properties, and frequency allows the designer to decrease the heat intensity in the adjacent areas of the workpiece and therefore avoid their undesirable heating. In contrast to locating concentrators only at the end areas of the coil (Figure 4.12), the application of a C-shaped flux concentrator, which warps around the coil will lead to some coil efficiency improvement. However, this will of course increase the cost of the design.

In some heat treatment applications, several coils are involved. Because of the relatively small distances between coils, strong magnetic ties can form between coils. This can lead to some negative electromagnetic effects (i.e., power transfer between coils). From another perspective, the stray flux might cause an undesirable temperature profile in the workpiece. In these applications, magnetic concentrators can be used as electromagnetic shields, which will allow undesirable coil interactions and their negative results to be eliminated. In general, the effectiveness of magnetic flux shields depends on various parameters such as frequency, magnetic field intensity, material properties, and the geometry of the induction system. Care should be taken at the corners of flux concentrators because of their tendency to overheat due to electromagnetic end effects.

Manufacturers of IHT equipment have found the development and use of flux control technology increasingly important in reducing the size and improving the quality of IHT systems. Before beginning a project, detailed mathematical modeling and laboratory tests should be conducted to determine the cost-effectiveness of using flux concentrators in a particular application. Different applications may call for different flux concentrator materials (i.e., Laminations, Alphaflux, Fluxtrol, Ferrotron, Alphaform, etc.). Interested readers will find an analysis of the use of various materials for flux concentrators in Refs. [141–152]. Here we give only a short description of the features of using flux concentrators based on the materials obtained from Fluxtrol Manufacturing Inc., IHS, HWG, Alpha 1, and many years of experience using flux concentrators at INDUCTOHEAT Inc.

The choice of the concentrator material depends on several factors where the higher the value, the better the situation, including relative magnetic permeability, electrical resistivity, thermal conductivity, Curie point, saturation flux density, and ductility [152]. Other important factors rely on lower values for a better situation including hysteresis losses and eddy current losses [1].

Additional factors that should be considered include the ability to be cooled and to withstand high temperatures; resistance to chemical attack by quenchants; machinability; formability, ease of installation and removal and cost, which is dependent on the concentrator material, frequency, power density, and geometry of the heat-treating system.

In heat treating, the materials typically used as magnetic flux concentrators are soft magnetic in nature, meaning that they are magnetic only when an external magnetic field is applied. In an electromagnetic field these materials can change their magnetization rapidly without much friction. They are characterized by a tall and narrow hysteresis loop of small area.

Magnetic materials that are soft magnetic usually have a uniform structure, low anisotropy, and magnetic domains that are randomly arranged. A random arrangement corresponds to a minimum energy configuration when the magnetic effects of the domains cancel each other. Therefore, the overall result is zero magnetization. The magnetic domains can be easily rearranged by applying an external magnetic field. The direction of domain rearrangement corresponds to the direction of the applied field. In this case, the magnetic materials behave as a temporary magnet [40,44–46].

As mentioned above, magnetic materials used for flux concentrators should have both a high slope of magnetic permeability and a high saturation flux density. Besides the magnetic permeability there are other important material properties of the concentrator material such as electrical resistivity and thermal conductivity. Magnetic materials with high electrical resistivity and reduce the eddy current losses of the flux concentrator, thereby reducing its temperature increase. High thermal conductivity flux concentrators usually have a longer life because they are not subject to local overheating. Local overheating can be caused by heat radiation from the heated workpiece or the high-density flux in certain areas of the magnetic concentrator.

One of the most important magnetic properties of concentrator materials is hysteresis loss. This quality is derived from the magnetization curve [40]. A typical magnetization curve, representing the magnetization process of a magnetic material, shows

1. Cycle of magnetization in one direction
2. Reversal of the applied magnetic field, which results in demagnetization of previously magnetized material and its magnetization in the opposite direction
3. Another reversal process resulting in magnetization in the original direction

Hysteresis loss is characterized by the conversion of electromagnetic energy into thermal energy due to the rearrangement of the magnetic domains during the hysteresis cycle. This loss should be as small as possible because it signifies a temperature rise in the flux concentrator, which can cause a loss of its magnetism, and therefore a reduction in coil efficiency.

Hysteresis loss is proportional to the area of the hysteresis loop and the frequency. Materials used for flux concentrators should have a coercive force as small as possible. A perfect flux concentrator with maximum efficiency would have no magnetization remaining after the external magnetic field has decreased to zero. A wide opening in the magnetization curve and a high frequency correspond to a high value of hysteresis loss. The flux concentrator properties can be determined from the manufacturer's data sheet or can be measured with the appropriate test equipment.

The materials most commonly used in IHT for flux concentrators and flux diverters are of the following types:

1. Laminations
2. Electrolytic iron-based materials
3. Carbonyl iron-based materials
4. Pure ferrites and ferrite-based materials
5. Soft formable materials

Laminations have been adapted for use in heat treatment from the motor and transformer industry. Grain-oriented magnetic alloys used in laminations are nickel–iron alloys and cold-rolled and hot-rolled silicon–iron alloys. Packets of laminated steel punchings are used effectively from line frequency to 30 kHz. However, there are cases where laminations have been successfully used at higher frequencies (i.e., 100 kHz plus). Laminations must be electrically isolated from each other and used at the proper frequency. Laminations are insulated with mineral and organic coatings. The thickness of the individual laminations should be held to a minimum to keep eddy current losses in the concentrators low. Generally, laminations are 0.06- to 0.8-mm thick. Thin laminations are used for higher frequencies. Laminations with a thickness greater than 0.5 mm are typically used for frequencies below 3 kHz. Compared to most available magnetic flux concentration materials, laminations have the highest relative magnetic permeability and saturation flux density. This is considered an important advantage. When laminations are applied, there are some problems that can occur. Laminations are particularly sensitive to aggressive environments such as quenchants, which leads to rust and degradation problems. Degradation of the magnetic properties of laminations are caused by an increase of coercive force and hysteresis loss. If the laminations are not firmly clamped, the punchings could start vibrating, resulting in mechanical damage, noise, and subsequent failure of the coil or process. Care should be taken with the corners of laminations because of their tendency to overheat due to electromagnetic end effects.

One of the main advantages of using laminations is that they are relatively inexpensive and can withstand high temperatures better than other materials. Lamination packets can also be used to support the induction coil while remaining insulated from it.

Electrolytic iron-based materials [143] were developed in the 1980s and early 1990s specifically for IHT applications. They can be machined by conventional methods, come in different sizes, and are available in two types of alloys with permeability up to 56. Some alloys are rated for higher frequencies (50–450 kHz) and others for lower frequencies (50 Hz to 50 kHz). They do not significantly degrade over time and can be easily removed and replaced for coil repairs.

Carbonyl iron-based materials were adapted to the induction industry from the products developed for the radio industry in the 1960s. They are easily machined but are available in only a few sizes having a low permeability ($\mu = 15$). These products were developed for the higher frequencies (200–450 kHz) and are easy to machine because of their high plastic content [143]. They cannot be soft soldered and must be acid etched. They have temperature characteristics similar to those of the electrolytic products but usually produce case depths about 10–20% less than the electrolytic products and laminations.

Other materials are pure ferrites or ferrite-based. Ferrites are dense ceramic structures made by mixing iron oxide (Fe_2O_3) with oxides or carbonates of one or more metals such as nickel, zinc, or magnesium. They are pressed, then fired in a kiln at high temperature and machined to suit the coil geometry. In relatively weak magnetic fields ferrites have very high magnetic permeability ($\mu = 2000$ plus). Ferrites are quite brittle materials and this is their main drawback. Other disadvantages of ferrites deal with the low saturation flux density (typically 3000–4000 G), low Curie point (approximately 220°C or 450°F), poor machinability, and the inability to withstand thermal shocks. Because of their high resistance, ferrite-based magnetic concentrators are particularly attractive for use in high power density magnetic fields or with high frequencies (50 kHz and higher).

Some concentrator materials are provided in a soft formable state (e.g., Alphaform [145]) that can be easily molded to a desired shape for developmental purposes and later machined, if desired, to exact tolerances. Alphaform is an advanced composite of insulated iron microparticles, space-age polymers, and a thermally sensitive catalyst.

In some applications, magnetic flux concentrators can be made from a single material. Others may be constructed of several materials. For example, in a split-return coil, laminations can be located in the middle area of the coil and iron- or graphite-based materials can be placed at the coil ends. Such designs are cost-effective and electrically efficient because they take into account the field distortion due to the electromagnetic end effect that results in additional losses within laminations at the coil ends.

As stated earlier, the properly used magnetic flux concentrators delivers substantial profits to the modern heat treater, such as:

- Reducing the operating power levels required to obtain the desired heating of the workpieces
- Improving the electrical efficiency of the process and decreasing the amount of energy used
- Making it possible to selectively heat specific areas of the workpiece
- Obtaining a superior heat pattern (i.e., more consistent hardness of the workpiece or more uniform heat patterns) and improving the physical and metallurgical properties
- Minimizing geometric distortion of the workpiece
- Preventing undesirable heating and annealing of adjacent parts
- Reducing the number of rejected parts, rework, and scrap
- Improving equipment life
- Reducing cycle time
- Eliminating the negative biological effects of electromagnetic field exposure on humans

Adding a flux concentrator to an existing inductor will result in an additional expense. A common saying among people who deal with induction is that if a good part can be produced without a flux concentrator, there is no reason to add to the coil cost.

One of the major concerns regarding the use of magnetic flux concentrators is the reliability of their installation. Typically, flux concentrators are soldered or screwed or sometimes simply glued to the induction coil. With time, due to different factors, there can be an unexpected shifting or movement of the concentrator to an improper position. Usually, magnetic concentrators are located in the areas with high magnetic flux density. Because of this, concentrators are affected by electromagnetic forces. Over time, as a result of those forces, the concentrator can become loose.

In addition to electromagnetic forces, this looseness can take place due to unstable temperature conditions. During the operation cycle, the flux concentrator could be heated to 250°C (482°F); then during quenching it could be cooled to the ambient temperature. In typical surface-hardening applications, heating–cooling cycles can be repeated up to 200–600 times per hour. After a series of such heating–cooling cycles an expansion–reduction of the volume of the flux concentrator takes place. Unstable temperature conditions can also result in the flux concentrator moving or relocating itself in an improper position. As a result, the heating and hardening patterns will change.

Unexpected changes in the hardening pattern can cause very serious damage. For example, in the automotive industry this can result in the car being recalled for replacement of the defective part. To prevent such a situation, the flux concentrators should be examined from time to time and repaired if necessary. In some cases, special monitoring systems can be installed to indicate changes in concentrator operation; however, such monitoring systems substantially increase the cost of the equipment.

In order to choose the right concentrator material, a heat treater can be confused by a wide variety of offerings available in today's market. Some manufacturers of magnetic flux concentrators may claim that their products are better suited to a certain induction heat-treating application than another.

The following examples illustrate some of the practical benefits and features of using flux concentrators in various heat treatment applications [142].

4.5.15.1   Application 1. Single-Shot Hardening of Long Shafts

Purpose: To increase the torsional strength of the shaft by case hardening to a depth of one fourth of the shaft diameter (one half the radius).

Single-shot applications can increase productivity by heating the entire hardened area at once. This application will require a larger power supply. When the power input is too low, the heat is conducted too deep; when power is too high, the surface overheats. By using a flux concentrator, the power requirement is minimized and better control is established over process parameters such as cycle time and pattern termination at the ends of the heated zone. Lower power also increases the life of the inductor. The shaft shown in Figure 4.85 (a dummy for test purposes only) was heated with less than 200 kW. Quench blocks have been removed to permit viewing of the heated area.

4.5.15.2   Application 2. Channel Coil for Continuous Processing

Purpose: To anneal parts that are moving through a long channel on a chain conveyor.

In multiple-station processes where conveyor speed is constant through all stations, the induction processing operation can be tuned to the constraints of the system by adjusting either the power input or the length of the inductors. Applications such as shell case and spark plug annealing are aided by flux concentrators on the channel coils (Figure 4.86), which increase the amount of energy that actually reaches the workpieces. Thus both the power requirement and coil length can be reduced.

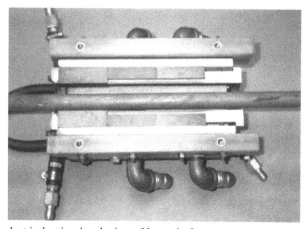

FIGURE 4.85 Single-shot induction hardening of long shafts.

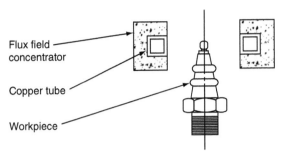

Flux field concentrator

Copper tube

Workpiece

FIGURE 4.86 The use of flux concentrators in channel-type induction coils, multistation steps.

4.5.15.3 Application 3. Single-Shot Hardening of the Drive Stem

Purpose: To increase strength of torque-transmitting component; to harden into fillet area and up to, but not into, the snap-ring groove.

Traditional scanning methods for induction hardening have created several problems, including (1) overheating of the corner of the large diameter near the fillet; (2) formation of a shallow and weak pattern in the fillet; and (3) a long, tapered pattern terminating below the snap-ring groove.

The single-shot inductor enhanced heating characteristics but still did not consistently harden to specified depths over the entire part. Hot bands were observed at each end of the stem, and underheating occurred on the corner of the large diameter, in the fillet, and in the shank. With the addition of flux concentrators, the shank, large diameter, and fillet between them are heated at the same rate to produce the required hardness pattern. With the flux concentrators placed correctly at the top of the inductor, the pattern termination near the snap-ring groove is sharper and more controllable. The improved heating pattern was clearly visible in the parts as they were heated. Figure 4.87a shows the cut samples, the effects of improper heating (through heating at the top of the stem), and the improved pattern produced by the inductor after adding flux concentrators.

As inductor designers gain experience and familiarity with families of workpiece configurations, inductors can be designed with flux concentrator materials installed during initial fabrication (Figure 4.87b). Minor changes are all that may be required for a final pattern development. Typical heat time for short stems like this, at 10 kHz, is 6 s at 100 kW.

4.5.15.4 Application 4. Hardening a Valve Seat

Purpose: To improve surface wear to avoid marking by the valve.

Using a magnetic concentrator in this case turned an almost impossible development task into a fairly reasonable one. There were two obvious complexities. First, the diameters adjacent to the seat are the same size as the seat's O.D. The heat pattern produced by a simple inductor small enough to pass through these bores will not reach the O.D. on the surface of the valve seat. Second, the tube inside the seat must not be hardened, yet the field intensity in a round inductor is greatest within the bore of the coil.

Magnetic flux concentrator material applied in the center and top of the inductor protects the center tube. It causes the heated pattern to flatten out across the seat and pushes the pattern across the entire surface. The pattern shown was accomplished with a 400-kHz unit at 6.0 kW in 4.0 s. Figure 4.88 shows the placement of the powdered iron concentrator.

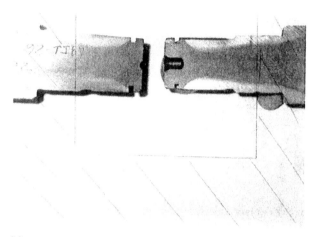

(a)

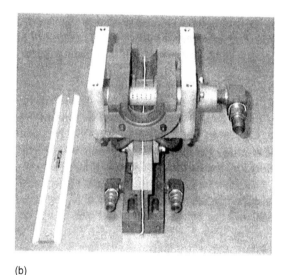

(b)

FIGURE 4.87 Single-shot hardening of the drive stem.

4.5.15.5   Application 5. Surface Hardening of Rocker Arm Tip

Purpose: To improve surface wear on nodular iron rocket arm.

The previous methods of heating rocker arm tips usually involved a circular inductor that heated the entire end of the rocker arm. The pattern was not optimal, and the time required to conduct heat to the center of the wear surface was excessive. A split-return type of inductor improved results, and the use of flux concentrators allows the wear surface to reach the hardening temperature before the remainder of the workpiece is appreciably heated. Heating time for this application was 1.0 s at 27 kW at a frequency of 25 kHz.

From the examples described above, one can conclude that when choosing a magnetic flux concentrator for a particular application, the selection factors discussed here should be carefully considered. At the same time, major attention should be given to the location of the concentrator, its shape, and applied frequency. When many factors are involved in obtaining the required heating pattern, the computational ability of the IHT manufacturer becomes an ultimate advantage over companies that rely on intuition and the experience of past mistakes.

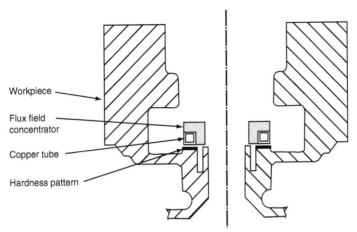

FIGURE 4.88 Placement of powdered iron-based concentrator for induction heat treatment of complex-shaped part.

4.6 SUMMARY

To briefly summarize the chapter, IHT may be viewed as a very specialized area of induction heating. Well-known basic principles of heat transfer and electromagnetics that apply to IHT are discussed, although no attempt was is made to provide an exhaustive analysis because of space limitations.

Modern technology requires a deeper study of the processes involved in IHT. The use of the most advanced computational methods and software allows engineers to obtain a more detailed knowledge of the process. This knowledge helps them to generate innovative ideas, discover new processes, and better understand the unique industrial technologies.

As mentioned in the introduction, one of the goals of this chapter has been to emphasize the interrelated features of electromagnetic and heat transfer phenomena. However, the metallurgical aspect is the third but no less important part of the heat treating. We specifically did not discuss the metallurgical aspect here because it is discussed in previous chapters and in [1].

The variety of IHT applications necessitates the use of many different combinations of power supplies, load matching, process control, and monitoring equipment. Some induction coils can be very effective with certain inverters and ineffective with others. Therefore, the optimal design of a modern IHT system must take into consideration not only the features of the physical stand-alone process but also a combination of the inductor, load-matching station, and inverter. The operational characteristics of the power supply must properly match the coil requirements to obtain the desired results. These subjects are discussed in Chapter 5.

Should further details be required or questions arise regarding particular applications, the authors welcome any reader inquiries or suggestions at INDUCTOHEAT Inc.

REFERENCES

1. V. Rudnev, D. Loveless, R. Cook, and M. Black, *Handbook of Induction Heating*, Marcel Dekker, New York, 2003.
2. C.A. Tudbury, *Basics of Induction Heating*, Rider, New York, 1960.
3. M.G. Lozinskii, *Industrial Applications of Induction Heating*, Pergamon Press, London, 1969.
4. E.J. Davies, *Induction Heating Handbook*, McGraw-Hill, New York, 1979.
5. A.E. Slukhotskii and S.E. Ryskin, *Inductors for Induction Heating*, Energy Publ., St. Petersburg, Russia, 1974 (in Russian).

6. S.L. Semiatin and D.E. Stutz, *Induction Heat Treatment of Steel*, ASM International, Metals Park, OH, 1986.

7. ASM International, *Induction Heating*, Course 60, ASM International, Metals Park, OH, 1986.

8. M. Orfeuil, *Electric Process Heating*, Battelle Press, Columbus, OH, 1987.

9. S. Zinn, *Elements of Induction Heating: Design, Control, and Applications*, ASM International, Metals Park, OH, 1988.

10. E.J. Davies, *Conduction and Induction Heating*, Peter Peregrinus, London, UK, 1990.

11. A.D. Demichev, *Surface Induction Hardening*, St. Petersburg, Russia, 1990 (in Russian).

12. G.F. Golovin and M.M. Zamjatin, *High-frequency Induction Heat Treating*, Mashinostroenie, St. Petersburg, Russia, 1990 (in Russian).

13. D.F. Mellon, Contour gear hardening using induction heating with RF and thermographic control, *Industrial Heating*, July, 1988.

14. INDUCTOHEAT Bulletin, *Contour Gear Hardening Eliminates Carburizing Problems*, November, 1988.

15. D.F. Mellon, Induction contour gear hardening for efficient in-line heating, *Precision Metals*, June, 1988, pp. 49–53.

16. V. Rudnev, D. Loveless, B. Marshall, K. Shepeljakovskii, and N. Dyer, *Gear Heat Treating by Induction*, Gear Technology, March, 2000.

17. G. Parrish and D. Ingham, *The Submerged Induction Hardening of Gears*, Heat Treatment of Metals, Vol. 2, 1998.

18. *American Metal Treating Co.*, General Presentation, 1999.

19. ASM, Heat treating, in *Metals Handbook*, 9th ed., Vol. 4, ASM, Cleveland, OH, 1991.

20. V. Rudnev, *Intricacies of Induction Hardening of Powder Metallurgy Parts*, Heat Treating Progress, ASM International, December, 2003.

21. K. Shepeljakovskii, *Induction Surface Hardening of Parts*, Mashinostroenie, Moscow, 1972.

22. *TSH Steels and Advances in Induction Hardening of Pinions*, ERS Engineering, 2002.

23. INDUCTOHEAT Bulletin, *Three-stage Hardening of Transmission Parts Shifts into Gear*, 1992.

24. INDUCTOHEAT Bulletin, *Harden and Temper ID and OD of Cams*, 1992.

25. INDUCTOHEAT Bulletin, *Annealing: Marine Engine Propeller Draft Shafts*, 1992.

26. B. Gebhart, *Heat Transfer*, McGraw-Hill, New York, 1970.

27. S. Patankar, *Numerical Heat Transfer and Fluid Flow*, Hemisphere, New York, 1980.

28. F.P. Incropera and D.P. Dewitt, *Fundamentals of Heat Transfer*, Wiley, New York, 1981.

29. R.F. Myers, *Conduction Heat Transfer*, McGraw-Hill, New York, 1972.

30. J.A. Adams and D.F. Rogers, *Computer Aided Analysis in Heat Transfer*, McGraw-Hill, New York, 1973.

31. W.M. Rohsenow and J.P. Hartnett, *Handbook of Heat Transfer*, McGraw-Hill, New York, 1973.

32. R. Siegel and J.R. Howell, *Thermal Radiation Heat Transfer*, 2nd ed., McGraw-Hill, New York, 1980.

33. J.A. Wiebelt, *Engineering Radiation Heat Transfer*, Holt, Rinehart and Winston, New York, 1966.

34. E.M. Sparrow and R.D. Cess, *Radiation Heat Transfer*, Wadsworth, Englewood Cliffs, NJ, 1966.

35. N.M. Beljaev and A.A. Rjadno, *Methods of the Heat Transfer*, Part 1–2, Vysshaja Schola, Moscow, 1982 (in Russian).

36. J.R. Howell, *A Catalog of Radiation Configuration Factors*, McGraw-Hill, New York, 1982.

37. ASM, *Metals Handbook*, Vol. 2, *Properties and Selection: Nonferrous Alloys and Pure Metals*, ASM, Cleveland, OH, 1979.

38. ASM, *High Temperature Property Data: Ferrous Alloys*, ASM, Metals Park, OH, 1988.

39. L.V. Bewley, *Flux Linkages and Electromagnetic Induction*, Dover, New York, 1964.

40. *Encyclopedia of Physics*, 2nd ed., VCH Publishers, Weinheim, Germany, 1991.

41. R.W. Chabay and B.A. Sherwood, *Electric and Magnetic Interactions*, Wiley, New York, 1995.

42. P. Hammond, *Electromagnetism for Engineers*, Pergamon Press, New York, 1978.

43. M.A. Plonus, *Applied Electromagnetics*, McGraw-Hill, New York, 1978.

44. I.E. Tamm, *Fundamentals of the Theory of Electricity*, Moscow, Russia, 1981 (in Russian).

45. W.H. Hayt Jr., *Engineering Electromagnetics*, McGraw-Hill, New York, 1981.

46. H.G. Booker, *Energy in Electromagnetism*, Peter Peregrinus, London, UK, 1982.

47. N.N. Rao, *Elements of Engineering Electromagnetics*, Prentice-Hall, Englewood Cliffs, NJ, 1987.

48. R. Ehrlich, J. Tuszynski, L. Roelofs, and R. Storner, *Electricity and Magnetism Simulations: The Consortium and Upper-level Physics Software*, Wiley, New York, 1995.

49. V.S. Nemkov, B.S. Polevodov, and S.G. Gurevich, *Mathematical Modeling of High-Frequency Heating Equipment*, 2nd ed., St. Petersburg, Russia, 1991 (in Russian).

50. V.I. Rudnev and K.L. Schweigert, Designing induction equipment for modern forge shops, *Forging*, Winter 1994, pp. 56–58.

51. V.S. Nemkov and V.B. Demidovich, *Theory of Induction Heating*, Energoatomizdat, St. Petersburg, Russia, 1988 (in Russian).

52. D.J. Griffiths, *Electrodynamics*, Prentice-Hall, Englewood Cliffs, NJ, 1989.

53. V.I. Rudnev, Mathematical Simulation and Optimal Control of Induction Heating of Large-Dimensional Cylinders and Slabs, Ph.D. thesis, Department of Electrical Technology, St. Petersburg Electrical Engineering University, Russia, 1986 (in Russian).

54. G.D. Smith, *Numerical Solution of Partial Differential Equations: Finite Difference Methods*, 3rd ed., Oxford University Press, Oxford, UK, 1985.

55. K.J. Binns, P.J. Lawrenson, and C.W. Trowbridge, *The Analytical and Numerical Solution of Electric and Magnetic Fields*, Wiley, New York, 1992.

56. A.A. Samarskii, *Theory of Finite Difference Schemes*, Moscow, Russia, 1977 (in Russian).

57. S.V. Patankar and B.R. Baliga, A new finite-difference scheme for parabolic differential equations, *Num. Heat Transfer 1*: 27–30 (1978).

58. K.S. Demirchian and V.L. Chechurin, *Computational Methods for Electromagnetic Field Simulations*, Moscow, 1986 (in Russian).

59. J.C. Stirkwerda, *Finite Difference Schemes and Partial Differential Equations*, Wadsworth & Brooks, Belmont, CA, 1989.

60. M.V.K. Chari, Finite Element Analysis of Nonlinear Magnetic Fields in Electric Machines, Ph.D. dissertation, McGill University, Montreal, PQ, Canada, 1970.

61. M.V.K. Chari and P.P. Silvester, Finite element analysis of magnetically saturated DC machines, *IEEE Trans. PAS 90*: 2362 (1971).

62. J. Donea, S. Giuliani, and A. Philippe, Finite elements in solution of electromagnetic induction problems, *Int. J. Num. Methods Eng. 8*: 359–367 (1974).

63. W. Lord, Application of numerical field modeling to electromagnetic methods of nondestructive testing, *IEEE Trans. Magn. 19*(6): 2437–2442 (1983).

64. W. Lord, Development of a Finite Element Model for Eddy Current NDT Phenomena, Ph.D. thesis, Electrical Engineering Department, Colorado State University, Ft Collins, CO, 1983.

65. C. Marchand and A. Foggia, 2D finite element program for magnetic induction heating, *IEEE Trans. Magn. 19*(6): 2647–2649 (1983).

66. P.P. Silvester and R.L. Ferrari, *Finite Elements for Electrical Engineers*, Cambridge University Press, New York, 1983.

67. W. Muller, C. Kramer, and J. Krueger, Calculation of 2- or 3-dimensional linear or nonlinear fields by the CAD-program PROFI, *IEEE Trans. Magn. 19*(6): 2670–2673 (1983).

68. R.K. Livesley, *Finite Elements: An Introduction for Engineers*, Cambridge University Press, New York, 1983.

69. N. Ida, 3-D Finite Element Modeling of Electromagnetic NDT Phenomena, Ph.D. thesis, Colorado State University, Ft Collins, CO, 1983.

70. J.D. Lavers, Numerical solution methods for electroheat problems, *IEEE Trans. Magn. 19*(6): 2566–2572 (1983).

71. D.A. Lowther and P.P. Silvester, *Computer Aided Design in Magnetics*, Springer, Berlin, 1986.

72. W. Lord, Y.S. Sun, S.S. Udpa, and S. Nath, A finite element study of the remote field eddy current phenomenon, *IEEE Trans. Magn. 24*(1): 435–438 (1988).

73. A. Muhlbauer, S.S. Udpa, V.I. Rudnev, and A.F. Sutjagin, Software for modeling induction heating equipment by using finite elements method, *Proceedings of the Tenth All-Union Conference on High-Frequency Application*, St. Petersburg, Russia, 1991, Part 1, pp. 36–37 (in Russian).

74. E.J.W. ter Maten and J.B.M. Melissen, Simulation of inductive heating, *IEEE Trans. Magn. 28*: 1287–1290 (1992).

75. J.M. Jin, *The Finite Element Method in Electromagnetics*, Wiley, New York, 1993.
76. S. Mandayam, L. Udpa, S.S. Udpa, and Y.S. Sun, A fast iterative finite element model for electrodynamic and magnetostrictive vibration absorbers, *Proceedings of the Ninth Conference on Computational Electromagnetic Fields*, COMPUMAG, Miami, 1993, pp. 8–10.
77. T. Dreher and G. Meunier, A 3D line current model of voltage driven coils, *Proceedings of the Ninth Conference on Computational Electromagnetic Fields*, COMPUMAG, Miami, 1993, pp. 50–52.
78. M. Hano, An improved finite element analysis of eddy current problems connected to voltage source, *IEEE Trans. Magn. 29*(2): 1491–1494 (1993).
79. E. Thimpson, The finite element method, Class Notes CE-665, Colorado State University, Progress in 1978–1990.
80. O.C. Zienkiewicz and R.L. Taylor, *The Finite Element Method*, 4th ed., Vol. 1, *Basic Formulation and Linear Problems*, McGraw-Hill, New York, 1989.
81. L.J. Segerlind, *Applied Finite Element Analysis*, Wiley, New York, 1976.
82. C.S. Desai, *Elementary Finite Element Method*, Prentice-Hall, Englewood Cliffs, NJ, 1979.
83. S.J. Salon and J.M. Schneider, A hybrid finite element–boundary integral formulation of the eddy current problem, *IEEE Trans. Magn. 18*(2): 461–466 (1982).
84. T.H. Fawzi, K.F. Ali, and P.E. Burke, Boundary integral equations analysis of induction devices with rotational symmetry, *IEEE Trans. Magn. 19*(1): 36–44 (1983).
85. S. Cristina and A. Di Napoli, Combination of finite and boundary elements for magnetic field analysis, *IEEE Trans. Magn. 19*(6): 2337–2339 (1983).
86. C.A. Brebbia, *The Boundary Element Method for Engineers*, Pentech Press, London, 1978.
87. K.R. Shao and K.D. Zhou, Boundary element solution to transient eddy current problems, *Eng. Anal. 1*: 182–187 (1984).
88. M.H. Lean, Electromagnetic Field Solution with the Boundary Element Method, Ph.D. thesis, University of Manitoba, Winnipeg, MB, Canada, 1981.
89. Y.B. Yildir, A Boundary Element Method for the Solution of Laplace's Equation in Three-Dimensional Space, Ph.D. thesis, University of Manitoba, Winnipeg, Canada, 1985.
90. T. Inuki and S. Wakao, Novel boundary element analysis for 3-D eddy current problems, *IEEE Trans. Magn. 29*(2): 1520–1523 (1993).
91. K.F. Wang, S. Chandrasekar, and H.T.Y. Yang, Finite element simulation of induction heat treatment and quenching of steels, *Trans. NAMRI/SME XX*: 83–90 (1992).
92. V.I. Rudnev, Characteristics of transverse electromagnetic edge effect in induction heating of magnetic and nonmagnetic slabs, in *The Study of Electrothermal Processes*, Cheboscary, Russia, 1985, pp. 30–34 (in Russian).
93. V.I. Rudnev, D.L. Loveless, M.R. Black, and P.J. Miller, Progress in study of induction surface hardening of carbon steels, gray irons and ductile (nodular) irons, *Industrial Heating*, March, 1996.
94. G. Roen, Metallurgical considerations in induction hardening, *Proceedings of the Sixth International Induction Heating Seminar*, Nashville, September, 1996.
95. INDUCTOHEAT Bulletin, *UNICOOL, Closed Loop Water and Recirculating System*, 1992.
96. INDUCTOHEAT Bulletin, *Water Recirculating System, Plate Type Closed Loop*, 1991.
97. INDUCTOHEAT Bulletin, *Quick Change Coil Adapters*, 1993.
98. INDUCTOHEAT Bulletin, *UNISCAN-1*, 1991.
99. INDUCTOHEAT Bulletin, *UNISCAN-II*, 1992.
100. INDUCTOHEAT Bulletin, *UNISCAN-IV*, 1992.
101. INDUCTOHEAT Bulletin, *STATISCAN VSS-20*, 1991.
102. INDUCTOHEAT Bulletin, *Horizontal Scanning Machine*, 1991.
103. INDUCTOHEAT Bulletin, *Single-Shot Hardening*, 1991.
104. V.I. Rudnev, *The Art of Computation of the Induction Heating Process*, INDUCTOHEAT Booklet, INDUCTOHEAT, Inc., Madison Heights, MI, 1994.
105. V.I. Rudnev, New induction heat technology in Russia, in *Heat Treating: Equipment and Processes*, Conference Proceeding, ASM International, 1994, pp. 209–213.
106. INDUCTOHEAT Bulletin, *UNIPOWER UPF, Self Contained Induction Heating System*, 1990.
107. V.I. Rudnev, R.L. Cook, and J. LaMonte, Induction heat treating: keyways and holes, *Metal Heat Treating*, March–April, 1996.

108. V.I. Rudnev, D.L. Loveless, and R.L. Cook, Striping phenomena, *Industrial Heating*, November, 1995.
109. Gear Research Institute, Review of Literature on Induction Hardening of Gears, Project A-1051 (C553), 1994.
110. American Gear Manufacturers Association, *Gear Materials and Heat Treatment Manual*, ANSI/ AGM 2004-B29, Section 5.2, Flame and Induction Hardening, AGMA, 1989.
111. AGMA, *Design Guide for Vehicle Spur and Helical Gears*, ANSI/AGMA 6002-B93, AGMA, Section 3.
112. K. Namiki, T. Urita, I. Machida, and T. Takagi, The application of hardenability assured cold forging medium carbon steels to CVJ outer race, SAE Tech. Paper 930965, SAE Int.
113. J.M. Storm and M.R. Chaplin, Dual frequency induction gear hardening, *Gear Technol. 10*(2): 22–25 (1993).
114. Y. Matsubara, M. Kumagawa, and Y. Watanabe, Induction hardening of gears by dual frequency induction heating, *J. Jpn. Soc. Heat Treatment 29*(2): 92–98 (1989).
115. V.I. Rudnev and R.L. Cook, Bar end heating, *Forging*, Winter 1995, pp. 27–30.
116. INDUCTOHEAT Bulletin, *Bar Heating*, INDUCTOHEAT, Inc., 1994.
117. V.I. Rudnev and W.B. Albert, Continuous aluminum bar re-heating prior to reducing mill, *33 Metal Producing*, January, 1995, p. 50.
118. V.S. Nemkov, V.B. Demidovich, V.I. Rudnev, and O. Fishman, Electromagnetic End and Edge Effects in Induction Heating, UIE Congress, Montreal, Canada, 1991.
119. V.I. Rudnev and D.L. Loveless, Induction slab, plate and bar edge heating for continuous casting lines, *33 Metal Producing*, October, 1994, pp. 32–34, 43–44.
120. V.I. Rudnev and D.L. Loveless, Longitudinal flux induction heating of slabs, bars and strips is no longer "black magic," Part 1, *Industrial Heating*, January, 1995, pp. 29–34.
121. V.I. Rudnev and D.L. Loveless, Longitudinal flux induction heating of slabs, bars and strips is no longer "black magic," Part 2, *Industrial Heating*, February, 1995, pp. 46–50.
122. R.M. Baker, Transverse flux induction heating, *Elec. Eng.*, October, 1950, pp. 922–924.
123. R.M. Baker, Transverse flux induction heating, *Trans. AIEE 69*(2): 711–719 (1950).
124. M. Lamourdedieu, Continuous heat treatment of aluminum alloy strip, *Metal Prog.*, October, 1951, pp. 88–92.
125. M. Lamourdedieu, Continuous heat treatment of aluminum alloys of the Duralumin type, *80*: 335–338 (1951/1952).
126. R.C. Gibson and R.H. Johnson, High efficiency induction heating as a production tool for heat treatment of continuous strip metal, *Sheet Metal Ind.*, December, 1982, pp. 889–892.
127. R. Waggott, D.J. Walker, R.C. Gibson, and R.H. Johnson, Transverse flux induction heating of aluminum alloy strip, *Metals Technol. 9*: 493–498 (1982).
128. J. Blacklung, Induction heating in rolling mills, new ideas and applications, *Rev. Metall., Cah. Inf. Tech. 84*: 67–71 (1987).
129. N.V. Ross, R.H. Kaltenhauser, and G.A. Walzer, Transverse Flux Induction Heating of Steel Strip, Electroheating Congress, Toronto, Canada, 1992, pp. 110–119.
130. T. Yamagishi, Y. Kitajima, K. Nagahama, H. Ishii, and H. Ikeda, TFX induction heating CAL for rolled aluminum sheet, Japan Light Metal Association, Tokyo, April 1988.
131. R.C.J. Ireson, Induction heating with transverse flux in strip metal process lines, *IEE Power Eng. J. 3*, 1989.
132. R.C. Gibson, W.B.R. Moore, and R.A. Walker, TFX—an induction heating process for the ultra rapid heat treatment of metal strip, ASM Heat Treatment and Surface Engineering Conference, Amsterdam, May, 1991.
133. R.C.J. Ireson, Developments in transverse flux (TFX) induction heating of metal strip, *Metallurgia 58*: 68 (February 1991).
134. S. Wilden, Inductive high-capacity annealing furnaces for the continuous treatment of strips, *Metallurgia 43*: (1989).
135. I, Oku, M. Inokuma, and K. Awa, Application of induction heating to continuous treatment for aluminum alloy strip, *J. Jpn. Inst. Light Metals 40*: 1990.

136. S.B. Lasday, Work processing for continuous annealing of sheet by transverse flux induction heating at steel plant, *Induction Heating*, October, 1991, pp. 43–45.
137. K. Standford, Transverse flux induction heating, *Eng. Digest*, September, 1987, pp. 23–25.
138. P.W. Ainscow, Inductive heating for production of galvannealed steel strip, *Steel Times*, April, 1992, pp. 173–175.
139. E. Balle, J. Calas, and A.B. Wilhelm, Lacquer coating line for tin mill products, *Iron Steel Eng.*, May, 1991, pp. 36–38.
140. W. Kolakowski, Economical production of hot strip with the compact strip production process, *Improved Technologies for the Rational Use of Energy in the Iron and Steel Industry*, NEC Birmingham, UK, 1992.
141. V.I. Rudnev and R.L. Cook, Magnetic flux concentrators: myths, realities, and profits, *Metal Heat Treating*, March/April, 1995, pp. 31–34.
142. J.S. LaMonte and M.R. Black, How flux concentrators improve inductor efficiency, *Heat Treating*, June, 1989.
143. R.S. Ruffini, Production and concentration of magnetic flux for eddy current heating applications, Report presented to 1993 International Federation of Heat Treatment and Surface Engineering, Beijing, China, 1993.
144. R.S. Ruffini, Induction heating with magnetic flux technology, *Modern Applied News*, 1991.
145. Alpha 1 Induction Service Center, General Presentation, *Industrial Heating*, pp. 54–55.
146. *Handbook of Experimental Stress Analysis*, Wiley, New York, 1963.
147. R.D. Cook, *Finite Element Modelling for Stress Analysis*, Wiley, New York, 1995.
148. A.J. Fletcher, *Thermal Stress and Strain Generation in Heat Treating*, Elsevier Science, London, 1989.
149. K. Weiss, In-line tempering on induction heat treating equipment, *Proceedings of the First International Induction Heating Seminar*, São Paulo, Brazil, 1995.
150. General Presentation of HWG, Germany, 1993.
151. K. Weiss, In-line tempering on induction heat treating equipment relieves stresses advantageously, *Industrial Heating*, December, 1995.
152. V.I. Rudnev, R.L. Cook, and D.L. Loveless, Keeping your temper with magnetic flux concentrators, *Modern Application News*, November, 1995.

5 Induction Heat Treatment: Modern Power Supplies, Load Matching, Process Control, and Monitoring

Don L. Loveless, Raymond L. Cook, and Valery I. Rudnev

CONTENTS

5.1 POWER SUPPLIES FOR MODERN INDUCTION HEAT TREATMENT

Induction heating power supplies are frequency changers that convert the available utility line frequency power to the desired single-phase power at the frequency required by the induction heating process. They are often referred to as converters, inverters, or oscillators, but they are generally a combination of these. The converter portion of the power supply converts the line frequency AC input to DC, and the inverter or oscillator portion changes the DC to single-phase AC of the required heating frequency.

Many different power supply types and models are available to meet the heating requirements of a nearly endless variety of induction heating applications [1–10]. The specific application will dictate the frequency, power level (Figure 5.1), and other inductor parameters such as coil voltage, current, and power factor (cos φ) or Q factor.

For purposes of efficient communication, it is best at this point to provide a brief glossary of terms commonly used by induction heat treatment specialists that may be encountered in this section.

Buttons/Taps/Studs: The individual terminal connections on the load-matching capacitors above the porcelain insulators. The connection of a button involves installing a conducting copper strap or washer between an existing copper bus and a capacitor terminal, thus making an electrical connection between a section of the capacitor and the bus.

Cans: Load-matching capacitors.

Efficiency: Ratio of the actual power output of the power supply or coil to the required input power.

kVAR: Kilovolt ampere reactive (kVAR) is a rating term for capacitors that specifies the amount of reactive volt–amperes that the capacitor can supply to the circuit when run at a specified voltage and frequency.

Power factor: The cosine of the phase angle between the voltage and the current of a given circuit. This is most often of interest to the user of induction heating as the ratio of kilowatts to the actual volt–ampere product required to produce that power. A higher power factor (cos φ) generally means a lower volt–ampere product required from the selected frequency converter.

Reactance and Q factor: Inductive reactance is the opposition to the flow of AC current by a pure inductance. A mechanical analogy would liken the inertia of a flywheel to the inductive reactance of a coil while the friction of the shaft and bearing of the flywheel is

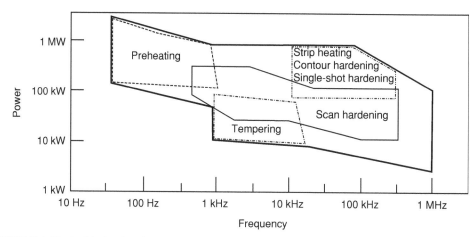

FIGURE 5.1 Typical induction heat treatment applications.

like the resistance of the coil. Using the same analogy, a high-Q coil is like a large flywheel with little shaft friction, and a low-Q coil is like a small flywheel with much friction. Q is the ratio of inductive reactance to the resistance of the coil.

Resonance: For a series circuit, resonance is the frequency at which the vector sum of capacitive and inductive reactances is equal to zero, the power factor (cos φ) is equal to 1, and the impedance is at its minimum value. For a parallel circuit containing resistance in either or both of the circuit paths, resonance may be defined as the frequency at which the vector sum of capacitive and inductive reactance is equal to zero, cos $\varphi = 1$, or the impedance is at its maximum value. For a low-Q circuit, these points may be slightly different from one another, but at higher Q values they are essentially the same. For practical purposes, in induction heating it is possible to consider them equivalent.

Over the hump: Refers to a condition where a swept frequency power supply is operating beyond the resonant frequency of the oscillating circuit.

Tank circuit: The components in the resonant portion of the induction heating circuit, usually consisting of capacitors, transformers, and the induction coil. These components may be connected in series or parallel, depending on the type of power supply used.

Trombone bus: A bus that is designed to slide in and out of a specified receiving bus. As the trombone bus is moved in and out of the receiving bus, the circuit inductance will vary to allow adjustment of the running frequency, voltage, and power.

Tuning or matching the load: The process of adjusting the circuit capacitance, inductance, and impedance in order to heat the workpiece most effectively with the hardware available.

Diode or rectifier: A semiconductor device that allows the current to flow in only one direction. Its function and symbol are illustrated in Figure 5.2.

Thyristor or silicon-controlled rectifier (SCR): A semiconductor device that allows current to flow in one direction but only after a firing signal or positive voltage is applied to its gate terminal. It turns off or blocks the current flow, only after the current supplied to it has stopped for a certain minimum period of time. This period is called the turn-off time. Its function and symbol are illustrated in Figure 5.3.

Transistor: A semiconductor device that will control current flow from its input terminal to its output terminal by the application of a control signal to its gate or base terminal. The transistor can be switched on or off at any time. Its function and symbol are illustrated in Figure 5.4.

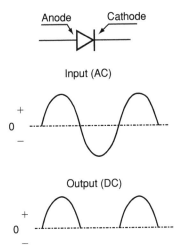

FIGURE 5.2 Symbol and waveshapes of diode.

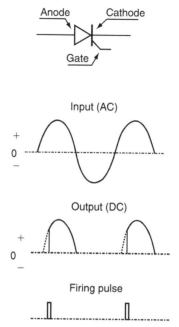

FIGURE 5.3 Symbol and waveshapes of thyristor.

Frequency is a very important parameter in induction heat treatment because it is the primary control over the depth of current penetration and therefore of the depth and shape of the resulting heat pattern. Frequency is also important in the design of induction heating

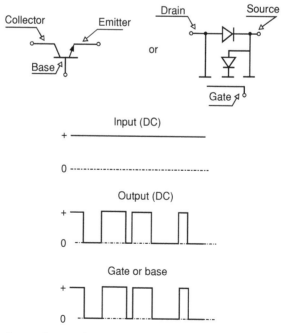

FIGURE 5.4 Symbol and waveshapes of transistor.

power supplies because the power components must be rated for operation at the specified frequency. The power circuit must ensure that these components are operated with adequate margin to yield high reliability at this frequency. The inverter circuits that convert DC to AC use solid-state switching devices such as thyristors (SCRs) and transistors. For high power and lower frequencies, large thyristors are commonly used. For frequencies above 25 kHz, or low power, transistors are used because of their ability to be turned on and off very fast with low-switching losses.

Vacuum-tube oscillators have been used extensively for many years at frequencies above 300 kHz. However, the tube oscillators have a low conversion efficiency of typically 50–60% compared to 83–95% for an inverter using transistors. Power vacuum tubes have a useful life of 2,000–4,000 h and are therefore a costly maintenance item. The high voltage (over 10,000 V) required for tube operation is more dangerous than 1,000 V or less present in typical transistorized inverters. These negative features of tube oscillators have brought about a dramatic move toward the use of transistorized power supplies in most heat treatment applications that require a frequency of less than 1 MHz.

Figure 5.5 shows in graphical form the various power and frequency combinations that are covered by thyristor, transistor, and vacuum-tube power supplies. There are obviously large areas of overlap where more than one type of power supply can be used.

The power required for a given application depends on the volume and kind of metal to be heated, the rate of heating, and the efficiency of the heating process. Small areas heated to a shallow depth may require as little as 1 or 2 kW, whereas heating wide, fast-moving steel strips to temperatures above their Curie point may require many megawatts of induction heating power. It is therefore necessary to define the process and its power requirements by using the numerical techniques described in the computation section of chapter 4 or by a careful extrapolation from similar applications.

The part and coil geometry and the electrical properties of the material to be heated determine the specific coil voltage, current, and power factor. Defining these parameters is necessary to ensure that the output of the power supply is capable of matching the requirements of the coil. Most induction heating power supply systems have the ability to match a reasonable range of heating-coil parameters.

Physical constraints imposed by the environment, in which the induction heat treatment is to be carried out, can also play an important part in the selection or application of the power supply. Each type of power supply, described in detail later in this section, has specific advantages that may directly affect its suitability in the overall heat treatment system.

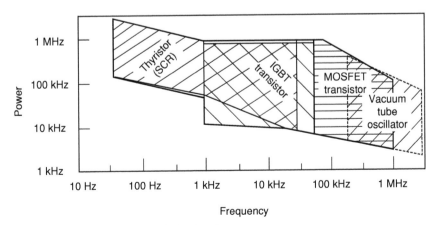

FIGURE 5.5 Modern inverter types for induction heat treatment.

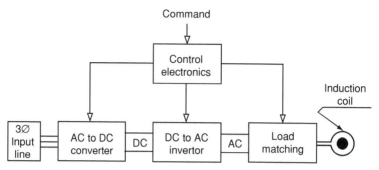

FIGURE 5.6 Induction heat treatment power supply, basic block diagram.

Floor space, machine design, and plant layout are important factors in many induction heat treatment installations. For example, in highly automated machines with a number of hardening and tempering stations, the very compact unitized construction of a transistor-based power supply with self-contained load-matching transformer and capacitors is a definite advantage. On the other hand, for installations requiring a long distance between the power supply and the work coil, the heat station or load-matching portion should be separated from the rest of the power supply and located at the work coil.

5.1.1 TYPES OF INDUCTION HEATING POWER SUPPLIES

An understanding of the various types of power supply circuits used for induction heating is necessary to select the best type for a given application or to assess the suitability of an available power supply to the application. A very basic block diagram that applies to nearly all induction heating power supplies is shown in Figure 5.6. The input is generally three-phase 50 or 60 Hz at a voltage that is between 220 and 575 V. The first block represents the AC to DC converter or rectifier. This section may provide a fixed DC voltage, a variable DC voltage, or a variable DC current. The second block represents the inverter or oscillator section, which switches the DC to produce a single-phase AC output. The third block represents the load-matching components, which adapt the output of the inverter to the level required by the induction coil. The control section compares the output of the system to the command signal and adjusts the DC output of the converter, the phase or frequency of the inverter, or both to provide the desired heating.

5.1.1.1 Full-Bridge Inverter

The most common inverter configuration is the full bridge as shown in Figure 5.7. Often referred to as an H bridge, it has four legs that each contains a switch. The output is located in the center of the H so that when switches S1 and S2 are closed, current flows from the DC supply through the output circuit from left to right. When switches S1 and S2 are opened and

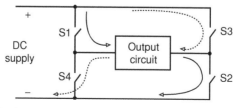

FIGURE 5.7 Basic full-bridge inverter.

switches S3 and S4 are closed, current flows in the opposite direction, from right to left. As this process is repeated, an AC is generated at a frequency determined by the rate at which the switches are opened and closed.

5.1.1.2 Half-Bridge Inverter

The half-bridge inverter, as its name implies, requires only two switches and two-filter capacitors to provide a neutral connection for one side of the output circuit as shown in Figure 5.8. The other side of the output circuit is then switched between positive DC supply by S1 and the negative supply by S2, thus generating an AC voltage across the output. This configuration is used in place of the full bridge where lower output voltage or output power is desired.

Many books and technical papers have been written about the detailed design and theory of operation of the various types of induction heating power supplies [1–10,24]. Inclusion of such detail would likely to be of little help to the heat treater. Therefore, the following paragraphs will only categorize the commonly used power circuit and control combinations. This will give some insight into the advantages and disadvantages of modern induction heating power supplies and their applications and features.

Figure 5.9 shows the principal design features of the inverter configurations most commonly used in induction heating power supplies. The two major types are the voltage-fed and the current-fed. The figure further subdivides each of these by the DC source (fixed or variable), the mode of inverter control, and the load-circuit connection (series or parallel).

5.1.1.3 Direct-Current Section

All of the power supplies outlined in Figure 5.9 have a converter section that converts the line frequency AC to DC [2,3]. Nearly all induction heating power supplies use one of the three basic converters. The simplest is the uncontrolled rectifier shown in Figure 5.10. The output voltage of this converter is a fixed value relative to the input line-to-line voltage, and no control of the output is provided by the converter section. The uncontrolled rectifier must therefore be used with an inverter section capable of regulating the power supply output.

The phase-controlled rectifier, shown in Figure 5.11, has thyristors that can be switched on in a manner that provides control of the DC output relative to the input line voltage. This relatively simple converter can be used to regulate the output power of the inverter by controlling the DC supply voltage. The circuit has two disadvantages. First, the input line power factor is reduced to values that are not acceptable to modern power quality specifications when the DC output voltage of the converter is less than maximum [11,12]. Second, the control response time is necessarily slower because it is not able to respond faster than the frequency of the input line it is acting upon. There are, however, schemes that require additional power components for alleviating both of these disadvantages [2,11].

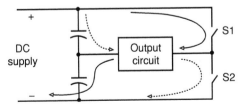

FIGURE 5.8 Basic half-bridge inverter.

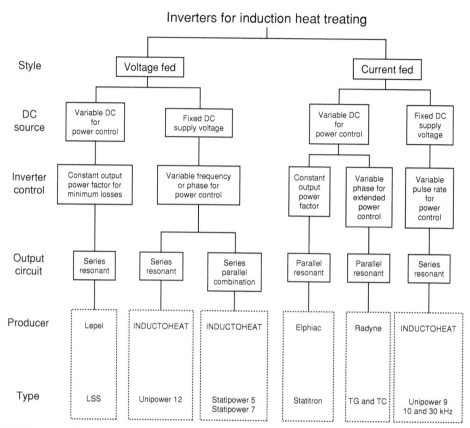

FIGURE 5.9 Induction heat treatment inverters.

The third converter type has an uncontrolled rectifier followed by a switch mode regulator as shown in Figure 5.12. The switch mode regulator shown in the diagram is one of the simplest forms and is called a buck regulator [2]. The level of DC voltage or current at the output is regulated by rapidly switching the pass transistor on and off. The greater the on-time/off-time ratio, the higher the output voltage or current. This converter can therefore regulate the output power of the inverter by controlling the supply of DC. The input line power factor is maximum at all power levels, and the response time can be very fast due to the relatively high switching rate of the buck regulator. Therefore, this converter overcomes the disadvantages of the simple controlled rectifier while remaining more complex, more costly, and slightly less efficient [2].

All of the converters just discussed draw non-sine wave current from the input AC line. This means that there are harmonics or multiples of the line frequency present in the current

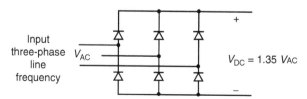

FIGURE 5.10 Uncontrolled rectifier.

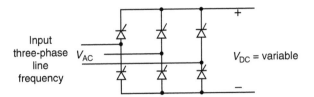

FIGURE 5.11 Phase-controlled rectifier.

waveshape. The harmonic distortion of the current waveshape can adversely affect supply transformers and other electronic equipment connected to the same line. In most heat treatment situations where the power supply rating is less than 600 kW and the plant power distribution system provides a low-source impedance or stiff line, a six-pulse rectifier as described above is acceptable. For higher power systems or where utility requirements require reduced harmonic content, a 12-pulse rectifier, which requires a six-phase input and 12 rectifiers, can be used. The table below compares the typical line current harmonics as a percent of the fundamental for these rectifier configurations.

Harmonic	6-Pulse (%)	12-Pulse (%)
5th	17.5	2.6
7th	11.1	1.6
11th	4.5	4.5
13th	2.9	2.9
17th	1.5	1.5

As shown in the table, the 5th and 7th harmonics are nearly eliminated in the 12-pulse case, resulting in a dramatic reduction of the total harmonic distortion of the line current. Use of higher pulse configurations such as 18 or 24 would lead to a further reduction but at a considerable expense.

5.1.1.4 Inverter Section

5.1.1.4.1 Voltage-Fed Inverters with Simple Series Load
Voltage-fed inverters are distinguished by the use of a filter capacitor at the input of the inverter and a series-connected output circuit as shown in the simplified power circuit of Figure 5.5. The voltage-fed inverter is used in induction heating to generate frequencies from 90 Hz to as high as 1 MHz. Thyristors, which are also called SCRs, can be used to switch the current at frequencies below 10 kHz. Below 50 kHz, insulated gate bipolar transistors

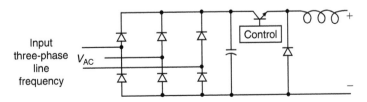

FIGURE 5.12 Uncontrolled rectifier with switch mode regulator.

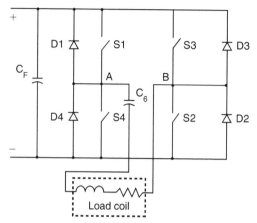

FIGURE 5.13 Voltage-fed series-connected output.

(IGBTs) are commonly used. Above 50 kHz power, MOSFET transistors are chosen for their very fast switching speeds.

The voltage-fed inverter can be switched below resonance as illustrated by the bridge output voltage (Figure 5.14, trace 1) and the output current waveshape (trace 2). This must be the case when thyristor switches are used because diode conduction must follow thyristor conduction for sufficient time to allow the thyristor to turn off. This minimum turn-off time requirement limits the practical use of thyristors to frequencies below 10 kHz. The INDUC-TOHEAT Statipower 6 is an example of this type of inverter [4].

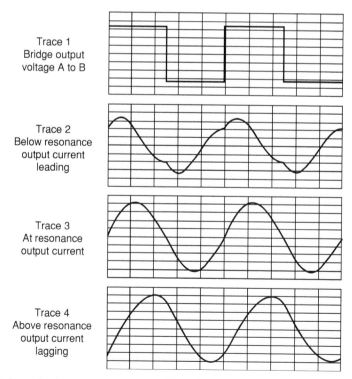

FIGURE 5.14 Voltage-fed inverter waveshapes with series-connected output.

Transistors do not require turn-off time and therefore can be operated at resonance as illustrated by the output current waveshape (Figure 5.14, trace 3). In this case, there is little or no diode conduction, and the transistor is switched while the current is at zero, thus minimizing switching losses and maximizing inverter power rating and efficiency. Operation at resonance means that the output power factor is unity and maximum power is transferred from the DC source to the load. To regulate power in this case, this DC supply voltage must be controlled. The Lepel Solid State (LSS) family of induction heating power supplies (produced by Lepel Corp.) is an example of this type; they are operated at resonance with power controlled by variable DC supplied by a switch mode regulator [5].

Transistors can also be switched above resonance as illustrated by trace 4 in Figure 5.14. In this case, the conducting switches (S3 and S4) are turned off before the current reaches zero. This forces the current to flow in the diodes (D1 and D2) that are across the non-conducting switches (S1 and S2). These switches (S1 and S2) can then be turned on and will conduct as soon as the load current changes direction. This mode of operation minimizes transistor- and diode-switching losses while allowing the inverter to operate off resonance to regulate power. The regulation of output power by control of the inverter frequency relative to the natural resonant frequency of the load will be discussed in Section 5.2.

The voltage-fed inverter supplies a square wave voltage at the output of the bridge, and the impedance of the load determines the current drawn through the bridge to the series load circuit. In nearly all heat treatment applications, an output transformer is required to step up the current available from the inverter to the higher level required by the induction heating coil. The secondary of this transformer is connected directly to the heating coil when the heating frequency is 30 kHz or less and the coil voltage is less than 250 V. In higher frequency applications where the coil voltage is necessarily greater, the series resonant capacitor is usually placed in the secondary circuit of the transformer and in series with the heating coil.

Figure 5.15 shows Lepel's model of a solid-state portable RF power supply. This low-power (15 kW), high-frequency (HF) (50–200 kHz) unit is specifically designed for such heat treatment applications as preheating, soft soldering, brazing, shrink fitting, annealing, through hardening, and epoxy curing of small parts. While capable of functioning as a low-power, in-place unit, the generator's light weight enables it to be hand-carried to wherever it is needed. A built-in electronic digital timer allows automatic timing of the heat cycle to a preset interval and system shutoff.

The salient features of the voltage-fed inverter with a simple series resonant induction heating load are compared to those of the current-fed bridge inverter and summarized in Figure 5.16.

5.1.1.4.2 Voltage-Fed Inverter with Series Connection to a Parallel Load
A popular variation of the voltage-fed inverter for induction heating has an internal series-connected inductor and capacitor that couple power to a parallel resonant output or tank circuit as shown in Figure 5.17. The values of the internal series inductor and capacitor are selected to be resonant above the operating or "firing" frequency of the inverter with an impedance at this firing frequency that will allow sufficient current to flow from the bridge to permit full-power operation. A very important feature of this style of inverter is that the internal series circuit isolates the bridge from the load. This protects the inverter from load faults caused by shorting or arcing and from badly tuned loads, making it one of the most robust thyristor-based induction power supplies available for heat treatment.

A second feature of this series–parallel configuration is realized when the internal series circuit is tuned to the third harmonic of the firing frequency. The power supply is then capable of developing full power into the parallel tank circuit tuned to either the fundamental firing frequency or the third harmonic. For example, the INDUCTOHEAT Statipower 5

FIGURE 5.15 Air-cooled portable radio-frequency solid-state induction heating power supply.

family of induction heat-treating power supplies are produced in three dual-frequency models, 1 and 3 kHz, 3.2 and 9.6 kHz, and 8.3 and 25 kHz [6] with a power range of 10–1500 kW (Figure 5.18). Because load current is not used for commutation, this system can be operated with the output shorted for easy troubleshooting. Solid-state accuracy ensures output power regulation of $\pm1\%$ with a line variance of up to $\pm10\%$. Reliability is further enhanced by placement of 95% of all circuitry on one control board that is accessible without entering the high-voltage section of the power supply.

The voltage-fed inverter with series connection to a parallel load commonly uses thyristors for power switching in the bridge and has an unregulated DC input supply. The regulation of output power is accomplished by varying the firing frequency relative to the parallel load

Bridge inverter features	
Voltage fed	Current fed
DC filter capacitor Square wave voltage Sine wave current Series resonant output Load current = output I. voltage x Q Best for low-Q loads	DC inductor Sine wave voltage Square wave current Parallel resonant output Load voltage = output V. current x Q Best for high-Q loads

FIGURE 5.16 Bridge inverter features.

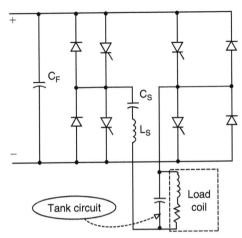

FIGURE 5.17 Voltage-fed inverter with series connection to parallel load.

resonant frequency. The waveshapes present in this style of inverter are shown in Figure 5.19. Trace 1 shows the voltage waveshape at the output of the bridge. Trace 2 shows the bridge current to the load. Trace 3 is the load current and the current when the load is tuned to the fundamental or firing frequency. The corresponding waveshapes for operation with the load tuned to the third harmonic of the firing frequency are shown in Figure 5.19 (trace 4).

5.1.1.4.3   Current-Fed Inverters
Current-fed inverters are distinguished by the use of a variable voltage DC source followed by a large inductor at the input of the inverter bridge and a parallel resonant load circuit at the

FIGURE 5.18 Solid-state induction heating power supply Statipower BSP 5 and 7.

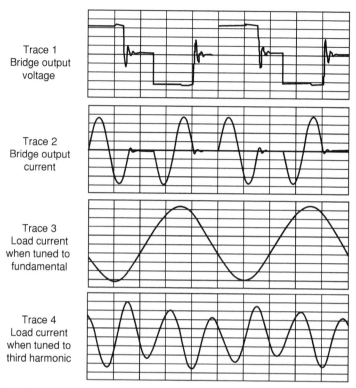

FIGURE 5.19 Waveshapes of voltage-fed inverter with series connection to parallel load.

output as shown in the simplified power circuit of Figure 5.20. Current-fed inverters are available in models that cover the entire 90 Hz to 1 MHz range of frequencies used for induction heat treatment. Thyristors are commonly used below 10 kHz, whereas transistors are chosen for the higher frequencies.

When the power switching is done with thyristors, the current-fed inverter must be operated above the resonant frequency of the parallel resonant load. As illustrated by the waveshapes of Figure 5.21, the voltage across the output of the bridge is a sine wave (trace 1) and the current (trace 2) is a square wave. It is interesting to note that this is just the reverse of the voltage-fed inverter, where the voltage is a square wave and the current is a sine wave. The DC bus voltage across the bridge after the large inductor L_{DC} (trace 3) resembles a full-wave rectified sine wave. The bus voltage is forced negative from the time the bridge is switched until the load voltage reaches zero. This time must be sufficiently long to provide turn-off

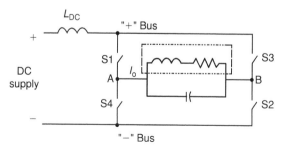

FIGURE 5.20 Current-fed full-bridge inverter.

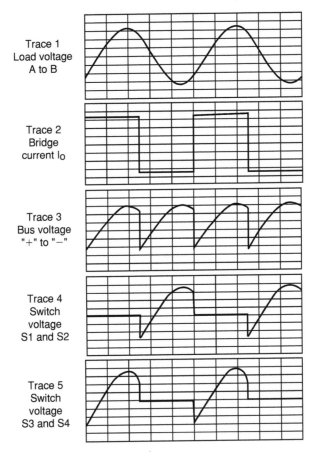

Trace 1
Load voltage
A to B

Trace 2
Bridge
current I$_o$

Trace 3
Bus voltage
"+" to "−"

Trace 4
Switch
voltage
S1 and S2

Trace 5
Switch
voltage
S3 and S4

FIGURE 5.21 Current-fed inverter waveshapes above resonance.

time to thyristors that are no longer conducting. The voltage across the thyristor switches is shown in traces 4 and 5 of Figure 5.21 with the negative portion of the waveshape noted as turn-off time. TG are very high power 1.5 to 4 megawatt current-fed thyristor power supplies. TC are medium power 100–3000 kW current-fed thyristor power supplies. The TG and TC family of induction heat treatment power supplies (produced by Radyne Limited, U.K.) are of this design (Figure 5.22) and have been in use since 1970 [7].

For operation of the current-fed inverter at frequencies above 10 kHz, transistors are used in the inverter bridge because they can be switched very fast and do not require turn-off time. In this case, the inverter can be operated at the resonant frequency of the parallel resonant tank circuit as shown in Figure 5.23. One diagonal of the bridge containing transistors T1 and T2 is turned on as transistors T3 and T4 of the other diagonal are turned off. This switching or commutation is done at a time when the voltage across the load, inverter bus, and transistors is zero. The inverter waveshapes obtained in this mode of operation are shown in Figure 5.24. Switching at zero voltage minimizes the switching losses in the transistors and therefore allows for higher frequency operation. When the inverter frequency is locked to the natural resonant frequency of the load, the output power must be regulated by controlling the input current to the inverter. This is accomplished by using one of the variable voltage DC supplies described earlier. The Statitron 3 (produced by Inducto Elphiac, Belgium) uses MOSFET transistors in a current-fed inverter configuration for heat-treating at frequencies from 15 to 800 kHz with power levels up to 1 MW [8].

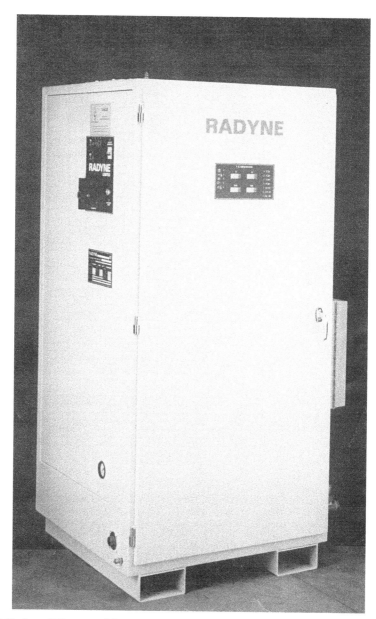

FIGURE 5.22 Radyne TG-type solid-state power supply.

Another inverter configuration that has been used extensively for heat-treating at 10 and 30 kHz uses only one thyristor and is referred to as a chopper or quarter bridge. Figure 5.25 shows a simplified circuit diagram. It is classified as a current-fed inverter because it has a large inductor in series with the DC supply to the inverter. Unlike the conventional full-bridge current-fed inverter, the chopper has a series-connected output circuit. When the thyristor is switched on, current flows both from the DC source through the large inductor and from the series load-tuning capacitor, discharging it through the load coil. The resulting load current pulse (Figure 5.26, trace 20) is nearly sinusoidal, with the first half-cycle of current passing through the thyristor and the second half-cycle through the diode. During this part of the period, current is rising in the input inductor. When the current stops flowing in the diode, the

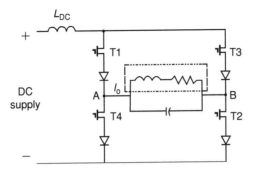

FIGURE 5.23 Current-fed full-bridge transistor inverter.

energy stored in the input inductor causes DC to flow in the output circuit, recharging the series load-tuning capacitor. The frequency of the output sine wave is determined by the series capacitor and the load coil inductance. It is this frequency that determines the penetration depth of the induction heating current. The firing rate of the inverter regulates the output power, and therefore a simple fixed voltage DC source may be used. The INDUCTOHEAT Unipower 9 and Uniscan induction scan hardening machine both make use of this simple inverter [2,9,13–15].

Operational considerations that impact on the suitability of each type of power supply include initial cost, operating cost or overall efficiency, reliability, maintainability, flexibility, cooling water availability, and the power supply's impact on utility power quality.

The initial cost is important but should be a deciding factor only when all of the inverter types considered meet the other operational requirements. In general, the chopper or quarter-bridge power supply has the lowest purchase price. For power levels below 750 kW, the voltage-fed inverter with series resonant load is the next choice based on cost. The current-fed inverter has a low cost per kilowatt when high power at low frequency is required. The most expensive is

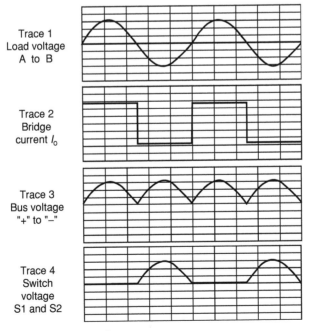

FIGURE 5.24 Current-fed inverter waveshapes at resonance.

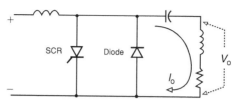

FIGURE 5.25 Current-fed chopper or quarter bridge.

usually the voltage-fed inverter with a series connection to a parallel load. It has more power components per kilowatt than any other type of inverter in its frequency range but is the most robust and flexible for induction heating applications.

Operating cost, which is usually determined by the power conversion efficiency, is also a consideration. Modern semiconductor-based induction heat treatment power supplies, however, all have reasonably high conversion efficiency compared to their motor generator and vacuum-tube predecessors. Most of them have a conversion efficiency of 80–93% when running at rated output power. The conversion efficiency referred here is that of the power supply from the input power connection to the output terminals and therefore does not include, in some cases, the output-matching transformer and load-tuning capacitors. The measurement and the specification of power conversion efficiency can be accomplished in many ways with differing results. At one extreme, only the losses in the inverter portion are used in the calculation of efficiency. At the other extreme, all the losses from line to load are used by taking the ratio of the output power delivered to a calorimeter load to the input line power to the system. This method includes the losses in the inductor coil, which can be relatively high, resulting in a much lower stated efficiency. It is therefore essential to know specifically what portions of the system are included in the specified efficiency to make direct comparisons of power supply efficiency.

The reliability and the maintainability and a power supply's tolerance to input and load perturbations is more a function of power component design margin and control circuit design than the general type of power supply circuit used. Without carrying out a detailed analysis of a power supply, it is very difficult to assess its reliability. Without this analysis, the best guide to equipment reliability is an assessment of the manufacturer's reputation, how

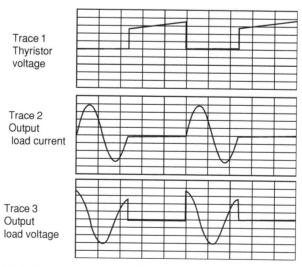

FIGURE 5.26 Current-fed chopper or quarter-bridge waveshapes.

long it has been in the business of producing induction heating power supplies, and the amount of its equipment in field use. Maintainability is affected by many features of power supply design, including the level of self-diagnostics provided, accessibility of components for inspection and measurement, and ease of component and subassembly removal and replacement. When power components, subassemblies, and control boards are interchangeable without adjustment or modification, troubleshooting and repair can be quickly and effectively accomplished by electrical maintenance personnel with only minimal training. Self-diagnostic systems can be very helpful in locating failures in a power supply. However, the inclusion of diagnostic circuitry, which can also fail, has a negative impact on reliability, and therefore a balance between the level of fault diagnostics and power supply reliability is necessary. A very reliable power supply design should require only very basic fault indicators, while more failure-prone designs should be equipped with more extensive diagnostics to speed up the repair process even though an incremental decrease in reliability will result.

The flexibility or the ability of a power supply to operate under varying load conditions or in different applications is an important factor in some situations. If the heat treatment machine is a general purpose equipment such as a scan-hardening machine used in a job shop, the ability to match a wide range of coils at more than one frequency is attractive if not essential. In this case, a dual-frequency power supply with a versatile load-matching system, including both transformer tap switches and dual-frequency capacitor banks, is recommended. The relatively new transistorized power supplies with external transformer tap switching are also attractive where their small size, light weight, and minimal cooling water requirements allow them to be portable and used by multiple machines. The Unipower 12 shown in Figure 5.27 is an example of such a multiple-application power supply [10].

FIGURE 5.27 Multiple application solid-state power supply Unipower 12.

5.2 LOAD MATCHING

5.2.1 PRELUDE TO DISCUSSION OF LOAD MATCHING

A very important facet of induction heat treatment that is often overlooked in the initial design stages is the ability to successfully deliver to the workpiece the maximum available power from a given power supply at the minimum cost. Circumstances do not always allow for an optimal design of a complete induction heat treatment system in which the power supply design is based on the application including the specific induction coil parameters. Quite often, the induction coil is designed to achieve the desired induction heat treatment pattern without regard for the power supply that will be used. When this is the case, a flexibile interface is required to match the output characteristics of the power supply to the input characteristics of the induction coil and workpiece combination [11]. If this match is not provided, the power supply will not be able to develop its rated power if the coil requires more voltage or current than the supply can deliver.

There are many factors involved, any of which can cause complications in arriving at the stated goal. Variable ratio transformers, capacitors, and sometimes inductors are connected between the output of the power supply and the induction coil. The adjustment of these components is commonly referred to as load matching or load tuning.

5.2.2 FOUR STEPS IN UNDERSTANDING LOAD MATCHING FOR SOLID-STATE POWER SUPPLIES

5.2.2.1 Step One

The most common example of matching a power source and load would be a simple lighting circuit application, where a 6-V light bulb is available for use on a 120-V_{AC} power line (Figure 5.28). Obviously, there is a need for some type of interface hardware to prevent 120 V_{AC} from destroying the light bulb. This would commonly be accomplished by inserting a transformer between the light bulb and the power line.

Induction heating circuits have not only a resistive element but also considerable inductance. As part of the electric circuit, any inductor can be introduced as a combination of resistance and reactance (inductance). Both the resistance and the reactance of the inductor are nonlinear functions of several parameters such as coil–workpiece geometry, material properties, and

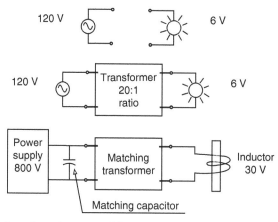

FIGURE 5.28 Load tuning—impedance matching.

frequency. Furthermore, the electrical resistivity and magnetic permeability of metals are non-linear functions of the temperature (see Figure 4.8 and Figure 4.9 of Chapter 4). As shown in Chapter 4, electrical resistivity and magnetic permeability vary during the heating cycle. In addition, modern metalworking processes require that workpieces of different sizes are heated in the same inductor. The combinations of production mix and variations of material properties result in changing coil resistance and reactance, which affects the tuning and performance of the power supply. Generally speaking, a change in coil resistance and reactance results in a change of the phase angle between the coil voltage and coil current of a given circuit. Such a change can be characterized by the coil power factor, which refers to the cosine of the phase angle (cos φ). Power factors of different types of inductors are affected differently by the various factors. At the same time, for different frequencies, the power factor can be significantly different (i.e., from cos $\varphi = 0.02$ up to cos $\varphi = 0.6$), which makes the Q-factor range from $Q = 50$ down to $Q = 1.7$.

In addition to these factors, the process itself usually requires that the part be heated at some frequency other than the line frequency. In conventional heat treatment, the applied frequency typically ranges from 200 Hz to 400 kHz. Since a relatively large current is required to successfully heat a workpiece, it is necessary to build power sources with extremely high output current capability or to use a simple resonant circuit to minimize the actual current or voltage requirement of the frequency converter. A simple example may help at this point.

Example: Given a work coil that requires 100 kW, 40 V, and 10,000 A at 10 kHz and a power source that is rated at 100 kW, 440 V, and 350 A, are the two incompatible?

By using an isolation transformer, we might select a ratio of 440/40 or 11:1 to match the work coil to the power source. This would leave us with a current requirement of 10,000/11 or 909 A, which is too high for the given power source.

By the addition of a specific capacitance to the load circuit, it is possible to lower the current requirement and still accomplish the heating task. The addition of sufficient capacitance to tune the circuit to unity power factor (cos $\varphi = 1$) would result in a required current from the power source of 100 kW/440 V or 227 A, well within the limitations of our selected power source. This relaxes the requirements not only on the power source but also on interconnecting cables, contactors, and transformers operating in the area of the improved power factor.

As shown in the previous section, there are two basic types of resonant frequency converters that use parallel and series resonant circuits. Figure 5.29 shows the characteristics of series and parallel resonant circuits. Looking first at the parallel circuit, it is easy to see that if the capacitor value is equal to zero, then a given voltage applied to the circuit at a fixed

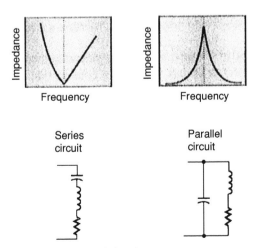

FIGURE 5.29 Resonance at series and parallel circuits.

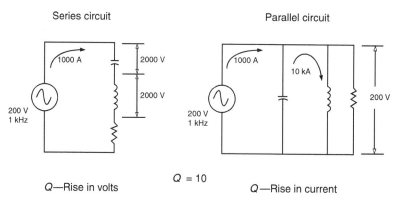

FIGURE 5.30 Series and parallel circuits.

frequency will result in a specific amount of power dependent on the circuit impedance. When sufficient capacitance is added to the circuit to tune the load circuit near resonance, the circuit impedance rises dramatically and the amount of current drawn from the power source falls off dramatically. The circuit voltage required to achieve a specific power level is the same as it was in the initial case of zero capacitance, but now the higher current required by the load is supplied by the capacitors rather than the power source.

In a parallel-tuned load circuit, we have a Q rise in current in the tank circuit compared with the input line from the power source (Figure 5.30). This analogy can be repeated for the case of the series circuit to realize that with the calculated change in circuit impedance; the circuit current will be much higher for a given input voltage when the circuit is tuned near the resonant frequency because the impedance is approaching zero. The load coil current required for a given power is the same for the given load circuit regardless of whether the connection is series or parallel, but because the overall impedance has fallen and the required current is fixed, the required driving voltage is approximately a factor of Q lower than the coil voltage. Hence, we have a Q rise in current in the parallel circuit and a Q rise in voltage with the series-connected circuit (Figure 5.29). It is therefore imperative to have an understanding of what type of circuit connection exists in order to understand the effect that tuning changes will have on the power source and workstation components.

5.2.2.2   Step Two

Turning now to look at the output power characteristic for a given load circuit versus the circuit-operating frequency (Figure 5.31), it is easy to see that if we begin at a low frequency and gradually increase the operating frequency to the point of resonance, we will have an increase in output power. Beyond the point of resonance, an increase in the operating

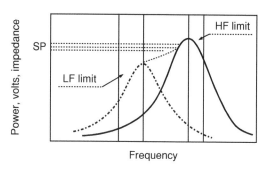

FIGURE 5.31 Load resonant.

frequency will result in a decrease in the output power. This characteristic is often used to accomplish the required regulation mode for the power source. The goal in tuning the workstation is to deliver the required power to heat the workpiece without exceeding any of the power source parameters. Figure 5.32 illustrates a typical workstation with capacitors, an autotransformer, and an isolation transformer.

A useful way to approach the tuning of a workstation is first to determine the ratings of power source and record the available workstation hardware. The next step would be to make an estimate of the required coil voltage for the desired output power. This may be done with previous

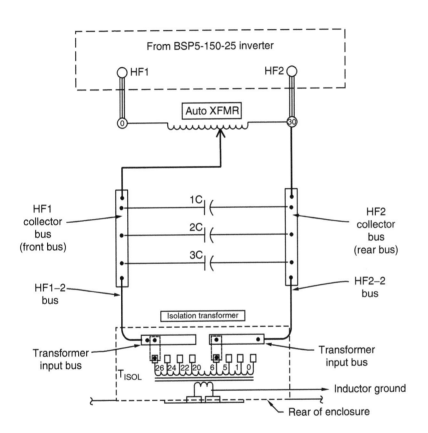

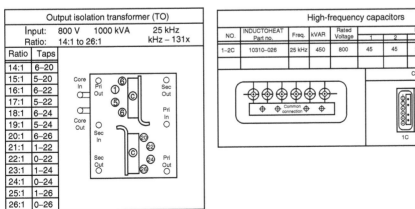

FIGURE 5.32 Typical load-tuning component interconnection.

FIGURE 5.33 Load frequency analyzer.

data or rule-of-thumb extrapolations. Then select an isolation transformer ratio to yield the proper voltage match. The next step would be to set up the load in the work coil as it will be heated and use a load frequency analyzer or a signal generator to determine the resonant frequency of the load circuit [12]. Figure 5.33 shows the load frequency analyzer, which is a solid-state portable instrument that quickly determines the resonant frequency of the induction heater without heating the workpiece. This analyzer eliminates hours of setup time and prevents the waste of production parts. Instead of guesswork, the load frequency analyzer can easily and precisely determine the resonant frequency for any induction heating or heat treatment system.

After obtaining a resonant frequency, it is necessary to add or subtract capacitance from the circuit to match the tuned frequency of the workstation to the rated frequency of the power source. At this time, it is useful to run a cycle and record data for use in extrapolating to the desired setup. Figure 5.34 gives a simple Mathcad sheet for use in extrapolations when tuning the workstation. Variables are presented for power, voltage, coil length, coil turns, frequency, and coil diameter. The variables P1, V1, L1, N1, F1, and D1 are the initial values recorded. The remaining variables are the desired values except for one variable that is chosen as the one to solve for. The equation can be solved manually for any of the given variables. An example at this point might prove helpful.

From previous data:

$$P1 = 100\,\text{kW}; \quad V1 = 30\,\text{V}; \quad F = 9.6\,\text{kHz}$$

$$N1, L1, D1 = 1$$

Desired: What voltage is required for the same power at 1.2 kHz?

$$P2 = 100\,\text{kW}; \quad V2 = ? \quad F = 1.2\,\text{kHz}$$

$$N2, L2, D2 = 1(\text{same as above})$$

Solve for V2.

Solver routine for the extrapolation equation: RLC 4/5/94

To use this program:
(1) Input all values that are known.
(2) Input 1's for all remaining variables.

P1 = 100   P2 = 100

(3) Insert the name of the variable that
    you are solving for in the Find(*x*)
    statement below.

V1 = 30   V2 = 1

F1 = 9600   F2 = 1200

(4) The answer is displayed below.

L1 = 1   L2 = 1

Note: To input variables, point with your
mouse to the number you want to
change and click the left button.
Then delete or enter number as
required. Click on the next position
or press return.

N1 = 1   N2 = 1

D1 = 1   D2 = 1

Given

Standard known voltages
for single-turn inductor

$$\frac{P1}{P2}=\left(\frac{V1}{V2}\right)^2 \cdot \left(\frac{F2}{F1}\right)^{0.674} \cdot \frac{L1}{L2} \cdot \left(\frac{N2}{N1}\right)^2 \cdot \frac{D2}{D1}$$

Volts	Freq.	kW
30	9.6 kHz	100
20	3 kHz	100
15	1.2 kHz	100

Answer = Find (V2)

Answer = 14.886

FIGURE 5.34 Solver routine.

Answer: 15 V (approximately).

Note: For most load-matching calculations, an answer within 10% is close enough to accomplish the job. Although not pleasing scientifically, it is very practical and will save considerable time probing for exact answers.

It may be profitable at this point to consider the effect of changes in various components or component values in the workstation. The induction coil may be varied in shape and size with respect to the workpiece that is heated. The effect of changing a simple cylindrical coil and workpiece is shown in Figure 5.35. The value of circuit Q is directly affected by the part-to-coil coupling and will cause a need for increased capacitive volt–amperes to balance the inductive portion of the circuit. The larger the gap between the part and the coil, the greater the required value of heat station capacitance (Figure 5.35).

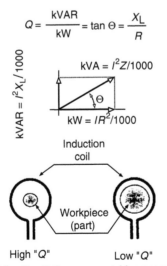

$$Q = \frac{kVAR}{kW} = \tan \Theta = \frac{X_L}{R}$$

$kVA = I^2 Z/1000$

$kVAR = I^2 X_L/1000$

$kW = IR^2/1000$

Induction coil

Workpiece (part)

High "Q"   Low "Q"

The higher the Q, the more matching capacitors required

FIGURE 5.35 Definition of the parameter Q.

The heat station isolation transformer can also be varied by physically changing tap connections to change the turns ratio. A change in this transformer will produce a significant change in the circuit inductance reflected to the primary side of the circuit and subsequently the amount of capacitance required to balance the load property.

The amount of capacitance connected in the circuit can also be changed. Increasing the circuit capacitance will lower the resonant frequency of the circuit, and a decrease in capacitance will result in an increase in the load resonant frequency. There is sometimes confusion in recording the actual amount of capacitance in a given load circuit due to the use of the terms kVAR and microfarads to describe the quantity of capacitance. The term kVAR can be expressed by the following formula:

$$kVAR = \frac{2\pi FCV^2}{1000}$$

which is obviously a function of the frequency F, microfarad value C, and operating voltage V of the capacitor. For a given capacitor, the kVAR is at the nameplate frequency and voltage. If the capacitor is used at another voltage or frequency, the actual required microfarad value (C) will change according to the formula mentioned above.

One might ask what the real benefit of the kVAR term is. It can facilitate calculations when transformers are installed between the coil and the capacitor bank. As shown in Figure 5.36, if we have a load that has a Q value of 8, the required value of the load or tank capacitor is approximately the Q value multiplied by the required kilowatts. For a 150 kW unit, this load would require 1200 kVAR kilovolt Ampere Reactive of heat station capacitance at the rated voltage and frequency. This is true regardless of the position of the capacitor in the circuit. If the microfarad value were used here, then the effective microhenry value of the coil would have to be calculated through each of the transformer connections and then the capacitor microfarad value would be calculated.

One of the major causes of misapplication of this information is that most users assume that the kVAR value of the capacitor is constant, but, in fact, it changes its value as the operating frequency and voltage change. It is then necessary to specify the kVAR at the operating voltage and select the capacitors so that they provide the required kVAR value at the desired voltage and operating frequency. For example, using the information above, if the circuit requires 1200 kVA at 600 V and 10 kHz, then the nearest standard capacitor that would be acceptable would probably be a 2133 kVAR, 800 V, and 10 kHz capacitor.

Manufacturers for many years have standardized on 220, 400, 440, and 800 V operating voltages to reduce the variety of capacitors produced and to match the existing standard output ratings of motor generators.

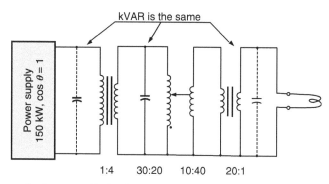

FIGURE 5.36 Schematic diagram of induction system.

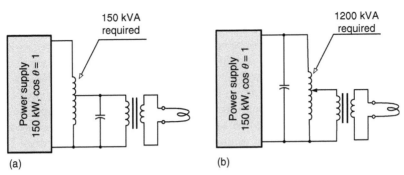

FIGURE 5.37 Features of the reactive power compensation due to different autotransformer location.

Another component that can be changed in the matching process is the autotransformer. A change in the tapping of this transformer will, for all practical purposes, affect the output voltage of the circuit but not the operating frequency. This is stated with the qualifier "for all practical purposes" because with solid-state power supplies the leakage reactance of the auto-transformer may sometimes be significant. Care should be taken in choosing the placement of the autotransformer because it can greatly affect the required kVAR of the transformer. As shown in Figure 5.37a, if the transformer is installed between the power supply and the capacitor bank it is operating at the power factor of the power source with relatively low current. If this same transformer is installed between the capacitor bank and the isolation transformer or work coil (Figure 5.37b), the current is much higher and the required kVAR of the transformer is much higher, roughly Q times the kilowatt-operating point of the power supply.

5.2.2.3 Step Three

Historically, for the motor generator set, tuning was approached by trying to add enough capacitance to read unity power factor or zero-phase angle on the panel meters. With solid-state power supplies, the power source often operates at less than unity power factor, and any inductance added in the transmission lines becomes more of a factor. Often a reactance located in the power supply must be considered part of the tuned circuit. To complicate matters further, each type of power source has a variety of limiting conditions that could prevent delivery of maximum power to the workpiece. It is advisable, before purchasing a power supply, to check with the manufacturer as to how much reserve capacity is available in the power supply. More than one user has been cut short by buying a 150 kW power supply only to find out that the maximum power it will deliver into the load circuit is 90 kW. Figure 5.31 shows a typical tuning curve for a swept frequency power supply operating into a parallel tank circuit. This type of power supply most often begins at a lower frequency, called the low frequency limit (LF in Figure 5.31) and begins sweeping up in frequency until the present power level is attained or a limit is reached. Typical limits would include a HF limit, a phase or low impedance limit, output voltage limit, output current limit, maximum power limit, etc.

One complication that can arise as a result of mistuning the load is that the frequency may be increased beyond the resonant frequency of the tuned circuit. This results in confusing the control circuit, which is normally in a mode of increasing the frequency to increase the output power. Since the power will decrease for an increase in frequency beyond the load resonant frequency, the power supply frequency will continue to increase until it reaches the HF limit. This condition is referred to as "going over the hump." The remedy is generally to reduce the value of the heat station capacitor or vary the inductance to produce an increase in the load resonant frequency.

It should be noted that on power supplies with an output series capacitor this "over the hump" condition will result when the series capacitance is too small relative to the kilowatt rating of the power supply. By adjusting the tank circuit capacitor and transformer or coil turns, it is possible to shift the curve in Figure 5.31 to the left or right to move away from expected limiting conditions. It should be noted that the power delivered is that shown on the curve and that if the curve is shifted too far to the left, a higher power than that desired may be delivered with no apparent control by the power potentiometer. Some solid-state converters will not run at a zero power output even though the potentiometer may be set to zero.

A current-fed power supply operating into a parallel tank circuit operates at a fixed frequency by phase locking itself to the tank circuit resonant frequency. One might think that this would eliminate the need for tuning. Unfortunately this is not the case. The problem of matching impedances still exists. On many current-fed units, the maximum allowable current is only slightly higher than that calculated for unity power factor at full voltage. This means that unless the impedance of the tank circuit is exactly right to deliver full current at full voltage, the power supply will deliver less than full power to the load. This sometimes requires the insertion of a special tuning bus to adjust the impedance for the correct value. If the current as a percent of maximum is higher than the percent of voltage, more inductance is required in the circuit. If the voltage is higher than the current, then inductance must be removed from the circuit. Another solution provided by the control of some current-fed inverters is to operate the inverter above the resonant frequency of the tank circuit. This reduces the load impedance to better match the output of the inverter.

The effort spent on load matching can be reduced by using a power supply that has more rated capacity than required or one that can demonstrate the capacity to run at 120% of its power rating. This will ensure ease of tuning when the applications call for 100% power or less.

Another general guide in load matching is to aim approximately for a voltage match as outlined in the following table:

% Voltage	% Power
10	1
20	4
30	9
40	16
50	25
60	36
70	49
80	64
90	81
100	100

Since the power varies as the square of the output voltage, striving for these values will give a setting that will allow easy adjustment to higher or lower values without continual limiting conditions.

5.2.2.4 The Final Step

A final caveat in load matching has to do specifically with the transmission lines from the power source to the load-matching (or heat) station and those from the load-matching capacitors or output transformer to the heating coil. Large inductances in these areas can cause considerable problems because much of the voltage generated by the power supply is dropped across the high-inductance elements of the circuit and not across the load itself. This can result in a considerable reduction in allowable output power and possibly in the inability to complete the desired heating task. This inductance is particularly critical in the higher kVA portion of the circuit (between the matching capacitors and the coil), especially at higher frequencies and higher currents. A good practice is to minimize the transmission line inductance within the required cost and size constraints.

5.2.3 Medium- and High-Frequency Transformers for Modern Heat Treatment

A transformer is an important part of the induction heating machine. Different types of transformers are used in inverters and heat stations. The total efficiency of the power supply is primarily affected by the transformer's efficiency. Years ago, when motor generators were widely used, the design of isolation transformers was a straightforward process. Some basic information, such as frequency, kilowatts, kilowatt–amperes, and input–output voltages, was all that was required. Today, with many different types of solid-state inverters and heat stations, the task of designing efficient transformers becomes more complex. The successful design of contemporary transformers should involve such features as the current–voltage waveforms, which can be square, sinusoidal, or sawtooth and often contain harmonics.

In 1955, the Jackson Transformer Company first started to manufacture transformers for induction heating applications. Since that time, thousands of different transformers have been developed to match the variety of applications. Here, we briefly introduce some basic types of medium- and high-frequency transformers for modern heat treatment based on the materials provided by the Jackson Transformer Company [16,17].

The transformer's main purpose is to change one voltage to another, making it possible to operate a great variety of loads at suitable voltages. In a transformer, the turns of the primary and secondary coils are coupled closely together so that their respective turns ratios determine very closely the output voltages and volt–ampere characteristics. The coils are usually wound on a laminated core of a magnetic material, and the transformer is then known as an iron-core transformer. Sometimes, as in many radio-frequency (RF) transformers, there is no magnetic core; the transformer then may be described as an air-core transformer.

Transformer manufacturers provide the induction heating industry with a wide range of transformers and other magnetic products from line frequency to 800 kHz, from a few volt–amperes to over a megawatt, and with water- and air-cooled designs. Products include isolation, auto, current, potential, and RF transformers, along with AC–DC reactors and integrated magnetic devices. As a general rule, most of the magnetic devices are water-cooled. This is because of size limitations, cost factors, power requirements, and frequency ranges.

5.2.3.1 AC–DC Reactors

AC reactor designs take into consideration any DC component, from a few hertz to several hundred kilohertz, water-cooled or dry, open construction or encapsulated. The legs of the inductors have distributed gaps to minimize flux leakage and to reduce noise. The legs are normally encapsulated to minimize vibrations. They are available from a few microhenries to several millihenries, and from a few amperes to several thousand amperes.

DC reactors are designed to handle an AC ripple component that may be present. They are available in both dry and water-cooled designs. Typically, they have a shell-type construction, whereby the gaps are distributed in the center leg to reduce fringing. The center leg is encapsulated to minimize noise and vibrations. They are available from a few microhenries to several millihenries and from a few amperes to several thousand amperes.

5.2.3.2 Variable Impedance Transformer

The variable impedance transformer (VIT) is an integrated magnetic device in which the primary windings and the magnetic amplifier windings are placed on common cores. Its purpose is to provide stepless power to electric furnaces that have silicon carbide elements, vacuum furnaces for deposition of metal, plating power supplies, and load banks. The VIT is a current-control device that requires minute signals to control a large amount of power. The VIT can operate with large unbalanced loads or with an open-circuit phase. The VIT can withstand short circuits for prolonged periods of time without incurring component failure. VITs are available in single-, two-, or three-phase designs that provide from a few hertz to 200 kHz, from a few volts to 2,000 V up to 500 kVA.

5.2.3.3 Heat Station Transformers

Jackson heat station transformers such as 52V1, 51V1, and 531V1 have been used in the heating, hardening, and annealing industry. The 52V1 transformers are normally used where the voltage needs to be stepped down anywhere from 5:1 to 22:1 or from 5:2 to 22:2 or other ratio combinations depending on customer requirements. The input voltages are anywhere from 220 to 1,200 V, and frequencies from 500 Hz to 10 kHz. The kVA values can range from 50 to over 10,000 kVA.

The construction of the windings can be either opened or epoxy-encapsulated. The output connections (secondary terminals) are generally referred to as fishtails. The input side of the transformer, which is referred to as the primary winding, is tapped to cover the required turns ratio. The windings use rectangular copper tubing of the thin wall because of the skin effect of the current at medium frequency. A typical profile of the tubing used would be $0.25 \times 1 \times 0.048$ wall. The primary and secondary windings are of an interleaved design to take advantage of the shape of the tubing and to reduce the resistance and impedance of the transformer. One of the unique features of this design is that the losses in the primary and secondary windings are equal. For a typical 22:1 ratio transformer, there are 22 primary turns in a series and there are 22 secondary turns connected in parallel in a one-turn construction. Therefore, the secondary resistance reflected to the primary is 22^2 times the secondary resistance. The total resistance is the primary resistance plus the reflected secondary resistance.

The construction of the core uses thin electrical steel (0.006 or 0.007 in. thick) of EE- or EI-type laminations. The core is water-cooled by means of copper-cooling plates sandwiched between the steel laminations. It has been concluded after many tests that the flux generated by the ampere turns in the magnetic circuit flows along the inside legs of the laminations just as current in a circuit takes the least resistive path. Therefore, the width of the outside leg of a shell-type transformer operated at medium frequencies can be less than one-half the tongue (center leg), as is required for low-frequency designs. The core loss of the transformer varies as the square of the input voltage, inversely as the square of the input turns, and approximately as the fourth root of the frequency.

5.2.3.4 Toroidal Transformers

Typically toroidal transformers are totally encapsulated and are used in through hardening, tempering, forging, and annealing. Normally, the output voltages are higher in the heat station transformers, and in many instances the output voltage is equal to or much higher than the applied voltage. Input voltages can be from 100 to 2000 V or higher. The output voltage can range from 50 V to several thousand volts. Taps are provided within the voltage range. The frequencies can be anywhere from 200 Hz to 10 kHz. The kVA can range from 50 to 3000 kVA or higher. They are more efficient than laminated transformers and have virtually no air gaps. A disadvantage of remaining encapsulated is that they are not easily repaired and in most cases must be replaced.

Toroidal autotransformers usually are smaller in size and have lower exciting current, better regulation, and higher efficiency than an isolation transformer. The reason for this is that in an isolation transformer all of the kVA is transferred to the secondary, whereas in an autotransformer only a portion of the total kVA is transformed, the rest flowing directly from the primary to the secondary without transformation. The windings in an autotransformer are wound around the same core and are used to step up or step down the input voltage. The core of toroidal transformers consists of thin steels wound in a cylindrical or toroidal form. Water-cooled copper heat sinks are used on the flat surface of the cores to carry away the heat generated by the core. Without water cooling, the physical size of the core would increase drastically. The windings are hand-wound over the core, using round copper tubing, and its size determined by the design current.

5.2.3.5   Integrated Magnetic Transformers

Jackson Transformer Company has developed and patented a method of combining a transformer and inductor in a single package whereby the inductor and the primary of the transformer have a common core. This product is referred to as a transinductor and can be designed to provide a fixed inductance in the primary or secondary or both. Variable ratios can be provided on the transformer portion. By combining the two components the size of the product is reduced, the overall efficiency is increased, and the leakage flux of the magnetic device is minimized.

5.2.3.6   Rectangular (C-Core) Transformers

The construction of a rectangular transformer uses a C-core and interleaved windings. Normally, the unit is epoxy-encapsulated. The design of the rectangular core transformer is usually at low to medium frequencies with input voltages from a few hundred to a few thousand volts, output voltages from a few hundred to a few thousand volts, and input power up to several thousand kilovolt–amperes. Specific requirements for this type of transformer are low-leakage inductance and high efficiency.

5.2.3.7   Narrow-Profile Transformers

Narrow-profile transformers are designed to deliver high power at medium frequencies within narrow physical constraints. A typical example is the induction heating bearing surfaces of an engine crankshaft. The construction of this style of transformer uses the interleaved winding design and ferrite cores and is epoxy-encapsulated. This allows the transformer to achieve its narrow-profile, high-efficiency, low-leakage inductance and be completely protected from its harsh environment and physical abuse.

5.2.3.8   Ferrite-Core Transformers

Ferrite-core transformers are similar to heat station transformers in that they have an interleaved-winding construction. One of the differences is that in place of the steel lamination used in the core, ferrite material is used. Ferrites offer advantages over steel laminations in that they have low eddy currents and high permeability over a wide frequency range. Having a homogenous ceramic structure and inherent low core loss, the ferrites become very attractive at frequencies above 10 kHz for transformer applications. In some cases, even though the ferrite core loss is low, they may still need to be water-cooled because of the frequencies at which they are used. In applications, when the output power from the power supply is fairly low and the frequency may be under 10 kHz, ferrites are more advantageous than steel because of the lower loss of the ferrite.

5.2.3.9 Radio-Frequency Transformers

The RF transformer is normally referred to as a current transformer and is designed without any core material. The critical element in the design is to obtain the highest current transfer ratio from primary to secondary. Generally, the primary winding is encapsulated silicon rubber, which is a moisture-resistant material. This is required because of the high dielectric strength needed between the primary and the secondary windings and also to protect the windings from the environment. Great care must be taken in the construction and the selection of the material. A clean room environment is highly recommended.

5.2.3.10 Maintenance, Sizing, and Specification of Transformers

As a general rule, when transformers are water-cooled, most failures occur because of a break-down of the insulation between the windings. Normally, this is due to lack of water, poor-quality water, too high water inlet temperature, or operation of the transformer outside its designed rating. Sometimes, insulation breaks down because of the harsh environment to which the transformer is subjected. Another failure that commonly occurs is the melting of the output connection (fishtail), which is caused by improper tightening or poor maintenance of the inductor (e.g., dirty and oxidized surfaces on the mating inductor or fishtail). Sometimes the core fails. Again, this may be due to lack of water, poor-quality water, too many input volts per turn (voltage per turn exceeds core loss temperature limitation), and improper use of frequency. A well-designed water system will pay for itself with reductions in component failures and downtime. A proper maintenance of the inductor and transformer connections will also help greatly.

To properly size or specify a transformer, the following information is generally required: input voltage to the transformer, power source wattage, the frequency range of the power source and the frequency at which the transformer will operate, the turns ratio or the output voltage required at full load (or no load), the input kVA at the minimum and maximum turns ratios, and expected efficiency (based on the kilowatt rating of the power source), or loss of the transformer. It is also helpful to know any unique characteristics of the power source, type of waveform, and if any DC will be present. The more information the designer has available, the more assurance the customer has of getting the proper, most efficient transformer.

5.3 PROCESS CONTROL AND MONITORING

5.3.1 PRELUDE TO DISCUSSION OF PROCESS CONTROL AND MONITORING

One of the most important features of a modern induction heat treatment machine is that it has an effective control and adequate monitoring system. The monitoring system provides an operator with information about what is actually happening during the process and whether the heat treatment of the workpiece has been successful [18,19].

In earlier years, controls as simple as dual set-point meters were used to determine whether a given parameter was running between two preset points at the time the circuit was interrogated. With the advent of the Programmable logic controller (PLC), a much larger number of points could be monitored in real time during the heat treatment cycle. In the early 1980s, HWG, in Germany (this company is now a part of the INDUCTOHEAT Group), developed the coil signature system, which was sold on many commercial machines and has been effectively applied for repeatable heat treatment processes.

The general idea of the signature concept is rather simple and can be described as follows. The monitoring system observes one of the unregulated variables related to the process and stores the most important parameters during the machine cycle. These values are compared to set points stored in the information bank within the PLC (ideal signature), and an output indication is given on the graphics display. In normal performance, all subsequent signatures of cycles are compared to the ideal one and must remain between the upper and lower limits (Figure 5.38). If any signature goes outside the area of limits, then the operator can see exactly during which part of the heat treatment cycle the signature was not repeated and the process exceeded the set limits. The operator can determine immediately what the problem is and what should be done to adjust the machine to get the cycle signature back into the correct setup. It is not necessary for the operator to know in detail the electromagnetic, heat transfer, or quenching features of the process. The operator merely needs to know how to adjust the machine to get the signatures back to the ideal shape.

As microprocessor technology continued to grow, this signature technology was employed by manufacturers of strain-measuring equipment and eddy current testers. Modern signature systems will allow the measurement of many variables at once with corresponding real-time graphing of the function within the preset set points as well as any required SPC analysis.

The question always remains as to which are the most significant variables to monitor to ensure that the process has been successfully repeated and what the correlation is between the variables and the heat treatment process. A list of variables for induction hardening would include the following:

Workpiece chemistry	Workpiece geometry
Coil geometry	Coil material
Workpiece-to-coil location	Frequency
Power, voltage, current	Cycle time
Workpiece temperature	Quench flow
Quench data (media, temperature, purity, concentration	

One manufacturer, who was purchasing a number of machines, took the refreshing approach of performing a design of experiments (DOE) for his particular process. The DOE determined that there were relatively few parameters that significantly affected the process. The decision was made to control the significant parameters rather than focusing on the larger number of insignificant items usually pursued.

During workpiece induction heating, there is a large variation in the material properties, including electrical resistivity and the relative magnetic permeability of ferromagnetic metals. These changes result in a significant redistribution of heat sources within the workpiece

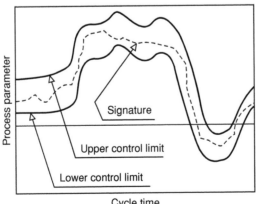

FIGURE 5.38 Control limits and a sample of a signature.

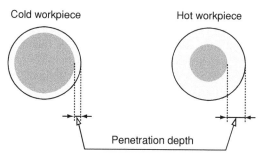

FIGURE 5.39 Depth of current penetration in cold and hot steel.

during heat treatment. Variation in electrical resistivity and magnetic permeability during the heating cycle leads to changes in the reference depth of current in the workpiece. These changes can be observed by monitoring the coil voltage, current, and phase angle (cos φ) at the induction coil (Figure 5.39 and Figure 5.40).

Although at first glance the changes may appear to be very dramatic, there are some factors that may mask the expected effect. For a mass heating system running at relatively low field intensity (i.e., induction heating before forging, rolling, or extrusion), the relative magnetic permeability of the workpiece surface may change from 200 to 1 during the transition from cold to hot. For a surface-hardening application running at 10 kW/in.[2] or more, the change of μ will typically be only from 8 to 1. Since the reference depth varies as the inverse square root of the permeability changes, the actual change in inductance from cold to hot load may be relatively small. In many cases of induction hardening, the actual change in inductance and impedance is relatively small and is greatly reduced for stronger magnetic fields and higher power densities (see Figure 4.16 of Chapter 4).

Different types of monitoring systems are available in the market. The choice of a particular monitoring system is a matter of operational features of the process, cycle time, technological requirements, and cost. In some applications, a relatively simple energy monitor will be sufficient. Other applications, however, may require advanced signature-monitoring devices.

5.3.2 ENERGY MONITOR

The simple energy monitor is shown in Figure 5.41 [18,20]. This monitor measures and displays the actual energy delivered to the induction coil in kilowatt–seconds. It is a relatively

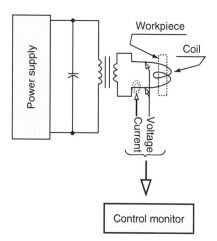

FIGURE 5.40 Monitoring of induction heat treatment system.

FIGURE 5.41 Energy monitor.

inexpensive device. Once a workpiece-heating pattern is developed and the correct power and heating time are established, this information is preset into the monitor. The acceptable lower and upper kilowatt–second limits are then entered by the user. If insufficient or excessive energy is applied to the load, the display will show REJECT/UNDER or REJECT/OVER, respectively. Auxiliary contacts can be used to reject the part in automated lines or sound an alarm in manual operations. If the count falls within the present range, the ACCEPTABLE indicator will be displayed. The energy monitor can be used as an induction process controller to turn off the power automatically when the desired amount of energy is delivered to the load. The energy monitor circuitry accurately measures and displays the output of the power supply. While most-earlier RF monitoring was on the input, fiber optics make it possible to monitor the high voltages and frequencies on the output safely.

5.3.3 ADVANCED MONITORING

In many cases, energy or coil monitors can monitor the heating cycle effectively. However, in some applications, these monitors give only a partial picture because the quenching phase is as critical to the proper hardening of the part as the heating phase. It would be desirable to have more advanced monitoring equipment available. This equipment could monitor several parameters simultaneously in real time and indicate which parameters may cause an improperly hardened part.

The Stativision concept has been used extensively in developing advanced monitoring devices [21,22]. Monitors that use this concept have a real-time interface that monitors the energy into a workstation and checks it against the ideal system set points. The signature of load parameters is polled, varying from 30 ms to 2 s, and 75 readings are taken. The ideal signature, which is based on previous successful laboratory developments and the history of operations, is stored online. During each successive machine cycle, a new signature is compared to the stored ideal signature. The difference in the two is displayed along with the actual running signature. If the two do not match, a fault is logged, and a signal is sent to the machine control.

Operator screens also display process data. The main screen has several graphs representing major process parameters such as power, frequency, and current. The fault screen displays a list of system problems, including faults as well as limits. From this screen, the operator or maintenance personnel can move the cursor to the fault or limit displayed. A keystroke will bring up a help screen developed specifically for the item indicated. These help screens provide user-friendly diagnostic routines that improve system-up time and provide the tuning needed for a particular load.

It is wise to remember that each parameter that is monitored adds cost to the machine. For a great majority of induction hardening applications, four signatures are sufficient to define the process [19].

Scanning induction surface hardening typically requires the following signatures: scan speed, quench pressure, rotation speed, and load power. As an example, scan speed and load power signatures are shown in Figure 5.42.

Single-shot heat treatment processes will require parameters such as rotation speed, quench pressure, load power, and quench flow. Since the part is not moving during the heating cycle, the scan speed is irrelevant to the process, although monitoring these parameters might indicate a tendency toward a failure of the machine index. Therefore, for maintenance scan speed could be monitored.

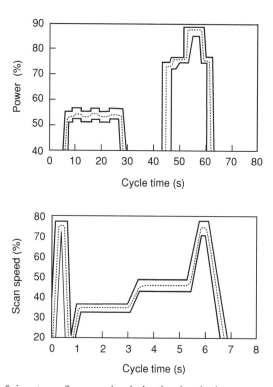

FIGURE 5.42 Samples of signatures for scanning induction hardening.

Other parameters that are advisable to be monitored with a signature system are shown in Figure 5.43. Workpiece temperature can be monitored during the cycle using an infrared monitor. In some applications, in addition to the parameters shown in Figure 5.42, signatures can involve other parameters that are essential to particular processes and change during each cycle.

An advanced commercial system such as INDUCTOHEAT's QA Ultra 8000 (Figure 5.44) is capable of producing four or eight simultaneous process parameter signatures. This system verifies all machine settings to provide confidence in the quality of processing for the part. It can detect high/low power, high/low rotation speed, excessive/slack quench, high/low heat time, etc. The features and benefits of this and other similar monitoring systems are:

- Machine repeatability on every component
- Real-time parameter sample monitoring
- Cost savings through reduced destructive testing
- Immediate test results
- Easy setup and operation with user-friendly software
- Help in troubleshooting, allowing identification of the fault condition and the point in the cycle where the problem occurred
- Trend analysis
- Data collection for SPC software

Advanced monitoring systems provide precise verification of the real process parameters. As with most processes, it is advantageous to use a number of techniques to manage the process. The combination of a reliable power supply, an efficient induction coil, and a full process monitoring system ensures a proper hardening profile.

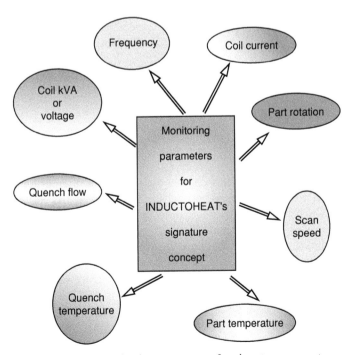

FIGURE 5.43 INDUCTOHEAT's monitoring parameters for signature concept.

FIGURE 5.44 Quality assurance monitor (signature monitor) QA Ultra 8000.

5.4 SUMMARY

The unique viewpoint presented in this chapter is that of the integration of modern power supply selection, load matching, process control and monitoring, and contemporary inductor design techniques. Current information has been presented on the use of different types of solid-state power supplies with their basic block diagrams, waveshapes, and resonant characteristics as well as photographs of practical applications. An attempt has been made to bridge the gap between induction coil design information and power supply, load matching, and process control and monitoring. This information is truly useful to the modern heat treatment engineer.

Should further details be required or questions arise regarding different types of induction heat treatment power supplies in particular applications, the authors welcome any reader inquiries or suggestions at INDUCTOHEAT, Inc.

REFERENCES

1. C.A. Tudbury, *Basics of Induction Heating*, Vol. 2, Rider, New York, 1960.
2. D.L. Loveless, R.L. Cook, and V.I. Rudnev, Considering nature and parameters of power supplies for efficient induction heat treating, *Ind. Heating*, June 1995, 33–37.
3. D.L. Loveless, Solid state power supplies for induction heating prior to rolling, forging and extrusion, *33 Metal Producing*, August 1995.
4. INDUCTOHEAT Bulletin, *Statipower 6*, 1991.
5. General presentation of activity of Lepel Corp., 1990.
6. INDUCTOHEAT Bulletin, *Statipower 5*, 1991.
7. General presentation of activity of Radyne Ltd., U.K., 1992.
8. General presentation of activity of Elphiac, Belgium, 1990.
9. INDUCTOHEAT Bulletin, *Unipower 9*, 1991.
10. INDUCTOHEAT Bulletin, *Unipower 12*, 1993.
11. R.L. Cook, D.L. Loveless, and V.I. Rudnew, Load matching in modern induction heat treating, *Ind. Heating*, September 1995.
12. INDUCTOHEAT Bulletin, *Load Frequency Analyzer*, 1991.
13. INDUCTOHEAT Bulletin, *Uniscan-I*, 1991.
14. INDUCTOHEAT Bulletin, *Uniscan-II*, 1992.
15. INDUCTOHEAT Bulletin, *Uniscan-IV*, 1992.
16. W.E. Terlop and S. Cassagrande, Special transformer technology for medium and high frequency applications, in *Proceedings of 1st International Induction Heating Seminar*, São Paulo, Brazil, 1995.
17. General presentation of Jackson Transformer Company, 1993.
18. R.L. Cook, R.J. Myers, and V.I. Rudnev, Process monitoring for more effective induction hardening control, *Mod. Appl. News*, August 1995.
19. R.J. Myers, Induction control with smoke and mirrors, in *Proceedings of International Heat Treating Conference: Equipment and Processors*, ASM International, Materials Park, Ohio, 1994, pp. 295–297.
20. INDUCTOHEAT Bulletin, *Energy Monitor: Quality Control Energy Monitor*, 1993.
21. INDUCTOHEAT Bulletin, *QA Ultra 8000*, 1992.
22. INDUCTOHEAT Bulletin, *Stativision*, 1993.
23. D. Oxbrough, Current fed inverters, in *Proceedings of 6th International Induction Heating Seminar*, Nashville, Tennesee, September 1996.

6 Laser Surface Hardening

Janez Grum

CONTENTS

6.1   INTRODUCTION

Lasers represent one of the most important inventions of the 20th century. With their development it was possible to get a highly intensive, monochromatic, coherent, highly polarized light wave [1,2]. The first laser was created in 1960 in Californian laboratories with the aid of a resonator from an artificial ruby crystal. Dating from this period is also the first industrial application of laser which was used to make holes into diamond materials extremely difficult to machine. The first applications of laser metal machining were not particularly successful mostly due to low capabilities and instability of laser sources in different machining conditions. These first applications, no matter how successful they were, have, however, led to a development of a number of new laser source types. Only some of them have met the severe requirements and conditions present in metal cutting. As the most successful among them, CO_2 laser should be mentioned. The high-intensity CO_2 lasers have proved to be extremely successful in various industrial applications from the point of view of technology as well as economy. A great number of successful applications of this technique have stimulated the development of research activities which, since 1970, have constantly been increasing.

Laser is becoming a very important engineering tool for cutting, welding, and to a certain extent also for heat treatment. Laser technology provides a light beam of extremely high-power density acting on the workpiece surface. The input of the energy necessary for heating up the surface layer is achieved by selecting from a range of traveling speeds of the workpiece and/or laser-beam source power.

The advantages of laser materials processing are as follows [3–7]:

- Savings in energy compared to conventional surface heat treatment welding or cutting procedures.
- Hardened surface is achieved due to self-quenching of the overheated surface layer through heat conduction into the cold material.
- As heat treatment is done without any agents for quenching, the procedure is a clean one with no need to clean and wash the workpieces after heat treatment.
- Energy input can be adapted over a wide range with changing laser source power, with having focusing lenses with different focuses, with different degrees of defocus (the position of the lens focus with respect to the workpiece surface), and with different traveling speeds of the work-piece and/or laser beam.
- Guiding of the beam over the workpiece surface is made with computer support.
- It is possible to heat-treat small parts with complex shapes as well as small holes.
- Optical system can be adapted to the shape or complexity of the product by means of different shapes of lenses and mirrors.
- Small deformations and/or dimensional changes of the workpiece after heat treatment.
- Repeatability of the hardening process or constant quality of the hardened surface layer.
- No need for/or minimal final machining of the parts by grinding.
- Laser heat treatment is convenient for either individual or mass production of parts.
- Suitable for automation of the procedure.

The use of laser for heat treatment can be accompanied by the following difficulties:

- Nonhomogeneous distribution of the energy in the laser beam
- Narrow temperature field ensuring the wanted microstructure changes
- Adjustment of kinematic conditions of the workpiece and/or laser beam to different product shapes
- Poor absorption of the laser light in interacting with the metal material surface

Engineering practice has developed several laser processes used for surface treatment

- Annealing
- Transformation hardening
- Shock hardening
- Surface hardening by surface layer remelting
- Alloying
- Cladding
- Surface texturing
- Plating by laser chemical vapor deposition (LCVD) or laser physical vapor deposition (LPVD)

For this purpose, besides CO_2 lasers, Nd:YAG and excimer-lasers with a relatively low power and a wave-length between 0.2 and 1.06 μm have been successfully used. A characteristic of these sources is that, besides a considerably lower wavelength, they have a smaller focal spot diameter and much higher absorption than CO_2 lasers.

In heat treatment using laser light interaction, it is necessary to achieve the desired heat input, which is normally determined by the hardened layer depth. Cooling and quenching of the overheated surface layer is in most cases achieved by self-cooling as after heating stops, heat conducted into the workpiece material is so intensive that the critical cooling speed is achieved and thus also the wanted hardened microstructure. Figure 6.1 illustrates the dependence of power density and specific energy on laser light interaction time and on the

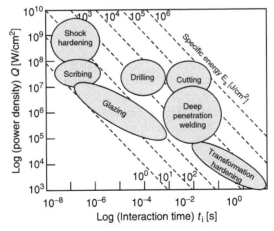

FIGURE 6.1 Dependence of power density, specific energy, and interaction time at laser metalworking processes.

work-piece surface in order to carry out various metalworking processes. Diagonally there are two processes, i.e., scribing and hardening, for which quite the opposite relationship between power density and interaction time has to be ensured [5–6]. In scribing material vaporization at a depth of a few microns has to be achieved, ensuring the prescribed quality and character resolution. On the other hand, for hardening a considerably lower power density per work-piece surface unit is required, but the interaction times are the longest among all the mentioned metalworking processes.

Thus different power densities and relatively short interaction times, i.e., between 10^{-1} and 10^{-3} s, are related to transformation hardening and remelting of the material. This group thus includes the processes in which either the parent metal alone or the parent metal and the filler material are to be melted. The highest power densities are applied in the case that both the parent metal and the filler material are to be melted. Such processes are laser welding and alloying. With cladding, however, the lowest power densities are required as the material is to be deposited on the surface of the parent metal and only the filler material is to be melted. The power density required in both welding and alloying of the parent metal, however, depends on the materials to be joined or alloyed. In hardening of the surface layers by remelting it is necessary, in the selection of the power density, to take into account also the depth of the remelted and modified layer. In laser cutting a somewhat higher power density is required than in deep welding. In the cutting area the laser-beam focus should be positioned at the workpiece surface or just below it. In this way a sufficient power density will be obtained to heat, melt, and evaporate the workpiece material. The formation of a laser cut is closely related to material evaporation, particularly flowing out of the molten pool and its blowing out due to the oxygen auxiliary gas, respectively.

Lasers produce a collimated and coherent beam of light. The almost parallel, single-wavelength, light rays that make up the collimated laser beam have a considerably higher power density profile across the diameter of the beam and can be focused to a spot size diameter on the workpiece surface [8–10].

Various types of lasers are classified with reference to a lasing medium [15–19]. In laser treatment processes, the most frequently used gaseous lasers are CO_2 lasers [20].

The transverse power density distribution emerging from the laser source is not uniform with reference to the optical laser axis. It depends on the active lasing medium, the maximum laser power, and the optical system for transmission and transformation of the laser beam [10–14].

Figure 6.2 shows different mode structures of the laser beam, i.e., TEM_{00} (a, Gaussian beam), and multimode beam structures TEM_{01} (b), TEM_{10} (c), TEM_{11} (d), and TEM_{20} (e), respectively [11].

The transverse power density distribution of the laser beam is very important in the interaction with the workpiece material. The irradiated workpiece area is a function of the focal distance of the convergent lens and the position of the workpiece with reference to the focal distance. The transverse power density distribution of the laser beam is also called the transverse electromagnetic mode (TEM). Several different transverse power density distributions of the laser beam or TEMs can be shaped. Each individual type may be attributed a different numeral index. A higher index of the TEM indicates that the latter is composed of several modes, which makes beam focusing on a fine spot at the work-piece surface very difficult. This means that the higher the index of the TEM the more difficult it is to ensure high power densities, i.e., high energy input. For example in welding, mode structures TEM_{00}, TEM_{01}, TEM_{10}, TEM_{11}, TEM_{20}, and frequently combinations thereof are used. Some of the laser sources generate numerous mode structures, i.e., multimode structures [8–12].

Projection
of laser beam
power density

Power density
profile across
diameter of beam

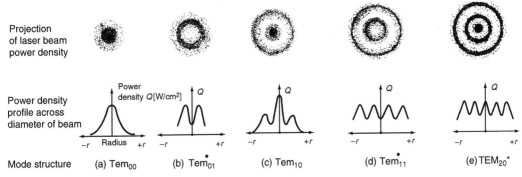

Mode structure   (a) Tem$_{00}$ (b) Tem$_{01}^{\bullet}$ (c) Tem$_{10}$ (d) Tem$_{11}^{\bullet}$ (e) TEM$_{20}^{*}$

FIGURE 6.2 Basic laser beam mode structures. (From Dawes, C. *Laser Welding*; Ablington Publishing and Woodhead. Publishing in Association with the Welding Institute: Cambridge, 1992; 1–95.)

The most frequently used lasers are continuous lasers emitting light with the Gaussian transverse power density distribution, TEM$_{00}$. A laser source operating 100% in TEM$_{00}$ is ideal for cutting and drilling. A TEM$_{00}$ beam can be focused with a convergent lens to a very small area thus providing a very high power density. In practice TEM$_{01}$, having energy concentrated at the periphery of the laser beam with reference to the optical axis, is used as well. This mode is applied primarily to drilling and heat treatment of materials because it ensures a more uniform elimination of the material in drilling and a uniform through-thickness heating of the material in heat treatment. For welding and heat treatment, very often mixtures of multimode structures giving an approximately rectangular, i.e., top-hat-shaped, energy distribution in the beam are used [13–20].

Heat treatment requires an adequate laser energy distribution at the irradiated work-piece surface. This can be ensured only by a correct transverse power density distribution of the laser beam. The power density required for heat treatment can be achieved by the multimode structure or a built-in kaleidoscope or segment mirrors, i.e., special optical elements.

Heating of the material for subsequent heat treatment requires an ideal power density distribution in the laser beam providing a uniform temperature at the surface, and below the surface to the depth to which the material properties are to be changed.

Figure 6.3 shows different shapes of transverse power density distribution which can be accomplished by both the defocusing of the laser beam and the laser-beam transformation through the mirrors or differently shaped optical elements [21]. When the focused laser beam with the Gaussian power density distribution is used, the irradiated spot (a) is smaller than the one obtained with the defocused beam (b). To fulfill the requirements of a uniform power density distribution, the top-hat mode of the power density should be ensured. Figure 6.3 shows two solutions both giving a rather uniform power density distribution. Such a power density distribution can be accomplished by the unidirectional (c) or bidirectional (d) oscillating laser beam that is obtained by being guided through appropriately shaped optical elements showing adequate kinematics [21]. It is characteristic of the two basic shapes of the laser beam that they are suitable for heat treatment and can be applied to heating of larger areas. Thus with the unidirectional oscillating laser beam, heating of the material is relatively fast whereas with the bidirectional oscillating beam heating is less intense, but it provides a larger and rectangular heated surface. Both types of laser beam are very suitable for fast

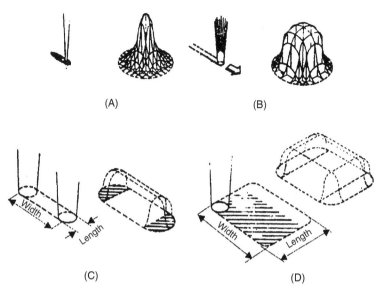

FIGURE 6.3 Basic shapes of transverse power density distribution. (A) Focused laser beam suitable for deep penetration welding and cutting, (B) defocused laser beam suitable for heat treatment, (C) unidirectional oscillating laser beam for wide paths and intensive heating, and (D) bidirectional oscillating laser beam for wide paths and medium heating. (From Seaman, F.D.; Gnanamuthu, D.S. Using the industrial laser to surface harden and alloy. In *Source Book on Applications of the Laser in Metalworking*; Metzbower, E.A., Ed.; American Society for metals: Metals Park, Ohio, 1981; 179–184.)

hardening of larger areas. In any case it should be taken into account that such a transformation of the laser beam is feasible only with a laser source of sufficiently high maximum power.

6.2 LASER SURFACE HARDENING

6.2.1 LASER HEATING AND TEMPERATURE CYCLE

A prerequisite of efficient heat treatment of a material is that the material shows phase transformations and is fit for hardening. Transformation hardening is the only heat-treatment method successfully introduced into practice. Because of intense energy input into the workpiece surface, only surface hardening is feasible. The depth of the surface-hardened layer depends on the laser-beam power density and the capacity of the irradiated material to absorb the radiated light defined by its wavelength. Laser heating of the material being characterized by local heating and fast cooling is, however, usually not suitable to be applied to precipitation hardening, steroidization, normalizing, and other heattreatment methods [3–6,22].

Kawasumi [23] studied laser surface hardening using a CO_2 laser and discussed the thermal conductivity of a material. He took into account the temperature distribution in a three-dimensional body taken as a homogeneous and isotropic body. On the basis of derived heat conductivity equations, he accomplished numberous simulations of temperature cycles and determined the maximum temperatures achieved at the workpiece surface.

Figure 6.4 shows temperature cycles in laser heating and self-cooling. A temperature cycles was registered by thermocouples mounted at the surface and in the inside at certain

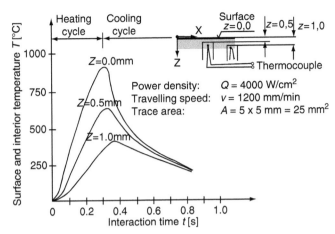

FIGURE 6.4 Temperature cycle on the workpiece surface and its interior versus interaction time. (From Kawasumi, H. Metal surface hardening CO₂ laser. In *Source Book on Applications of the Laser in Metal working*; Metzbower, E.A., Ed.; American Society for Metals: Metal Park, Ohio, 1981; 185–195.)

workpiece depths [23]. In this case the laser beam with its optical axis was traveling directly across the centers of the thermocouples inserted in certain depths. A thermal cycle can be divided into the heating and cooling cycles. The variations of the temperature cycles in the individual depths indicate that:

- Maximum temperatures have been obtained at the surface and in the individual depths.
- Maximum temperature obtained reduces through depth.
- Heating time is achieved at the maximum temperature obtained or just after.
- The greater the depth in which the maximum temperature is obtained the longer is the heating time required.
- In the individual depths the temperature differences occurring are greater in heating than in cooling.
- Consequently, in the individual depths the cooling times are considerably longer than the heating times to obtain, for example, a maximum temperature.

Figure 6.5 shows two temperature cycles in laser surface heating [23]. In each case the maximum temperature obtained at the surface is higher than the melting point of the material; therefore remelting will occur. The remelting process includes heating and melting of the material, fast cooling, and material solidification. The maximum temperature at the surface being higher than the melting point, a molten pool will form in the material surrounding the laser beam. Because of a relative travel of the laser beam with reference to the workpiece, the molten pool travels across the workpiece as well, whereas behind it the metal solidifies quickly. The depth of the remolten material is defined by the depth in which the melting point and the solidification temperature of the material have been attained.

The depth of the remelted layer can be determined experimentally by means of optical microscopy or by measuring through-depth hardness in the transverse cross section.

Figure 6.6 shows the dependence of the maximum temperature attained at the workpiece surface in laser heating on the given power density and different traveling speeds v [23]. The four curves plotted indicate four different power densities Q_i, i.e., 2, 4, 6, and 8 kW/cm². With the lowest power density Q_1, transformation hardening can be performed if the traveling

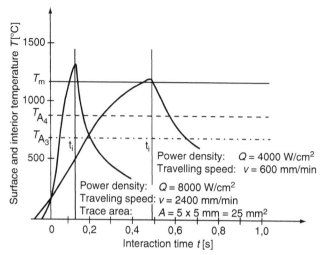

FIGURE 6.5 The effect of laser light interaction time on the temperature cycle during heating and cooling at various power densities and traveling speeds. (From Kawasumi, H. Metal Surface hardening CO_2 laser. In *Source Book on Applications of the Laser in Metal working*; Metzbower, E.A., Ed.; American Society for Metals: Metal Park, Ohio, 1981; 185–195.)

speed is varied between 0.3 and 0.5 m/min. With the highest power density Q_4, hardening by remelting of the thin surface layer can be efficiently performed as the traveling speed v of the workpiece does not exceed 2.0 m/min. With the power densities between the highest and the lowest, Q_2 and Q_3, it is the transformation hardening which can be efficiently performed because the wide range of the traveling speeds, i.e., from 0.4 up to 2.0 m/min, makes it possible to obtain a hardened microstructure. The wide range of traveling speeds of the workpiece and/or the laser beam enables hardening of the surface layers of different thicknesses.

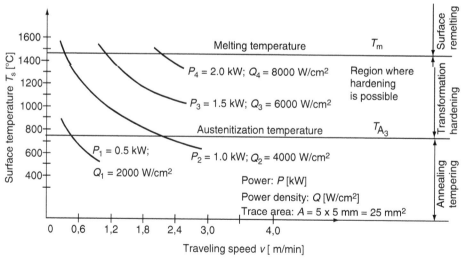

FIGURE 6.6 The effects of power density and traveling speed on hardening. (From Kawasumi, H. Metal Surface hardening CO_2 laser. In *Source Book on Applications of the Laser in Metal working*; Metzbower, E.A., Ed.; American Society for Metals: Metal Park, Ohio, 1981; 185–195.)

6.3 METALLURGICAL ASPECT OF LASER HARDENING

Prior to transformation hardening, an operator should calculate the processing parameters at the laser system. The procedure is as follows. Some of the processing parameters shall be chosen, some calculated. The choice is usually left to operators and their experience. They shall select an adequate converging lens with a focusing distance f and a defocus z_f taking into account the size of the workpiece and that of the surface to be hardened, respectively. Optimization is then based only on the selection of power and traveling speed of the laser beam. The correctly set parameters of transformation hardening ensure the right heating rate, then heating to the right austenitizing temperature T_{A_3}, and a sufficient austenitizing time t_A. Consequently, with regard to the specified depth of the hardened layer, in this depth a temperature a little higher than the transition temperature T_{A_3} should be ensured. Because of a very high heating rate the equilibrium diagram of, for example, steel does not suit; therefore it is necessary to correct the existing quenching temperature with reference to the heating rate. Thus with higher heating rates a higher austenite transformation temperature should be ensured in accordance with a time–temperature-austenitizing (TTA) diagram. The left diagram in Figure 6.7 is such a TTA diagram for 1053 steel in the quenched-and-tempered state (a) whereas the right diagram is for the same steel in the normalized state (b) [24]. As the steel concerned shows pearlitic–ferritic microstructure, a sufficiently long time should be ensured to permit austenitizing. In fast heating, austenitizing can, namely, be accomplished only by heating the surface and subsurface to an elevated temperature. For example with a heating time t of 1 s, for total homogenizing a maximum surface temperature T_s of 880°C should be ensured in the first case and a much higher surface temperature T_s, i.e., 1050°C, in the second case. This indicates

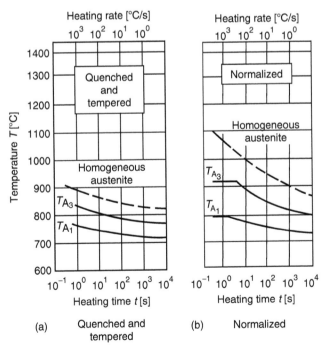

FIGURE 6.7 TTA diagram of steel 1053 for various states. (From Amende, W. Transformation hardening of steel and cast iron with high-power lasers. In *Industrial Applications of Lasers*; Koebner. H., Ed.; John Wiley & Sons: Chischester, 1984; 79–99; Chapter 3.)

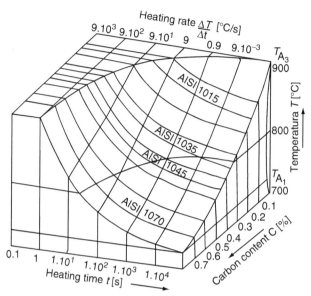

FIGURE 6.8 Influence of heating rate and carbon content on austenitic transformation temperature. (From Amende, W. Transformation hardening of steel and cast iron with high-power lasers. In *Industrial Applications of Lasers*; Koebner. H., Ed.; John Wiley & Sons: Chischester, 1984; 79–99; Chapter 3.)

that around 170°C higher surface temperature ΔT_s should be ensured in the second case (normalized state) than in the first case (quenched-and-tempered state).

Figure 6.8 shows a space TTA diagram including numerous carbon steels with different carbon contents. The TTA diagram gives particular emphasis to the characteristic steels, i.e., 1015, 1035, 1045, and 1070 steels, and their variations of the transition temperature T_{A_3} with reference to the given heating rate and the corresponding heating time [24].

When the laser beam has stopped heating the surface and the surface layer, the austenitic microstructure should be obtained. Then the cooling process for the austenitic layer begins. To accomplish martensite transformation, it is necessary to ensure a critical cooling rate that depends on the material composition. Figure 6.9 shows a continuous cooling transformation (CCT) diagram for EN 19B steel including the cooling curves. As carbon steels have different carbon contents, their microstructures as well show different contents of pearlite and ferrite. An increased carbon contents in steel decreases the temperature of the beginning of martensite transformation T_{M_s} as well as of its finish T_{M_F}. Figure 6.10 shows the dependence between the carbon content and the two martensite transformations. Consequently, the increase in carbon content in steel results in the selection of a lower critical cooling rate required. In general, the microstructures formed in the surface layer after transformation hardening can be divided into three zones, that is:

• A zone with completely martensitic microstructure
• A semimartensitic zone or transition microstructure
• A quenched-and-tempered or annealed zone with reference to the initial state of steel

Sometimes, particularly in the martensitic zone, retained austenite occurs too due to extremely high cooling rates and the influence of the alloying elements present in steel.

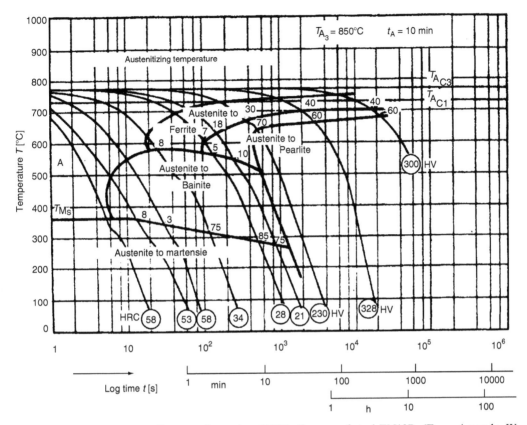

FIGURE 6.9 Continuous cooling transformation (CCT) diagram of steel EN19B. (From Amende, W. Transformation hardening of steel and cast iron with high-power lasers. In *Industrial Applications of Lasers*; Koebner. H., Ed.; John Wiley & Sons: Chischester, 1984; 79–99; Chapter 3.)

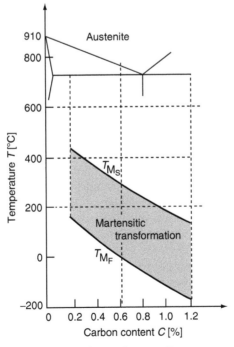

FIGURE 6.10 Influence of carbon content in steels according to start and finish temperature of martenzitic transformation.

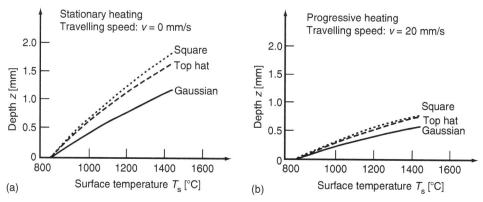

FIGURE 6.11 Calculated hardening depths for C45 steel at stationary and progressive laser beam acting for a square beam of 5×5 mm, a uniform distributed beam of 5 mm, and a Gaussian beam $w = 5$ mm. (From Meijer J.; Kuilboer, R.B.; Kirner, P.K.; Rund, M. Laser beam hardening: transferability of machining CIRP Seminar on Manufacturing Systems—LANE' 94. In *Laser Assisted Net shape Engineering*; Geiger, M., Vollertsen, F., Eds.; Meisenbach-Verlag: Erlangen, Bamberg, 1994; 234–252.)

Figure 6.11 shows three calculated depths of the hardened layer of C45 heat-treatment steel obtained with three different shapes of the laser beam, i.e., square, tophat, and Gaussian distributions [25]. Different energy inputs selected permitted to achieve different maximum temperatures at the material surface. In the calculatiions two cases of laser heating, i.e., a steady beam, the traveling speed ν being 0 mm/s (Fig. 6.11a), and a beam traveling with a speed ν of 20 mm/s (Fig. 6.11b), were taken into account. The greatest depth of the hardened layer was obtained with the square power density distribution in the beam cross section. It was followed by the depths of the hardened layer with the top-hat and Gaussian distributions. With the stationary hardening, the depth of the hardened layer was by approximately 30% greater than with the top-hat distribution. In kinematic or scan hardening, i.e., when the laser beam is traveling at 20 mm/s, and with the square energy density distribution, a 2.3-times smaller depth of the hardened layer was obtained at the maximum surface temperature T_{Smax}, i.e., 1420°C, a temperature just below the melting point of the given steel.

Figure 6.12 shows a through-depth variation of the maximum temperature of transformation-hardened SAE 1045 steel [26]. Two heating conditions were selected, they were defined by different power densities and different traveling speeds of the laser beam across the material surface. It is true in both cases that with the same transformation temperatures, T_{A_1} in T_{A_3}, different depths of the hardened layer and different widths of the transition zone consisting of the hardened microstructure and the microstructure of the base metal were obtained. The results of measurements indicated that the depth of the transition zone amounted to around 20% of the depth of the hardened layer. This relationship between the depth, i.e., the width of the transition zone and the depth of the hardened layer depends primarily on thermal conductivity of the material. The materials with higher thermal conductivity require a higher energy input so that the same maximum temperature in the same depth may be obtained, which, however, results in greater depths of the hardened layers and a larger transition zone. Figure 6.13 shows a complete diagram including possible processing parameters and the depths of the hardened layers obtained in transformation hardening [26]. The data are valid only for the given mode structure of the laser beam (TEM), the given area of the laser spot (A), and the selected absorption deposit. In this case the processing parameters are selected from the laser-beam power P [W] and the traveling speed of the laser beam ν [mm/s]. The upper limit is a power P of 8 kW. This is the limiting energy input

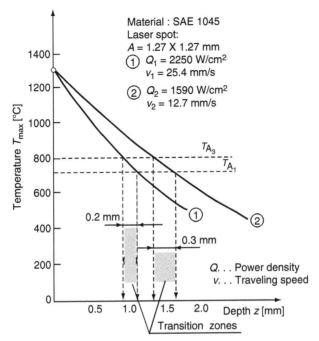

FIGURE 6.12 Influence of processing parameters on hardening depth in laser surface transformation hardening. (From Belforte, D.; Leirtt M., Eds.; *The Industrial Laser Handbook*, Section 1, 1992–1993 Ed. Springer Verlag: New York, 1992; 13–32.)

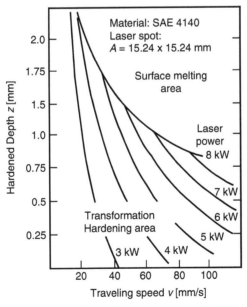

FIGURE 6.13 Influence of laser power and traveling speed on depth of hardened layer at a given laser spot. (From Belforte, D.; Leirtt M., Eds.; *The Industrial Laser Handbook*, Section 1, 1992–1993 Ed. Springer Verlag: New York, 1992; 13–32.)

which permits steel melting. Although the data collected are valid only for a very limited range of the processing parameters, conditions of transformation hardening with the absorptivity changed due to the change in the deposit thickness of the same absorbent, or other type of absorbent, can efficiently be specified as well. Greater difficulties may occur in the selection of the laser trace that can be obtained in different ways and can be varied too. In case the size of the laser spot changes due to optical conditions, it is recommended to elaborate a new diagram of the processing conditions of transformation hardening.

6.4 MICROSTRUCTURAL TRANSFORMATION

In transformation hardening of steel we start from its initial microstructure, which is ferritic–pearlitic, pearlitic–ferritic, or pearlitic. In steel heating, transformation into a homogeneous austenitic microstructure should be ensured.

We have thus a controlled diffusion transformation of the microstructure in the surface layer of the work-piece material of a certain thickness. The thickness of the hardened layer required can be obtained if the heating process is well known. Heating is, namely, defined by the heating rate, the maximum temperature attained at the surface, heating to the depth required, and the time of heating of the material above the austenitizing temperature T_{A_3}. The time required for homogenizing of austenite depends on the type of the initial microstructure and the grain size. It is a basic prerequisite to ensure a homogeneous distribution of carbon in austenite, which results in homogeneous martensite showing a uniform hardness after quenching and self-cooling, respectively. Account should be taken that the martensite transformation is a nondiffusion transformation occurring during the cooling process only at a sufficiently high cooling rate.

Figure 6.14 shows two thermal cycles, i.e., the one at the surface, $z = 0.0$ mm, and the one below the surface, $z = 0.5$ mm, in laser heating of C45 heat-treatment carbon steel [25]. Figure 6.14a shows two thermal cycles of a laser beam with a spot diameter D of 6.0 mm traveling at a speed v of 25 mm/s. Figure 6.14b shows the operation of a static laser beam of the same diameter.

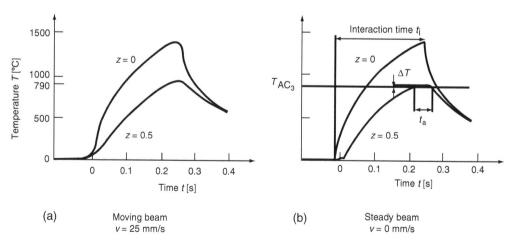

FIGURE 6.14 Temperature cycles during laser heating and cooling of steel C45 at the same interaction time. (From Meijer J.; Kuilboer, R.B.; Kirner, P.K.; Rund, M. Laser beam hardening: transferability of machining CIRP Seminar on Manufacturing Systems—LANE' 94. In *Laser Assisted Net Shape Engineering*; Geiger, M., Vollertsen, F., Eds.; Meisenbach-Verlag: Erlangen, Bamberg, 1994; 234–252.)

For efficient homogenizing, the temperature difference ΔT is important. It is defined as a difference between the maximum temperature T_{max} and the austenitizing temperature T_{A_3} occurring in a depth z of 0.5 mm. Such a temperature difference ensures, with regard to the heating and cooling conditions of the specimen, the time required for austenite homogenizing t_a in the given depth.

Figure 6.15 shows a shift of the transformation temperature, which ensures the formation of inhomogeneous and homogeneous austenite within the selected interaction times [25]. A shorter interaction time will result in a slightly higher transformation temperature T_{A_1} and also a higher transformation temperature T_{A_3}. To ensure the formation of homogeneous austenite with shorter interaction times, considerably higher temperatures are required. Figure 6.15a shows a temperature–time diagram of austenitizing of C45 steel. The isohardnesses obtained at different interaction times in heating to the maximum temperature that ensures partial or complete homogenizing of austenite are plotted. Figure 6.15b shows the same temperature–time diagram of austenitizing of 100 Cr6 hypereutectoid alloyed steel. The diagram indicates that with short interaction times, which in laser hardening vary between 0.1 and 1.0 s, homogeneous austenite cannot be obtained; therefore the microstructure consists of austenite and undissolved carbides of alloying elements that produce a relatively high hardness, i.e., even up to 920 HV$_{0.2}$. After common quenching of this alloyed steel at a temperature of homogeneous austenite, a considerably lower hardness, i.e., only 750 HV$_{0.2}$, but a

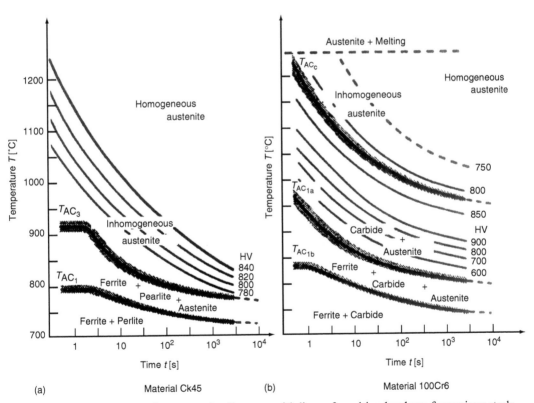

(a)   Material Ck45   (b)   Material 100Cr6

FIGURE 6.15 Temperature–time–austenite diagrams with lines of resulting hardness for various steels. (From Meijer J.; Kuilboer, R.B.; Kirner, P.K.; Rund, M. Laser beam hardening: transferability of machining CIRP Seminar on Manufacturing Systems—LANE' 94. In *Laser Assisted Net Shape Engineering*; Geiger, M., Vollertsen, F., Eds.; Meisenbach-Verlag: Erlangen, Bamberg, 1994; 234–252.)

relatively high content of retained austenite was obtained. Retained austenite is un-wanted as it will produce unfavorable residual stresses and reduce wear resistance of such a material.

6.5 MATHEMATICAL MODELING

6.5.1 MATHEMATICAL PREDICTION OF HARDENED DEPTH

The classical approach to modeling the heat flow induced by a distributed heat source moving over the surface of a semi-infinite solid starts with the solution for a point source, with integrations over the beam area [27]. This method requires numerical procedures for its evaluation. These solutions are rigorous, so their computations are complex and the results difficult to be applied. Bass [28] gives an alternative approach by presenting temperature field equations for various structural beam modes. Analytical results show good response to various materials. Ashby and Easterling [29] and Li et al. [30] developed further analytical approach and an approximate solution for the entire temperature field. Comparison of the analytical results with numerical calculations shows adequate description of laser transformation hardening. Variability of laser transformation hardening parameters and changing material properties with temperature result in some scatter. With the use of dimensionless parameters to simplify computation, results general to all materials are obtained. There exist many such examples in the analysis of welding [31] and in laser surface treatment [32–34].

In mathematical modeling the following assumptions were considered:

- The surface absorptivity A is constant.
- The latent heat of the α- to γ-transformation is negligible.
- The thermal conductivity λ and the thermal diffusivity of steel are a constant.
- The eutectoid temperature T_{A_1} is as given by the phase diagram.
- The radius of Gaussian beam r_B is the distance from the beam center to the position where the intensity is $1/e$ times the peak value.

The origin of the coordinate system is the beam center. The laser of total power P moves in the x direction with traveling speed v, with the y axis across the track, and z axis is the distance below the surface.

The temperature field equation from Ashby and Easterling [29] is valid for the Gaussian line source:

$$T - T_0 = \frac{Aq}{2\pi\lambda v[t(t+t_0)]^{1/2}} \cdot \exp\left(-\frac{1}{4a}\right)\left[\frac{(z+z_0)^2}{t} + \frac{y^2}{(t+t_0)}\right]$$

The equation contains two reference parameters, defined by $t_0 = r_B^2/4a$ and z_0 is a characteristic length, as a function to limit the surface temperature.

Shercliff and Ashby [33] define the following dimensionless parameters:
$T^* = (T - T_0)/(T_{A1} - T_0)$ is the dimensionless temperature rise,
$q^* = Aq/r_B\lambda\,(T_{A_1} - T_0)$ is the dimensionless beam power,
$v^* = vr_B/a$ is the dimensionless traveling speed,
$t^* = t/t_0$ is the dimensionless time, and $(x^*, y^*, z^*) = (x/r_B, y/r_B, z/r_B)$ is the dimensionless x,y,z coordinates.

The distance z_0 is normalized as follows:

$$z_0^* = z_0/r_B$$

and dimensionless temperature parameter is then

$$T^* = \frac{(2/\pi)\,(q^*/v^*)}{[t^*(t^*+1)]^{1/2}} \exp - \left[\frac{(z^*+z_0^*)^2}{t^*} + \frac{y^{*2}}{(t^*+1)}\right]$$

The time-to-peak temperature t_p^* at an (x^*, y^*, z^*) position is found by differentiating with respect to the time.

$$t_p^* = \frac{1}{4}\left[2(z^*+z_0^*)^2 - 1 + [4(z^*+z_0^*)^4 + 12(z^*+z_0^*)^2 + 1]^{1/2}\right]$$

At stationary laser beam of uniform intensity Q produces a peak surface temperature given by Bass [28]:

$$T_p - T_0 = \frac{2Aq}{\pi^{1/2}\lambda}(a\tau)^{1/2}$$

The average intensity of a Gaussian beam is $Q = q/\pi r_B^2$ so the previous equation can be rewritten:

$$T_p - T_0 = \frac{2Aq}{\pi^{3/2}r_B^2\lambda}(a\tau)^{1/2}$$

or in dimensionless form τ being $2r_B/v$ and

$$(T_p^*)_{z^*=0} = (2/\pi)^{3/2}q^*/(v^*)^{1/2}.$$

The conditions for the first hardening $T_p = T_{A_1}^*$ or the onset of melt $T_p = T_m$ are thus defined by a constant value of the process variables as follows:

$$q^*/T_p^*(v^*)^{1/2} = (\pi/2)^{3/2} = \text{constant}$$

with T_p^* taking the value appropriate to the peak temperature of interest T_{A_1} or T_m.

Bass [28] gives a more general solution for the peak surface temperature at stationary Gaussian beam acting in dimensionless form

$$(T_p^*)_{z^*=0} = (1/\pi)^{3/2}q^* \tan^{-1}(8/v^*)^{1/2}$$

A constant value of a single dimensionless parameter defines the first hardening and the onset of melt as follows for all v^*:

$$(q^*/T_p^*)\tan^{-1}(8/v^*)^{1/2} = \pi^{3/2} = \text{constant}.$$

Four dimensionless parameters define laser hardening with a Gaussian beam. The aim is to produce a diagram from which process variables may be readily selected. A convenient plot is of the dependent variables z_c^*, v^*, q^*, and $T_p^* = 1$ which give $T_p = T_{A_1}$ as shown in Fig. 6.16a. As surface melting is generally undesirable, the contours are dashed if the surface melts $T_p^* = (T_m - T_0)/(T_{A_1} - T_0)$ at $z^* = 0$. Figure 6.16a shows the depth at which surface melt commences for a 0.4-pct carbon steel at $T_p^* \approx 2.12$.

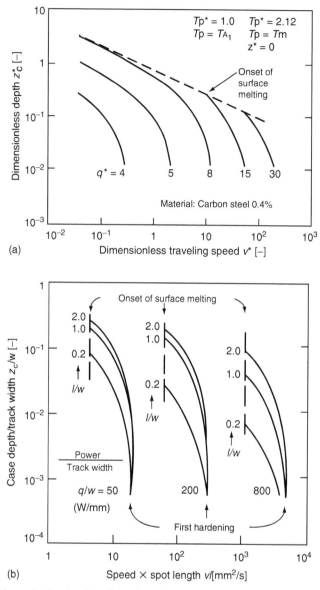

FIGURE 6.16 (a) Dimensionless hardened depth Z_a^* against laser beam traveling speed v^* with curves of constant laser beam power q^* for Gaussian power density cross section. (b) Dimensionless hardened depth Z_c/w against laser beam traveling speed parameter vl with curves of $q/w = $ const. for rectangular power density and spot ratio $l/w = $ const. (From Ashby, M.F.; Easterling K.E. *Acta Metall.* 1984, *32*, 1933–1948.)

A non-Gaussian source may be simulated by super-posing a number of Gaussian sources. The total power is shared between the Gaussian sources, to give the best fit to true energy profiles. For particular location and given time the temperature rises due to each source contribution to heating. This is valid if the thermal properties of the material are independent of the temperature, meaning that the differential heat flow according to the equation is linear. This method will be presented by the same authors for laser-beam heating with rectangular sources defined as laser beam spot ratio. A laser-beam ratio R is defined by length l in travel direction x, and the width w, of track in cross direction y ($R = l/w$). In practice fixed width is normally used and the length is varied, so it is sensible to normalize the process variables using the beam width as follows:

$$q_R^* = Aq/w\lambda q\ (T_A - T_0)$$
$$v_R^* = vw/a$$
$$z_R^* = z/w$$

where the subscript R refers to rectangular laser beam.

Figure 6.16b shows a dimensionless diagram for medium carbon steel according to various ratio power to track width q/w [W/mm] and spot ratio $R = l/w$ [−].

The diagram shows the nondimensional ratio of the hardened-layer depth to the laser-spot width for a rectangular laser beam as a function of a product of the traveling speed and the laser-trace length. Three characteristic power densities per unit of laser-spot width, i.e., $q/w = 50$, 200, and 800 W/mm, were chosen, which equals a ratio of 1 to 4 to 16. In the diagram there are curves plotted for individual power densities q/w valid with certain ratios of the laser-spot dimensions of the rectangular beam, i.e., $l/w = 0.2$, 1.0, and 2.0. In the individual cases laser-hardening conditions and boundary conditions of laser hardening, i.e., laser remelting, were known too. The individual curves in the diagram indicate that by increasing the laser-beam power density per unit of laser-spot width the boundary conditions will be achieved with smaller depths of the hardened layer. The diagram shown is of general validity and permits the determination, i.e., prediction, of the hardened-layer depth based on the variation of the power density per unit of laser-spot width and the traveling speed.

It can be summarized that the approximate heat flow model of Ashby and Easterling [29] for laser transformation hardening has been presented, describing Gaussian and non-Gaussian sources over a wide range of process variables.

The following advantages can be noted:

- Simplification by choosing dimensionless parameters.
- Surface temperature calibration, extending the approximate Gaussian solution to all laser-beam traveling speeds.
- High traveling speed solution was found to be acceptable for rectangular, uniform sources.
- General Gaussian solution and enabling extension of the model to non-Gaussian sources.
- Identification of constant dimensionless parameters containing all of the process variables for both sources, which determine the position of the first hardening or the onset of melting.
- Process diagrams for rectangular sources allow the choice of the process variables such as the beam power, track width, traveling speed, and spot dimensions.

6.5.2 MATHEMATICAL MODELING OF MICROSTRUCTURAL CHANGES

Ashby and Easterling [29] presented their results as laser-processing graphs, which show the microstructures and hardnesses regarding process variables.

In their experiments they studied two steels, i.e., a Nb microalloyed and a medium-carbon steel. They varied the laser power P, the beam spot radius r_B, and its traveling speed v.

They carried out laser surface heat treatments using a 0.5- and a 2.5-kW continuous wave CO_2 laser using Gaussian and "top hat" energy profiles.

The microhardness of individual martensitic and ferritic regions in the low-carbon steel could be estimated fairly well from a simple rule-of-mixtures

$$H_p(\text{mean}) = f_m H_m + (1 - f_m)H_f$$

where H_m is the mean microhardness of the martensite and H_f that of the ferrite at that depth.

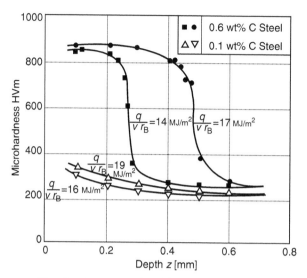

FIGURE 6.17 Hardness profiles for the two steels with different energy densities, $E = q/vr_B$ and interaction time, $t_i = r_B/v$. (From Ashby, M.F.; Easterling K.E. *Acta Metall.* 1984, *32*, 1933–1948.)

Figure 6.17 shows the hardness profiles for the two steels with different energy densities q/vr_B and interaction time r_B/v. It can be noted that worse hardening is obtained when the carbon content of the steel is low. For the high-carbon steel, complete carbon redistribution occurs within the austenitization process, which gives a uniform high hardness.

They developed simple models for pearlite dissolution, austenite homogenization, and martensite formation.

The heat cycle $T(t)$ at the depth causes microstructural changes if high enough temperature is achieved. Some of the microstructure changes are diffusion controlled such as the transformation of pearlite to austenite and the homogenization of carbon in austenite. The microstructure changes depend on the total number of diffusive jumps that occur during the temperature cycle. It is measured by the kinetic strength, I, of the temperature cycle, defined by

$$I = \int_0^\infty \exp\left(-\frac{Q_A}{RT(t)}\right) dt$$

where Q_A is the activation energy for the typical microstructural transformation and R is the gas constant. It is more convenient to write it as

$$I = \alpha\tau \exp\left(-\frac{Q_A}{RT_p}\right)$$

where T_p is the maximum temperature and τ is the thermal constant. The constant α is well approximated by

$$\alpha = \sqrt[3]{\frac{RT_p}{Q_A}}.$$

At rapid heating, the pearlite first transforms to austenite, which is followed by carbon diffusion outward and increases the volume fraction of high-carbon austenite.

If the cementite and ferrite plate spacing within a colony distance is λ, it might be thought that lateral diffusion of carbon to austenite occurs. In an isothermal heat treatment, this would require a time t given by $\lambda^2 = 2Dt$, which is

$$\lambda^2 = 2D_0 \alpha \tau \exp\left(-\frac{Q_A}{RT_p}\right)$$

where D is the diffusion coefficient for carbon. In a temperature cycle $T(t)$ the quantity Dt is where the maximum temperature (T_p) is found to cause the transformation.

The most important to the understanding of laser transformation hardening is the modeling of the carbon redistribution in austenite. When a hypoeutectoid, plain-carbon steel with carbon content c is heated above the T_{A_1} temperature, the pearlite transforms instantaneously to austenite. The pearlite transforms to austenite containing $c_e = 0.8\%$ carbon and the ferrite becomes austenite with negligible carbon content c_f. Thereafter, the carbon diffuses from the high to the low concentration regions, to an extent which depends on temperature and time. On subsequent cooling of steel from austenitic temperature, the austenite with carbon content greater than the critical value of 0.05 wt% C transforms to martensite and the rest of austenite with carbon content less than 0.05 wt% C transforms into pearlite.

The volume fraction occupied by the pearlite colonies is as follows:

$$f_i = \frac{c - c_f}{0.8 - c_f} \approx \frac{c}{0.8}$$

where c_f is the carbon content of the ferrite.

The volume fraction of the martensite is as follows [29]:

$$f = f_m - (f_m - f_i)\exp - \left[\frac{12f_i^{2/3}}{\sqrt{\pi g}}\ln\left(\frac{c_e}{2C_c}\right)\sqrt{Dt}\right]$$

where g [m] is mean grain size.

The hardness of the transformed surface layer depends on the volume fraction of martensite and its carbon content. The authors [29] calculated the hardness of the martensite and ferrite mixture by using a rule of mixtures

$$H = fH_m + (1 - f)H_f$$

and suggested the following formula for calculating hardness

$$H = 1667c - 926c^2/f + 150.$$

Figure 6.18 shows three measured hardness profiles for different energy parameters (full lines) and the calculated profiles (dashed lines). Figure 6.19 shows a laser-processing diagram for a 0.6 wt% plain carbon steel. The horizontal axes present energy density q/vr_B and beam spot radius r_B. These variables determine the temperature cycle in the transformed layer. The vertical axis is the depth below the surface. Within the shaded region, melting occurs and outside the transformation hardening process. The diagram also shows the contours of martensite volume fraction. The volume fraction and carbon content of the martensite are used to calculate the hardness HV after laser surface transformation hardening.

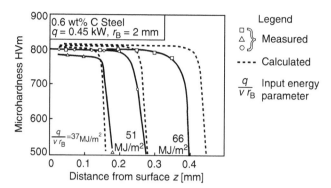

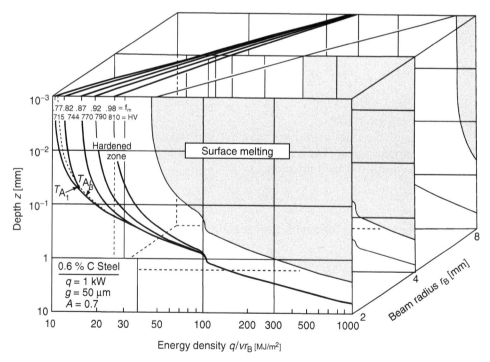

FIGURE 6.18 Measured and calculated hardness profiles for the 0.6 wt% carbon steel compared with those predicted. (From Ashby, M.F.; Easterling K.E. *Acta Metall.* 1984, *32*, 1933–1948.)

The author [29] showed that:

- Steels with a carbon content below about 0.1 wt% do not respond to transformation hardening.
- Optimum combination of process variables gives maximum surface hardness without surface melting.

The method could be used for laser glazing and laser surface alloying.

FIGURE 6.19 Laser processing diagram for 0.6 carbon steel. (From Ashby, M.F.; Easterling K.E. *Acta Metall.* 1984, *32*, 1933–1948.)

6.6 COMPUTING METHOD FOR CALCULATING TEMPERATURE CYCLE

Several methods exist to solve the heat conduction equations for various conditions; interesting descriptions are given by Carslaw and Jaeger [27]. Most of the computing method to calculate temperature cycles are based on one of the many cases which are modified to suit the particular case [35].

Gregson [36] discussed a one-dimensional model using a semi-infinite flat-plate solution for idealized uniform heat source which is constant in time. Expressions used for temperature profile separated for heating and cooling are as follows.

Heating temperature–time profile:

$$T(z, t) = \frac{\varepsilon z Q_{AV}}{\lambda} \sqrt{at} \cdot \text{ierfc} \left\{ \frac{z}{2\sqrt{xt}} \right\}, \quad Q(t) = \begin{cases} Q & \text{for } t > 0 \\ 0 & \text{for } t < 0 \end{cases}.$$

Cooling temperature–time profile:

$$T(z, t) = \frac{2Q_{AV}\sqrt{a}}{\lambda} \left\{ \sqrt{t} \cdot \text{ierfc} \frac{z}{2\sqrt{at}} - \sqrt{t - t_L} \cdot \text{ierfc} \left(\frac{z}{2\sqrt{a(t - t_L)}} \right) \right\}$$

$$f(t) = \begin{cases} Q & \text{for } 0 < t < tL \\ 0 & \text{for } 0 > t > tL \end{cases}$$

where T [°C] is the temperature, z [cm] is the depth below the surface, t [s] is the time, $\varepsilon \cong 1$ is the emissivity, Q_{AV} [W/cm^2] is the average power density, λ [W/cm °C] is the thermal conductivity, a [cm^2/sec] is the thermal diffusivity, t_0 [s] is the time start for power on, t_L [s] is the time for power off, and ierfc is the integral of the complementary error function.

These equations for description of laser heating and cooling process are valid if the thickness of the base material is greater than $t \geq \sqrt{4at}$ and they could be approximately described for the hardened layer.

These one-dimensional analyses may be applied to laser transformation hardening process with idealized uniform heat sources, which are produced by using optical systems such as laser-beam integrator or high-power multimode laser beam with top-hat power density profile. These equations present one-dimensional solutions and provide only approximate temperature–time profile. For better description of thermal conditions, a two- or three-dimensional analysis considering actual input power density distribution and variable thermophysical properties treated material is required.

Sandven [37] presented the model that predicts the temperature–time profile near a moving ring-shaped laser spot around the periphery of the outer or inner surface of a cylinder. This solution can be applied to the transformation hardening processes using toric mirrors. Sandven [37] developed his model based on a flat-plate solution and assumed that the temperature time profile $T(t)$ for cylindrical bodies can be approximated by

$$T = \theta I$$

where θ depends on workpiece geometry and I is the analytical solution for a flat plate. The final expression for a cylindrical workpiece, which is derived from this analysis, is

$$T \approx (1 \pm 0.43\sqrt{\phi})\frac{2Q_0 a}{\pi\lambda\nu}\int_{x-B}^{x+B} e^u \cdot K_0(z^2+u^2)^{1/2}du$$

where the + sign means the heat flow into a cylinder, the − sign means the heat flow out of a hollow cylinder, Q_0 is the power density, ν is the laser-beam traveling speed in the x direction, K_0 is the modified Bessel function of the second kind and 0 order, u is the integration variable, $2b$ is the width of the heat source in the direction of motion, and z is the depth in radial direction,

$$B = \frac{\nu_b}{2a}, \quad Z = \frac{\nu_z}{2a}, \quad X = \frac{\nu_x}{2a}$$

where $\phi = at/R^2$, and R is the radius of the cylinder.

Sandven [37] provided graphical solutions for $Z=0$ for various values of B. To estimate an approximate depth of hardness, maximum temperature profile across the surface layer is the only item to be interested.

Cline and Anthony [38] presented most realistic thermal analysis for laser heating. They used a Gaussian heat distribution and determined the three-dimensional temperature distribution by solving the equation:

$$\partial T/\partial t - a\nabla^2 T = Q_{AV}/c_p$$

where Q_{AV} is the power absorbed per unit volume and c_p is the specific heat per unit volume.

They used a coordinate system fixed at the workpiece surface and superimposed the known Green function solution for the heat distribution; the following temperature distribution is

$$T(x, y, z) = P(c_p a r_B)^{-1} f(x, y, z, \nu)$$

where f is the distribution function

$$f = \int_0^\infty \frac{\exp(-H)}{(2\pi^3)^{1/2}(1+\mu^2)}d\mu \quad \text{and} \quad H = \frac{\left(X+\frac{\tau\mu^2}{2}\right)^2 + Y^2}{2(1+\mu^2)} + \frac{Z^2}{2\mu^2}$$

where $\mu^2 = 2at'/r_B$, $\tau = \nu r_B/a$, $X=x/r_B$, $Y=y/r_B$, $Z=z\,r_B$, P is total power, r_B is the laser-beam radius, t' is the earlier time when laser was at (x', y'), and ν is the traveling speed.

The cooling rate can be calculated as follows:

$$\partial T/\partial t = -\nu[x/\gamma^2 + \nu/2a(1+x/\gamma)]T$$

where $\gamma = \sqrt{x^2+y^2+z^2}$.

The given cooling rate is calculated only when point heat source is used.

This three-dimensional model is a great improvement over one-dimensional models because it includes temperature-dependent thermophysical properties of the material used for numerical solutions.

Grum and Šturm [39] obtained a relatively simple mathematical model describing the temperature evolution $T(z,t)$ in the material depending on time and position, where we distinguished between the heating cycle and the cooling cycle.

For reasons of simplifying the numerical calculations, it is necessary to make certain assumptions:

- Latent heat of material melting is neglected.
- Material is homogeneous with constant physical properties in the solid and liquid phase. So we assume that material density, thermal conductivity, and specific heat are independent of temperature.
- Thermal energy is transferred only through transfer into the material; thermal radiation and transfer into the environment are disregarded.
- Laser light absorption coefficient to workpiece material is constant.
- Limiting temperatures or transformation temperatures are assumed from phase diagrams.
- Remelted surface remains flat and ensures a uniform heat input.

Thus we obtained a relatively simple mathematical model describing the temperature evolution $T(z,t)$ in the material depending on time and position, where we distinguished between the heating cycle and the cooling cycle.

1. The heating cycle conditions in the material can be described by the equation:

$$T(z,\ t) = T_0 + \frac{AP}{2\pi A \nu_B \sqrt{t(t_i + t_0)}} \left[e^{-\left(\frac{(z+z_0)^2}{4at}\right)} + e^{-\left(\frac{(z-z_0)^2}{4at}\right)} \right] \mathrm{erfc}\left(\frac{z+z_0}{\sqrt{4at}}\right)$$

for $0 < t < t_i$

2. The cooling cycle conditions in the material can be expressed by the equation:

$$T(z,\ t) = T_0 + \frac{AP}{2\pi A \nu_B \sqrt{t(t_i + t_0)}} \left[e^{-\left(\frac{(z+z_0)^2}{4at}\right)} + e^{-\left(\frac{(z-z_0)^2}{4at}\right)} - e^{-\left(\frac{(z-z_0)^2}{4a(t-t_1)}\right)} \right]$$
$$\mathrm{erfc}\left(\frac{z+z_0}{\sqrt{4at}}\right)$$

for $t > t_i$

where variable t_0 represents the time necessary for heat to diffuse over a distance equal to the laser-beam radius on the workpiece surface and the variable z_0 measures the distance over which heat can diffuse during the laser-beam interaction time [29]. C is a constant, in our case defined as $C = 0.5$. Figure 6.20a presents the time evolution of temperatures calculated according to equations at specific depth of the material in the nodular iron 400–12 at a laser-beam traveling speed $\nu_B = 12$ mm/s. Figure 6.20b illustrates the variation of heating and cooling rates during the process of laser remelting in the remelted layer and in deeper layers of the material. The temperature gradient is at the beginning of the laser-beam interaction with the workpiece material, i.e., on heating up, very high, on the surface achieving values as high as 48,000°C/s. The results show that the highest cooling rate is achieved after the beam has passed by half the value of its radius r_B across the measured point. Knowing the melting and austenitization temperatures, they can successfully predict the depth of the remelted and modified layer (see Figure 6.21). Considering the fact that on the basis of limiting temperatures it is possible to define the depth of particular layers and that these can be confirmed by microstructure analysis, we can verify the success of the proposed mathematical model for the prediction of remelting conditions. Thus a comparison is shown in Figure 16.22a and Figure

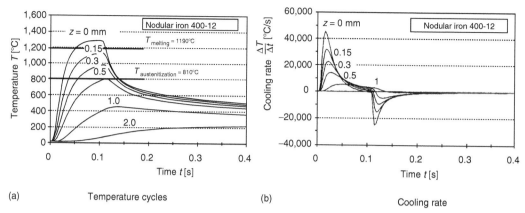

FIGURE 6.20 Temperature cycles and cooling rate versus time at various depths. (From Grunm, J.; Štrum, R. Calculation of temperature cycles heating and quenching rates during laser melt-hardening of cast iron. In *Surface Engineering and Functional Materials*, Proceedings of the 5th European conference on Advanced Materials, Functionality and Design; Maastricht, NL., Sarton, L.A.J.L., Zeedijk, H.B., Eds.; Published by the Netherlands Society for Materials Science Aj Zwijondrecht; 3/155–3/159.)

16.22b between the experimentally obtained results for the depth of particular zones of the modified layer and the results calculated according to the mathematical model. We can see that the calculated depths of the remelted and hardened zones correlate well with the experimentally measured values. Large deviations in the depth of the modified layers are found only in gray iron at insufficient workpiece traveling speeds, and they are probably due to the occurrence of furrows on the workpiece surface.

6.7 LASER LIGHT ABSORPTIVITY

With the interaction of the laser light and its movement across the surface, very rapid heating up of metal workpieces can be achieved, and subsequent to that also very rapid cooling down

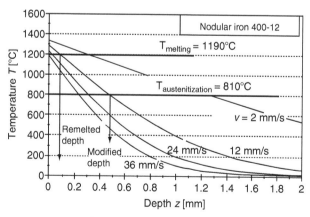

FIGURE 6.21 Maximum temperature drop as a function of depth in nodular iron 400–12. (From Grum, J.; Štrum, R. Calculation of temperature cycles heating and quenching rates during laser melt-hardening of cast iron. In *Surface Engineering and Functional Materials*, Proceedings of the 5th European Conference on Advanced Materials, Functionality and Design; Maastricht, NL., Sarton, L.A.J.L., Zeedijk, H.B., Eds.; Published by the Netherlands Society for Materials Science Aj Zwijondrecht; 3/155–3/159.)

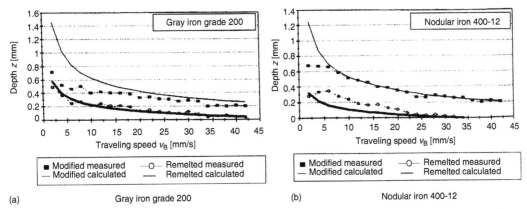

(a)   Gray iron grade 200   (b)   Nodular iron 400-12

FIGURE 6.22 Comparison of experimentally measured, remelted, and hardened zone depths with those calculated with the mathematical model. (From Grum, J.; Štrum, R. Calculation of temperature cycles heating and quenching rates during laser melt-hardening of cast iron. In *Surface Engineering and Functional Materials*, Proceedings of the 5th European Conference on Advanced Materials, Functionality and Design; Maastricht, NL., Sarton, L.A.J.L., Zeedijk, H.B., Eds.; Published by the Netherlands Society for Materials Science Aj Zwijondrecht; 3/155–3/159.)

or quenching. The cooling speed, which in conventional hardening defines quenching, has to ensure martensitic phase transformation. In laser hardening the martensitic transformation is achieved by self-cooling, which means that after the laser light interaction the heat has to be very quickly abstracted into the workpiece interior. While it is quite easy to ensure the martensitic transformation by self-cooling, it is much more difficult to deal with the conditions in heating up. The amount of the disposable energy of the interacting laser beam is strongly dependent on the absorptivity of the metal. The absorptivity of the laser light with a wavelength of 10.6 μm ranges in the order of magnitude from 2 to 5%, whereas the remainder of the energy is reflected and represents the energy loss. By heating metal materials up to the melting point, a much higher absorptivity is achieved with an increase of up to 55%, whereas at vaporization temperature the absorptivity is increased even up to 90% with respect to the power density of the interacting laser light.

Figure 6.23 illustrates the relationship between laser light absorptivity on the metal material surface and temperature or power density [40,41]. It is found that, from the point of view of absorptivity, laser-beam cutting does not pose any problems, as the metal takes the liquid or evaporated state, and the absorptivity of the created plasma can be considerably increased. Therefore it is necessary to heat up the surface, which is to be hardened, onto a certain temperature at which the absorptivity is considerably higher and enables rapid heating-up onto the hardening temperature or the temperature which, for safety reasons, is lower than the solidus line. This was successfully used in heat treatment of camshafts [42].

The heating-up of the workpiece surface material by the laser beam is done very rapidly. The conditions of heating-up can be changed by changing the energy density and relative motion of the workpiece and the laser beam.

In surface hardening this can be achieved without any additional cooling and is called self-quenching. The procedure of laser surface hardening is thus simpler than the conventional flame or induction surface hardening as no additional quenching and washing is required. Absorptivity always depends on the wavelength of light and the surface preparation of the material that interacts. In Table 6.1 typical values of reflectivity of polished surfaces of a titanium alloy, aluminum, 304 austenitic stainless steel, and soft steel in interaction with CO_2 laser light are stated [43]. The reflectivity of the polished surface of the given materials

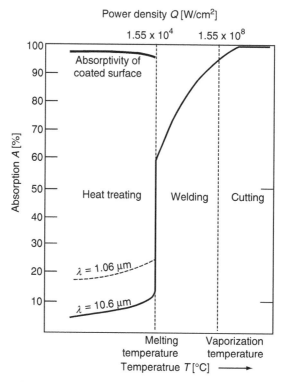

FIGURE 6.23 The effect of temperature on laser light absorptivity. (From Tizian, A.; Giordano, L., Ramous, E. Laser surface treatment by rapid solidification. In *Laser in Materials Processing*. Metzbower, E.A., Ed.; Conference Proceedings; American Society for Metals: Metals Park, Ohio, 1983; 108–115. Engel S.L. Section IV, Surface hardening—basics of laser heat treating. In *Source Book on Applications of the Laser in Metalworking*; Metzbower, E.A., Ed.; American Society for Metals: Metals Park, Ohio, 1981; 149–171.)

depends on the temperature and wavelength of the incident light. Bramson [44] defined the dependence between electric resistance and emissivity ε_λ (T) for the light radiation striking the material surface at a right angle.

$$\varepsilon_\lambda(T) = 0.365\left(\frac{\rho_r T}{\lambda}\right)^{1/2} + 0.0667\left(\frac{\rho_r T}{\lambda}\right) + 0.006\left(\frac{\rho_r T}{\lambda}\right)^{3/2}$$

where ρ_r [Ω cm] is the electrical resistivity at a temperature $T[°C]$, ε_λ (T) is the emissivity at $T[°C]$, and λ [cm] is the wavelength of incident radiation. Reflectivity depends on the incident angle of the laser beam with reference to the polarization plane and the specimen surface [45–53]. Figure 6.24 shows the reflectivity of CO_2 laser light from a steel surface at different incident angles and different temperatures [43,53,54]. The diagram combines experimental data (plotted dots) on reflectivity and absorptivity, and theoretically calculated values or reflectivity (uninterrupted lines). The variations of absorptivity indicate that absorptivity strongly increases at elevated temperatures due to surface oxidation and at very high temperatures due to surface plasma absorption.

Figure 6.25 shows the influences exerted on absorptivity of CO_2 and Nd: YAG laser light in the interaction with specimens made of Ck45 steel [53]. The steel specimens were polished,

TABLE 6.1
Typical Values of Surface Reflectivity for Various Materials and Surface States

Material	Surface State	Reflectivity R [%], $\lambda = 10.6\ \mu m$
Titanium 6A1-4V	Polished 300C	85
Aluminum	Polished	98
Stainless steel 304	Polished	85
Steel, mild	Polished	94
	Roughened with sand paper	
	to 1 μm	92
	to 19 μm	32

Source: From Stern, W.M. Laser cladding, alloying and melting. In *The Industrial Laser Annual Handbook 1986*; Belforte, D., Levitt, M., Eds.; Penn Well Books, Laser Focus: Tulsa Oklahoma, 1986, 158–174.

ground, turned, and sandblasted. To Ck45 heat-treatment carbon steel, various methods of surface hardening, and particularly laser hardening, are often applied.

From the column chart it can be inferred that absorptivity of steel specimens subjected to different machining methods is considerably lower with CO_2 laser light than with Nd:YAG laser light. The lowest absorptivity was obtained with the polished specimens. It varied between 3 and 4% with reference to the laser-light wavelength. Absorptivity was slightly stronger with the ground and then turned surfaces. It turned out in all cases that the

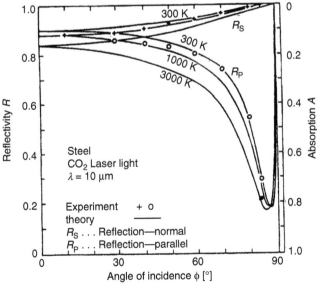

FIGURE 6.24 Variation of reflectivity with angle and plane of polarization. (From Wissenbach, K.; Gillner, A.; Dausinger, F. *Transformation Hardening by CO₂ Laser Radiation, Laser und Optoelectronic*; AT-Fachferlach, Stuttgart, 1985; Vol. 3, 291–296.)

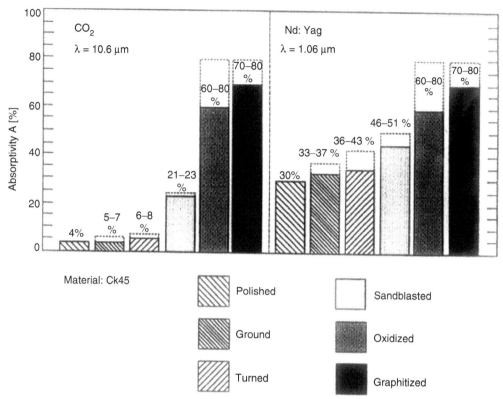

FIGURE 6.25 Influence of various steel Ck45 treatments on absorption with CO_2 or Nd:YAG laser light. (From Wissenbach, K.; Gillner, A.; Dausinger, F. *Transformation Hardening by CO_2 Laser Radiation, Laser und Optoelectronic*; AT-Fachferlach, Stuttgart, 1985; Vol. 3, 291–296.)

absorptivity of Nd:YAG laser light is 7 times that of CO_2. If absorptivity of the two wavelengths is considered, smaller differences may be noticed with the sandblasted surfaces. The oxidized and graphitized surfaces showed the same absorptivity of laser light regardless of its wavelength. The latter varied between 60 and 80%.

6.8 INFRARED ENERGY COATINGS

To increase laser-beam absorptance at metal surfaces various methods were used:

- Metal surface painted with absorbing coatings followed by laser processing
- Chemical conversion coatings
- Uncoated metal surfaces processed by a linearly polarized laser beam [55]

Infrared energy coatings having high absorptance must have the following features for increased efficiency during laser heating at heat treatment:

- High thermal stability
- Good adhesion to metal surface
- Chemically passive to material heat conduction from coating to material
- Proper coating thickness referring to type of material and the way of heat treatment

- High heat transfer coefficient for better heat conduction from coating to material
- Easily applied and removed
- As lower expenses for coatings as possible

6.8.1 PAINT AND SPRAY COATINGS

Several commercially available paints exhibit low normal spectral reflectance for CO_2 laser with wavelength $\lambda = 10.6$ μm. These paints in general contain carbon black and sodium or potassium silicates and are applied to metal surfaces by painting or spraying. Thicknesses of paint coatings range from 10 to 20 μm.

6.8.2 CHEMICAL CONVERSION COATINGS

Chemical conversion coatings, such as manganese, zinc, or iron phosphate, absorb infrared radiation. Phosphate coatings are obtained by treating iron-base alloys with a solution of phosphoric acid mixed with other chemicals. Through this treatment, the surface of the metal surface is converted to an integrally bonded layer of crystalline phosphate. Phosphate coatings may range in thickness from 2 to 100 μm of coating surface. Depending on the workpiece geometry, phosphating time can range from 5 to 30 min regarding temperature and concentration of solution. Phosphate coatings on the metal surface can be prepared with a fine or coarse microstructure.

In terms of chemical passiveness and ease of coating application on metal surfaces, silicates containing carbon black are more effective than the phosphate coatings. Figure 6.26 schematically illustrates the reaction of manganese phosphate with metal surface and subsequent formation of low melting compounds, which can penetrate along the grain boundaries over several grains below the surface of metal material. This reaction can be prevented by using chemically inert coatings.

6.8.3 LINEARLY POLARIZED LASER BEAM

Metals have lower reflectance for linearly polarized electromagnetic radiation. The basis of this optical phenomenon has been applied to the CO_2 laser heat treatment of uncoated iron-base alloys. An unpolarized laser beam can be linearly polarized by using proper reflecting optical elements. Figure 6.27 shows unpolarized laser beam with specific incident angle referring to metal mirror and reflected beam is linearly polarized. This angle of incidence is called the polarizing angle. When the laser beam is linearly polarized, the dominant vibration

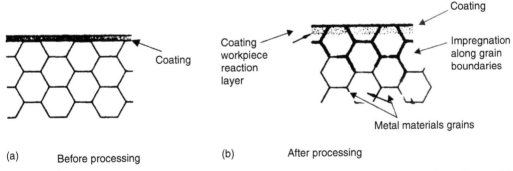

(a) Before processing (b) After processing

FIGURE 6.26 Potential reaction of infrared energy-absorbing coating after laser-treated metal material. (From Guanamuthu, D.S.; Shankar, V. Laser heat treatment of iron-base alloys. In *Laser Surface Treatment of Metals*. Drapper, C.V., Mazzoldi, P., Eds.; NATO ASI, Series-No. 115, Martinus Nijhoff Publishers. Dordrecht, 1986; 413–433.

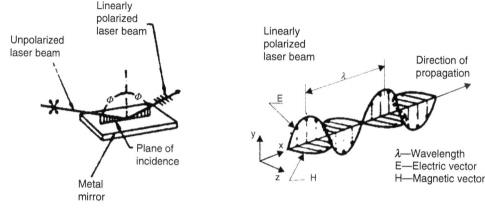

Φ—Angle of incidence

FIGURE 6.27 Conversion of an unpolarized laser beam to a linearly polarized beam by reflection at a specific angle. (From Guanamuthu, D.S.; Shankar, V. Laser heat treatment of iron-base alloys. In *Laser Surface Treatment of Metals*. Drapper, C.V., Mazzoldi, P., Eds.; NATO ASI, Series-No. 115, Martinus Nijhoff Publishers. Dordrecht, 1986; 413–433.)

direction is perpendicular to the plane of incidence. The plane of incidence is defined as the plane that contains both the incident laser beam and the normal to the reflecting surface.

The electric vector **E** of the linearly polarized beam has components parallel E_p and perpendicular E_s to the plane of incidence. Figure 6.28 illustrates absorptance as a function

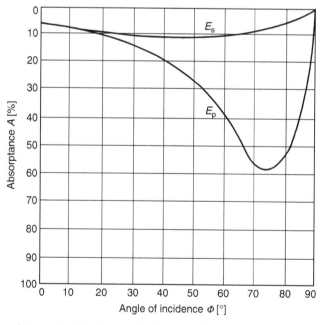

FIGURE 6.28 Effect of the angle of incidence of a linearly polarized laser beam on absorptance by iron-base alloys. (From Guanamuthu, D.S.; Shankar, V. Laser heat treatment of iron-base alloys. In *Laser Surface Treatment of Metals*. Drapper, C.V., Mazzoldi, P., Eds.; NATO ASI, Series-No. 115, Martinus Nijhoff Publishers. Dordrecht, 1986; 413–433.

of the angle of incidence for iron [55]. At an angle of incidence between 70° and 80°, the absorptance is between 50 and 60% for E_p and 5 to 10% for E_s. Thus by directing a linearly polarized laser beam at an angle of incidence greater than 45°, substantial absorptance by iron-base alloys is possible. Possible weakness of this method is the important laser-beam power loss during conversion of an unpolarized to a linearly polarized laser beam.

6.9 INFLUENCE OF DIFFERENT ABSORBERS ON ABSORPTIVITY

Rothe et al. [56] determined absorptivity of metal surfaces on the basis of calorimetric measurements. Kechemair et al. [57] studied the influence of the graphite absorption deposit on the surface of the specimens of 35 NCD 16 steel.

Woo and Choo [58] analyzed the general effects of coating thickness of absorber on the depth and width of hardened layer. Hardening experiments were carried out at various laser energy input. Trafford et al. [59] studied absorptivity in the case when the traveling speed of the laser beam across the workpiece surface varied. They studied various absorption deposits, e.g., carbon black and colloidal graphite, zinc and manganese phosphates, which were deposited on carbon steel with a carbon content of 0.4%. The first two absorbents behaved very favorably at lower surface temperatures, but at higher temperatures their effect weakened. Another deficiency of depositing carbon black or colloidal graphite is that in the manipulation of articles the deposit will get damaged and its effect will therefore be somewhat poorer. Consequently, it can show in a nonuniform depth of the hardened layer.

Phosphate deposits are wear and damage resistant and quite easy to apply to the surface. Consequently, they are considered to be better and more useful. They absorb both the light of the IR spectrum and visible light; therefore these deposits look black as well. When hardening was performed only at a limited area of the workpiece surface, it could be assessed from the rest of the absorption deposit at the boundary between the hardened and unhardened surfaces how the heating process proceeded.

Figure 6.29 shows the results obtained in calorimetric measurements of absorptivity shown by steel with the carbon content of 0.4%, when coated with the respective four deposits, as a function of the traveling speed of the laser beam. The other laser-hardening parameters were a square power density distribution of the laser beam Q of 4.5×10^3 W/cm^2,

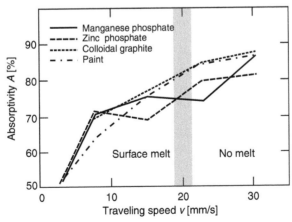

FIGURE 6.29 Influence of different absorbers and traveling speeds on absorptivity. (From Trafford, D.N.H.; Bell, T.; Megaw, J.H.P.C.; Bransden, A.S. heat treatment using a high power laser. *Heat Treatment'79*; The Metal Society: London, 1979; 32–44.)

which allows both transformation hardening and hardening by remelting. A difficulty encountered in hardening by remelting is that temperatures occurring are very high and the absorption deposit will get damaged, which will deteriorate the absorption effect. In the figure there is a hatched area separating low traveling speeds of the laser beam across the workpiece surface that in connection with the other parameters ensures surface hardening by remelting. The boundary zone of the traveling speeds is comparatively wide, i.e., $v = 20 \pm 1.5$ mm/s, which confirms how difficult it is to ensure the repeatability of the depth of hardening in both transformation hardening and hardening by remelting. The zone at the extreme left end of the hatched area corresponds to low traveling speeds that produced temperatures at the surface exceeding the melting point suitable for remelting. The traveling speed can thus be used to control the depth of the remelted and hardened layer. The zone at the extreme right corresponds to the temperatures exceeding the limiting speed with which no remelting but only transformation in the solid state could be obtained. The traveling speed defines the through-depth heating conditions and thus determines the depth of the hardened layer and the size of the transition zone. The size of the transition zone and the variation of hardness in it are extremely important as they permit to fulfil requirements set for dynamically loaded parts.

Inagaki et al. [60] studied various types of deposits in order to increase the depth of the hardened layer and reduce laser-beam energy losses by means of improved absorption in the interaction. Relevant tests were performed with a CO_2 laser with a maximum power of 2.0 kW and the square power density distribution ensuring a uniform depth of the hardened layer after the interaction. A quantitative analysis of absorptivity and the conditions of transformation hardening showed that the most fit-for-purpose absorption deposit was a blend of mica and graphite powder, which provided a 1.8-mm-deep and 10.0-mm-wide single hardened spot. The following substances were used: graphite, mica, and various oxides, e.g., SiO_2, TiO_2, Al_2O_3, which were added to graphite. The binder was acrylic resin. The different ratios between the absorbent and the binder were tested too. A ratio between the contents of "powder" and "binder" was selected among the ratios of 2 : 1, 1 : 1, 1 : 2, 1 : 3, and 1 : 4. The authors focused their studies on a relationship between the size of the graphite particles and its content and the content of binder. The graphite particles were defined by their average size, which amounted to around 0.5 μm, around 4.0 μm, and around 10.0 μm.

With the graphite-oxide type of deposit, the particles were very fine and showed an average size of graphite particles of 0.5 μm and an average size of oxides of 1.0 μm. In this case, the grains being very tiny, the ratio between the content of the two substances and the binder was very limited. Thus only the ratio of 2 : 1 was selected.

Figure 6.30 shows the testing of graphite deposited on SK3 tool steel with a carbon content of 1.1%, taking into account all the above variables in the deposit composition [60]. It turned out that the absorbent having smaller graphite grains had a better effect, while a higher ratio between the graphite particles and the binder had no particular influence and the largest particles even diminished the effect. At the extreme right end the effect of zinc phosphate as absorbent on the depth of the hardened layer is shown. The tests showed that with lower traveling speeds of the laser beam, i.e., with a higher surface temperature, the zinc-phosphate-based deposit was more efficient than the graphite one.

Figure 6.31 shows the effect of the second type of absorbents consisting of graphite particles with an average grain size of 0.5 μm and an oxide with an average grain size of 1.0 μm [60]. The ratio of the two constituents was 1 : 1 and the ratio between the solid substances and the binder 2 : 1. Testing of the absorptivity of laser light was performed with different graphite deposits with oxides or mica deposited on the same steel surfaces. The depth and the area of the obtained hardened layer, respectively, were somewhat greater with graphite than with the mica-based deposit. The absorptivity was somewhat lower with the deposits consisting of different oxides. Testing was performed using SK3 tool steel with a

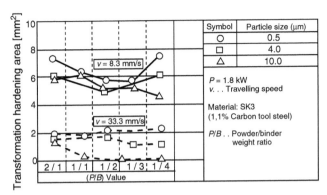

FIGURE 6.30 Effect of graphite particle size and powder/binder weight ratio on the transformation hardening area in cross section. (From Inagaki, M.; Jimbou, R.; Shiono, S. Absorptive surface Coatings for CO_2 Laser Transformation Hardening; *Proceedings of the 3rd International Colloquium on Welding and Melting by Electrons and Laser Beam*, Organized by Le Commisariat a l'Energie Atomique l'Institut de Soudure; Contre, M., Kunceirc, M., Eds., Lyon, France, 1983; Vol. 1, 183–190.)

carbon content of 1%, a laser-beam power of 1.7 kW, and different traveling speeds v of the laser beam, i.e., 33.8 and 5 mm/s.

Figure 6.32 shows a comparison of the depth of a single hardened trace obtained with different surface preparations, i.e., with mica and graphite, the binder, and a zinc-phosphate coating [60]. As this comparison of absorptivity was made using the same tool steel, it can be stated that in laser transformation hardening the best results were obtained with the graphite deposit with the addition of mica.

Grum et al. [61] studied microstructures of various aluminum–silicon alloys after casting and the influence of laser hardening by remelting on the changes in microstructure and hardness of a modified surface layer. Their aim was to monitor changes in the thin, remelted surface layer of a specimen material in the form of thin plates with regard to remelting conditions. The authors gave special attention to the preparation of specimen surface by an absorbent.

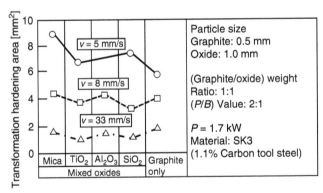

FIGURE 6.31 Effect of adding various oxides to graphite on the transformation hardening area. (From Inagaki, M.; Jimbou, R.; Shiono, S. Absorptive surface Coatings for CO_2 Laser Transformation Hardening; *Proceedings of the 3rd International Colloquium on Welding and Melting by Electrons and Laser Beam*, Organized by Le Commisariat a l'Energie Atomique l'Institut de Soudure; Contre, M., Kunceirc, M., Eds., Lyon, France, 1983; Vol. 1, 183–190.)

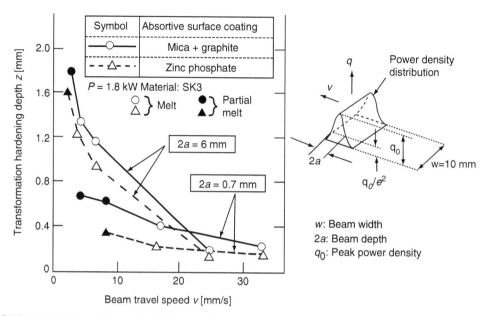

FIGURE 6.32 Effect of beam irradiating conditions on the transformation hardening depth for various absorbers. (From Inagaki, M.; Jimbou, R.; Shiono, S. Absorptive surface coatings for CO_2 Laser Transformation Hardening; *Proceedings of the 3rd International Colloquium on Welding and Melting by Electrons and Laser Beam*, Organized by Le Commisariat a l'Energie Atomique l'Institut de Soudure; Contre, M., Kunceirc, M., Eds., Lyon, France, 1983; Vol. 1, 183–190.)

Aluminum and its alloys absorb laser light poorly; therefore it is needed that the surface of the aluminum alloys be suitably prepared prior to laser remelting. This can be done by surface treatment or by deposition of a suitable absorbent. To this end a high-temperature coating medium based on silicon resin with the addition of metal pigment was applied. It was a sluggish resin. The silicon resin containing pigment was applied to the surface as an about 10-μm-thick layer. The absorbing coating permitted a considerable increase in laser light absorption and energy input, which almost agrees with the calculated values [62–65].

Hardening of the coating medium was carried out at different temperatures, i.e., at the ambient temperature (T_{AM}) and temperatures of 100, 150, 200, and 250°C. In accordance with recommendations of the manufacturer, a single hardening time was selected, i.e., 1 h at the temperature selected. The aim of the different hardening conditions for the thermo-stable resin coating was to study its absorptivity regarding laser light and thus establish the influence on the size of the remelted layer after remelting. The surface appearance seemed not to change after hardening. In accordance with the manufacturer's assurance, the absorbent evaporates during the remelting process and does not affect formation of the remelted layer.

Figure 6.33 shows the influence of hardening temperature of absorbent on the width and depth of the remelted layer, which is different with different alloys. The largest remelted trace is obtained after absorbent hardening at a temperature between 100 and 150°C. With alloys AlSi12 and AlSi8Cu the largest trace is obtained, provided the absorbent has been hardened, at a temperature of around 100°C and with alloys AlSi5 and AlSi12CuNiMg at a hardening temperature of the absorbent of around 150°C.

Gay [66] conducted an investigation on the absorption coefficients obtained with various surface treatment processes, including different absorbents and different fine machining processes.

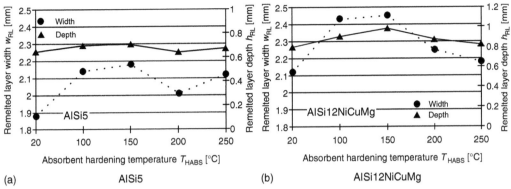

(a) AlSi5   (b) AlSi12NiCuMg

FIGURE 6.33 Size of the remelted layer obtained after laser remelting AlSi5 (a) and AlSi12NiCuMg (b) versus hardening temperature of the absorber in the given remelting conditions: $P = 1500$ W, $\nu_b = 350$ mm/min and $z_s = 9$ mm. (From Grum, J.; Božič, S.; Šturm, R. Measuring and analysis of residual stresses after laser remelting of various alluminium alloys. In Proc. of the 7th Int. Seminar of IFHT, Heat Treatment and Surface Engineering of Light Alloys, Budapest, Hungary; Lendvai, J., Reti, T., Eds.; Hungarian Scientific Society of Mechanical Engineering (GTE); 507–516.)

The investigations were conducted on SAE 1045 heat-treatment steel. An absorption coefficient was determined by the calorimetric method [67,68].

Guangjun et al. [69] studied exacting laser transformation hardening of precision V-slideways. Experiments with conventional hardening and laser hardening were performed. For this purpose four steels were selected, i.e., 20, 45, 6Cr15, and 18Cr2Ni4WA. Table 6.2 refers to all the steels concerned and both heat-treatment methods and gives detailed data on the preliminary heat treatment and execution of surface hardening. Data on the hardness achieved with the two heat-treatment methods are given as well. In order to improve the absorptivity of laser light by the individual steels, carbon ink and manganese phosphate $Mn_3(PO_4)_2$ were used as coatings.

Laser irradiation of metal surfaces is a very complex phenomenon, which is described by the three-dimensional thermal conductivity in the material. There are several equations available for simple calculations of the effects of heat conduction in terms of the temperature obtained inside the hardened layer or the heat-affected zone. It is known from the theory of

TABLE 6.2
Comparison between Laser-Hardened and Various Conventional Heat-Treated Structural Steel Hardness

Material	Laser Hardened	Conventional Heat-Treated Hardness Structure	Heat Treatment prior to Laser Transformation Hardening
Steel C20	547-529 HK, 51–54 HRC	<40–45 HRC	Annealing
Steel C45	712-889 HV, 60–66 HRC	HRC 45–50 (oil-quenched) HRC 52–60 (water quenched)	
Steel 6Cr15	880-939 HK, 66–68 HRC	HRC 64–66 (oil-quenched)	Quenching and tempering
Steel 18Cr2Ni4WA	524-620 HV, 51–56~HRC	HRC 37–39 (air-quenched) HRC 41–42 (oil-quenched)	

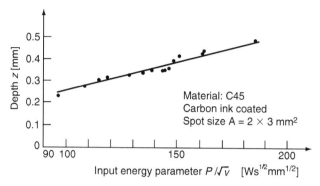

FIGURE 6.34 Dependence of laser beam power and traveling speed on hardened depth of steel C45. (From Guangjun, Z.; Qidun, Y.; Yungkong, W.; Baorong, S. Laser transformation hardening of precision v-slide-way. In *Proceedings of the 3rd International Congress on Heat Treatment of Materials*, Shanghai, 1983. Bell, T., Ed.; The Metals Society London: 1984; 2.9–2.18.)

thermal conductivity that the heating temperature of the irradiated material is, in approximation, dependent on the incident laser-beam power and the irradiation time. Thus it can be presumed that the depth of the hardened layer depends on parameter $P/\sqrt{D_B \nu}$, which is valid only in the case that no melting of the surface occurs. In the equation, P is the power, D_B is the laser-spot diameter at the specimen surface, and ν is the laser-beam traveling speed. The laser-beam diameter is defined as a diameter with which the power is exponentially reduced by a factor of $1/e$. It turned out that the hardness achieved was in direct proportion to $H\alpha P\sqrt{\nu\nu}$.

Figure 6.34 indicates that the strongest influence on surface heating and the temperature variation was exerted by the incident laser-beam power, and a reverse influence by the laser-beam traveling speed [69]. Both parameters of laser transformation hardening affect the changes of microstructure and, consequently, changes in hardness. The dependence shown in the diagram can generally be valid only with the same defocus and the same surface preparation preceding laser hardening. Meijer et al. [70] studied the possibility of hardening smaller workpieces or smaller workpiece surfaces by a CO_2 laser delivering a continuous energy, (CW) of low power in the TEM_{00} mode structure. Two ways of line hardening, i.e., the one with parallel hardened traces and the other with crossing hardened traces including an unhardened space, were selected. Hardening was applied to C45 heat-treatment carbon steel by (1) transformation hardening and (2) hardening by remelting. The authors [70] discussed the behavior of the colloidal graphite as an absorbent from the viewpoint of

- Deposits of different thickness, i.e., ranging between 2 and 14 μm
- Different traveling speeds of the laser beam
- Influence of the absorbent condition, i.e., wet (immediately after deposition), partially dried, and dry

Based on calorimetric measurement a relation between absorptivity and the thickness of the absorbent deposited as well as different traveling speeds of the beam defining surface temperature was determined. Figure 6.35 shows the absorptivity achieved with reference to the deposit thickness and the temperature obtained at the specimen surface [70]. The results obtained in measurements made after line transformation hardening confirmed that the differences in absorptivity were generally amounting up to 20%, that the optimum deposit thickness of colloidal graphite was 8 μm with the higher traveling speeds ν, i.e., 4 or 8 mm/s, and as much as 14 μm with the lower speeds. The optimum absorbent deposit with lasers of

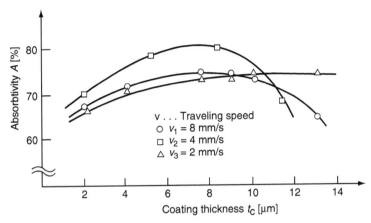

FIGURE 6.35 Absorptivity of a sprayed colloidal graphite coating, measured during line hardening. (From Meijer, J.; Seegers, M.; Vroegop, P.H.; Wes, G.J.W. Line hardening by low-power CO_2 lasers. In *Laser Welding, Machining and Materials Processing*. Proceedings of the International Conference on Applications of Lasers and Electro-Optics "ICALEO'85", San Francisco, 1985; Albright, C., Eds.; Springer-Verlag, Laser Institute of America: Berlin, 1986; 229–238.)

greater powers amounted even up to 50 μm, which meant that the deposit had to be made in several layers, depending on the absorbent used, to obtain the thickness required.

For the determination of the deposit thickness they used the optical method, i.e., by measurement of the light transmitted through the absorbent and glass to a photocell. This was accomplished simultaneously with the deposition of absorbent on the specimen and glass under the assumption that a uniform deposit thickness was obtained. Figure 6.36 shows the results of the transmission of light of a certain intensity (U_0) through the absorbent and glass (U) showing a dependence $U = U_0 e^{-\alpha t}$ [70]. For colloidal graphite, it follows that $1/\alpha = 1.06$ μm, which is valid during or after drying of the deposited absorbent.

Arata et al. [71] state that phosphate coating from zinc-phosphate—$Zn_3(PO_4)_2$—and manganese phosphate—$Mn_3(PO_4)_2$—have outstanding absorptive properties, extreme heat resistance, and good adhesion with a constant thickness of the deposit on the base material. Therefore phosphate coatings are widely used in industry. In testing phosphates in terms of absorption, the optic conditions were chosen so that, using a laser source $P = 1.5$ kW, the spot ran in the direction of the y axis and had a size of $Dy = 3$ mm.

The absorptivity was analyzed in the air and argon atmosphere in different kinematic conditions defined by the traveling speed of the workpiece or the laser beam. The results have confirmed that the absorptivity is almost independent of the atmosphere, which immediately brought about a simplification of the conditions to be maintained during hardening.

The absorptivity dependence on the workpiece traveling speed was more significant. The tests were carried out at traveling speeds of 1 and 8 m/min. The results have confirmed that at lower traveling speeds the absorptivity is less due to heat transfer to the cold workpiece material and also into the surrounding area. When the traveling speed is increased from 1 to 8 m/min, an absorptivity increase from 40 to almost 70% is achieved (Figure 6.37).

Arata et al. [71] also studied the effects of optical conditions on phosphate absorptivity. The starting point for this study was the spot size in the direction of the y axis, which was denoted by Dy. The spot size was changed from 1 to 6 mm for the purpose of studying the absorptivity dependence on different workpiece traveling speeds. Thus at the spot size of $Dy = 6$ mm and traveling speed of 1 m/min, the absorptivity was $A = 65\%$; at the traveling

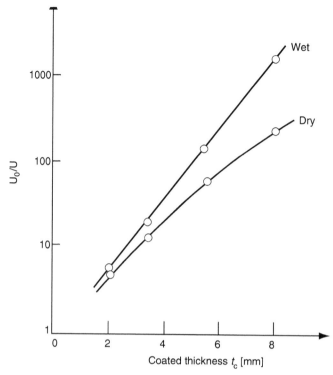

FIGURE 6.36 Light transmittance (inverse) as a measure of coating thickness. (From Meijer, J.; Seegers, M.; Vroegop, P.H.; Wes, G.J.W. Line hardening by low-power CO_2 lasers. In *Laser Welding, Machining and Materials Processing*. Proceedings of the International Conference on Applications of Lasers and Electro-Optics "ICALEO'85", San Francisco, 1985; Albright, C., Eds.; Springer-Verlag, Laser Institute of America: Berlin, 1986; 229–238.)

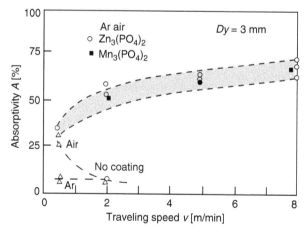

FIGURE 6.37 The absorptivity of specimens coated with zinc and manganese phosphates was measured in air and argon atmospheres. (From Arata, Y.; Inoue, K.; Maruo, H.; Miyamoto, I. Application of laser for material processing—heat flow in laser hardening. In *Plasma, Electron and Laser Beam Technology, Development and Use in Materials Processing*: Arata, Y., Eds.; American Society for Metals: Metals Park, Ohio, 1986; 550–557.)

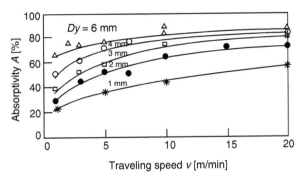

FIGURE 6.38 The spot size was changed from 1 to 6 mm in order to study the absorptivity dependence on different workpiece traveling speeds. (From Arata, Y.; Inoue, K.; Maruo, H.; Miyamoto, I. Application of laser for material processing—heat flow in laser hardening. In *Plasma, Electron and Laser Beam Technology, Development and Use in Materials Processing*: Arata, Y., Eds.; American Society for Metals: Metals Park, Ohio, 1986; 550–557.)

speed of 8 m/min, it was $A = 80\%$ (Figure 6.38). The absorptivity of metal surfaces coated with phosphates is essentially bigger than that of metal surfaces that are only polished.

Research by Grum and Žerovnik [72] has confirmed that laser-beam hardening can be successfully carried out despite the laser source's extremely low power.

The effects they had chosen to study include the following:

- Influences of alloying elements on heat treatment results
- Influences of different kinds of absorbers on heat treatment results
- Influences of energy input on the microstructure and on the magnitude of residual stresses and their nature
- Influences of various laser-beam paths on heat treatment results and the magnitude of residual stresses

Therefore four different types of surfaces prepared in various ways were chosen:

- Surface machined by grinding
- Surface treated with absorber A
- Surface treated with absorber B
- Surface treated with $(Zn_3PO_4)_2$

Absorber A—Miox, a PVK medium is made from polymer base with additions. It was applied by submersion to a thickness of not more than 20 µm, which was achieved by hardening at room temperature.

Absorber B—Melit Email was deposited by submersion, and hardening was carried out at a temperature between 120 and 150°C for 30 to 40 min. This product is made from an alkaline base and alumina resins that have to be air-dried for 10 to 15 min before hardening for the solvents to evaporate and the paint deposit to set. The instructions given by the manufacturer must be strictly observed, and special attention should be paid to the cleanliness of the surface, paint deposition technique, and drying method [72].

The zinc-phosphate absorber—$Zn_3(PO_4)_2$—was deposited on the workpiece surface simply and efficiently. The specimens for laser hardening were first degreased in alcohol or trichloroethylene and then submerged in a zinc phosphate bath at a temperature of 50°C

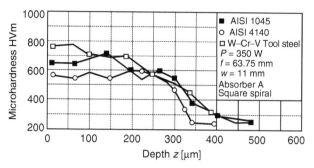

FIGURE 6.39 The microhardness profiles for given steels after using absorber A and a square-spiral laser path. (From Grum, J.; Žerovnik, P. Laser hardening steels, Part 1. Heat treating. Vol. 25–7, July 16–20, 1993.)

(122°F) for 4 to 5 min. This was followed by flushing in hot water and drying. We obtained a uniform and high-quality zinc phosphate coating with good absorptivity.

Grum and Žerovnik [72] show in Figure 6.39 and Figure 6.40 differences in microhardness at a given measurement position for all kinds of steel, using absorber A or B after hardening with a laser beam traveling along a square spiral line.

Figure 6.39 shows the microhardness profiles subsequent to heating with an energy input of the laser source with a power of $P = 350$ W and workpiece traveling speed $v = 1000$ mm/min. The optical conditions are defined by laser-beam diameter ($d = 8$ mm) before the focusing lens, distance to the focus ($f = 63.5$ mm) and by the defocusing degree ($w = 11$ mm). The defocusing degree describes the distance of the lens focus from the workpiece surface. The microhardness was measured in the middle of the hardened track and into the depth until the hardness of the base material is reached. As expected the highest hardness profile is displayed by W–Cr–V tool steel with the highest proportion of carbon and alloying elements having a positive effect on hardenability or hardness. The lowest hardness on the surface, ranging between 500 and 600 HVm, was achieved on AISI 4142 steel with the shallowest hardened track. This surprising finding points to the disturbances in heating, which may have been caused by the laser source or the poor quality of the absorber deposit. The steels were chosen carefully to permit the study on the effects of alloying elements in terms of their kind and amount. According to heat treatment instructions, the chosen steels have equal hardening temperatures; therefore the effects due to these differences might be excluded. Heat conduct-

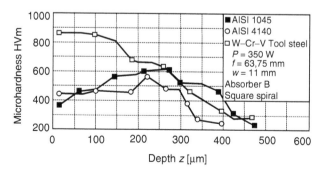

FIGURE 6.40 The microhardness profiles for steels after using absorber B and a square-spiral laser path. (From Grum, J.; Žerovnik, P. Laser hardening steels, Part 1. Heat treating. Vol. 25–7, July 16–20, 1993.)

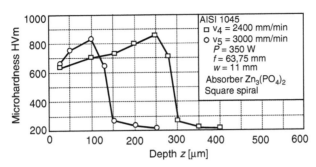

FIGURE 6.41 The microhardness for AISI · 1045 after using a zinc-phosphate absorber and a square-spiral laser path.

ivity might have exerted a certain influence on the surface as well as on the differences in the depth and width of the hardened track.

Figure 6.41 shows the microhardness profile for different hardness track depths on AISI 1045 steel using the zinc-phosphate absorber, with the beam traveling via a square spiraling path. The workpiece traveling speeds were considerably higher than when absorber A or B was used. These tests have shown that the workpiece traveling speed when using zinc phosphate was up to three times higher than those with absorber A or B.

Gutu et al. [73] studied transformation hardening of large gears made of alloyed 34MoCrNi15 steel and two heat-treatment steels, OLC45 and OLC60, respectively. The absorbents used were carbon black and colloidal graphite. The effects of the absorbents used were monitored by measuring the depth and the through-depth hardness of the relevant hardened layers.

Ursu et al. [74] studied the influence of the irradiation of a copper surface by CO_2 laser light. During irradiation, the temperature of the specimen was measured with a thermocouple positioned at the back side of the specimen. The irradiation test was performed twice on the same specimen. Prior to and after the irradiation of the same specimen, the absorptivity, A_0 and A_1, was determined and analyses of the surface with transmission electron microscope (TEM) and scanning electron microscopes (SEMs) were made. Some zones at the specimen were chosen for crystallographic analysis by x-ray diffraction as well.

6.10   LASER-BEAM HANDLING TECHNIQUES

At laser transformation hardening the laser beam passes over the surface so fast that melting cannot occur [75]. Traveling speed or scan speed is one of the important laser transformation hardening parameters. There are two scanning methods:

- Linear traverse using a defocused beam and
- Transversely oscillating a focused beam and moving to the direction of traverse.

When the laser beam is defocused to a spot dimension according to the surface of the workpiece, a compromise between hardened depth and traveling speed at surface hardening has to be achieved [75]. Figure 6.42a shows the general shapes of the cross section of the hardened tracks produced in the metal by this method. The center of the hardened track represents the deepest area at the cross section for laser beam with Gaussian energy distribution. To increase the width of the heat-treated single track, multiple overlapping tracks may be used. However, as shown in Figure 6.42b this method still does not give uniform

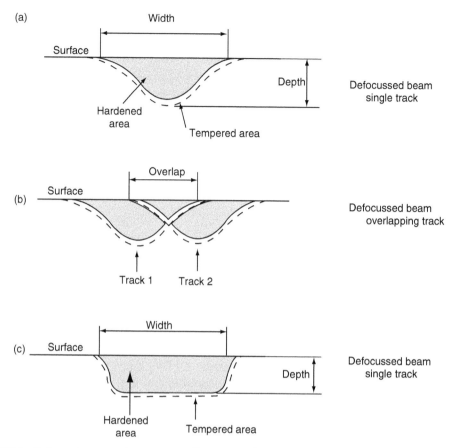

FIGURE 6.42 Laser heat treating patterns. (From Engel, S.L. Basics of laser heat treating. In *Source Book on Applications of the Laser in Metalworking. A Comprehensive Collection of Outstanding Articles from the Periodical and Reference Literature*; Metzbower, E.A. Ed.; American Society for Metals: Metals Park, Ohio, 1979; 49–171.)

hardened depth and interface soft streaks named heat-affected zone (HAZ) may develop at the overlap region [75]. Figure 42°c shows the desirable heat-treated shape [75]. The hardened depth is uniform where the surface width represents the width of the hardened track and the HAZ is comparatively small.

The method of oscillating a laser beam requires more hardware, the added expense, and the increase in the number of process parameters. Figure 6.43 illustrates the general principle of the laser-beam oscillating method [75]. In the simplest case, the beam is oscillated transverse to the direction of part or beam traverse, as shown in the right side of the figure. The temperature time diagram is the combination of the heating and cooling cycle as superimposed on transformation curve. The laser beam is focused to produce power densities of at least 10^5 W/in.^{2}. This power density assures high heating rate or fast temperature rise times. The oscillating frequency and traveling speed are selected so that the temperature of a given laser-beam spot on the surface varies between the melting temperature (T_m) and the austenitic transformation temperature (T_{A_3}). This condition is allowed to exist until the required volume of metal below the surface reaches the transformation temperature. At transformation temperature the bulk of the material provides the necessary heat sinking to transform austenite to a hardened matrix of martensite. Uniform hardened depth is achieved when

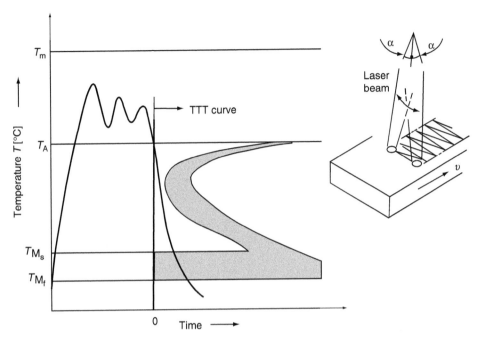

FIGURE 6.43 General concept of laser heat treating by the use of oscillating laser beam. (From Engel, S.L. Basics of laser heat treating. In *Source Book on Applications of the Laser in Metalworking. A Comprehensive Collection of Outstanding Articles from the Periodical and Reference Literature*; Metzbower, E.A. Ed.; American Society for Metals: Metals Park, Ohio, 1979; 49–171.)

the beam oscillates so fast that the metal, because of its relatively slow conductivity, sees the oscillating tracks as a solid line of energy. Appropriate overlapping of the single tracks will produce a minimum of overtempered zone in the hardened depth, as shown in Figure 6.42c [75].

If the oscillating frequency is too slow at transformation surface hardening for a given laser power density, the surface temperature will be driven into the melting region [75]. If the power density is too low for a given frequency, the material will not be heated up to its transformation temperature and no hardening will be produced. If the beam is allowed to oscillate too long in a given area in an attempt to increase hardened depth, the temperature in the surface layer will be raised to the point where efficient self-quenching no longer takes place. The cooling curve will then cross the "nose" of the transformation curve resulting in the transformation of austenite into pearlite (annealed, soft microstructure) rather than martensite.

6.11   PRESENTATION OF CO_2 LASER MACHINING SYSTEMS

Considering numerous practical applications of laser in electrical, metal working, and chemical industry, we can see that laser technology offers several advantages over the previously known technologies. Let us just mention one of the special plus factors which is contactless machining where there is no tool in the conventional sense but just the laser beam that never wears off. Thus frequent problems caused by tool wear such as workpiece deformation, surface layer damage, and damage of vital machine tool elements have all been avoided. However, there are other requirements that have to be considered in laser machining, as for example the quality of the laser source, modes of guidance of the workpiece or laser beam. All this requires interdisciplinary research and development work if we want to meet the needs of end users. The high investments required for machining systems have to be justified by high-quality products, increased productivity and possibilities of integration into manufacturing cells. The need to integrate laser machining systems into a manufacturing cell requires further development of mechatronic and electronic systems which would, together with computer and systems engineering, enable a more intensive introduction of this technology into the factories of the future. Considering the difficulties encountered in microdrilling, cutting, welding, and heat treatment, laser technology is the only technology that can, in the near future, as part of an integrated system, provide all sorts of manufacturing possibilities: from the blank to the end product. Laser technology will no doubt play an important role in the development of the factory of the future. This has already been recognized by a number of leading producers, e.g., in car industry, where it is impossible to imagine the assembly of car body parts without laser cutting and welding [76,77].

Up to now the development of laser technology has dealt with the following technological innovations:

- Guidance of the laser beam to different working locations in the space
- Guidance of the laser beam over larger distances
- Possibility of dividing the laser beam into several working places
- Timing of the laser beam
- Kinematic conditions within the workpiece laser–beam relation

These technological innovations offer extreme possibilities of adaptation and flexibility and allow machining of geometrical elements that previously could not be machined by conventional technologies and processes. Among the machining advantages we can mention:

- Possibility of welding thin parts
- Possibility of drilling small holes
- Adaptation of heat conditions to parts geometry in heat treatment

In addition to that, the high flexibility of laser technologies is supported also by high machining capabilities of the laser beam on a variety of materials with very different properties, that is:

- Possibilities of machining plastics
- Possibilities of machining ceramics
- Possibilities of machining metal parts
- Possibilities of machining different composite materials

Figure 6.44 shows the main components of a CO_2 laser machining system manufactured by Toshiba, type LAC 554 with a power of 3 kW [76]. By guiding the laser beam and by guiding the workpiece in x–y plane we can achieve a great variety of machining by cutting and produce very different geometric elements. The maximum length of the laser beam is 18 m and is led through an optic system of prisms and mirrors along the longitudinal and traverse direction. A given configuration of optic and kinematic systems allows machining to go on only on one place. Previous laser machining systems only allowed the possibility of moving the operating table or the workpiece, which also required very large working areas, higher investment, and resulted in lower product quality. Technologies based on "time sharing method" were also known allowing the possibility of guiding the laser beam to different working places where activities are going on (Figure 6.45) [77].

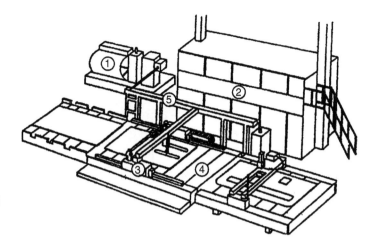

LASER TYPE LAC 554

1. Laser oscillator
2. System monitoring room
   laser control board
   NC equipment for laser
   system console
3. Beam scanner
4. Work transporting conveyor
5. Beam transmission route (18m)

FIGURE 6.44 A Toshiba laser machining system with a laser source power of 3 kW. (From Toshiba CO_2 Laser Machining System; Toshiba Corp. Shiyodo-ku: Tokyo, 18 pp.)

Figure 6.46 shows the production line equipped with transport belts supplying parts with a device for prepositioning and positioning within the reach of the laser beam [78]. This is followed by removal, deposition, and further transport on the belt until the next station. The whole equipment enables continuous running of the machining process. In Figure 6.46, we can see a laser machining system illustrating a special aspect of guidance of the laser beam through a prism to the working place. The working place rotates, in this way enabling that the work is done in sequences, e.g., inserting, laser machining, cooling, taking off, and deposition on the conveyer belt. To achieve a higher utilization rate of the laser system and increase of productivity, it is possible to introduce another working place by redirecting the laser beam.

Another technological change is introducing robots for guiding the laser beam. New types of devices for dividing/guiding the laser beam have been developed known as "Robolasers." Type 1 in Figure 6.47 illustrates a robot for feeding the workpiece/product under a fixed laser

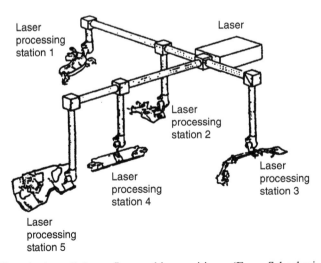

Laser
processing
station 1

Laser

Laser
processing
station 2

Laser
processing
station 3

Laser
processing
station 4

Laser
processing
station 5

FIGURE 6.45 Guiding the laser light to five working positions. (From Schachrei, A.; Casbellani, M. Application of high power lasers in manufacturing. Keynote papers. Ann. CIRP 1979, *28*, 457–471.)

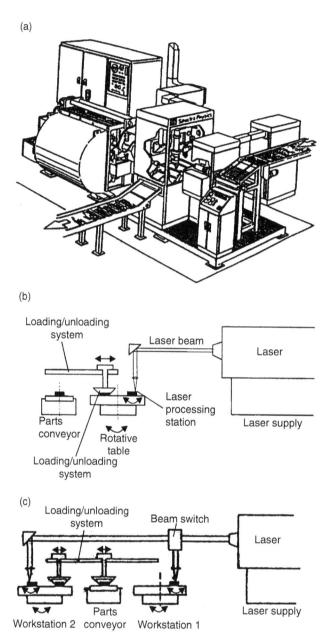

FIGURE 6.46 (a) Laser machining system with components for guiding the laser beam and the workpiece. (b) Workstation with loading/unloading system and laser processing station. (c) Laser machining system and parts handling/feeding double workstation. (Spectra Physics; From Carroz, J. Laser in high rate industrial production automated systems and laser robotics. In *Laser in Manufacturing*, Proceedings of the 3rd International Conference, Paris; IFS: France, Bedford, 1986, Quenzer, A., Springer-Verlag, Berlin, 1986; 345–354.)

beam [78]. In order to achieve a constant focal distance from the workpiece surface, the laser system has to have a suitable degree of freedom with a corresponding software provided by mechatronic and electronic systems. Type 2 (Figure 6.47) shows a robotized system which separately guides the laser beam and the nozzle for guiding the assistant gas. The third type,

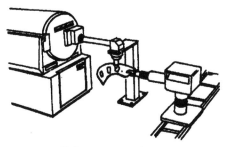

Type 1: Robot moves part
under fixed beam

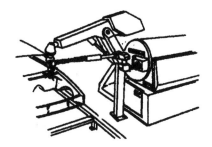

Type 2: Robot moves beam
delivery system

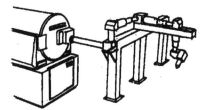

Type 3: Robot has built-in
beam delivery system

FIGURE 6.47 Three types of robolasers (laser + robot). (From Carroz, J. Laser in high rate industrial production automated systems and laser robotics. In *Laser in Manufacturing*, Proceedings of the 3rd International Conference, Paris; IFS: France, Bedford, 1986, Quenzer, A., Springer-Verlag, Berlin, 1986; 345–354.)

by using a special design, ensures a coordinate motion of the laser beam and auxiliary gas to the working place.

In Figure 6.48 we can see a system for guiding the laser beam, manufactured by Spectra-Physics, having the base plate fixed at the exit of the system/laser beam [78]. The laser beam

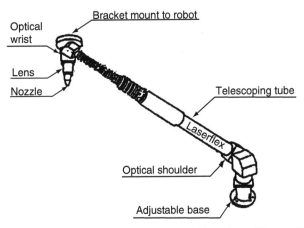

FIGURE 6.48 Laserflex for guiding the laser beam to the working place. (Spectra Physics; From Carroz, J. Laser in high rate industrial production automated systems and laser robotics. In *Laser in Manufacturing*, Proceedings of the 3rd International Conference, Paris; IFS: France, Bedford, 1986, Quenzer, A., Springer-Verlag, Berlin, 1986; 345–354.)

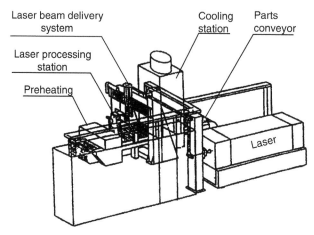

FIGURE 6.49 Laser system for line-surface hardening of cam-shafts. (From Mordike, S.; Puel, D.R.; Szengel, H. Laser over-flächenbehandlung - ein Productionsreifes Verfahren für Vielfältige Anwendungen. In *New Technologies in Heat Treating of Metals*; Croatian Society for Heat Treatment: Zagreb, Croatia, Liščić, B., Ed.; 1–12.)

travels through the optic shoulder to the optic joint and further to the focusing lens and goes out through the symmetric nozzle. On the upper side of the robot there is a gas tube through which the assistant gas is led from the gas station. The whole system is highly upgraded and flexible, enabling a very successful machining on very complex products as well as on large products. Because it is so highly flexible the system got the name Laserflex.

Figure 6.49 shows a laser machining system enabling a line-surface hardening of cam-shafts for the car industry [79]. Because of problems with absorptivity, the laser system was added in place of induction heating or pre- heating of parts on to the temperature of 400°C. This is followed by heating up with the laser beam to the hardening temperature. After hardening follows the cooling down of parts in the cooling chamber and their deposition on the conveyer belt leading to the next working place.

Figure 6.50 shows a line-beam of laser light, obtained from a system of optic elements such as a prism and a spherical, concave mirror [79]. A convex mirror ensures a constant distance of the optic system focus from the object surface/camshaft. The entire operation of surface heat treatment takes 2 to 5 s for one camshaft assembly and at the maximum 29 s for one camshaft depending on the size and kind of the latter.

A most important place in building laser machining systems with computer support is now given to mechatronic and electronic systems as the quality of the whole system depends on them [80]. Modern highly flexible laser technology represents a universal tool suitable for small and large products of simple or complex shape. Another fact which strongly favors this new technology is the possibility of guiding the laser beam to different, even very distant working places. Finally, using the same tool but in different working conditions it is possible to carry out different heat processes from heat treatment to welding. Thus we can establish that the advent of laser technology will finally solve the problems of conventional heat treatment and welding and will ease the transition to the manufacturing systems structure necessary for the factory in the future [81–83].

6.11.1 POSSIBILITIES OF KALEIDOSCOPE USE FOR LOW-POWER LASERS

Laser surface hardening is desirable to utilize the available energy distribution for heating and microstructural changes; however, the Gaussian distribution with large differences in energy

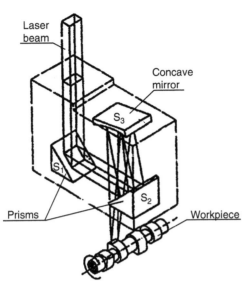

Laser beam

Concave mirror

S₃

Prisms

S₁

S₂

Workpiece

FIGURE 6.50 Optic system for redirecting the laser beam to the working place. (From Mordike, S.; Puel, D.R.; Szengel, H. Laser over-flächenbehandlung - ein Productionsreifes Verfahren für Vielfältige Anwendungen. In *New Technologies in Heat Treating of Metals*; Croatian Society for Heat Treatment: Zagreb, Croatia, Liščić, B., Ed.; 1–12.)

distribution across the laser beam does not provide this possibility. In practice, therefore, we tend to use other laser-beam energy distributions, namely, the square top-hat and rectangular. A very simple device, kaleidoscope, is therefore suitable for practical use. A kaleidoscope is a device in the form of a chimney that is a hollow body with a square or rectangular cross section [84,85]. It is necessary for the beam to reflect at least two or three times along the length of the kaleidoscope. To achieve this high reflectivity, a material of high heat conductance and mirror-smoothed material should be used. Reflectivity depends on the wavelength of the laser light and the material from which the kaleidoscope is made [86].

To assess the uniformity of beam intensity, the relative change in beam power density is considered irrespective of the effect of interference. Figure 6.51a shows the influence of kaleidoscope length and cross section on the relative change in beam intensity [84]. Figure 6.51b [84] shows the uniformity of beam power density distribution for different kaleidoscope lengths.

It is evident from the figure that with the increase in kaleidoscope length l and decrease in cross-dimensions "axa" or "axb," the uniformity of the beam increases. In the case of the rectangular cross section, it then follows that the kaleidoscope length must be chosen according to the magnitude of the longer side, which means larger length of the kaleidoscope than in the case of the square cross section. The size of the kaleidoscope depends also on the focal length of the lens at its entrance side. Figure 6.52 shows the dependence of focal length of the entrance lens and the kaleidoscope dimensions $1/a$ on the uniformity of the laser-beam power density at the exit side. From the focal length of the entrance lens, the kaleidoscope dimensions can therefore be determined and this also ensures the highest uniformity of the beam power density at the exit side.

In the experiments, a longer than the minimum length necessary for the square cross section was chosen in order to enable the use of the same kaleidoscope for both types of cross section [87]. A kaleidoscope with a length of 75 mm was applied, which is more than the recommended length of the rectangular cross section (Figure 6.53). This proved disadvantageous later as the interference raster was not distinct. The too long kaleidoscope for the

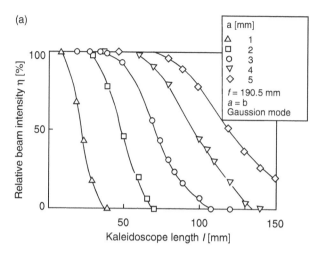

(a)

(b)

FIGURE 6.51 (a) Influence of kaleidoscope length and cross section on the relative change in beam intensity. (b) Power density distribution of laser beam versus kaleidoscope length. (From Shono, S.; Ishide, T.; Mega, M. Uniforming of Laser Beam Distribution and Its Application to Surface Treatment, Takasago Research & Development Center, Mitsubishi Heavy Industries, Ltd, Japan; Institute of Welding; IIW-DOC-IV-450-88, 1988; 1–17.)

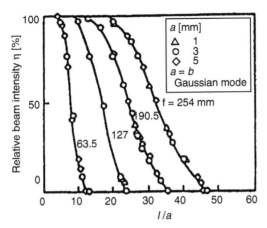

FIGURE 6.52 Influence of local length of entrance lens on the relative change in beam power density. (From Shono, S.; Ishide, T.; Mega, M. Uniforming of Laser Beam Distribution and Its Application to Surface Treatment, Takasago Research & Development Center, Mitsubishi Heavy Industries, Ltd, Japan; Institute of Welding; IIW-DOC-IV-450-88, 1988; 1–17.)

square cross section of 3×3 mm^2 enabled an interference of most of the elementary laser light beams despite the fact that the optical axes of the laser system and the kaleidoscope may not have fitted due to the assembly error.

On the upper side of the kaleidoscope with a cross section of 4×2 mm^2, an opening was made for the laser light to enable the best possible interference of elementary laser beams. The kaleidoscope consists of perpendicular copper walls, polished to a high gloss, thus enabling the creation of mirroring surfaces reflecting laser light. The walls are fastened to one another by bolts, which enables dismounting of the kaleidoscope and repeated polishing.

In the experimental part of the research, Grum [87] tried to choose such laser surface hardening conditions that may have shown the advantages of using a kaleidoscope in laser

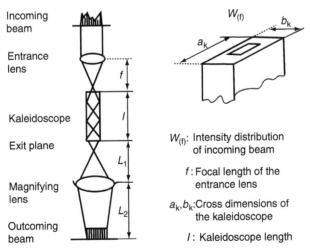

FIGURE 6.53 Kaleidoscope structure. (From Shono, S.; Ishide, T.; Mega, M. Uniforming of Laser Beam Distribution and Its Application to Surface Treatment, Takasago Research & Development Center, Mitsubishi Heavy Industries, Ltd, Japan; Institute of Welding; IIW-DOC-IV-450-88, 1988; 1–17.)

heat treatment with a low-power source. The main feature of the kaleidoscope is that it considerably lowers the traveling of the workpiece, still granting the desired uniformity of heating and also uniformity of hardening depth.

Figure 6.54 presents the hardened tracks and the measured microhardness on heat-treatable carbon steel AISI 1042. The hardening was carried out with the aid of a kaleidoscope with a defocusing degree defined by the size of the bright spot on the workpiece surface, which was 1.65 mm. The workpiece traveling speeds changed in the same way as those in the treatment of tool steel, i.e., 0.10, 0.12, and 0.14 m/min [87].

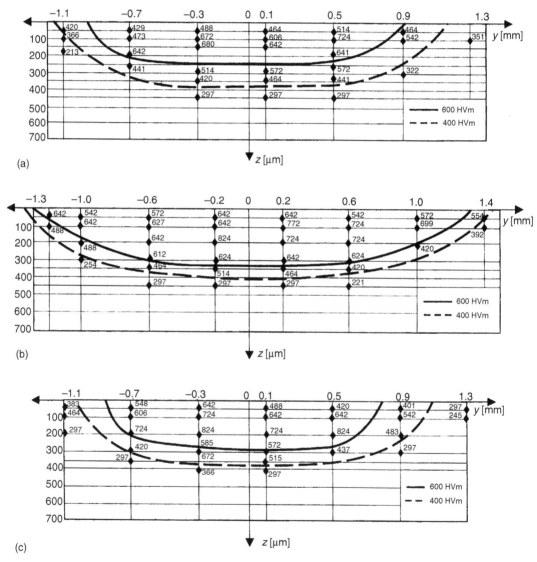

FIGURE 6.54 Measured hardnesses in laser surface transformation hardening 1042 AISI (0.46% C, 0.65% Mn) steel, using a kaleidoscope, traveling speed $\nu = 0.10$ m/min (a), $\nu = 0.12$ m/min (b), $\nu = 0.14$ m/min (c), and beam dimension $a = b = 1.65$ mm. (From Grum, J. Possibilities of kaleidoscope use for low power lasers. In *Conference Proceedings Heat Treating: Equipment and Processes*; Totten, G.E., Wallis, R.A., Eds.; ASM International: Materials Park, Ohio, 1994; 265–274.)

6.12 RESIDUAL STRESS AFTER LASER SURFACE HARDENING

From the technology viewpoint, various surface hardening processes are very much alike as they all have to ensure adequate energy input and the case depth required. In the same manner, regardless of the hardening process applied, in the same steel, the same microstructure changes, very similar microhardness variations, and similar variations of residual stresses within the hardened surface layer may be achieved.

Normally, the surface transformation hardening process also introduces compressive stresses into the surface layers, leading to an improvement in fatigue properties. Hardened parts—always ground because of their high hardness—require a minimum level of final grinding. This can only be achieved by a minimum oversize of a machine part after hardening, thus shortening the final grinding time and reducing cost of the final grinding to a minimum. With an automated manufacturing cell, one should be very careful when selecting individual machining processes as well as machining conditions related to them. Ensurance of required internal stresses in a workpiece during individual machining processes should be a basic criterion of such a selection. In those cases when the internal stresses in the workpiece during the machining process exceed the yield stress, the operation results in workpiece distortion and residual stresses. The workpiece distortion, in turn, results in more aggressive removal of the material by grinding as well as a longer grinding time, higher machining costs, and a less-controlled residual stress condition. The workpiece distortion may be reduced by subsequent straightening, i.e., by material plasticizing, which, however, requires an additional technological operation, including appropriate machines. This solution is thus suited only to exceptional cases when a particular machining process produces the workpiece distortion regardless of the machining conditions. In such cases the sole solution seems to be a change of shape and product dimensions so that material plasticizing in the machining process may be prevented.

It is characteristic of laser surface hardening that machine parts show comparatively high compressive surface residual stresses due to a lower density of the martensitic surface layer. The compressive stresses in the surface layer act as a prestress which increases the load capacity of the machine part and prevents crack formation or propagation at the surface. The machine parts treated in this way are suited for the most exacting thermomechanical loads as their susceptibility to material fatigue is considerably lower. Consequently, much longer operation life of the parts can be expected.

Residual stresses are the stresses present in a material or machine part when there is no external force and/or external moment acting upon it. The residual stresses in metallic machine parts have attracted the attention of technicians and engineers only after manufacturing processes improved to the level at which the accuracy of manufacture exceeded the size of deformation, i.e., distortion, of a machine part.

The surface and subsurface layer conditions of the most exacting machine parts, however, are monitored increasingly by means of the so-called surface integrity. This is a scientific discipline providing an integral assessment of the surface and surface layers and defined at the beginning of the 1960s. More detailed information on the levels of surface integrity description may be obtained [88,89].

6.13 THERMAL AND TRANSFORMATION RESIDUAL STRESSES

Yang and Na[90] proposed the model for determination of thermal and transformation residual stresses after laser surface transformation hardening. Thermal stresses were induced due to thermal gradient and martensitic phase transformation causing prevailing residual stresses. The dimensions of the specimen for simulation were 6 mm in length, 2 mm. in

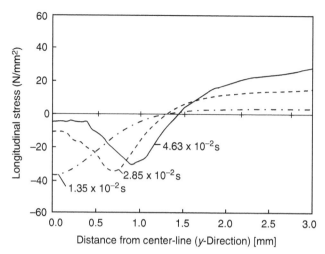

FIGURE 6.55 Longitudinal stress distributions at the surface during laser heating. (From Yang, Y.S.; Na, S.J. *Proc. Inst. Mech. Eng.* 1990, *204*, 167–173.)

thickness, and the width of the sliced domain was 0.2 mm. To carry out heat treatment laser power $P = 1$ kW, traveling speed $v = 25$ mm/s, absorptivity $A = 0.15$, and characteristic radius of heat flux $r = 1$ mm were chosen. At rapid heating austenitic transformation temperature started at 830°C and ended at 950°C. Cooling rate was also high enough in the heated material and austenite transformed to martensite in the temperature range from 360 to 140°C. In the simulation the volume changes at phase transformation base microstructure to austenite at heating ($\alpha_{F+P \rightarrow A} = 2.8 \times 10^3$) and then also volume changes at phase transformation of austenite to martensite at cooling ($\alpha_{A \rightarrow M} = 8.5 \times 10^{-3}$) were considered. Figure 6.55 shows longitudinal stress profiles in the direction of the laser-beam travel at various time moment at heating, namely, at $t_1 = 1.35 \times 10^{-2}$ s, $t_2 = 2.85 \times 10^{-2}$ s, and $t_3 = 4.65 \times 10^2$ s [90]. From the graph it can be seen that the compressive stresses are in immediate vicinity of the laser beam. Nevertheless, the thermal stresses at heating change to tensile stresses according to the distance from the laser beam.

As the laser provides more energy according to the heating time, the temperature increases, so that the compressive thermal stresses become progressively lower in the vicinity of the laser-beam center.

Figure 6.56 shows longitudinal stress profiles in the direction of the laser-beam travel at various cooling times during laser surface hardening [90]. The residual stress in laser surface hardening is induced from temperature gradient and phase transformations. During cooling process longitudinal stresses are always tensile and change to compressive during phase transformation. This means that longitudinal residual stress is always of compressive nature. As shown in Figure 6.56 the compressive residual stress was generated in the hardened depth within 0.5 mm from the center line of the exposed laser beam.

6.14 MATHEMATICAL MODEL FOR CALCULATING RESIDUAL STRESSES

Li and Easterling [91] developed a simple analytical model for calculating the residual stresses at laser transformation hardening. Following the temperature cycle of the laser beam travel, a certain volume of the heat-affected material expands as a result of the martensitic transform-

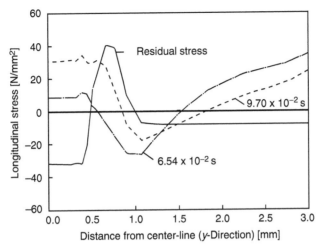

FIGURE 6.56 Longitudinal stress distributions and residual stress distributions at the surface during cooling in laser surface hardening. (From Yang, Y.S.; Na, S.J. *Proc. Inst. Mech. Eng.* 1990, *204*, 167–173.)

ation. The created martensite microstructure has higher specific volume than matrix microstructure, which causes residual stresses in the surface layer.

The magnitude of this residual stress is calculated as a function of the laser input energy and the carbon content of the steel.

To simplify the calculations, the following assumptions can be made:

- Stresses induced by thermal expansion near the surface can be neglected.
- Plastic strains caused by the martensitic transformation can be considered negligible.
- Dilatation in the HAZ is completely restrained along the x and y directions during laser transformation hardening.

From the author's equations, different factors and parameters are influential on residual stress calculations, despite the simplifying assumptions made. These factors include:

- The dimensions of the specimen
- The properties of the material and composition
- The laser processing variables
- The residual stress σ_{xx} should equal

$$\sigma_{xx} = -\frac{\beta f E}{1-\nu} + \frac{1}{bc}\int_0^c dz \int_{-b/2}^{+b/2} \frac{\beta f E}{1-\nu}\,dy + \frac{12(0.5c-z)}{bc^3}\int_0^c dz \int_{-b/2}^{+b/2} \frac{\beta f E}{1-\nu}(0.5c-z)\,dy$$

This equation is a complete description of the residual stress in the x direction, where the first term is negative and decreases with depth below the surface, and the second and third terms are both positive. It shows that these terms resulted in a compressive residual stress at the surface and changed to a tensile residual stress at some depth below the surface.

The y component of the residual stress σ_{yy} can be expressed as

$$\sigma_{yy} = -\frac{\beta f E}{1-\nu}$$

As several laser runs overlap in the y direction there is some stress relief in the overlapping zone. The maximum σ_{yy} given by the equation is still valid.

Assuming that expansion in the z direction proceeds freely, the z component of residual stress σ_{zz} is zero

$$\sigma_{zz} = 0$$

It is impossible to change the volume because of phase transformation, therefore shearing strains do not occur, and the shear components of residual stresses are zero

$$\sigma_{xy} = \sigma_{yz} = \sigma_{zx} = 0.$$

The given equations present a simplified way of calculating residual stresses at laser transformation hardening.

Figure 6.57 shows the effect of carbon content on σ_{xx} through depth residual stress profile [91].

It can be seen that the value of the compressive residual stress is very dependent upon carbon content, increasing by about 200 N/mm^2 for each 0.1% C. The crossover from compressive to tensile residual stress in the x direction occurs approximately at the depth where the steel has been heated to the T_{A_1} temperature. The figure shows also the profile for σ_{yy} in the steel with 0.44% C. As can be seen σ_{yy} lies at slightly higher compressive stresses than the corresponding profile for the x direction. The effect of changing the input energy density of the laser process on residual stress profile σ_{xx} is shown in Figure 6.58. It can be seen that, for the 0.44% C steel, changes in laser input energy do not substantially affect the value of the compressive residual stress but do affect the depth d_0 of the compressive layer. Figure 6.59 shows changes in compressive residual stress σ_{xx} and depth of compression laser d_0 as a function of laser input energy density. In practical laser transformation hardening, a surface

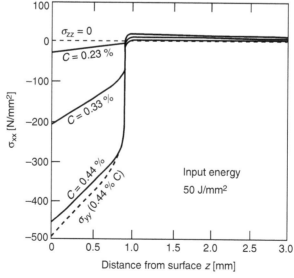

FIGURE 6.57 Predicted residual stress distribution σ_{xx} as a function of depth below surface and steel carbon content C. (From Li, W.B.; Easterling, K.E. *Surf. Eng.* 1986, 2, 43–48.)

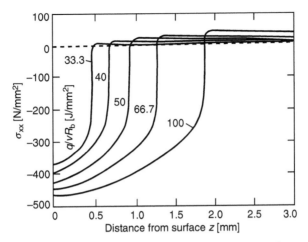

FIGURE 6.58 Predicted residual stress σ_{xx} as a function of depth below surface and laser input energy density q/vR_b for 0.44% C steel. (From Li, W.B.; Easterling, K.E. *Surf. Eng.* 1986, *2*, 43–48.)

may be hardened not by a single laser track but by a number of overlapping tracks. This obviously results in local stress relaxation in the y directions in the overlapping zones, although the residual stresses calculated at the center of each laser track will not be affected and remain as presented above. If the overlapping tracks cover the entire surface, end effects, as discussed in conjunction with σ_{xx}, should be included in the calculation of σ_{yy} for completeness.

6.15 INFLUENCE LASER SURFACE TRANSFORMATION HARDENING PARAMETERS ON RESIDUAL STRESSES

Solina [92] presented residual stress distribution below the surface after laser surface transformation hardening at various heating and cooling conditions of different steels.

For experimental laser surface transformation hardening they used a CO_2 laser. The absorptivity of the material for the laser light with a wavelength of $\lambda = 10.6$ μm was increased by an absorption coating of Zinc phosphate. In the tests two types of steel were used, i.e.,

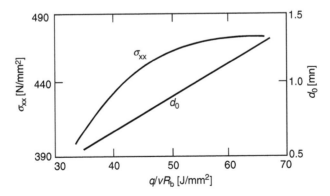

FIGURE 6.59 Predictions of surface compressive residual stress σ_{xx} and depth d_0 as a function of laser input energy density for 0.44% C steel. (From Li, W.B.; Easterling, K.E. *Surf. Eng.* 1986, *2*, 43–48.)

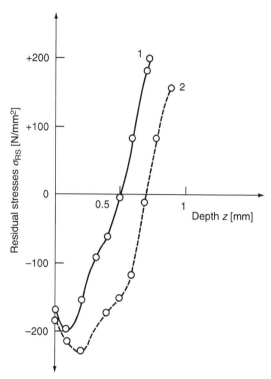

FIGURE 6.60 Residual stress distribution below the surface for given specimens: 1: C43, $P = 27.5$ W/mm^2, $t_i = 0.56$ s, cooling mode 1.2: 40CoMo4, $P = 27.5$ W/mm^2, $t_i = 0.56$ s, cooling mode 1. (From Li, W.B.; Easterling, K.E. *Surf. Eng.* 1986, 2, 43–48.)

carbon steel AISI 1045 and a low-alloy steel AISI 4140, both in normalized state. The specimen was self-cooled (mode 1) and quenched with liquid jet (mode 2).

The ranges for laser hardening parameters were 10 to 38 W/mm^2 s for the energy density and from 0.024 to 1.12 s for the beam–metal interaction time. From residual stress profiles in the surface layer, it is found that the residual stresses in the more superficial layers may vary, with changing laser hardening conditions, from compressive residual stress values $+200$ MPa to strong tensile residual stress values up to $+800$ MPa.

Figure 6.60 shows the residual stress profiles as a function of the depth below the surface in two specimens of carbon steel AISI 1045 (C43) and AISI 4140 (40CoMo4) alloyed steel [92]. The specimens were heated with the same power density $Q = 27.5$ W/mm^2 and at interaction time $t_i = 0.56$ s at beam spot area 14×14 mm^2 and self-quenched (mode 1). The residual stress profiles for both steels show the same laser surface hardening results in very similar changing residual stresses from the hardened layer. Maximum values of the compressive residual stress are about -200 N/mm^2, and very similarly residual stress droops to neutral plane with small shift for alloyed steel:

Figure 6.61 shows residual stress profiles through the surface layer after laser surface hardening of alloyed steel AISI 4140 at very high power density $Q_1 = 121.9$ W/mm^2, at interaction time $t_{i1} = 0.16$ s and $Q_2 = 418.7$ W/mm^2, at interaction time $t_{i2} = 0.024$ s, with cooling in liquid jet (cooling mode 2) [92]. In both laser surface conditions very drastic tensile residual stress profiles are obtained. Residual stress profiles differ only in absolute values. At laser surface hardening conditions for both steels tensile residual stresses were found on the surface layer from $+200$ to $+450$ N/mm^2.

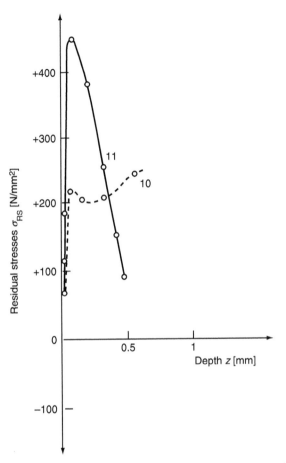

FIGURE 6.61 Residual stress distribution below the surface for given specimens: 10: 40CoMo4, $P = 121.9$ W/mm², $t_i = 0.16$ sec, cooling mode 2. 11: 40CoMo4, $P = 418.7$ W/mm², $t_i = 0.024$ s, cooling mode 2. (From Solina, A. *J. Heat Treat.* 1984, *3*, 193–203.)

Com-Nogue and Kerrand [93] analyzed the through-depth variations of residual stresses and microhardness of a laser-hardened layer on S2 chromium steel. Heating was carried out by a laser beam with a power P of 2 kW and a traveling speed v of 7 mm/s. A semielliptical power density distribution in the laser beam with defocusing was chosen so that a constant depth of the hardened layer d, i.e., 0.57 mm, was obtained with a width W of 10.8 mm. Figure 6.62 shows the through-depth variations of the measured hardness and experimentally determined residual stress in the hardened layer. The variations of both hardness and residual stresses in the surface hardened layer depended on the microstructure of the latter. Microstructural changes can be affected by the heating conditions which, in laser heating, can be determined by the selection of the laser power and the area of the laser-beam spot diameter d_B or A_B with a given mode structure. The traveling speed of the laser beam was then suitably adapted. The measured average surface hardness was 725 HV. It slowly decreased to around 300 HV, i.e., hardness of steel in a soft state, in a depth d of 0.57 mm.

The variation of microhardness permitted an estimation that heating was correctly chosen to achieve the maximum surface hardness and a very distinct width of area with the transition microstructure (a mixture of hardened and base microstructure).

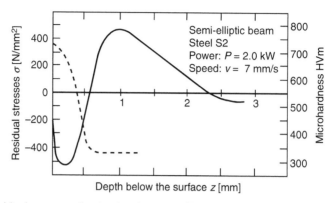

FIGURE 6.62 Residual stress and microhardness profiles through the laser surface hardened layer. (From Com-Nougue, J.; Kerrand, E. Laser surface treatment for electromechanical applications. In *Laser Surface Treatment of Metals*. Draper, C.W. Mazzoldi, P., Eds.; Martinus Nijhoff Publishers in Cooperation with NATO Scientific Affairs Division: Boston, 1986; 497–571.)

The through-depth variation of residual stresses of the hardened layer depended on the residual stresses occurring in the material prior to heating. Then followed relaxation of the existing stresses due to laser heating for further hardening. Finally is the variation of the residual stresses after laser heating affected also by the initial microstructure and the cooling rate of steel. A higher austenitizing temperature and an ensured homogeneity of carbon in austenite and as long as possible heating of the material for the given depth of the hardened layer ensured efficient tempering of the preceding residual stresses and introduced new low-tensile stresses. Thus after laser surface hardening, comparatively high compressive residual stresses could be ensured. They could be confirmed by computations as well. The through-depth variation of residual stresses of the hardened layer was, as expected, coordinated with the variation of hardness. It was very important that the maximum residual stress occurred just below the surface and amounted to -530 N/mm^2. Then it changed the sign in a depth of 0.57 mm and became the maximum tensile stress in a depth of 0.9 mm. In the range between the maximum compressive and compressive tensile residual stresses, i.e., between the depths of 0.57 and 0.9 mm, a linear variation of the residual stresses occurred.

Mor [94] studied measurement of residual stresses in laser surface hardened specimens made of AISI 4140 steel using the x-ray diffraction method. Figure 6.63 shows the results of the measurements performed with two power densities Q, i.e., 10 and 14 W/cm^2, and three respective times of interaction t_i, i.e., 2.0, 1.8, and 1.0 s. The residual stresses were measured in the direction perpendicular to the hardened traces. Microhardness of the base material, i.e., 200 HVm, increased up to around 700 HVm in the hardened trace and reduced to around 400 HVm in the zone of overlapping of two adjacent traces. The residual stresses in the hardened traces showed the compressive character due to an increased specific volume of the martensitic microstructure formed, but at the edge of the hardened traces their character gradually turned into the tensile one. Tensile residual stresses occurred also in the case of two parallel laser-hardened traces. In the hardened traces the maximum compressive residual stresses amounted to -300 N/mm^2, whereas in the zone of overlapping of the hardened traces they varied between 220 and 300 N/mm^2. At the edge of the hardened traces the tensile residual stresses amounted on the average up to 1 to 0 N/mm^2. The variations of residual stresses obtained in the hardened traces were as expected. However, the variation of residual stresses having the tensile character in the zone of overlapping of two adjacent traces, which may be a critical zone with dynamically loaded parts is surprising.

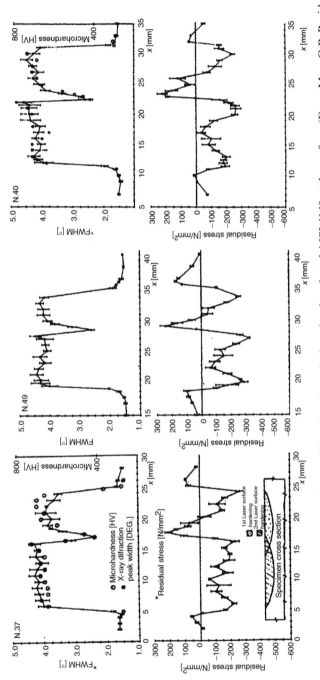

FIGURE 6.63 Microhardness, residual stress, and x-ray diffraction along the laser hardened steel AISI 4140 at the surface. (From Mor, G.P. Residual stresses measurements by means of x-ray diffraction on electron beam welded joints and laser hardened surfaces. In *Proceedings of the 2nd International Conference on Residual Stresses "ICRS2"*. Beck, G., Denis, S., Simon, A., Eds.; 696–702 Elsevier Applied Science: Nancy, London, 1988; 696–702.)

6.16   RESIDUAL STRESSES, MICROSTRUCTURES, AND MICROHARDNESSES

Ericsson et al. [95] studied residual stresses, the content of retained austenite, and the microstructure supported by microhardness measurements. The experiment consisted in making a laser-hardened trace on a cylinder with a diameter of 40 mm and a length of 100 mm. In the experiment AISI 4142 and AISI 52100 steels in quenched and tempered (320 HV) and fully annealed conditions (190 HV) were used. Laser surface hardening was carried out with a CO_2 laser with a power of 3 kW in continuous-wave mode. Several laser surface hardening parameters were chosen. The laser power, the laser-beam spot diameter, and the traveling speed of the workpiece, however, were varied. Figure 6.64 shows the results of calculations in laser surface hardening with reference to through-depth distribution of austenite in the heated layer after 24 s and then the through-depth variations of martensite and residual stresses in the hardened layer after quenching. In the calculations of the austenite and martensite contents and the variation of residual stresses in surface hardened AISI 4142 steel with 55.2 mm in diameter, a power density Q of 6.6 MW/m^2, a traveling speed v of 0.152 m/min, and width of hardened trace W of 8.175 mm were taken into account. After hardening, up to 35% of martensite was found in the surface layer to a depth of 0.5 mm. Then the martensite content was decreasing in a linear manner to a depth of 1.0 mm. Very similar was the through-depth variation of residual stresses of the compressive character with a maximum

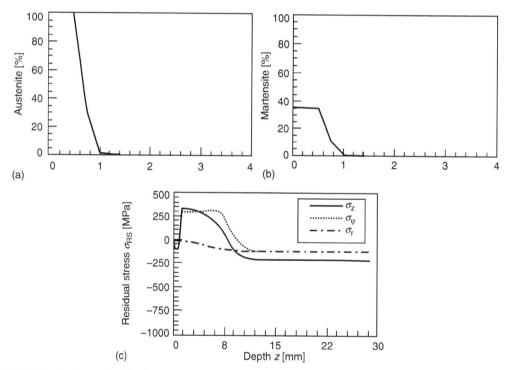

FIGURE 6.64 Austenite distribution after 24 s (a), martensite distribution at the end of cooling (b), and σ_z, σ_φ, and σ_r residual stress profiles (c). (From Ericsson, T.; Chang, Y.S.; Melander, M. Residual Stresses and Microstructures in Laser Hardened Medium and High Carbon Steels. In *Proceedings of the 4th International Congress on Heat Treatment of Materials*, Berlin, Int. Federation for the Heat Treatment of Materials; Vol. 2, 702–733.)

value of around -150 N/mm^2 in the axial and tangential directions. This is followed by a transition to tensile residual stresses in the unhardened layer amounting to 350 N/mm^2. This was followed by a transition to the central part of the specimen with a constant stress of around -200 N/mm^2. The radial residual stresses were almost all the time constant, i.e., $\sigma_r = 0$, from the surface to a depth z of 20 mm, then they increased to around -200 N/mm^2 in the center of the cylinder. An efficient indicator of the variation of residual stresses was the martensite content. The martensite content showed the variation of the residual stresses occurring in the axial and tangential directions, which means that the martensite transformation had a decisive role in the determination of the variation of residual stresses.

Cassino et al. [96] studied the variation of residual stresses after laser surface hardening of 55 CI heat-treatment carbon steel with a carbon content of 0.52%.

The investigations were conducted using flat specimens of $100 \times 80 \times 20$ mm in size and a square spot area of 5×15 mm. Heating was carried out with a laser power varying between 3.0 and 3.4 kW, and a traveling speed of the workpiece v of 5 mm/s. The specimen surfaces were coated with absorbents such as graphite and manganese phosphate that ensured absorptivity in the range from 65 to 85%. The specimens were subjected to forced cooling with argon (A) and water (W). Residual stresses were measured in the longitudinal and transverse directions using the x-ray diffraction method.

X-ray of 2A wavelength from synchrotron radiation was employed to obtain diffraction peaks from (211)-lattice planes of martensite at nine different ψ angles (the angle between the normal to the diffracting planes and the normal to the surface of the specimen) in the range of $-40.0°$ to $35.0°$. The strains were measured in the regions corresponding to the center of the laser traces and along the directions parallel and perpendicular to them. The elastic constants $E = 0.21$ TPa and $v = 0.28$ were used for stress calculations.

Table 6.3 shows residual stress values along longitudinal and transverse direction. For quenching of the workpieces with argon (A$_3$, A$_4$), the longitudinal compressive residual stress ranged between 540 and 610 to N/mm^2, whereas the transverse one ranged between 230 and 350 N/m^2. With forced quenching of the specimens with water (W$_2$, W$_4$, and W$_5$), the longitudinal compressive residual stress ranged between 200 and 400 N/mm^2 and the transverse one between 200 and 100 N/mm^2.

The dissipation of the measured longitudinal and transverse stresses was exceptionally strong and represented up to 50% of their measured value. The results of measurement of the residual stresses confirmed that in the given case a sufficient cooling rate was achieved by cooling with argon. It was necessary to use any more strong cooling medium.

TABLE 6.3
Residual Stress Values Along Longitudinal and Traverse Directions

Specimen	$\sigma\ [\phi = 0]$ MPa	$\sigma\ [\phi = \pi/2]$ MPa	Peak Width [°]
A3	-610 ± 230	-230 ± 160	-2.1 ± 0.6
A4	-540 ± 370	-350 ± 100	-2.3 ± 0.3
W2	-200 ± 130	-100 ± 150	-2.0 ± 0.2
W4	-300 ± 160	-60 ± 20	-2.2 ± 0.2
W5	-440 ± 1200	-100 ± 120	-2.2 ± 0.3

Source: From Cassino, F.S.L.; Moulin, G.; Ji, V. Residual stresses in water-assisted laser transformation hardening of 55C1 steel. In *Proceedings of the 4th European Conference on Residual Stresses "ECRS4"*; Denis, S. Lebrun, J.L., Bourniquel, B., Barral, M., Flavenot, J.F., Eds.; Vol. 2, 839–849.

6.17   SIMPLE METHOD FOR ASSESSMENT OF RESIDUAL STRESSES

Grevey et al. [97] proposed a simple method of assessment of the degree of residual stresses after laser hardening of the surface layer. The method proposed was based on the knowledge of the parameters of interaction between the laser beam and the workpiece material taking into account laser power and the traveling speed of the laser beam across the workpiece and the thermal conductivity of the material concerned. The authors of this method maintained that in the estimation of residual stresses with low-alloy and medium-alloy steels, the expected deviation of the actual variation from the calculated variation of residual stresses in the thin surface layer did not exceed 20%.

On the basis of the known and expected variation, the authors divided the residual-stress profiles through the workpiece depth into three areas as shown in Figure 6.65 [97]

- Areas I and II make up a zone from the surface to the limiting depth with compressive stresses defined in accordance with a TTA diagram (Orlisch diagram) for the given steel.
- Area III is the zone extending from the limiting depth, thus being an adjoining zone where the specific volume of the microstructure is smaller than that at the surface and, consequently, acts in the sense of relative contraction, which produces the occurrence of tensile stresses in this zone. The zone showing tensile residual stresses is strongly expressed in the specimens that were quenched and tempered at 400 or 600°C and then laser surface hardened. Because of a thermomechanical effect occurring during surface heating and fast cooling, certain residual stresses persist in the material after cooling as well. This thermomechanical effect was related to the state of the material prior to laser surface hardening.

Area III is known as the retempered zone of the material at the transition between the laser-hardened surface layer and the base metal, which is in a quenched-and-tempered state in the given case. The thermal effects in laser heating occurring in the retempered zone were calculated by the authors from the variation of the temperature in the individual depths of the specimen on a half infinite solid plate:

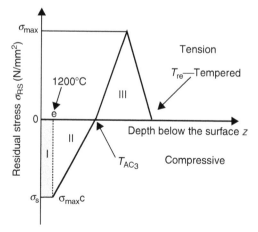

FIGURE 6.65 Residual stress profile, classified in typical areas. (From Grevey, D.; Maiffredy, L.; Vannes, A.B. A simple way to estimate the level of the residual stresses after laser heating. *J. Mech. Work. Technol.* 1988, *16*, 65–78.)

$$T(z) = T_s \left[1 - \text{erf}\left(\frac{z}{2\sqrt{t_i}}\right) \right]$$

where T_s [°C] is the surface temperature, $T(z)$ is the temperature at depth z, $T_s = \rho P_0 a^{1/2} t_i^{1/2}/S$ λ, ρ [%] is total energy efficiency, P_0 [W] is the average power required, S [cm^2] is the spot area of the beam, a [cm^2/s] is the thermal diffusivity, t_i [s] is the interaction time, and v [cm/s] is the traveling speed.

An error function was approximated with $(1-\exp(\sqrt{\pi u}))$, which, with $0.2 < u < 2.0$, provided a favorable agreement between the theoretical and experimental results. The depth in which the transition from compressive residual stresses to tensile residual stresses occurred was calculated using the following equation:

$$z = -\frac{4}{\pi}\sqrt{at_i} \ln \frac{T(z)v^{1/2}\pi \, r_0^{3/2}\lambda}{\rho a^{1/2} P_0}$$

The mathematical description of the variation of thermal cycle permitted them to predict the depth in which the austenite transformation T_{A_3} occurred and the temperature range in which retempering occurred.

The difficulty of the theoretical method consists in the requirement for knowledge of three parameters: thermal conductivity λ [Wcm/°C], thermal diffusivity a [cm^2/s], and total energy efficiency ρ [%].

Concerning the first two parameters they depend on the type of material used, on its initial microstructural state, and also on its temperature. They are linked by the relationship:

$$a = \lambda/dc_p$$

where $d = 7.8$ [g/cm^3] is the material density and c_p [J/gm/°C] is the specific heat which depends on the temperature.

In the calculations only the effective energy was taken into account; therefore the efficiency of material heating and the global coefficient ρ [%], including optical losses along the hardened trace and the reflection of the laser beam from the specimen, were taken into account. Thus they treated the display power P_0 and calculated, by means of the global coefficient of efficiency ρ, the so-called effective power, $P_e = \rho P_0$. The authors experimentally verified the effective power and confirmed the linear relationship as shown in Figure 6.66a and Figure 6.66b.

Figure 6.67 shows the results of calculations and measurements of longitudinal residual stresses. The deviations could be defined with reference to the depth of the transition of the compressive zone into the tensile zone ($\Delta h = h_{\text{EXP}} - h_{\text{EST}}$) and the deviation of the size of the maximum tensile stress in the subsurface ($\Delta\sigma_{\text{maxT}} = \sigma_{\text{maxEXP}} - \sigma_{\text{maxEST}} < 4\%$).

In the calculations the deviation between the experimental and estimated powers ($\Delta P = P_{\text{EXP}} - P_{\text{EST}} < 3\%$) was taken into account.

A very simple and practical method of determination of the through-depth variation of longitudinal residual stresses in the laser-hardened trace was proposed. The procedure was based on measurement of longitudinal residual stress at the surface σ_s and measurement of different characteristic depths by means of metallographic photos or from the through-depth hardness profile, which are defined by e, p, and h, that is:

$$\sigma_m = -\frac{p+e}{h-p}\sigma_s.$$

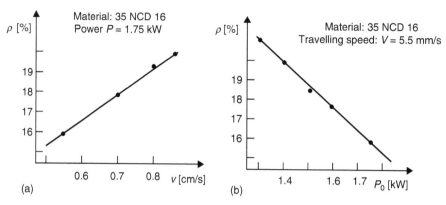

FIGURE 6.66 Variation of efficiency recording traveling speed (a) and laser beam power (b). (From Grevey, D.; Maiffredy, L.; Vannes, A.B. A simple way to estimate the level of the residual stresses after laser heating. *J. Mech. Work. Technol.* 1988, *16*, 65–78.)

6.18 INFLUENCE OF PRIOR MATERIAL HEAT TREATMENT ON LASER SURFACE HARDENING

Chabrol and Vannes [98] analyzed the variations of hardness and residual stresses in the thin surface layer of alloyed heat-treated steels, i.e., steel A (Cr–Ni), steel B(Cr–Mo), steel C(Cr–Ni–Mo), and micro-alloyed steel D (V). All steels were quenched and tempered at

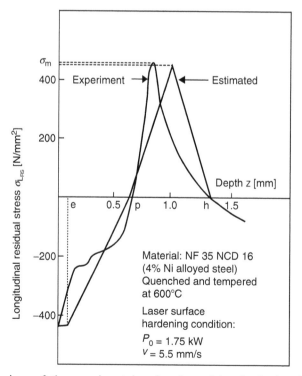

FIGURE 6.67 Comparison of the experimental and estimated longitudinal residual stress profiles. (From Grevey, D.; Maiffredy, L.; Vannes, A.B. A simple way to estimate the level of the residual stresses after laser heating. *J. Mech. Work. Technol.* 1988, *16*, 65–78.)

temperatures of 400 and 500°C. First measurements of hardness and residual stresses were made at a single trace. The specimens showed a variation of Vickers hardness $HV_{0.5}$ of steel B that was annealed and then laser transformation hardened in the first case, and quenched and tempered at 400 and 500°C and then laser surface hardened in the second case. They represent dissipation of hardness, which varied with different specimens between 610 and 750 $HV_{0.5}$, after laser surface hardening. The initial state of steel was soft.

In the second case steel was quenched and tempered and then laser surface treated, which produced a considerable change of variation of hardness in the surface layer. Dissipation of hardness in the thin surface layer was reduced too to vary between 640 and 695 $HV_{0.5}$. The depth of the hardened layer thus increased by approximately 30%. The variation of hardness at the transition between the hardened layer and the quenched-and-tempered base metal where softening, i.e., quenching and tempering, of the material occurred was interesting as well.

This study is based on the identification of residual stresses on the relaxation method, which consists of measuring specimen strain. The relaxation was induced by electro-chemical removal of the stresses surface layer, causing a breakdown in the existing equilibrium state. The restoration of the equilibrium is accompanied by specimen strains. The strains were measured by means of resistance strain gauges and calculated into residual stresses using a mathematical model.

Figure 6.68 shows the variations of the longitudinal and tangential residual stresses in the hardened trace on steel A that was preliminarily annealed. The two variations are very similar, that is:

- Compressive residual stresses of 330 N/mm^2 to a depth of 0.15 mm
- Then follows an almost linear variation of the residual stresses up to 200 N/mm^2 in a depth of 1.5 mm including a transition from the compressive zone to the tensile one occurring in a depth of 1.10 to 1.20 mm
- In the depths exceeding 1.5 mm the two stresses are of the tensile character and almost constant

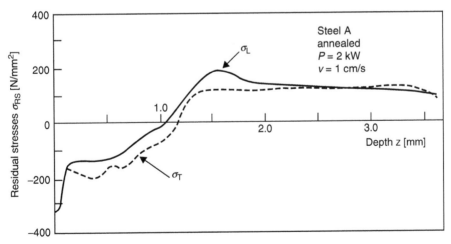

FIGURE 6.68 Longitudinal and traversal residual stress distribution in annealed steel A at given laser hardening parameters. (From Chabrol, C.; Vannes, A.B. Residual stresses induced by laser surface treatment. In *Laser Surface Treatment of Metals*; Draper, C.W. Mazzoldi, P., Eds.; Martinus Nijhoff Publishers in Cooperation with NATO Scientific Affairs Division: Boston, 1986; 435–450.)

The variation of residual stresses in multipass laser surface hardening was very similar to the one obtained with a single trace. The variations of the longitudinal and tangential residual stresses were very similar as well.

The measurement of residual stresses in the x-ray diffraction method showed surface stresses in the hardened traces as well as at the transition between the hardened traces.

Based on the experimental results, it is possible to make some conclusions on laser surface transformation hardening without melting of hypoeutectoid steels:

1. Microstructures observed in the hardened layer are martensitic, characteristic of conventional hardening but with some peculiarities:

   • Martensite is heterogeneous as a result of the nonuniform austenite. This heterogeneity is related to the initial microstructure and the transformation curves which are displaced to higher temperatures.
   • It is impossible to dissolve all the carbides with a heat treatment not involving melting process. Incomplete dissolution of carbides changes the hardenability. Nevertheless, undissolved carbides can be considered hard inclusions which are favorable in wear applications.

2. The analysis of residual stress profiles through the hardened layer confirms microstructural observations. Their profiles change with the type of steel and laser surface hardening conditions, but they correlate with microstructural transformations.

Compressive stresses, always observed in the laser surface hardened layer, improve the mechanical behavior. Tensile residual stresses that appear below the hardened layer are heterogeneous. Quenched-and-tempered steels, however, are more homogeneous and the tensile residual stresses much more pronounced and potentially dangerous because they facilitate crack initiation. Based on the microstructural and residual stress analysis, laser surface transformation hardening leads to better results in a plain medium carbon steel than in alloyed steel. Multipass laser surface hardening shows softening and tensile residual stresses at the surface. Both the softening effect and tensile residual stresses should be avoided in practice with a chosen higher overlapping degree.

6.19 INFLUENCE OF LASER SURFACE HARDENING CONDITIONS ON RESIDUAL STRESS PROFILES AND FATIGUE PROPERTIES

Ericsson and Lin [99] presented the influence of laser surface hardening on fatigue properties and residual stress profiles of notched and smooth specimens.

The authors studied the effect of laser surface hardening on two Swedish steels, SS 2225 (0.26% C, 1.13% Cr, 0.1% Mo) and SS 2244 (0.45% C, 1.02% Cr, 0.16% Mo). Both steels were delivered as hot rolled bars in quenched and tempered condition with a hardness of 300 HV. Smooth and notched fatigue specimens in quenched and tempered condition were laser surface hardened. By using x-ray diffraction technique residual stress profiles in the hardened layer were measured.

Figure 6.69 shows axial residual stress profiles in the center of the laser-hardened track. For laser-hardened smooth specimen of SS 2225 steel, compression residual stress profiles were found in both axial and tangential directions. Compressive residual stresses were also found in the hardened layer of smooth specimen of SS 2244 steel. The values of axial and tangential residual stresses are very similar. Figure 6.69 shows that residual stress profiles in

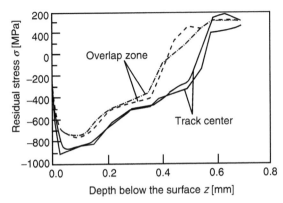

FIGURE 6.69 Axial residual stress profiles at the center of the track on smooth specimen of SS 2244 steel. (From Ericsson, T.; Lin, R. Influence of laser surface hardening on fatigue properties and residual stress profiles of notched and smooth specimens. In *Proceeding of the Conference "MAT-TEC 91"*, Paris; Vincent, L., Niku-Lari, A., Eds.; Technology Transfer Series, Published by Institute for Industrial Technology Transfer (IITT) Int.; Gowruay-Sur-Marne, France, 1991; 255–260.)

the track center were more compressive in comparison to those at the overlap track. The difference became much larger at depth between 0.4 and 0.6 mm, which could be explained by the difference in microstructures.

Figure 6.70 shows that the compressive residual stress profiles were found in the hardened notches specimen of SS 2225 steel and the some shape specimen of SS 2244 steel. The value of axial residual stress was higher than that of tangential residual stress especially for SS 2225 steel. This might be due to the different constraint to deformation at the notch. Although the input heat from hardening the second track had little tempering effect, relaxed the first track, and changed residual stress profiles. As shown in Figure 6.70, the value of residual stress was lower in the first track.

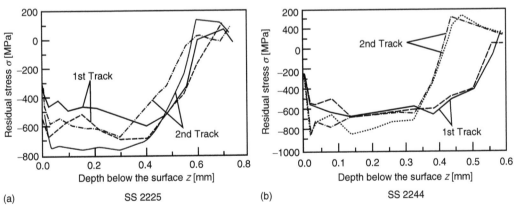

FIGURE 6.70 Residual stress profiles after laser surface hardening notched specimens. Full and open symbols refer to axial and tangential residual stress, respectively. (From Ericsson, T.; Lin, R. Influence of laser surface hardening on fatigue properties and residual stress profiles of notched and smooth specimens. In *Proceeding of the Conference "MAT-TEC 91"*, Paris; Vincent, L., Niku-Lari, A., Eds.; Technology Transfer Series, Published by Institute for Industrial Technology Transfer (IITT) Int.; Gowruay-Sur-Marne, France, 1991; 255–260.)

TABLE 6.4
Bending Fatigue Limits and Notch Sensitivity

Material	Specimen Type	Condition	Fatigue Limits [MPa]	Notch Sensitivity Index
SS 2225	Smooth	Q&T	480 ± 41	0.66
	Notched	Q&T	350 ± 62	
SS 2225	Smooth	Q&T + Laser hard	627 ± 10	0.16
	Notched	Q&T + Laser hard	575 ± 21	
SS 2244	Smooth	Q&T	481 ± 7	0.71
	Notched	Q&T	343 ± 10	
SS 2224	Smooth	Q&T + Laser hard	621 ± 33	0.13
	Notched	Q&T + Laser hard	581 ± 29	

Source: From Ericsson, T.; Lin, R. Influence of laser surface hardening on fatigue properties and residual stress profiles of notched and smooth specimens. In *Proceeding of the Conference "MAT-TEC 91"*, Paris; Vincent, L., Niku-Lari, A., Eds.; Technology Transfer Series, Published by Institute for Industrial Technology Transfer (IITT) Int.; Gowruay-Sur-Marne, France, 1991; 255–260.

As can be seen, the fatigue strength of quenched and tempered specimens was greatly improved after laser surface hardening. Both low and high cycle fatigue lives were increased. The evaluated bending fatigue limits are summarized in Table 6.4, together with the values of notch sensitivity index q. Notch sensitivity index q is a measure of the effect of a notch on fatigue and is defined as $(K_f - 1)/(K_t - 1)$, where K_f is the ratio of the fatigue limit of smooth specimen to that of notched specimen.

Figure 6.71 shows the S–N curves for smooth and notched specimens. The curves 1 and 2 represent fatigue strength for smooth specimens after quenched-and- tempered (Q&T) conditions, curves 3 and 4 represent fatigue strength for smooth and notched specimens after laser surface hardening. It can be concluded that:

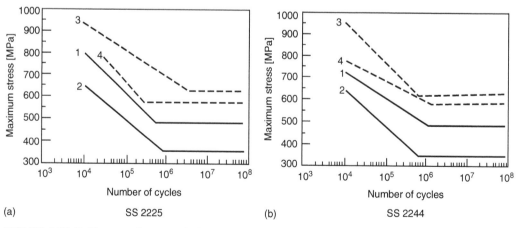

FIGURE 6.71 S–N curves for smooth (1,2) and notched (3,4) specimens. (From Ericsson, T.; Lin, R. Influence of laser surface hardening on fatigue properties and residual stress profiles of notched and smooth specimens. In *Proceeding of the Conference "MAT-TEC 91"*, Paris; Vincent, L., Niku-Lari, A., Eds.; Technology Transfer Series, Published by Institute for Industrial Technology Transfer (IITT) Int.; Gowruay-Sur-Marne, France, 1991; 255–260.)

- Bending fatigue limits of smooth specimens were increased by 31%for SS 2225 steel and 29% for SS 2244 steel, while those of notched specimens were increased by 64% for SS 2225 steel and 69% for SS 2244 steel.
- Notch sensitivity index q decreased from 0.66 to 0.16 for SS 2225 steel and from 0.71 to 0.13 for SS 2244 steel.
- Important increase in fatigue properties was caused by compressive residual stresses in the surface layer after laser surface hardening.

Bohne et al. [100] analyzed X 39 CrMo17 1 (1.4122) steel which was quenched and tempered at temperatures of 520 and 200°C. The hardened specimens were rapidly reheated with a Nd:YAG laser under such conditions that a maximum surface temperature ranging between 860 and 1390°C was obtained, and then quenched with a cooling rate ranging between 480 and 1160°C/s. Depending on the conditions of the preceding heat treatment and various parameters of laser processing, differences in the microstructure of a laser trace, in the hardness, and in the profile of residual stresses in the laser-hardened trace were obtained. Depending on the heating conditions, different contents of retained austenite, i.e., between 2 and 62%, were found in the laser-hardened traces. These great differences in the measured contents can be attributed to the height of the maximum surface temperature, the cooling rate, and the preceding heat treatment, i.e., tempering. It was found that the major influence on the formation of retained austenite was exerted by the tempering temperature in the preceding hardening so that a lower tempering temperature resulted in a higher content of retained austenite. The content of retained austenite, however, increased with an increase in the maximum temperature in laser surface heating as well as by an increase in the rate of laser heating.

In Table 6.5 the results of microhardness measurements obtained in a depth of 30 μm below the surface of the laser-hardened trace are stated in dependence of the maximum

TABLE 6.5
Microhardness and Retained Austenite (RA) Content after Laser Surface Hardening in Tempered Steel at 520°C

Specimen	T_{max} [°C]	T/t [°C/s]	RA [%]	Microhardness [$HV_{0.05}$]
Tempered at 520°C				
520-2r	860	620	<2	540 ± 14
520-3r	900	620	3	451 ± 7
520-3l	980	775	7	700 ± 18
520-4r	1020	670	9	706 ± 13
520-2l	1100	700	6	700 ± 17
520-1l	1125	920	24	653 ± 12
520-5r	1399	1160	62	636 ± 19
Tempered at 200°C				
200-6l	800	480	<2	397 ± 10
200-7l	840	5202	20	487 ± 14
200-7r	980	665	24	480 ± 9
200-6r	1399	1050	60	646 ± 14

Source: (From Bohne, C.; Pyzalla, A.; Reimers, W.; Heitkemper, M.; Fischer, A. *Influence of Rapid Heat Treatment on Microstructure and Residual Stresses of Tool Steels*. Eclat—European Conf. on Laser Treatment of Materials, Hanover, 1998; Werkstoff - Informationsgesellschaft GmbH: Frankfurt, 1998; 183–188).

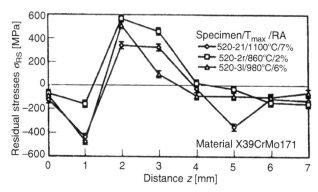

FIGURE 6.72 Longitudinal residual stress profiles after laser surface heat treating of quenched and tempered steel at 520°C. (From Bohne, C.; Pyzalla, A.; Reimers, W.; Heitkemper, M.; Fischer, A. *Influence of Rapid Heat Treatment on Microstructure and Residual Stresses of Tool Steels.* Eclat—European Conf. on Laser Treatment of Materials, Hanover, 1998; Werkstoff - Informationsgesellschaft GmbH: Frankfurt, 1998; 183–188.)

surface temperature T_{Smax} [°C] and the rate of laser heating, $\nu = \Delta T/\Delta t$ [°C/s]. It is worth mentioning the difference in the microhardness attained after steel tempering at the temperatures of 520 and 200°C, respectively. In the case of steel tempering from 520°C and then laser hardening, the microhardness decreased from 700 to 636 $HV_{0.05}$ with the increase in the content of retained austenite. In case of steel tempering from 200°C the microhardness increased with the increase in the content of retained austenite, which was to be attributed to precipitates that additionally hardened austenite and martensite.

Residual stresses were measured in the longitudinal and transverse directions in the central part of the hardened trace. The variation of the residual stresses is shown for the previously hardened steel that was tempered at 520°C and then rapidly laser heated to T_{Smax}, which provided lower contents of retained austenite, i.e., between 2 and 7%.

Figure 6.72 shows the variation of the residual stresses in the longitudinal direction with reference to the orientation of the hardened traces and Figure 6.73 the variation in the

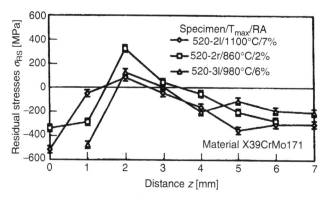

FIGURE 6.73 Transverse residual stress profiles after laser surface heat treating of quenched and tempered steel at 520°C. (From Bohne, C.; Pyzalla, A.; Reimers, W.; Heitkemper, M.; Fischer, A. *Influence of Rapid Heat Treatment on Microstructure and Residual Stresses of Tool Steels.* Eclat—European Conf. on Laser Treatment of Materials, Hanover, 1998; Werkstoff - Informationsgesellschaft GmbH: Frankfurt, 1998; 183–188.)

transverse direction. The specimens were, prior to surface hardening, quenched and tempered at 520°C. Both the longitudinal and transverse residual stresses show the compressive character at the surface and then get balanced through depth where they finally obtained the tensile character. The through-depth variations of the longitudinal and transverse residual stresses are similar. The compressive residual stresses can be attributed to austenite–martensite transformation whose magnitude and variation depend on the content of retained austenite and the precipitates that could additionally harden the surface layer. The magnitude and variation of the residual stresses were affected also by internal stresses occurring in the cooling process. If the latter were higher than the yield stress at the given temperature, plastic deformation and certain internal stresses accompanied by transformation stresses occurring in further cooling was obtained. The stresses occurring during the cooling process are called residual stresses after hardening. They are accompanied by specimen deformation. Account should be taken also of the residual stresses due to the preceding conventional hardening of the specimen.

Mordike [101] reported on several applications of laser transformation hardening, laser remelting, laser nitriding, and carbonitriding as well laser cladding and alloying. The results show the variation of residual stresses in the longitudinal and transverse directions in pearlitic gray cast iron, for two sets of conditions after laser remelting of the surface layer. Both through-depth profiles of the residual stresses confirmed great differences in the energy input obtained by varying the power density and the traveling speed of the laser beam. Although the two profiles of the residual stresses were very similar, it was found that the depths of the remelted layers differed, i.e., $d_1 = 0.2$ mm and $d_2 = 1.0$ mm. In both cases there were compressive residual stresses in the surface layer. They differed, however, in the depth in which the transition from compressive to tensile stresses occurred. In the first case the transition was outside the remelted depth whereas in the second case it was within the remelted zone. The compressive stresses at the surface amounted to around 400 N/mm^2 under the conditions of shallow remelting and to around 700 N/mm^2 in the subsurface, i.e., in a depth z of 0.3 mm, under the conditions of deep remelting.

Grum and Žerovnik [102] presented the experimental results of residual stress analysis after previously mentioned laser transformation hardening of various steels by using different absorbers. Their study based the identification of residual stresses on the relaxation method, which consists of measuring specimen strain. The relaxation was induced by electrochemical removal of the stressed surface layer, causing a breakdown in the existing equilibrium state. The restoration of the equilibrium is accompanied by specimen strains. The strains were measured by means of resistance strain gauges and calculated into residual stresses using a mathematical model.

Figure 6.74 through figure 6.76 show the variation of residual stresses as a function of depth, the distance of the observed location from the surface. In laser transformation hardening, tracks have very small widths and depths that are dependent on the optical and kinematic conditions and laser source power. In the investigation the specimen traveling speed was changed with respect to the absorber. Figure 6.74 shows three residual stress profiles of the investigated W–Cr–V tool steel, and 4140, 1045 structural steels with the laser beam traveling along a zigzag path and the ground surfaces were free from any of the absorbers. The residual stress profiles are very similar to one another, displaying the following characteristics:

- In all three cases, the surface is subjected to compressive residual stresses.
- At a depth between 300 and 600 μm, the compressive stresses transform themselves into tensile residual stresses.
- Tensile residual stresses are present in deeper layers and have an almost constant value around 500 N/mm^2.

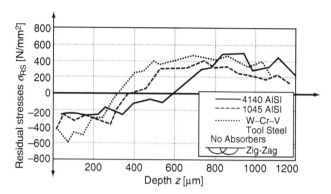

FIGURE 6.74 Residual stresses profiles in steels after laser transformation hardening at zigzag traveling path; no absorber. (From Grum, J.; Žerovnik, P. Laser Hardening Steels, Part 2. Heat Treating; Chilton Publication Company, August, 1993; Vol. 25, No. 8, 32–36.)

From the residual stress results, it can be seen that both the heat treatable steels have low compressive stresses on the surface and a less explicit transition into tensile stresses. The alloyed W–Cr–V tool steel has a considerably greater content of carbon than the heat treatable steels; therefore we can maintain that carbon plays a decisive role in residual stresses size and transformation.

Residual stresses W–Cr–V in tool steel, to which absorbers A and B were applied and the laser traveled a square-spiral path, are further illustrated in Figure 6.75. Thanks to light absorption, shorter interaction times are necessary, and as a result, the compressive residual stresses on the surface are lower. In addition, the depth of the transition from compressive into tensile stresses is lower using absorber A. The latter yields lower compressive residual stresses and a steeper transition into tensile stresses. Both features increase the sensitivity of parts to dynamic loads. Therefore the better choice is absorber B.

Figure 6.76 illustrates the variation of residual stresses with an intermittent laser trace [103]. The residual stresses in the compressive and tensile zone are low, a special feature being a thin surface layer with tensile stresses. Because after laser heat treatment grinding should be applied, in each of the discussed cases compressive stresses in the surface are a usual result.

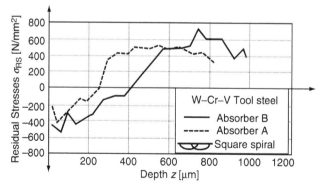

FIGURE 6.75 Residual stress profiles in steel S1 after laser transformation hardening at square spiral traveling path; absorbers A and B. (From Grum, J.; Žerovnik, P. Laser Hardening Steels, Part 2. Heat Treating; Chilton Publication Company, August, 1993; Vol. 25, No. 8, 32–36.)

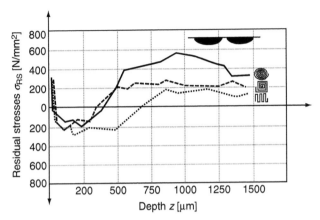

FIGURE 6.76 Residual stresses in W–Cr–V tool steel for inter-spaced hardening traces, concentric circle, square spiral, and zigzag hardening methods. (From Grum, J.; Žerovnik, P. Residual stresses in laser heat treatment of plane surfaces. In *Proceedings of the 1st International Conference on Quenching & Control of Distortion, Chicago, Illinois*; Totten, G., Ed.; ASM International: Materials Park, Ohio, 1992; 333–341.)

An interesting study was made by Yang and Na [104] in his paper titled "A study on residual stresses in laser surface hardening of a medium carbon steel" by using two-dimensional finite element model. By using the proposed model, the thermal and residual stresses at laser surface hardening were successively calculated. The phase transformation had a greater influence on the residual stress than the temperature gradient. The simulation results showed that a compressive residual stress region occurred near the hardened surface of the specimen and a tensile residual stress region occurred in the interior of the specimen. The maximum tensile residual stress occurred along the center of the laser track in the interior region.

The compressive residual stress at the surface of the laser-hardened specimen has a significant effect on the mechanical properties such as wear resistance, fatigue strength, etc.

The size of the compressive and tensile regions of the longitudinal residual stress for various spot ratios of the square beam mode is shown in Figure 6.77. It should be observed that with increasing beam width the compressive region becomes wide but shallow. From the comparison of the results, it is recommended that wide laser-beam spots are used for obtaining the desirable heat-treated region.

Figure 6.78 shows the sizes of the compressive and tensile regions of the longitudinal residual stress for various laser-beam power and traveling speed at a given input energy. Although the input energy is constant, the compressive residual stress region increases according to increased laser power and traveling speed. That means that it is desirable to use the high power beam and high traveling speed at laser surface hardening.

Estimation and optimization of processing parameters in laser surface hardening was explained by Lepski and Reitzenstein [105]. Optimum results were obtained if the processing was based on temperature cycle calculations, taking into account the material properties and input energy distribution. A user-friendly software is required in industry application of laser surface hardening. The software should fulfill the following criteria:

- Ability to check any given hardening problem
- Ability to estimate without experiments which laser power beam shaping or beam scanning system is selected
- Ability to predict the laser hardening results at a given application

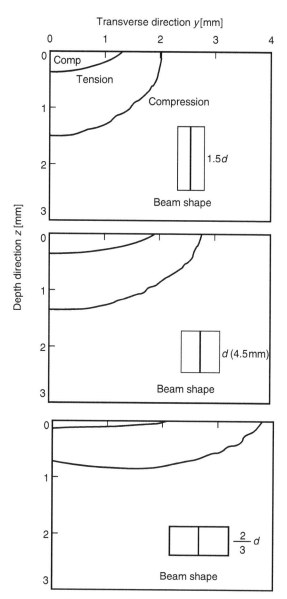

FIGURE 6.77 Compressive and tensile regions of the longitudinal residual stress for various laser beams of square beam structural mode at constant laser power. (From Yang, Y.S.; Na, S.J. *Surf. Coat. Technol.* 1989, *38*, 311–324.)

- Ability to calculate the processing parameters with minimum cost at desired hardening and annealing zone size
- Graphically present the relationship between the processing parameters and the hardening zone characteristics.

The integration of laser hardening in complex manufacturing systems requires hardening with high traveling speeds. In order to get a sufficient hardening depth neither the surface maximum temperature nor the laser interaction time must fall below certain limits even for

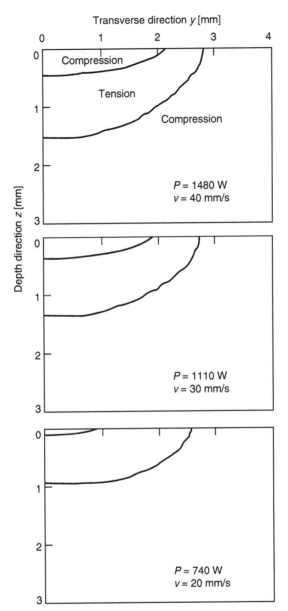

FIGURE 6.78 Compressive and tensile regions of the longitudinal residual stress for various laser powers and traveling speeds at a given input energy. (From Yang, Y.S.; Na, S.J. *Surf. Coat. Technol.* 1989, *38*, 311–324.)

high traveling speeds. This may be achieved to a certain degree by laser spot stretching along the traverse direction. In Figure 6.79 the track depth as well as the track width is represented for the steel C45 as functions of the spot axis ratio (SAR) (SAR $= y/x = 1 \ldots 10$) for various values of the traveling speeds ($v = 1$–10 m/min and a laser power of 10 kW). Values less than unity of the ratio SAR correspond to a spot stretched along the traverse direction.

Meijer et al. [106] presented a very interesting and practical subject of transferability of the same testing parameters in laser hardening using different laser systems used in several laboratories.

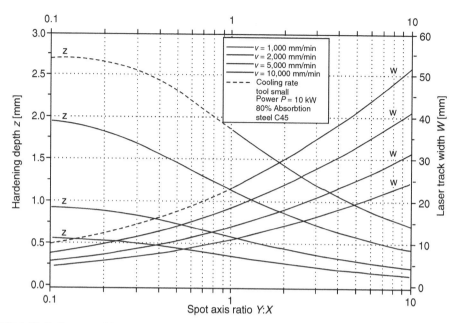

FIGURE 6.79 Influence of laser spot axis ratio and various traveling speeds on obtained depth and width of single hardened track. (From Lepski, D.; Reitzenstein, W. Estimation and optimization of processing parameters in laser surface hardening. In *Proceedings of the 10th Meeting on Modeling of Laser Material Processing*, Igls/Innsbruck; Kaplan A., Schnöcker D., Eds.; Forschungsinstitut für Hochcleistungsstrahltechnik der TüW Wien, 1995; 18 pp.)

The chosen experimental conditions show that the same material was tested, the same absorption deposit and the same working method in a shielding atmosphere were applied, but different degrees of defocus, small power densities, different interaction times, and different mode structures of the laser beam were used.

The deviations occurring are relatively strong; therefore it is dubious whether with different laser systems the same and reproducible results with the hardened surfaces can be obtained. As the machining conditions were uniformly selected, the same depths of the hardened traces could be assumed, yet this was not the case. The differences were exceptionally strong. They can be attributed to the chosen mode structure of the laser beam and other characteristics selected such as the degree of defocus and the travel speed of the laser beam. Consequently, the results of the investigation should serve only as a proof that the influence of energy input is strong. As far as the energy input is concerned, not only the mode structure of the laser beam but also the selected and interdependent parameters such as the travel speed and the degree of defocus should be taken into account.

Results of the second set of joint measurements are given in dependence of the width of a single hardened trace on the material surface and the depth of the hardened trace. The deviations occurring can be attributed only to the mode structures selected in the individual laboratories. The differences in the measured widths and depths of the hardened single traces are, as expected, smaller with shorter interaction times. They gradually increase with increasing interaction times.

The deviation occurring in hardening of carbon steel can be attributed to the available mode structures of the laser beam and the selection of different interaction times.

Kugler et al. [107] reported on the latest developments of powerful diode lasers permitting the manufacture of new laser hardening systems. Advantages of the diode laser are as follows:

- Optics adjustable and adaptable to the workpiece shape.
- Adjustable optics permits shallow or deep heating of the surface layers.
- Absorptivity of laser light with a double wavelength, $\lambda_1 = 0.808$ and $\lambda_1 = 0.94$, ranges, as assessed, between 50 and 70%, which means that no particular preparation of steel surface is required for heat treatment.
- It has a pyrometer mounted to measure the temperatures at the workpiece surface surrounding the laser beam.
- A computer-aided system for temperature setting makes a comparison with the reference temperature and automatically sets a higher or lower laser-source power in the very course of heat treatment.
- A robotic hand and optical transmission of laser light to the workpiece make it possible to perform heat treatment of large parts simply and efficiently.
- It is suitable for both small-scale and large-scale productions.

Such a way of temperature monitoring, i.e., by changing the power input, is extremely important at the workpiece edge, along the notches in the workpiece, or at the transition from a larger diameter to a smaller one.

Figure 6.80 shows the dependence between the power density distribution across the beam diameter, the specific energy, and the depth of the modified layer. In the diagram, different distributions, i.e., Gaussian (A), cylindrical (B), prismatic (C), and sharp-edged (D) distributions, are shown. The individual characteristic power density distributions across the beam diameter ensure, with the same heating conditions, small differences in the energy input, and great differences of the depth of the hardened layer attained. The lowest specific energy is

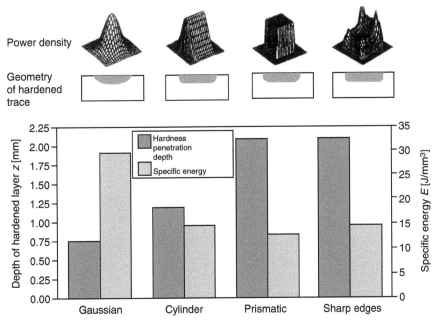

FIGURE 6.80 Influence of various laser beam shapes and specific energy on the depth of the hardened layer. (From Kugler, P.; Gropp, S.; Dierken, R.; Gottschling, S. Temperature controlled surface hardening of industrial tools—experiences with 4kW-diode-laser. In *Proceedings of the 3rd Conference "LANE 2001": Laser Assisted Net Shape Engineering 3, Erlangen;* Geiger, M., Otto, A., Eds.; Meisenbach-Verlag GmbH: Bamberg, 2001; 191–198.)

obtained with the cylindrical (B), prismatic (C), and sharp-edged (D) power density distributions. It ranges between 13.0 and 15.0 J/mm^3. The highest specific energy is attained with the Gaussian power density distribution. On the average, it is 100% higher than the lowest one. A more important piece of information is the one on the depth of the hardened layer attained. The latter equals 0.75 mm with the Gaussian power density distribution, 1.17 mm with the cylindrical distribution, and over 2.10 mm with the prismatic and sharp-edged distributions. This proves that the distributions most suitable for heat treatment are the prismatic and sharp-edged power density distributions.

6.20 PREDICTION OF HARDENED TRACK AND OPTIMIZATION PROCESS

Marya and Marya [108] reported about prediction hardened depth and width and optimization process of the laser transformation hardening. They used a dimensionless approach for Gaussian and rectangular sources to find laser heating parameters at given dimensions of the hardened layer.

The laser transformation hardening has been performed on a 0.45% carbon steel, coated with a carbon to maintain the surface absorptivity to about 70%. Figure 6.81 shows the author's results according to their model of predicted dimensions of the hardened layer and process optimization. The diagram shows that decreasing power and laser-beam traveling speed at Gaussian distribution of laser power density in cross section which defines interaction times are required to obtain specific data. Thus any useful combination of laser processing parameters must maximize heat diffusion to required depth of hardened layer. It is necessary that surface melting temperature is reached. Figure 6.81 and figure 6.82 show a low dimensionless traveling speed v^* that is necessary to allow the heat conduction in depth and achieve high dimensionless depth $Z_h^* = Z_h/R$ and dimensionless width $W_h^* = W_h/R$. The

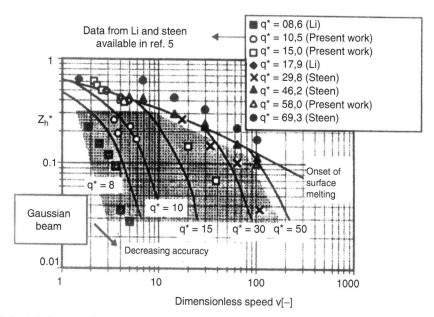

FIGURE 6.81 Influence of dimensionless power (q^*) and traveling speed (v^*) on hardened depth after (z_n^*) for Gaussian beam. (From Marya, M.; Marya, S.K. Prediction and optimization of laser transformation hardening. In *Proceedings of the 2nd Conference "LANE'97": Laser Assisted Net Shape Engineering 2, Erlangen;* Vollersten, F., Ed.; Meisenbach-Verlag GmbH: Bamberg, 1997; 693–698.)

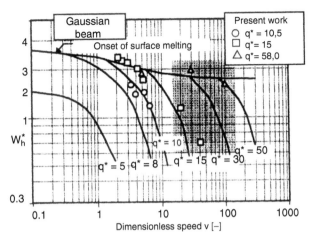

FIGURE 6.82 Influence of dimensionless power (q^*) and traveling speed (v^*) on hardened width (w^*) for Gaussian beam. (From Marya, M.; Marya, S.K. Prediction and optimization of laser transformation hardening. In *Proceedings of the 2nd Conference "LANE'97": Laser Assisted Net Shape Engineering 2, Erlangen;* Vollersten, F., Ed.; Meisenbach-Verlag GmbH: Bamberg, 1997; 693–698.)

dimensionless power parameter (q^*) is determined with respect to (v^*) to reach the melting on set.

If the laser beam moves faster, greater values (q^*) must be selected to reach surface melting. Similar calculations were realized for square laser-beam power density. Hardened widths should correspond rather well to beam spot diameter because the step energy gradient of the beam edge should produce an evenly steep temperature gradient.

Results in stationary laser-beam hardening show that melting cannot be achieved at dimensionless power below 7.6. Similar conclusions have already been drawn for Gaussian beams. Indeed, as the beam speed approaches zero, the beam profile contribution decreases as heat tends to dissipate more uniformly.

Figure 6.83 shows optimized beam spot dimensions at the surface which are very well predicted by the theoretical analysis. Moreover, the results show that hardened depth increases with heat input energy. They experimentally verified that melting conditions are proportional to the spot dimension. Although increasing powers and beam spot diameter produced wider hardened layer the heating rates decreased significantly. Figure 6.83 shows the variation in surface hardness according to the hardness of base material ($HV = 205$). In an optimization process, a compromise between a high quenched depth and a significant hardness increase must therefore be found.

6.21 MICROSTRUCTURE AND RESIDUAL STRESS ANALYSIS AFTER LASER SURFACE REMELTING PROCESS

Ductile iron is commonly used in a wide range of industrial applications because of its good castability, good mechanical properties, and low price. By varying the chemical and microstructure composition of cast irons, it is possible to change their mechanical properties as well as their suitability for machining. Ductile irons are also distinguished by good wear resistance, which can be raised even higher by additional surface heat treatment. With the use of induction or flame surface hardening, it is possible to ensure a homogeneous microstructure in the thin surface layer; however, this is possible only if cast irons have a pearlite matrix. If

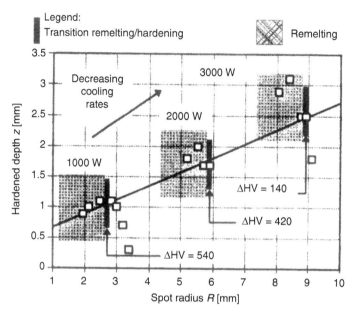

FIGURE 6.83 Influence of spot radius and laser-beam power on hardened depth. (From Marya, M.; Marya, S.K. Prediction and optimization of laser transformation hardening. In *Proceedings of the 2nd Conference "LANE'97": Laser Assisted Net Shape Engineering 2, Erlangen*; Vollersten, F., Ed.; Meisenbach-Verlag GmbH: Bamberg, 1997; 693–698.)

they have a ferrite–pearlite or pearlite–ferrite matrix, a homogeneous microstructure in the surface hardened layer can be achieved only by laser surface remelting.

After the laser beam had crossed flat specimen, a microstructurally modified track was obtained, which was shaped like a part of a sphere (Figure 6.84). To achieve a uniform thickness of the remelted layer over the entire area of the flat specimen (Figure 6.85), the kinematics of the laser beam were adapted by 30% overlapping of the neighboring remelted traces [109].

The microstructure changes in the remelting layer of the ductile iron are dependent on temperature conditions during heating and cooling processes. In all cases of the laser surface remelting process two characteristic microstructure layers were obtained, i.e., the remelted

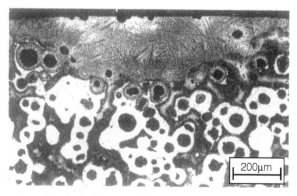

FIGURE 6.84 Cross section of a single laser-modified trace; remelting condition: $P = 1.0$ kW, $z_s = 22$ mm, and $\nu_b = 21$ mm/s. (From Grum, J.; Šturm, R. *J. Mater. Eng. Perform* 2001, *10*, 270–281.)

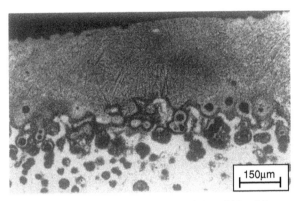

FIGURE 6.85 Laser surface modified layer at 30% overlap of the width of the remelted traces; remelting conditions: $P = 1.0$ kW, $z_s = 22$ mm, and $v_b = 21$ mm/s. (From Grum, J.; Šturm, R. *J. Mater. Eng. Perform*. 2001, *10*, 270–281.)

layer and hardened layer. Figure 6.86 shows that the microstructure in the remelted surface layer is fine grained and consists of austenite dendrites, with very fine dispersed cementite, together with a small portion of coarse martensite. X-ray phase analysis of the remelted layer showed the average volume percentages of the particular phases as follows: 24.0% austenite, 32.0% cementite, 39.0% martensite, and 5.0% graphite. Figure 6.87 shows the microstructure of the hardened layer consisting of martensite with a presence of residual austenite, ferrite, and graphite nodules. Graphite nodules are surrounded by ledeburite and/or martensite shells [109].

Grum and Šturm [109] showed the results (Figure 6.88) of the calculations of principal residual stresses in the flat specimen after laser remelting with the zigzag laser beam guiding at the 0 and 30% overlapping degree of the remelted layer. Different modes of guiding the laser beam over the specimen surface were selected, i.e., zigzag (A), square-shaped spiral toward the center (B), and square-shaped spiral away from the center (C), with the laser beam turning round outside the specimen to achieve more uniform thermal conditions in the material. In this way, it was possible to achieve different thermal conditions in the thin flat specimen during the remelting process as well as during cooling, which influence the preheating of the specimen prior to remelting and the tempering of the created modified microstructure. They stated the following:

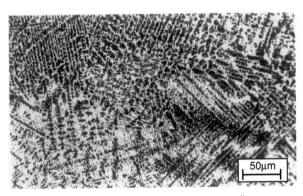

FIGURE 6.86 Microstructure of the remelted layer. (From Grum, J.; Šturm, R. *J. Mater. Eng. Perform*. 2001, *10*, 270–281.)

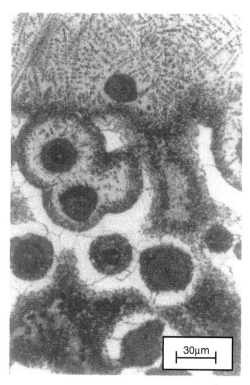

FIGURE 6.87 Microstructure of the hardened layer. (From Grum, J.; Šturm, R. *J. Mater. Eng. Perform.* 2001, *10*, 270–281.)

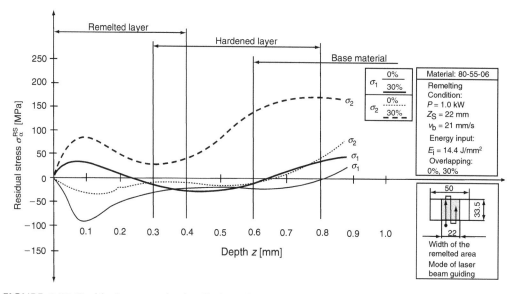

FIGURE 6.88 Residual stresses in ductile iron 80-55-06; zigzag laser beam guiding laser-remelting conditions. (From Grum, J.; Šturm, R. *J. Mater. Eng. Perform.* 2001, *10*, 270–281.)

- Residual stresses σ_1 are directed in the longitudinal direction and σ_2 directed in the transverse direction of the flat specimen.
- During the process of laser surface remelting, the specimen bends more in the longitudinal direction causing additional lowering of tensile residual stresses.
- At 0% overlapping of the remelted layer, the compressive residual stresses were achieved between $+100.0$ and -5.0 MPa. A 30% higher overlapping degree induces the occurrence of tensile residual stresses of $+90$ MPa in the remelted layer and compressive stress of 50 MPa in the hardened layer.

The same authors in Figure 6.89 showed the results of principal residual stresses in the thin surface layer with a laser beam guiding in the shape of a square spiral at a given laser-remelting conditions, without overlapping traces.

They showed the calculated principal residual stresses profiles through the depth and the direction angles of the principal residual stresses. The residual stresses were measured to the depth of 0.9 mm, which means that they were measured to the transition into the matrix.

From the results of calculated residual stress profiles the following conclusions can be drawn:

- Principal residual stresses σ_1 are directed perpendicularly to the direction of the laser path.
- Principal residual stresses σ_2 acting in the direction of laser remelting and being of tensile nature are much higher than the residual stresses σ_1.
- Principal residual stresses σ_1 are reduced because of the temperature field, and lowering the yield strength of the material.
- Residual stresses in the modified surface layer in the longitudinal direction of the specimen are compressive and range between 80.0 and 5.0 MPa when, in the transition area from the hardened layer to the matrix, they change into tensile residual stresses.

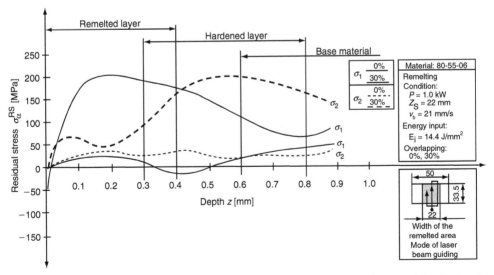

FIGURE 6.89 Residual stresses in ductile iron 80-55-06 with a circular laser beam guiding in the shape of a square spiral beginning on the edge of the remelted area and ending in the middle of the specimen at the above indicated laser-remelting condition. (From Grum, J.; Šturm, R. *J. Mater. Eng. Perform.* 2001, *10*, 270–281.)

- In the transverse direction of the specimen, residual stresses are always tensile and range from +50.0 to +200.0 MPa in the modified layer.

All of these laser-remelting conditions, each in its own way, change the amount of the energy input and can have an important effect on the size and quality of the modified layer and residual stresses. A greater amount of input energy into the specimen results in a higher increase of temperature in it and higher over-heating of the specimen, which, on the other hand, lowers the cooling rate in the modified layer and gives rise to the occurrence of small microstructure residual stresses in this layer.

6.21.1 DIMENSIONS OF THE REMELTED TRACK

Hawkes et al. [110] studied laser-remelting process on ferrite and pearlite gray iron at various melting conditions. The results are presented by microstructure, microhardness analysis, and size data of the remelted tracks. Figure 6.90 shows remelted depth according to traveling speed and beam diameter. Depth of remelting layer increased significantly by approaching the focussed beams at lower traveling speed of 100 mm/s due to the keyholing mechanism operating in the molten pool. This occurred at sufficient laser-beam power density to vaporize the metal under the beam center. The pressure of the expanding vapor held the cavity open forming a black body, thus transferring more energy deeper into the material.

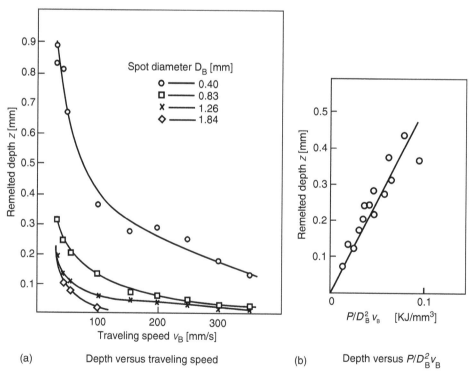

(a)   Depth versus traveling speed   (b)   Depth versus $P/D_B^2 v_B$

FIGURE 6.90 Remelting depth according to traveling speed v and input energy $P/D_B{}^2 v_b$. (From Hawkes, I.C.; Steen, W.M.; West, D.R.F. Laser Surface Melt Hardening of S.G. Irons. In *Proceedings of the 1st International Conference on Laser in Manufacturing*, Brighton, UK; Kimmit, M.F., Ed.; Co-published by: IFS (Publications), Bedford, UK, Ltd. and North-Holland Publishing Company: Amsterdam, 1983; 97–108.)

The remelting width was greater at smaller areas of beams at low traveling speeds due to the intense vapor and liquid convection. Figure 6.90a shows the dependence of melting depth according to traveling speed at various beam spot diameters. Figure 6.90b shows the depth of remelting, assuming depth/width is constant, proportional to $P/D_B{}^2V_B$.

Greater depth of remelting was achieved in the ferritic gray iron as compared to the pearlitic gray iron with higher beam spot diameters at lower traveling speeds. Figure 6.91 shows microhardness profiles across overlapping tracks at two different laser-remelting conditions. Laser track B shows a much wider HAZ at increased microhardness, and laser track A shows a greater HAZ at increased microhardness. The differences in HAZ width and microhardness profiles depend on the austenite decomposition temperature in the center of overlapping tracks.

6.21.2 MATHEMATICAL MODELING OF LOCALIZED MELTING AROUND GRAPHITE NODULE

Roy and Manna [111] described in their paper mathematical modeling of localized melting around graphite nodules during laser surface hardening of austempered ductile iron. Similar findings were presented by Grum and Šturm [112] at laser surface remelting in the transition zone, while heating at low power beam $P = 700$ W and traveling speed v of 60 mm/s, where dissolution of the graphite nodules at depth $z = 100$ μm below the surface occurred. At heating process in the region with dissolved graphite in austenite, lowering melting temperature was reached which resulted in local remelting. Figure 6.92 gives the variation of carbon concentration as a function of the graphite nodule distance from the surface. From the

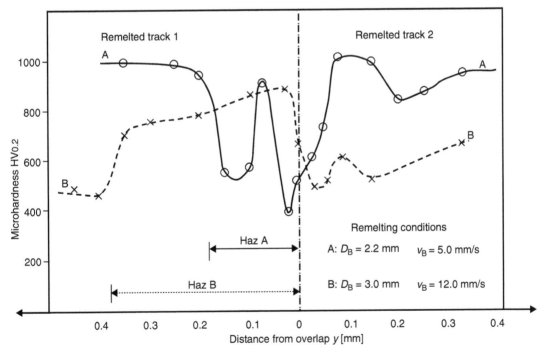

FIGURE 6.91 Microhardness profiles of tempered microstructures in overlap multitracks. (From Hawkes, I.C.; Steen, W.M.; West, D.R.F. Laser Surface Melt Hardening of S.G. Irons. In *Proceedings of the 1st International Conference on Laser in Manufacturing*, Brighton, UK; Kimmit, M.F., Ed.; Co-published by: IFS (Publications), Bedford, UK, Ltd. and North-Holland Publishing Company: Amsterdam, 1983; 97–108.)

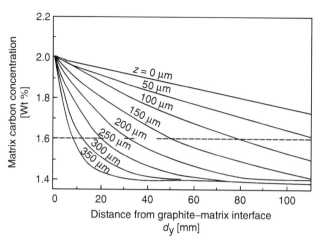

FIGURE 6.92 Variation of matrix carbon concentration as a function of distance from graphite matrix interface. (From Roy, A.; Manna, I. *Opt. Lasers Eng.* 2000, *34*, 369–383.)

diagram it can be concluded that carbon concentration in the matrix is higher on the surface and lower with decreasing depth below the surface. The dashed line shows 1.6 wt% carbon, which determines the minimum level of carbon enrichment and maximum width y_m of the localized remelted zone around the graphite nodule.

Figure 6.93 gives the matrix carbon concentration and maximum temperature as a function of graphite interface distance d_y, the dashed horizontal line denotes the effective temperature and maximum carbon solubility 1.6 wt% for austenite, and the vertical one represents the given remelt width (y_m) from the graphite surface.

Figure 6.94 shows the changing of remelt width (y_m) as a function of the depth below the surface (z) at given laser surface hardening conditions. Solid symbols in the diagram show theoretically predicted maximum melt width y_m around the graphite nodule as a function of

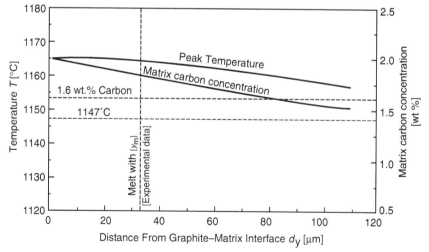

FIGURE 6.93 Variation of the peak temperature and matrix carbon concentration according to the distance from graphite matrix interface. (From Roy, A.; Manna, I. *Opt. Lasers Eng.* 2000, *34*, 369–383.)

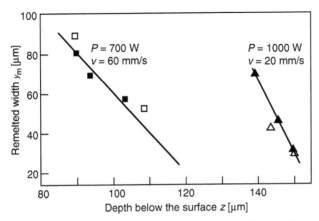

FIGURE 6.94 Variation of remelted width y_m according to remelting depth z. (From Roy, A.; Manna, I. *Opt. Lasers Eng.* 2000, *34*, 369–383.)

the depth below the surface. The open symbols note the experimental data at given laser hardening conditions and depth below the surface as well.

6.21.3 Transition between the Remelted and Hardened Layers

Grum and Sturm [112] studied rapid solidification microstructure in the remelting layer and microstructure changes in the hardened layer.

The application of laser surface remelting to nodular iron 400–12 causes the material to undergo microstructural changes. A newly created austenite–ledeburite microstructure with the presence of graphite nodules in the remelted layer and a martensite–ferrite micro-structure with graphite nodules in the hardened layer have been observed. Microscopy of the hardened layer was used to analyze the occurrence of ledeburite shells and martensite shells around the graphite nodules in the ferrite matrix. The thickness of the ledeburite and martensite shells was supported by diffusion calculations. The qualitative effects of the changed microstructures were additionally verified by microhardness profiles in the modified layer and micro-hardness measurements around the graphite nodules in the hardened layer.

The tests involved the use of an industrial CO_2 laser with a power of 500 W and Gaussian distribution of energy in the laser beam. The optical and kinematic conditions were chosen so that the laser remelted the surface layer of the specimen material. The specimens were made from nodular iron 400-12 with a ferrite–pearlite matrix that contained graphite nodules.

A characteristic of the transition area between the remelted and the hardened layer is that local melting occurs around the graphite nodules. Grum and Šturm [112] showed in Figure 6.95 a schematic representation of the sequence of processes occurring in the phases of heating and cooling:

- Matrix transforms into a nonhomogeneous austenite.
- Diffusion of carbon from graphite nodules into austenite.
- Increased concentration of carbon in the austenite around the graphite nodule lowered the melting-point temperature in local melting of part of the austenitic shell around the graphite nodule.
- After rapid cooling, a ledeburite microstructure is formed locally. This is then further surrounded by a martensite shell.

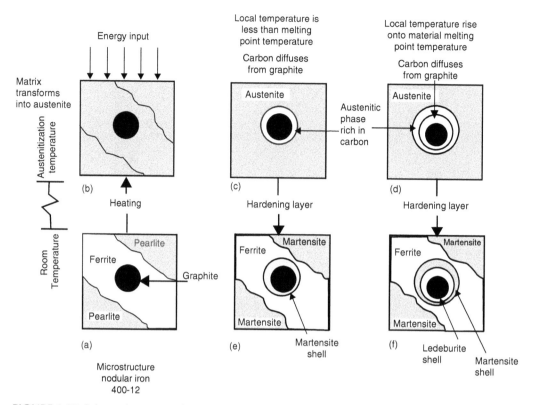

FIGURE 6.95 Schematic presentation of the microstructural changes in the transition area between the melted and the hardened zones. (From Grum, J.; Sturm, R. *Mater. Charact.* 1996, *37*, 81–88.)

The transition area is very narrow and very interesting from the microstructural point of view. However, it is not significant in determining the final properties of the laser-modified surface layer.

Figure 6.96 shows the measured and calculated thickness of martensite shells around graphite nodules with respect to their distance from the remelted layer. Our estimation is that the differences between the measured and the predicted martensite shell thickness are within the expected limits as the data on heat conductivity and diffusivity were chosen from the literature. The calculations confirmed the validity of the mathematical model for the determination of temperature T and time t remaining in a given temperature range, which enabled us to define the diffusion path of the carbon or, in other words, define the thickness of the martensite shell.

6.21.4 CIRCUMSTANCES FOR RAPID SOLIDIFICATION PROCESS OF CAST IRON

Three features are significant for the rapid solidification process after laser remelting [113]:

- The cooling rate $\dot{\varepsilon} = dT/dt$
- The solidification rate $R = dx/dt$, which characterizes crystal grain growth per unit time in the liquid–solid interface
- The temperature gradient $G = dt/dx$ across the liquid solid interface to a given location

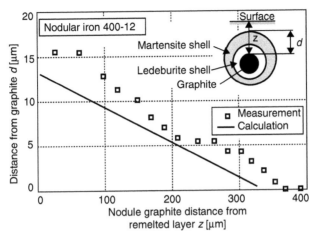

FIGURE 6.96 Comparison of measured and calculated thicknesses of martensite shells around graphite nodules in the hardened zone. (From Grum, J.; Šturm, R., *Master Charact.* 1996 *37*, 81–88.)

These parameters are connected by the equation

$$\dot{\varepsilon} = RG$$

Depending on the values of the above variables, different microstructures can be produced after solidification process. It is important that different microstructures can be obtained in the same material at various solidification conditions. In Figure 6.97, one set of parallel lines represent lines equal to G/R ratios and the other rectangular to the former one, which represents RG products (equal to $\dot{\varepsilon}$) [113].

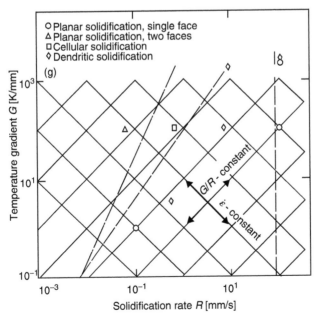

FIGURE 6.97 Variation of microstructure of a cast iron with solidification conditions. (From Bergmann, H.W. *Surf. Eng.* 1985, *1*, 137–155.)

The *G/R* parallel lines give similar solidification conditions before solidification increases with ε. Therefore an increasing nucleation frequency is obtained resulting in a finer microstructure of the same morphology. For the set of lines where ε is constant, the same grain size occurs, with changed solidification. At laser remelting it is difficult to verify experimentally the two limiting cases of solidification. Therefore special conditions must occur at rapid solidification:

- Substantial superheating of the melt which influences the heterogeneous nucleation
- Extreme temperature gradients which assure rapid, directional solidification
- Epitaxial growth on substrate crystals

In Figure 6.98, cooling rate, remelted depth, and dendrite arm spacing are correlated [113].

6.21.5 EVALUATION OF RESIDUAL STRESSES AFTER LASER REMELTING OF CAST IRON

Domes et al. [114] gave estimation of microstructure, microhardness, and residual stress distribution after laser surface remelting of gray cast iron and nodular iron.

The authors carried out experiments with an 18-kW CO_2 laser. The laser beam was shaped by a line integrating mirror to the laser spot area of 2×15 mm. The residual stresses were measured by x-ray diffractometer.

Figure 6.99 shows the microstructure of laser-remelted nodular iron consisting of primary austenite dendrites, which can solve more carbon at increasing solidification rate and eutectic ledeburite. Therefore with increasing solidification rate the ledeburitic portion decreases because of the formation of supersaturated austenite solid solution. A portion of austenite transforms, depending on the cooling rate, into martensite with residual austenite (RA), bainite, or pearlite. Figure 6.99a shows the microstructure of laser-remelted nodular iron after remelting with different preheating temperatures ($T = 500$ or $600°C$). The non-preheated specimens are characterized by a high content of residual austenite at preheating temperature, while after preheating to $400°C$ the microstructure consists mainly of martensite and bainite. Further raising of T_p generates higher portion of bainitic and/or pearlitic microstructure. The

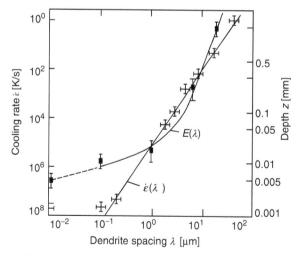

FIGURE 6.98 Variation of dendrite arm spacing with solidification parameters. (From Bergmann, H.W. *Surf. Eng.* 1985, *1*, 137–155.)

formation of residual austenite can be observed at preheating temperature up to a T_p of 500°C and at traveling speed of 1 m/min (Figure 6.99e). The residual austenite can be avoided either by further increasing T_p or by decreasing the traveling speed or by a post-heat treatment.

The surface hardness of laser-remelted nodular iron is about 700 to 800 HV. As shown in Figure 6.99b and Figure 6.99c, higher traveling speed and a lower preheating temperature T_p increase the microhardness. The microhardness of ledeburitic surface layers results from a law of mixture, where the hardness of the transformed austenite dendrites and eutectic ledeburite contributes at a given solidification rate. The microhardness in the remelted layer increases with raising temperature gradients and quenching solidification rate, respectively.

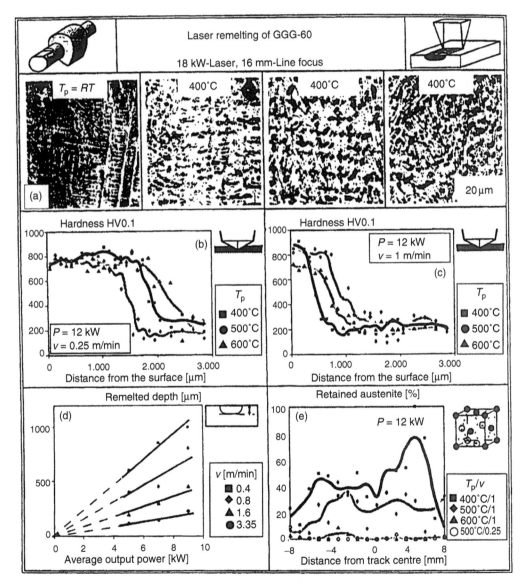

FIGURE 6.99 Microstructure and microhardness profiles after laser remelting of nodular iron at given laser-remelting conditions. (From Domes, J.; Müller, D.; Bergmann, H.W. Evaluation of Residual Stresses after Laser Remelting of Cast Iron. In Deutscher Verlag fuer Schweisstechnik (DVS), 272–278.)

The formation of cracks in the remelted layers depends strongly on the process parameters as well as on melt depth and preheating treatments, respectively. To achieve crack-free surfaces a martensitic transformation in the remelted or heat-affected zone has to be avoided. Figure 6.99e shows that preheating to more than 600°C at traveling speed $v = 1$ m/min is necessary to avoid retained austenite and therefore martensitic transformation. A smaller traveling speed allows a lower preheating temperature T_p, because the specimen is heated during surface remelting, so that the critical cooling rate is not achieved. A similar effect can be reached by preheating on temperature $T > T_{M_s}$ if post-heat treatment is added.

Figure 6.100a and Figure 6.100b show the influence of a preheat treatment on the longitudinal and transversal residual stress profiles at the surface after remelting of nodular iron. Because of high contents of retained austenite for low preheating temperatures only the residual stress after preheating to 500 and 600°C is analyzed. In both cases, compressive residual stresses in the modified layer can be observed, which reach higher values for higher preheating temperature. The reasons for this are as follows:

- The martensitic transformation in the HAZ can be avoided.
- The difference in heat expansion coefficient is rising because of the higher content of cementite in the remelted layer with increasing preheating temperature T_p.

The influence of the specimen geometry on the resulting longitudinal residual stress profiles in the modified layer is shown in Figure 100b and Figure 100c. After laser remelting of the flat specimens tensile residual in longitudinal direction can be found, while remelting of the web specimens leads to compressive residual stresses. A possible explanation for the difference in residual stress profiles is the stress relaxation during the laser-remelting process. The compressive residual stresses reach into the depth of the remelted layer and into the HAZ, which leads to tensile residual stresses (Figure 6.100d).

Figure 6.100e and Figure 6.100f show the deviations of the residual stresses in the center of the modified track. These specimens were preheated to $T_p = 500°$C and post-heat treated at the same temperature. For both types of specimens compressive residual stresses of about -100 MPa are measured at the surface. Differences between longitudinal and transversal residual stresses can only be observed for both types of specimens.

Pre- and post-heat treatment of the laser-treated materials has an influence on the portion of residual austenite in the layers as well as on the portion of martensite in the HAZ, which causes tensile residual stresses and the possible formation of cracks in the modified layer. To create a complete compressive residual stress profiles in the modified layer a suitable heat treatment has to be applied considering the specimen geometry.

Grum and Šturm [115] gave estimation of residual stress distribution after laser-remelting gray cast iron Grade 250 at different laser-beam guiding. Heating and cooling conditions in a relatively thin workpiece are very much dependent on laser-beam power, laser-beam diameter on the workpiece surface, interaction time, and the degree of overlapping of the remelted layer.

Heating and cooling conditions are also very much dependent on the paths along which the laser beam travels over the flat specimen surface. Figure 6.101 shows the variation of residual stresses as a function of the depth of the modified layer for gray iron Grade 250 after different paths of laser-beam travel [115,116]. The graphs present two measured curves, i.e., main stresses σ_1 and σ_2.

From the graphs we can conclude the following:

- Residual stresses have, in all cases of different laser-beam travel paths, a very similar profile differing only in absolute values (Figure 6.101a through Figure 6.101c). In the surface remelted layer, tensile residual stresses were found in the range 50 to 300 MPa.

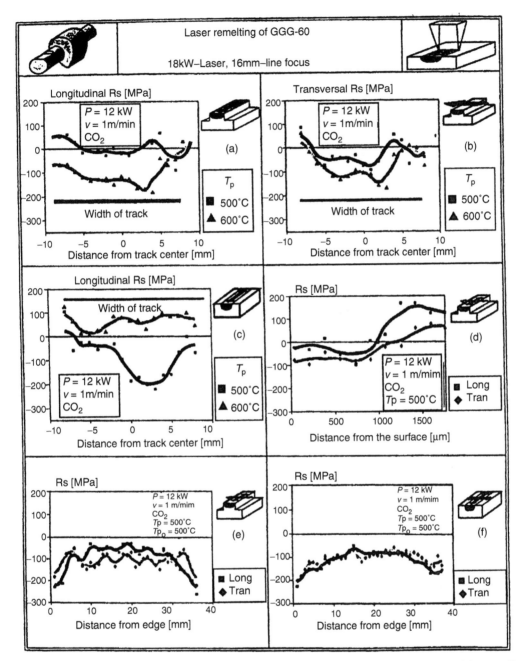

FIGURE 6.100 Residual stresses after laser remelting of nodular iron at given laser-remelting conditions. (From Domes, J.; Müller, D.; Bergmann, H.W. Evaluation of Residual Stresses after Laser Remelting of Cast Iron. In Deutscher Verlag fuer Schweisstechnik (DVS), 272–278.)

- The change from tensile to compressive residual stress takes place in the transition area between the remelted and hardened zones. Maximum compressive residual stress values were found in the middle of the hardened zone in the range 25 to 150 MPa. But there is one exception. In the case when the laser-beam travel path starts in the center of the workpiece (Figure 6.101c) no compressive stresses were found.

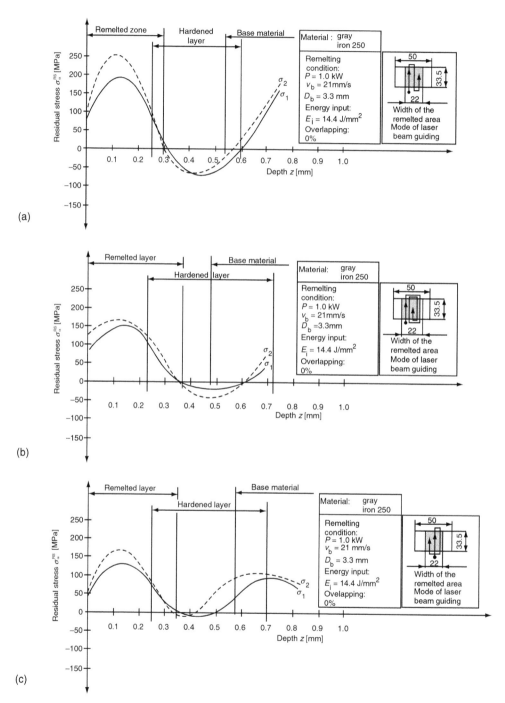

FIGURE 6.101 Residual stresses in gray iron grade 250 after different paths of laser beam travel: $P = 1$ kW; $v_B = 21$ mm/s; $D_b = 3.3$ mm; overlapping 0%; σ_1, solid line and open symbols; σ_2, broken line and closed symbols. (From Grum, J.; Šturm, R. *Mater. Sci. Technol.* 2001, *17*, 419–424.)

- By x-ray diffraction the amount of constituents present in the remelted layer was analyzed. From the above results it can be concluded that in the case when it is possible to produce the remelted surface layer without any residual austenite in the microstructure, tensile residual stresses are minimized. The higher the amount of martensite in the remelted layer the lower the tensile residual stresses.
- The remelted surface layer cracked in the direction of graphite flakes in the gray iron. It is believed that during the cooling process, owing to high temperature differences, extremely high tensile stresses were generated which exceeded the yield point of the material at increased temperature in the remelted surface layer.

With laser-remelting procedure a sufficient depth of the modified layer is achieved, in addition to desirable microstructure changes and good microhardness profiles of the modified layer. Surface remelting a cast microstructure with under-cooled graphite results in an extended HAZ in which the carbon diffuses into the ferrite. Success of the carbon penetration depends on the temperature or the distance from the surface. In the remelted layer crystals are supersaturated and grow initially epitaxially and with a planar front. This is then followed by dendritic growth of crystals. Depending on the cooling rate, carbide particles or carbide eutectic can remain between the dendrites. The graphite eutectic is completely dissolved in the short heating cycle.

After rehardening, the fine dendrites in the remelted layer at overlapping regions coalesce and the striped appearance of cementite is replaced by a random arrangement of martensite between the cementite regions. In the transition zone, the microstructure consists of finely dispersed carbides and martensite. Great attention has to be given to the selection of optimal laser remelting as different structure of the microstructure matrix and the size of flaky graphite can substantially affect the thermal condition of the material.

After laser remelting of gray and nodular irons, the measured residual internal stresses show a similar variation as well as absolute values at different kinds of guiding of the laser beam across the surface. Residual stresses are in all the cases of tensile type on the surface and then decrease to the depth of 0.3 mm where they transform into compressive stresses. The relatively great depths of transformation of tensile into compressive internal stresses confirm that even after finish grinding it is not possible to achieve the desired stress state in the material with more or less high compressive residual stresses on the surface.

6.22 ASSESSMENT OF DISTORTION AND SURFACE CHANGES

Careful assessment is being made of component distortion in view of the attractions of treating finish-machined components and of the stringent dimensional specifications. The present indications are that dimensional changes resulting from laser treatment are minimal, provided that a high-temperature stress relief is employed at the appropriate stage in machining and that the volume of transformed material is low.

Grum and Šturm [117] presented a new method of simultaneous measuring of the strain on the bottom side of the flat specimen during laser surface remelting process.

In the experiment, the chosen thickness of the specimen was 5.5 mm and was such that the maximum achieved temperature on the bottom side of the flat specimen was always less than 350°C. This temperature was dictated by strain measuring rosette used to measure the specimen strain.

Figure 6.102 shows the experimental system for continuous measurements of the strain and temperature on the bottom side of specimen during the process of laser remelting, going on the top side of the flat specimen. On the bottom side of the specimen were placed

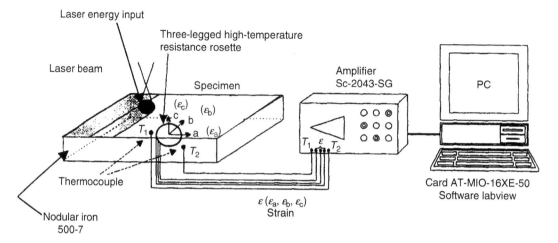

FIGURE 6.102 Experimental system for measuring the temperatures and strain of the flat specimen during the laser-remelting process. (From Grum, J.; Šturm, R. *J. Mater. Eng. Perform.* 2000, *9*, 138–146.)

a three-legged, 45° high-temperature self-compensating resistance-measuring rosette and two thermocouples.

The thermocouples, placed in the longitudinal direction of the specimen on the left and right sides next to the high-temperature resistance-measuring rosette, continuously measure the temperature and thus define the temperature cycles on the bottom side of the flat specimen induced on the specimen's top side. This kind of placement enables the continuous monitoring of the strain and specimen temperature during the remelting process.

The measurements of specimen strains during the remelting and cooling process and subsequent calculations of the main residual stresses were made on the nodular iron 500-7 (ISO) with a pearlite–ferrite matrix.

Figure 6.103 presents the results of the measured temperature cycles in the middle of the specimen bottom side during the remelting process for three different laser traveling ways. The temperature gives information on the temperature changes in heating and cooling of the material on the bottom side of the specimen. To measure the effects of different laser-beam traveling ways and the different number of laser-beam passes across the surface, thermocouples registered partial temperature during the heating process. In the phase of heating the specimen, partial temperature occurs with a period of laser-beam passage across the specimen surface. Partial temperature occurs with a period of laser-beam passage across the specimen surface. In the phase of heating, the highest peaks of partial temperature can be noted with the zigzag laser-beam traveling ways. The process of cooling can be described in general from the moment the maximum temperature in the specimen material on the bottom side was reached. It is possible to conclude that the time at maximum temperature depends on the laser-beam traveling way. For the chosen three laser-beam traveling ways at 0% overlap, the same amount of energy $E = 14.4$ J/mm^2 was provided. Each laser-beam passage across the specimen surface induces gradual heating of the material, the result of which is preheating of the material before the next laser-beam passage. The increased temperature of the specimen material makes the yield point of the material slightly lower, which may, with the given internal stresses, result in strain of the specimen. How much the specimen will preheat depends on laser-beam traveling way. Considering the three different laser-beam traveling ways across the specimen surface, the following can be stated:

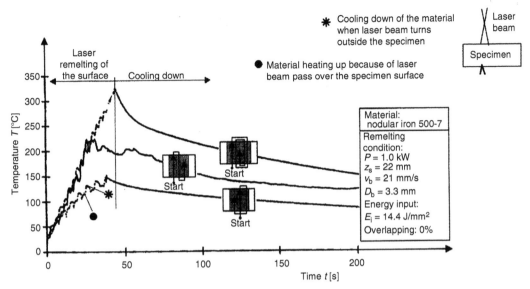

FIGURE 6.103 Temperature measured on bottom side of the specimen during laser surface remelting at various laser beam travels. (From Grum, J.; Šturm, R. *J. Mater. Eng. Perform.* 2000, *9*, 138–146.)

- Lowest maximum temperature is reached at circular beam traveling way, starting in the middle, of the remelted area. In this case, a considerable high cooling rate in the remelted layer is ensured, which has influence on the amount of the residual austenite.
- Higher temperatures are achieved at laser-beam traveling way. From known volume changes in phase transformations, a smaller amount of residual austenite has an influence on the size of tensile residual stress profiles in the modified layer.
- Maximum temperature in the material on the bottom side of the specimen is achieved at circular mode of laser-beam traveling way beginning on the edges of the remelted area of the specimen. On completion of cooling, the remelted layer contains a smaller proportion of residual austenite, which strongly lowers the residual stress profiles.

In cooling, the just-remelted surface layers solidify, and cooling is continued in solid state involving phase transformations, which are reflected in a characteristic microstructure in the lower area of the modified surface layer. As a result of microstructural changes in solid state, i.e., due to the austenite–martensite transformation, an increase in the volume of the layer takes place. The volume of the remelted layer decreases slightly, which causes the occurrence of tensile stresses in it during the process of cooling or at ambient temperature. The decrease in the volume of the remelted surface layer is greater than the increase in the volume of the hardened layer. Therefore the result is the strain of the specimen.

In Figure 6.104, results of strain measurements with different traveling directions are presented. The following can be seen:

- A very different progress of strain changes in the particular directions.
- Largest strain ε_a is found in the direction of the longer side of the specimen.
- In the direction of the shorter side of the specimen, i.e., in the transverse direction, the strain ε_c is at first of tensile nature but after six passes it changes into a compressive one.

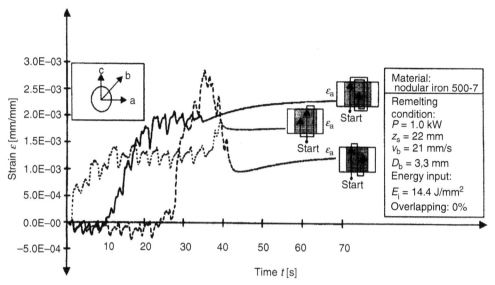

FIGURE 6.104 Time dependence of specimen strain. (From Grum, J.; Šturm, R. *J. Mater. Eng. Perform.* 2000, *9*, 138–146.)

A very important conclusion is that the size of tensile strain on the bottom side of the specimen is increasing with the increase in temperature of the specimen material.

Continuous measurement of specimen strain during laser surface remelting with a resistance-measuring rosette is a new method, not yet reported in the literature, of describing the strain events in the specimen. It has been found that the information about the time-dependent changes in specimen strain contributes to better knowledge of the conditions during laser remelting and thus better process of optimization.

6.23 OPTIMIZATION OF THE LASER SURFACE REMELTING PROCESS

For making decisions in determining the optimal conditions Grum and Šturm [118] worked out a set of descriptive evaluation criteria regarding the condition of the remelted layer, hardened layer, and the surface. This enabled them to define the highest and lowest possible specimen traveling speed which will still ensure a sufficient thickness of the modified layer of acceptable quality.

Optimization of the laser surface remelting process is related to the selection of the remelting conditions such as laser power, the spot diameter at the surface of the workpiece, and the relative speed of movement between the laser beam and the workpiece. If the power of laser radiation is directed to the surface of the workpiece, then the radiated area is defined with the spot diameter and power density. The calculated power density can represent a satisfactory relative comparison of the remelting processes, which occur at different remelting conditions. Actually, it is the energy input that is important for the evaluation of remelting, which is dependent on the power density as well as on the traveling speed of the laser beam. The effect of the input energy is related to material absorptivity.

Optimization of the laser surface-remelting process for a thin-surface layer can be accomplished in four ways, that is:

- Selection of heating conditions for a given remelting depth
- Selection of heating conditions and the mode of guiding the laser beam by minimizing the strain of the machine parts
- Selection of heating conditions for the formation of compressive or minimum tensile residual stresses at the surface and in the thin-surface layer
- Selection of heating conditions for the formation of compressive or minimum tensile residual stresses at the surface and in the thin-surface layer and a requirement for a minimum strain of the machine part

The optimization of the laser-remelting process has to be performed for the lamellar and nodular iron separately [119].

6.24 FATIGUE PROPERTIES OF LASER SURFACE HARDENED MATERIAL

The fatigue properties of ferritic S.G. iron were studied with the pull–pull test [113,120]. The mean tensile stress was 50% of the yield point, i.e., ≈ 150 N mm^2. A 10-Hz frequency of various amplitudes was then applied to the specimens and the fatigue limit determined. Untreated specimens were ground after machining. As there is no relevant data in the literature, optimized processing parameters for laser surface hardening had to be defined. This was done by keeping the depth of the hardened layer as a constant at about 10% of the thickness. Compared to the untreated S.G. iron, a decrease in fatigue limit was found at laser surface hardened specimens. In addition, a spiral laser track gives better results than a longitudinal multitrack. Grinding the surface, i.e., smoothing small amounts of roughness produced at laser surface hardening specimens, leads to a small decrease in fatigue limit as compared with untreated specimens.

The most favorable values (Figure 6.105) were found when the specimens were annealed at 240°C for 2 h after laser surface hardening with helium and subsequently ground [113,120]. It can be seen that the untreated specimens for a medium mean load can carry a higher amplitude. However, with increasing load the advantage of strengthening due to laser remelting allows operation at mean loads over the yield point of the untreated material.

In Figure 6.106 the corresponding Smith diagram gives the fatigue life as a function of the mean load [120]. Pure fatigue behavior was found in laser surface hardened specimens, which may partly depend on the surface finish and partly on residual stresses. Grinding at the

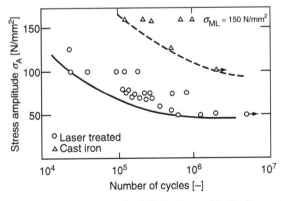

FIGURE 6.105 Fatigue behavior of laser-remelted S.G. iron with ferrite matrix. (From Bergmann, H.W. Laser surface melting of iron-base alloys. In *Laser Surface Treatment of Metals*; Draper, C.W., Mazzoldi, P., Eds.; Series E: Applied Science—No. 115, NATO ASI Series; Martinus Nijhoff Publishers: Dordracht, 1986; 351–368.)

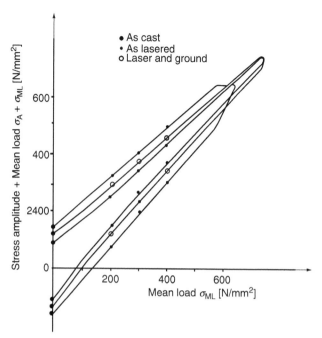

FIGURE 6.106 Smith diagram. (From Bergmann, H.W. Laser surface melting of iron-base alloys. In *Laser Surface Treatment of Metals*; Draper, C.W., Mazzoldi, P., Eds.; Series E: Applied Science—No. 115, NATO ASI Series; Martinus Nijhoff Publishers: Dordracht, 1986; 351–368.)

surface improves the fatigue behavior but is not able to give properties equivalent to the untreated specimens.

6.25   WEAR PROPERTIES OF LASER SURFACE HARDENED MATERIAL

The advantage of laser-remelted S.G. iron was demonstrated for rolls which run dry against each other with a fixed relative slip, one wheel being driven and the other partially braked (Figure 6.107). Various stresses were applied (e.g., 500 and 1000 N/mm^2) and the humidity was controlled. For comparison, ~1% slip was used. It is obvious that the wear properties of laser-remelted cast iron are superior to those of conventional hardened steels. Combinations of lasered irons and steels must be avoided, as significant deformation of the steel occurs. When laser treatments were carried out under He, better results were found than with other gases. Excellent results were obtained when laser-remelted S.G. irons were used in combination with TiN- or TiC-coated, hardened steels, the TiN/TiC surface exposed to the wear. After the tests, no deformation is visible on the ledeburitic surface. Good wear properties are obtained in the combination of laser-treated S.G. iron with nitrided or borided steels. A comparison between laser-treated and TIG-remelted ledeburitic surfaces favors the laser hardening process.

Grum and Šturm [121] presented research results of laser surface remelting on different qualities of gray and nodular irons. The effects of laser hardening are expressed in the qualitative changes in the microstructure resulting in higher hardness and better wear resistance of the modified machine part surface [122,123].

In the gray iron, an extremely high hardness around 1000 HV$_{0.1}$ on the surface of the remelted area can be noted, which is falling uniformly to the value 620 HV$_{0.1}$ in the transition

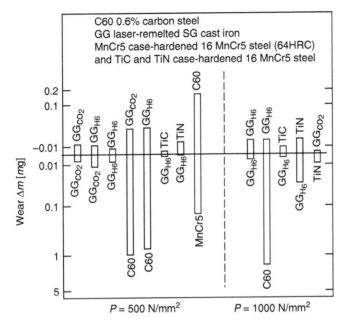

FIGURE 6.107 Comparison of wear properties of laser and TIG surface remelted S.G. iron with those of various steels, with and without surface treatments: $P =$ load on rolls during test. (From Bergmann, H.W. Laser surface melting of iron-base alloys. In *Laser Surface Treatment of Metals*; Draper, C.W., Mazzoldi, P., Eds.; Series E: Applied Science—No. 115, NATO ASI Series; Martinus Nijhoff Publishers: Dordracht, 1986; 351–368.)

layer. The drop in microhardness then reoccurs at a depth of 250 μm (depth of the remelted area), which is attributed to changes in size and concentration of the residual austenite. Then follows hardened layer with a predominantly martensitic microstructure ensuring an increased hardness ranging from 700 to 830 $HV_{0.1}$. On the other hand, in the nodular iron, the remelted area displays a rather homogeneous distribution of the austenite and ledeburite giving very uniform profile of the measured microhardnesses ranging from 730 to 880 $HV_{0.1}$. In the heat-affected or hardened layer there is a fine-grained martensitic microstructure with a ferrite surrounding, the microhardness ranging from 540 to 680 $HV_{0.1}$.

Figure 6.108 presents a comparison of microhardness values subsequent to surface remelting by a single trace and by several overlapping traces. The results of microstructure and the measured microhardness in the heat-affected layer confirm that in surface remelting by overlapping, the remelted traces of tempering martensite into the fine bainite microstructure in the nodular iron come into effect. This effect is seen in Figure 6.108b in the lowering of microhardness values. In the gray iron, on the other hand, the overlapping of the remelted traces leads to the effect of material preheating, and for this reason we can achieve greater depth of the heat-affected area. A comparison of the graphs in Figure 6.108a and Figure 6.108b shows that the effect of preheating is significant only in case of unfavorable flaky graphite.

In laser surface remelting, traces of heat treatment remain on the specimen. Their dimensions depend on laser-beam parameters, conditions of rotational and translatory motion of the specimen, and kind of material. Table 6.6 shows the average values of the measurements of the depth of the remelted and heat-affected layer measured from the workpiece surface, as shown on the right side of the table.

The wear resistance test was performed on an Amsler machine, the specimens being made of gray and nodular irons with the remelted surface in sliding lubricated contact with a

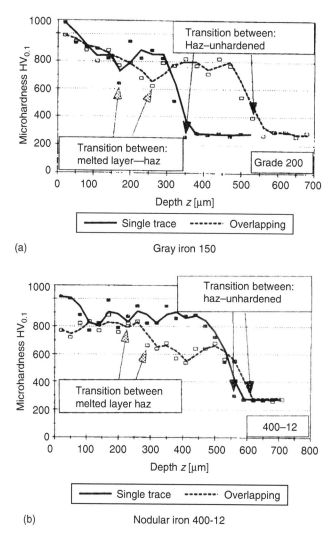

(a)   Gray iron 150

(b)   Nodular iron 400-12

FIGURE 6.108 Microhardness profiles in modified layer for single and overlapping traces. (From Grum, J.; Šturm, R. Laser surface melt-hardening of gray and nodular iron. In *Proceedings of the International Conference on Laser Material Processing*, Opatija; Croatian Society for Heat Treatment, 1995; 165–172.)

hardened heat-treatable steel C55 (ISO). The normal loading force was 700 N at a sliding pressure of 700 N/cm^2. The wear resistance of the discussed casts with a laser-remelted surface was determined by the measurements of mass losses and by defining the wear coefficient k, which is expressed as [124,125]:

$$k = \frac{W}{\mathrm{Fn}L} \,[\mathrm{mm}^3/\mathrm{Nm}]$$

where W [mm^3] is the volume of wear, Fn [N] is the applied load, and L[m] is the operating path.

Figure 6.109 is a graphical presentation of the results of cumulative loss of mass and calculated values of wear coefficient versus operating path. From the bar chart, we can see

TABLE 6.6
Measured Depths of Remelted and Modified Layer

	Depth z [mm]			
	Remelted Layer		Modified Layer	
Material	$Z_{R\ min}$	$Z_{R\ max}$	$Z_{M\ min}$	$Z_{M\ max}$
Grade 200	0.12	0.29	0.37	0.57
400-12	0.19	0.30	0.47	0.62

Source: From Grum, J.; Šturm, R. Laser surface melt-hardening of gray and nodular iron. In *Proceedings of the International Conference on Laser Material Processing*, Opatija; Croatian Society for Heat Treatment, 1995; 165–172.

that the nodular iron wears down more than the gray iorn and that the wear coefficient after the initial running-in period drops significantly in all materials Figure 6.109a. Considering the recommendation in the literature, we can, however, say that the wear coefficient is, in all the cases, substantially lower than the one allowable for machine parts ($k < 10^{-6}$ mm^3/Nm) [125]. This shows that the conditions of the discussed sliding pair are very favorable.

The cumulative loss of mass (Figure 6.109b) presents the growing loss of mass after different operating paths chosen as measuring spots. The remelted surface of the flake graphite gray iron has, by almost 100 HVm, higher hardness on the surface than the nodular iron, which is reflected, as the test results have shown, in the wear resistance. Therefore the loss of mass on nodular iron specimens is much bigger, even by two or three times. On the other hand, from the experimental results of microhardness measurements and structure analysis of the different qualities of nodular iron, we can note no differences in the quality of the remelted layer. However, it was found that a bigger loss of mass on nodular iron specimens may be due to the following reasons:

- A fall-out of graphite nodules of the surface of the remelted layer during wear test.
- A smaller amount of chemically bound carbon in the cementite or ledeburite, which results in a lower hardness of the remelted layer.
- A higher ductility under the remelted layer, which results in higher deformation of the surface layer in loaded condition during the wear test.

The experimental results have confirmed that with a low-power laser source it is possible to achieve a sufficient thickness of the modified layer if surface remelting is applied. Because of the morphology of its graphite, gray iron is a very demanding material for heat treatment even at suitably chosen machining conditions, the obtained surface was uneven, and had small craters, cracks, and gaseous porosity in the remelted layer. On the other hand, in nodular iron no such irregularities could be observed. Yet it should be mentioned that, because of incomplete dissolution of graphite nodules, a ledeburite microstructure is obtained with a slightly lower hardness and thus a lower wear resistance compared to gray iron.

Fukuda et al. [126] carried out the wear test on the pin-and-disc machine (Figure 6.110). It was found that laser-treated iron shows remarkable wear resistance by comparison with non-heat-treated steel, and also good wear resistance compared to induction surface hardened cast steel. There exists a quantitative relationship between the surface hardness of laser-treated nodular iron and its martensite portion. The surface hardness of laser-treated nodular iron is lower than that of induction surface hardened cast steel.

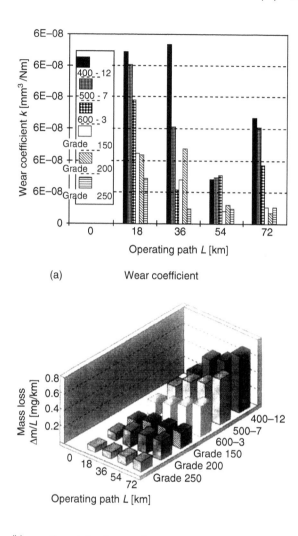

(a)   Wear coefficient

(b)   Cumulative losses of mass

FIGURE 6.109 Wear of gray iron 200 and nodular iron 400-12 with laser-remelted surfaces in sliding lubricated contact with a hardened heat-treatable C55 steel. (From Grum, J.; Šturm, R. Laser surface melt-hardening of gray and nodular iron. In *Proceedings of the International Conference on Laser Material Processing*, Opatija; Croatian Society for Heat Treatment, 1995; 165–172.)

Transformation hardening of steels or cast irons having the pearlitic microstructure is very efficient despite the comparatively short interaction time in laser heating. Cementite lamellae in the ferritic matrix disintegrate very fast. The change into the face-centered cubic lattice with the austenitic microstructure takes a very short time, too. Moreover, some time is available for the migration of carbon atoms. The migration of carbon atoms increases in alloyed steels with carbides of alloying elements and in gray or nodular cast irons having a larger content of carbon that will dissolve in austenite when heated and contribute to an increased hardness of the surface layer after cooling. Thus it often occurs that the hardness at the surface and in the surface layer obtained in transformation hardening is higher than that obtained after induction hardening of the same steel. Many authors monitored the variations of hardness in the hardened surface layer after laser hardening and induction surface hardening. It is extremely

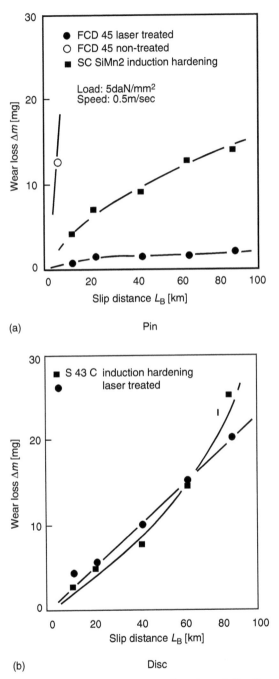

(a)   Pin

(b)   Disc

FIGURE 6.110 Wear loss on pin and disc machine as a function of slip distance. (From Fukuda, T.; Kikuchi, M.; Yamanishi, A.; Kiguchi, S. Laser hardening of spheroidal graphite cast iron. In *Proceedings of the Third International Congress on Heat Treatment of Materials, Shanghai 1983*; Bell, T., Ed.; The Metals Society: London, 1984; 2.34–2.44.)

TABLE 6.7
Conditions Used in the Dry Wear Tests

Type of Specimen	Pin Type	Load F [daN]	Surface Preparation	
As cast	Steel*	2.5	As received (ground)	*Hardness 927 HV grid 1000
Hardened	SI_3N4	5.0	Ground on SiC paper	
Remelted	SI_3N_4	5.0	As received (ground)	

Source: From Magnusson, C.F.; Wiklund, G.; Vuorinen, E.; Engström, H.; Pedersen, T.F. Creating tailor-made surfaces with high power CO_2-lasers. In *Proceedings of the 1st ASM Heat Treatment and Surface Engineering Conference Mater. Sci. Forum*; 1992, Vol. 102–104, 443–458.

important that in laser hardening a higher hardness of the hardened surface layer is achieved. Then follows a very fast variation of hardness in the transition zone from the hardened to the unhardened layer. In induction hardening this transition from the hardened to the unhardened zone is very gentle, which allows a better behavior of the material of dynamically loaded parts. This indicates that induction hardening provides more favorable hardness profiles in the hardened layer and thus also a better inoperation behavior of machine parts. It seems that the martensite surrounding the graphite nodules and in the matrix, which has been transformed by the laser hardening, prevents the ferrite and the graphite nodule from deforming under the wear.

Magnusson et al. [127] in his paper titled "creating tailor made surfaces with high power CO_2-lasers" connected various types of surface modification on wear resistance. They investigated laser hardening, remelting, and alloying of cast irons and the influence on microstructure, wear properties, and resistance to tempering.

Untreated specimens and specimens with laser-hardened single circular tracks were exposed to dry sliding wear in a conventional pin-on disc machine. Additional specimens were treated with overlapping hardened as well as remelted tracks. Test conditions are given in Table 6.7.

For laser-treated specimens, the wear rates are in all cases small but lower for surface remelted than for surface hardened specimens. The results of the wear tests are presented in Figure 6.111. For the laser-treated specimens, the wear process leads to tempering of the surface layer, which gives a decrease in hardness of 100 HV.

The wear rate as well as the wear mechanism depends on the type of cast iron. The lowest wear rate of the surface hardened and surface remelted condition was found out. On surface hardened gray cast iron and also in the case of ferritic–pearlitic nodular iron, the wear loss appears to take place by brittle fracture in the material at the graphite flakes or nodules.

The surface hardness of cast irons can be increased from about 200 to 500–800 HV by surface hardening, remelting, and alloying by a CO_2 laser. The wear rate of different cast irons will decrease by a factor of approximately 10 after surface hardening or remelting. Finally, results have shown that it is possible to combine wear resistance and tempering resistance by an appropriate selection of cast iron and laser hardening process.

6.26   REMELTING OF VARIOUS ALUMINUM ALLOYS

Coquerelle and Fachinetti [128] presented friction and wear of laser-treated surface, i.e., laser remelting at various aluminum alloys. Hypoeutectic, eutectic, and hypereutectic aluminum silicon alloys have been tested for identification of tribological properties at assuring the

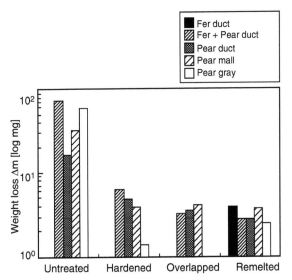

FIGURE 6.111 Wear test results for different gray cast and nodular irons. The untreated specimens were tested with a 2.5-daN load on a steel pin, the other with a 5-daN load on a SI3N4 pin. (From Magnusson, C.F.; Wiklund, G.; Vuorinen, E.; Engström, H.; Pedersen, T.F.; Creating tailor-made surfaces with high power CO_2-lasers. In *Proceedings of the 1st ASM Heat Treatment and Surface Engineering Conference Mater. Sci. Forum*; 1992, Vol. 102–104, 443–458.)

lowest possible thermal expansion coefficient. All mechanical properties were desired for such application as cylinder blocks and piston for automotive industry.

Before laser treatment the face of the flat specimens was phosphated or blacked with a graphite, so that the laser-beam absorption could be increased. The single track is larger at blacked surface conditions than after phosphatation. The remelted surface layer of the alloys when the power density was reached 10^5 W/cm^2 at only 1-mm laser-beam spot. The remelted depth increases with the portion of added silicon in the alloy at the same power density but decreases with higher traveling speed. The microstructure was determined with interdendritic space, which depends on the interaction time and chosen traveling speed. Figure 6.112 is the influence of the traveling speed on the interdendritic spaces in eutectic alloy (12% Si) presented. A substantial reduction of the silicon particle size has been achieved on hypereutectic alloy at traveling speed $v = 1$ cm/s below 5 μm, and at traveling speed $v = 100$ cm/s it was smaller than 1 μm.

Figure 6.113a and Figure 6.113b show the influence of laser-beam traveling speed on surface hardness after remelting alloy with 12% Si or 17% Si. For the tribological test the remelted surface at different degrees of overlapping was made. Table 6.8 shows the Vickers microhardness value at loading 3 N for eutectic aluminum silicon alloy after remelting with given conditions. The microhardness of AlSi12 alloy is 83 HV$_{0.3}$ before treatment and after single-track remelting is 160 HV$_{0.3}$. The hardness of the overlapping tracks is always lower than that of the single track.

Coquerelle and Fachinetti [128] conducted sliding and wear experiments by a pin-on-disk machine at ambient temperature. Figure 6.114 shows the wear rate for three aluminum silicon alloys before and after laser remelting. Wear rate depends on the silicon portion and silicon particle size. The wear rate is determined by volume mass loss of the particle [mm^3] on sliding path [km]. The wear rate in mm^3/km is about two times greater at soft material than at fine-grained microstructures after remelting in hypoeutectic alloy. With the increase of the silicon

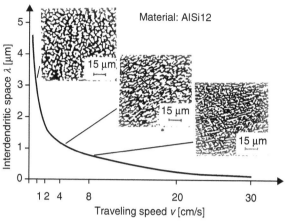

FIGURE 6.112 Influence of the traveling speed on interdendritic spaces (Si12 wt%). (From Coquerelle, G.; Fachinetti, J.L. Friction and wear of laser treated aluminium–silicon alloys. Paper Presented at the European Conf. on Laser Treatment of Materials, Bad Nauheim, 1986. In *Laser Treatment of Materials*; Mordike, B.L., Ed.; DGM Informationsgesellschaft Verlag: Oberursel, 1987; 171–178.)

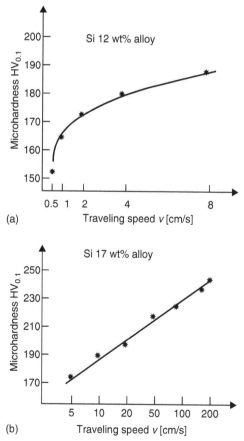

FIGURE 6.113 Influence of the traveling speed on the microhardness Al–Si alloys. (From Coquerelle, G.; Fachinetti, J.L. Friction and wear of laser treated aluminium–silicon alloys. Paper Presented at the European Conf. on Laser Treatment of Materials, Bad Nauheim, 1986. In *Laser Treatment of Materials*; Mordike, B.L., Ed.; DGM Informationsgesellschaft Verlag: Oberursel, 1987; 171–178.)

TABLE 6.8
Vickers Hardness Values

			Surface Generation		
Eutectic Alloy	Before Treatment	Single Scan	Mode a/2	Mode a/2	Mode β
Power: 3.0 kW	83 $HV_{0.3}$	160 $HV_{0.3}$	140 $HV_{0.3}$	100 $HV_{0.3}$	150 $HV_{0.3}$
Speed: $\nu = 1.0$ cm/s		93%	70%	20%	80%

Source: From Coquerelle, G.; Fachinetti, J.L. Friction and wear of laser treated aluminium–silicon alloys. Paper Presented at the European Conf. on Laser Treatment of Materials, Bad Nauheim, 1986. In *Laser Treatment of Materials*; Mordike, B.L., Ed.; DGM Informationsge sellschaft Verlag: Oberursel, 1987; 171–178.

portion in the alloy, wear rate decreased. Figure 6.115 shows the influence of silicon particle size on wear rate at various remelting conditions for hypereutectic alloy. If the silicon particle size decreases from 100 to 2 μm, the wear rate decreases 4 times. The authors presented also the significant wear mechanisms for various types and states of alloys, such as the following:

- A work-hardened layer can be observed on hypoeutectic alloys (5% Si) at soft state. The depth of this layer is about 25 μm and transverse cracks are often present.
- The wear mechanism in eutectic alloy is not the same as in hypoeutectic alloy. Before laser surface remelting, three phases—Si, $Al_8Si_6Mg_3Fe$, and Al_3Ni—are very large and harder than a matrix. Under Hertzian pressure, the harder phases were fractured and they additionally hardened the soft matrix in the surface layer. Subsequently, the wear rate is reduced. Meanwhile, the harder phases in contact with other surface react abrasively and the wear rate is still high.
- After laser remelting the size of the dendrites is lower than in the alloy or the soft state. For the eutectic alloy, almost all the phases of silicon have been melted again and after solidification the size of the particles is lower than 2 to 4 μm. The wear rate decreases by about 100%.

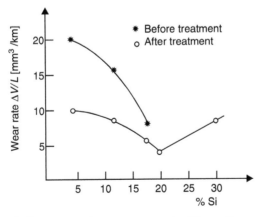

FIGURE 6.114 Influence of silicon percentage on wear rate. (From Coquerelle, G.; Fachinetti, J.L. Friction and wear of laser treated aluminium–silicon alloys. Paper Presented at the European Conf. on *Laser Treatment of Materials*, Bad Nauheim, 1986. In *Laser Treatment of Materials*; Mordike, B.L., Ed.; DGM Informationsgesellschaft Verlag: Oberursel, 1987; 171–178.)

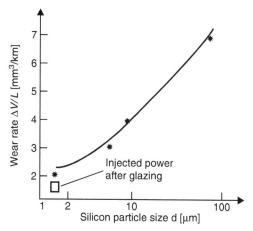

FIGURE 6.115 Influence of silicon particle size on wear rate (hypereutectic alloy). (From Coquerelle, G.; Fachinetti, J.L. Friction and wear of laser treated aluminium–silicon alloys. Paper Presented at the European Conf. on Laser Treatment of Materials, Bad Nauheim, 1986. In *Laser Treatment of Materials*; Mordike, B.L., Ed.; DGM Informationsgesellschaft Verlag: Oberursel, 1987; 171–178.)

For the hypereutectic alloy the mechanism is similar to the one observed in the eutectic alloy. The wear rate in hypereutectic alloy is lower than that in eutectic alloy. The wear rate increases with the size of the silicon particles because fracture is more often found on specimens at higher particle size.

Antona et al. [129] tested AlSi7Cu3 aluminum cast alloy at various remelting conditions. The surface layer of the individual flat specimens was assessed on the basis of metallographic inspection to point out the maximum depth, hardness, and microstructure. Figure 6.116 shows the depth of the remelted single laser track with three rectangular laser spot sizes and three amounts of traveling speeds. The remelted depths were measured at the center of the section taken at 20 and 40 mm from the starting point of the laser remelting. From microstructure analysis it can be concluded that the most remarkable effect of rapid solidification is the refinement of the dendritic microstructure. Figure 6.117 shows a micro-structure at a typical area of the remelted layer.

In general the authors distinguished three areas with different microstructure in the remelted surface layer as follows:

- An interface between remelted area and base material with partially coarse microstructure and microporosities
- An area with a very fine microstructure near the surface with typical equiaxial growth of the dendrites
- An intermediate area between the two previously mentioned areas with column type of dendrites

Luft et al. [130] reported on laser surface remelting of eutectic aluminum alloys.

In the past, surface remelting was researched only using eutectic or near eutectic Al–Si alloys for automotive pistons. It was observed that rapid remelting resulted in a very fine grained microstructure. Much research on rapid solidification has shown that many other alloys can have mechanical properties improved. In general nonferrous alloys were investigated in which the solubility in solid state was limited or where at high melting temperature point intermetallic compounds could be formed. Typical promising solute elements are Fe,

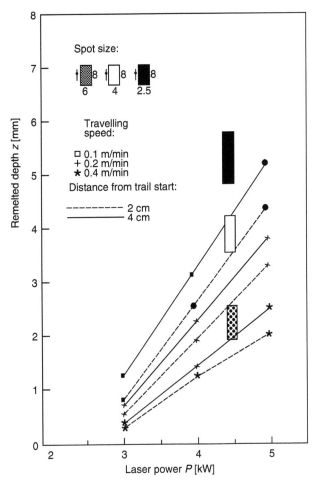

FIGURE 6.116 Remelting depth according to laser-remelting conditions. (From Antona, P.L.; Appiano, S.; Moschini, R. Laser surface remelting and alloying of aluminum alloys. Paper Presented at the European Conference on Laser Treatment of Materials, Bad Nauheim, 1986. In *Laser Treatment of Materials*; Mordike, B.L., Ed.; DGM Informationsgesellschaft Verlag: Oberursel, 1987; 133–145.)

Cr, Mn, and Ni, which have influence on the refining mechanical properties of alloys. Rapid solidification at remelting surface layer shows that the range of solid solubility can be extended. The second method for improving mechanical properties is precipitation hardening of remelted surface layer. The electron beam heat treatment must be carried out in a vacuum, which limits the size of the workpiece.

In Table 6.9 the composition of the investigated aluminum alloys after laser surface remelting process and the melting point of the intermetallic compounds are presented.

The laser conditions for the microstructural investigations were the same for all alloy: focused beam with 2-kW power and laser-beam traveling speed rate of 4 mm/min. The influence of the traveling speed on the remelted area for laser and electron beam is given in Figure 6.118. As would be expected, laser-beam heating is less efficient than electron beam heating.

The microstructures showed fractured intermetallic compounds aligned in the rolling direction in an aluminum matrix. At laser surface remelting process these depend on the size of the melting point of the interme-tallic compound and the interaction time, which have

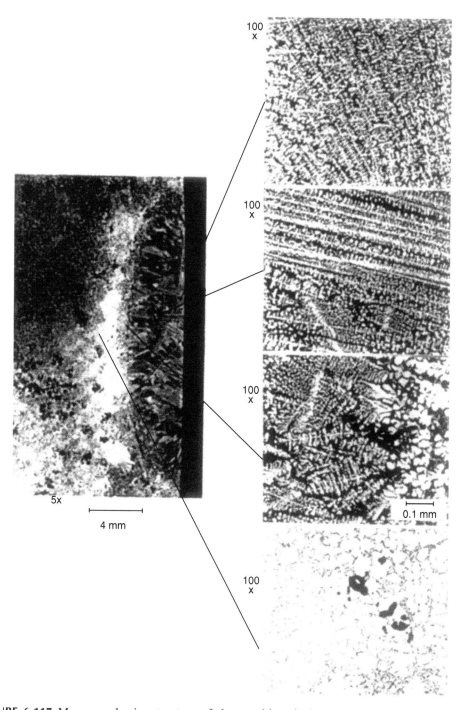

FIGURE 6.117 Macro- and microstructure of the remelting single track at a typical area. (From Antona, P.L.; Appiano, S.; Moschini, R. Laser surface remelting and alloying of aluminum alloys. Paper Presented at the European Conference on Laser Treatment of Materials, Bad Nauheim, 1986. In *Laser Treatment of Materials*; Mordike, B.L., Ed.; DGM Informationsgesellschaft Verlag: Oberursel, 1987; 133–145.)

TABLE 6.9
Various Compositions of Aluminium Alloys and Specific Data About Solubility, Type of Intermetallic Compounds, and Melting Temperature

Solute Element	Content [wt%]	Solubility [wt%]	Intermetallic Compounds	T_m [°C]
Cr	6	0.72	$CrAl_7$	725
			$CrAl_{11}$	940
			$CrAl_4$	1010
			$CrAl_3$	1170
Mn	6	1.40	$MnAl_6$	710
			$MnAl_4$	822
			$MnAl_3$	880
Fe	6 and 8	0.00	$FeAl_3$	1160
Ni	6	0.00	$NiAl_3$	854
Ti	1.5	0.20	$TiAl_3$	1340
Zr	1.5	0.28	$ZrAl_3$	1580
Si	11.8	1.65	—	—

Source: From Luft, U.; Bergmann, H.W.; Mordike, B.L. Laser surface melting of aluminium alloys. In *Paper Presented at the European Conference on Laser Treatment of Materials, Bad Nauheim, 1986*. Mordike, B.L., Ed.; Laser Treatment of Materials; DGM Informationsgesellschaft Verlag: Oberursel, 1987; 147–161.

influence on the degree of dissolving, and their redistribution depends on the size of the hardened layer.

In general, segregation lines can be observed, reflecting solidification conditions and the stepwise control of the specimen movement. The degree of solubility extends according to

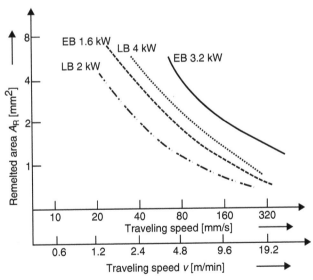

FIGURE 6.118 Remelted area as a function of traveling speed at constant power obtained by laser or electron beam. (From Luft, U.; Bergmann, H.W.; Mordike, B.L. Laser surface melting of aluminium alloys. In *Paper Presented at the European Conference on Laser Treatment of Materials, Bad Nauheim, 1986*. Mordike, B.L., Ed.; Laser Treatment of Materials; DGM Informationsgesellschaft Verlag: Oberursel, 1987; 147–161.)

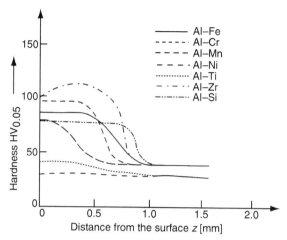

FIGURE 6.119 Microhardness profiles of laser-remelted layer of various Al alloys. (From Luft, U.; Bergmann, H.W.; Mordike, B.L. Laser surface melting of aluminium alloys. In *Paper Presented at the European Conference on Laser Treatment of Materials, Bad Nauheim, 1986.* Mordike, B.L., Ed.; Laser Treatment of Materials; DGM Informationsgesellschaft Verlag: Oberursel, 1987; 147–161.)

traveling speed at given remelting conditions. In Figure 6.119 the microhardness profiles across the remelted layer of various types of aluminum alloys are presented.

Based on given data for laser surface hardening by remelting various nonferrous alloys at given laser-remelting conditions the following can be concluded:

- The alloys of aluminum and transition metals such as Ti and Zr form intermetallic compounds with the highest melting point. The high melting point of $ZnAl_3$ or $TiAl_3$ prevents complete solution in the matrix. The small amount of precipitated intermetallic compounds on point boundary or segregation line does not offer an improvement of the hardness after laser remelting.
- The aluminum–cromium alloys after remelting and subsequent cooling include precipitates of $CrAl_7$ that took place in the matrix, which improve the hardness from 45 to 100 HV. The size of the dendrites was about 1 to 2 μm, but it is finest at the substrate.
- The aluminum manganese alloys appear to consist of a relatively coarse lamellar eutectic with a spacing in the range from 0.3 to 0.5 μm. Close to the base material intermetallic compounds $MnAl_7$ can be found as interdendritic eutectic phase. After laser surface remelting of the thin surface layer, a hardness of 120 HV was obtained.
- Dissolution of Al_3Fe in Al–Fe alloys at laser surface remelting was complete in the remelted layer; only near the base material, i.e., in the transition layer, could it be incompletely dissolved; fine intermetallic dendrites and interdendritic eutectic are found in the transition layer. The hardness of the remelted material was 90 to 100 HV and the hardness of the base material is about 40 HV. After laser remelting by overlapping single laser tracks a microstructure is formed in eutectic crystals of the size between 8 and 12 μm, which contains Al_3Fe particles of the size between 1 and 2 μm. Fine intermetallic dendrites and interdendritic eutectic are found. The hardness in the remelted layer is 90 to 100 HV and in the base material about 40 HV.
- The aluminum and nickel alloys could only be remelted using graphite absorber coatings. The microstructure consists of fine feathery eutectic with grain size from 4 to 10

μm and lamellar spacing of 0.3 μm. Measured hardness is approximately 80 HV of the remelted layer. After annealing of the remelted specimen at 500°C for 1 h there is no effect on changing the hardness profile through the remelted layer. After annealing of the remelted specimen at 580°C for 1 h in Al–Fe and Al–Ni alloys a reduction in hardness was obtained.

Vollmer and Hornbogen [131] investigated various aluminum silicon as follows:

- Hypoe alloy with 8.0% Si
- Eutec with 12.5% Si
- Hypereutectic alloy with 17.0% Si.

Remelting process was realized with two remelting conditions. At various interaction times, i.e., in the first case $t_i = 7.4 \times 10^{-3}$ sec and in the second case $t_i = 0.4 \times 10^3$ s. Rapid solidification occurred during subsequent cooling by a eutectic reaction, giving a much finer eutectic spacing as from the base material. This investigation has clearly shown that the laser-remelting process depends on energy density, on the duration of laser treatment, and, finally, on chemical composition and type of base material microstructure.

The various base microstructures may be exposed to different laser heating conditions as follows:

- The temperature at the liquid–solid interface may be higher than the melting temperature of eutectic $T_E = 577°C$.
- The interface temperature may be higher than the melting temperature of aluminum $T_{Al} = 660°C$.
- The interface temperature may be higher than the melting temperature of silicon $T_{Si} = 1410°C$.

Surface remelting is of special interest for possibilities of improving surface wear resistance. The main purpose of this research was to describe melting and rapid solidification with special attention to liquid–solid interface at various heating conditions. For various heating conditions, various temperature cycles were achieved, i.e., maximum temperature solidification rate.

However, if an alloy with primary crystallization of silicon or aluminum is exposed to a temperature above T_E but below the melting temperature of pure components, mixing of aluminum and silicon is required to acquire the low melting temperature of the eutectic. Consequently, long-range diffusion is necessary until the front of melting can propagate.

As a result of this the authors expected a high velocity V_m of the melting zone according to

$$v_m \sim \frac{D_f}{b}$$

for the first mentioned case, the mobility is determined principally by one atomic hop of about the atomic spacing b. The last mentioned case requires high velocity of the melting zone $V_m = D_f/S$ long-range diffusion, which implies a diffusion path S of the order of magnitude of the size of primary crystallized particles.

The second process occurred during the melting of the hypo- and hypereutectic alloy with subsequent laser treatment. As a consequence, an inhomogeneous liquid containing undissolved crystals has formed. Evidently, there is no formation of glasses and only small amounts of supersaturated homogeneous phase form surface are present by given laser treatment conditions and alloy compositions.

Grum et al.[132] studied microstructures of various aluminum–silicon alloys after casting and the influence of laser hardening by remelting on the changes in microstructure and hardness of a modified surface layer. Their aim was to monitor changes in the thin, remelted surface layer of a specimen material in the form of thin plates with regard to remelting conditions. With the selected remelting conditions for the thin surface layer, a sufficiently high energy input into the surface of individual specimen was ensured. It varied between 165 and 477 J/mm^2. A comprehensive study provides a good insight into the circumstances of laser remelting of aluminum alloys and permits an efficient prediction of the microstructure, hardness level, and residual stresses in the remelted surface layer of the specimen in different remelting conditions.

As it is known, the magnitude and variation of residual stresses exert a decisive influence on the operating performance of machine parts; therefore constructors very often set requirements regarding the magnitude of residual stresses in the most stressed surface layer. To this end it is necessary to study the influences producing residual stresses, particularly in the surface layer. In order to be able to ensure the wanted properties of the surface layer, the influences on the generation of residual stresses are to be known. In the hardening of the thin surface layers by laser remelting, the following should be additionally taken into account [132]:

- Control mode for the laser-beam travel across the workpiece surface.
- Separate influences of the laser-beam power density and travel speed across the workpiece surface are to be known. By changing each of the parameters the same energy input can be ensured. Energy input may influence the size of the remelted trace and, which is even more important, overheating of both the remelted and the non-remelted specimen parts.
- Because of a relatively small width of the remelted trace on the surface in comparison to the workpiece size, laser beam has to travel across the workpiece surface several times. With regard to the Gaussian distribution of energy in the beam, overlapping of the remelted traces has to be ensured so as to ensure also a uniform depth of the remelted layer across the entire workpiece. Consequently, in a relatively narrow range also the degree of overlapping between two neighboring remelted traces may be varied. In our case a 30% degree of overlapping of the two neighboring remelted traces was selected.

The graph in Figure 6.120 presents the main stresses and the depth of the remelted layer of AlSi12CuNiMg alloy, defined by measurements of strain in a given direction and by calculating the main stresses and defining the directions of the main axes.

In the conditions of unbalanced state, i.e., at higher cooling rates, copper remains dissolved in the aluminum with a concentration higher than the balanced concentration. After laser remelting, the aluminum crystals are oversaturated with atoms of copper, which results in the segregation of copper in the form of fine inclusions and thus higher hardness of the matrix.

Residual stresses are a result of temperature and microstructural stresses occurring in the workpiece material directly after the process of remelting a thin surface layer. During the process of rapid cooling, when the process of solidification is going on, the volume of machine parts contracts, resulting in temperature stresses. However, the variation and size of residual stresses in the remelted layer depend also on the composition and homogeneity of the melt and conditions of cooling. The cooling conditions are very important because, at higher

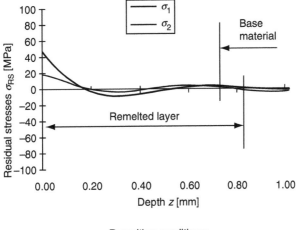

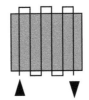

Remelting conditions:
$P = 1500$ W
$V_b = 350$ mm/min
$Z_S = 9$ mm

Way of laser beam traveling

FIGURE 6.120 Residual stresses versus depth of the modified layer after laser remelting of AlSi12Cu-NiMg alloy. (From Grum, J.; Božič, S.; Šturm, R. Measuring and Analysis of Residual Stresses after Laser Remelting of Various Aluminium Alloys. In *Proceedings of the 7th International seminar of IFHT, Heat Treatment and Surface Engineering of Light Alloys, Budapest, Hungary*; Lendvai, J., Réti, T., Eds.; 1999; 507–516.)

cooling rates, it is possible to achieve an ever finer distribution of silicon in the matrix, i.e., in the solid solution of aluminum and silicon.

Hardening of the thin surface layer may be influenced by:

• As fine as possible distribution of the silicon particles in the oversaturated solid solution of aluminum and silicon with other alloying elements
• As fine as possible and uniform distribution of intermetallic compounds such as Al_2Cu and Ni_3Al. Regarding the nature of laser-remelting process, in which there is a melt pool, which mixes due to hydrodynamic and electromagnetic forces, around the laser beam, a rather homogeneous melt and, after rather rapid cooling, quite uniform hardness in the remelted specimen layer may be established.

Measurements of residual stresses were made on the specimen on which laser beam was led at a 15% overlapping of the remelted layer. After laser surface remelting, residual stresses of a magnitude of 20 to 60 MPa were identified at a depth of 0.2 mm. In the depth greater than 0.2 mm from the specimen surface, compressive and tensile residual stresses were intermittently present ranging up to 10 MPa. The variation and size of residual stresses depend on the cooling rates or time necessary for the remelted layer to cool down to ambient temperature.

Microhardness of aluminum alloys prior to laser remelting ranged, depending on the content of silicon and other intermetallic compounds, between 52 $HV_{0.1}$ with AlSi5 alloy and 10 $HV_{0.1}$ with AlSi12NiCuMg alloy. Figure 6.121 shows the variations of microhardness as a function of the depth of the laser-remelted layer of the four aluminum alloys under the same remelting conditions [132].

The diagram indicates the following findings:

- The lowest microhardness in the remelted layer was obtained in AlSi5 alloy and ranged between 60 and 65 $HV_{0.1}$. The microhardness of the hypoeutectic alloy increased from 52 $HV_{0.1}$ in the soft state to 62 $HV_{0.1}$ in the hardened condition. The microhardness increased only by around 20% with reference to the initial soft state of the alloy.
- After remelting of AlSi12 and AlSi8Cu3 eutectic alloys, very similar variations of microhardness were obtained although the microhardnesses of the two alloys differed in the soft state, i.e., it was 63 $HV_{0.1}$ with AlSi12 alloy and 76 $HV_{0.1}$ with AlSi8Cu3 alloy. In the remelted layer a microhardness of around 100 $HV_{0.1}$ was achieved, i.e., the microhardness of the alloys increased by 30% to 40% with reference to their initial soft state. The important increase in microhardness is attributed to both fine distribution of silicon in the solid solution and Al_2Cu intermetallic compound.
- In AlSi12NiCuMg compound a maximum microhardness of up to 160 $HV_{0.1}$ was obtained after remelting of the thin surface layer. Other intermetallic compounds (Al_3Ni, Mg_2Si), which were present in the matrix in a relatively fine form, contributed to a major increase in microhardness as well.

The microhardness measurements performed at specimens of different materials confirmed that the increase in microhardness depended on the type of aluminum alloy. Laser

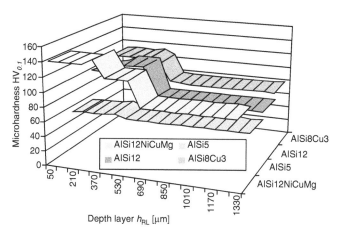

FIGURE 6.121 Through-depth variations of microhardness of four aluminum alloys; remelting conditions. $P = 1500$ W; $v_b = 400$ mm/min; $z_s = 8$ mm. (From Grum, J.; Božič, S.; Šturm, R. Measuring and Analysis of Residual Stresses after Laser Remelting of Various Aluminium Alloys. In *Proceedings of the 7th International seminar of IFHT, Heat Treatment and Surface Engineering of Light Alloys, Budapest, Hungary*; Lendvai, J., Réti, T., Eds.; 1999; 507–516.)

surface hardening by remelting of the thin surface layer provides a homogeneous fine-grained microstructure consisting of finely distributed silicon and intermetallic compounds in the solid solution of aluminum and silicon.

6.27 CORROSION PROPERTIES OF LASER SURFACE REMELTED IRON

During metallographical preparation it is a common feature that laser-remelted iron in the as-quenched condition etches less than the substrate. This indicates that the remelted material exhibits better corrosion properties [120].

From the current–density–potential curves one can derive that the remelted material is more noble than the unremelted one. The fact that this is true even for white cast substrates can only be interpreted by the fact that the higher supersaturation and small grain size prevent the $\gamma \rightarrow \alpha$ transformation almost completely so that the difference in the potential corresponds to the difference in γ- and α-iron [120]. This would explain the etching behavior in the as-quenched condition and is consistent with the fact that corrosion behavior does change when the laser-treated material is annealed. The second thing that is obvious is that for all potentials the current density is about a factor 3 to 6 times smaller than for the untreated material [120].

De Damborena et al. [133] studied the elimination of intergranular corrosion susceptibility of austenitic stainless steel 304 after laser surface remelting.

The typical composition of a given austenite steel is 18% Cr, 8% Ni, 0.03–0.20% C. The chromium content and crystal lattice are responsible for its good anticorrosion characteristics.

The use of this type of steel is limited by sensitization, which results in intergranular corrosion. The sensitization of stainless steel is due to the precipitation of chromium carbides at grain boundaries at temperatures of 450 to 900°C. The corrosion process is accelerated due to the formation of galvanic couples in affected areas, which has influence on the corrosion rate.

The solutions to this problem are as follows:

- Use of a steel with low carbon content
- Stabilization of the steel material with titanium and niobium
- Solution of the carbides at $\approx 1050°C$, followed by a cooling rate which assures a typical microstructure.

Before the laser-remelting application of the beam the specimens were shot peened to remove any impurities and to minimize laser-beam reflection from the metal surface.

The microstructures after laser remelting are a typical rapid solidification process, mainly composed of dendritic cellular growth of austenite.

Three typical types of solidified microstructure are stated:

- The dendritic cellular microstructure of remelted austenite.
- The precipitates of chromium carbides at austenitic grain boundaries in HAZ have disappeared in it.
- The last region contains a fully sensitized microstructure.

Figure 6.122 shows the polarization curve for the austenitic stainless steel 304 before laser remelting. The I_r/I_a quotient is 0.0031, which means that the given austenitic steel is immune to intergranular corrosion.

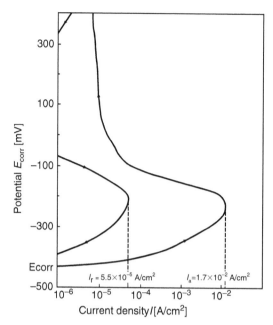

FIGURE 6.122 Polarization curve for 304 stainless steel as received. (From de Damborena, J.; Vazquez, A.J.; Gonzalez, J.A.; West, D.R.F. *Surf. Eng.* 1989, *5*, 235–238.)

Figure 6.123 shows curves obtained following laser surface remelting, showing that the remelted surface has better resistance to corrosion than that of the sensitized state of steel, and also better than the one in the received state.

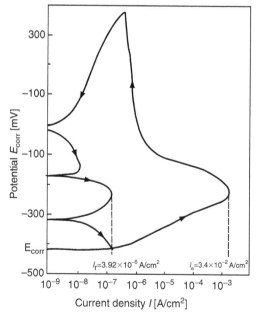

FIGURE 6.123 PPR test for material after laser surface melting. (From de Damborena, J.; Vazquez, A.J.; Gonzalez, J.A.; West, D.R.F. *Surf. Eng.* 1989, *5*, 235–238.)

TABLE 6.10
Results of Corrosion Tests

Condition	$E_{corr.}$ [mV]	I_r/I_a
As received	−429	0.0031
Sensitized	−432	0.60
Sensitized and laser treated	−400	6.6×10^{-5}
Single track	−420	0.0032
Double track: overlap 100%	−410	0.0011
Triple track: overlap 50%	−443	0.0011
Double track: overlap 75%	−405	0.0032

Source: From de Damborena, J.; Vazquez, A.J.; Gonzalez, J.A.; West, D.R.F. *Surf. Eng.* 1989, *5*, 235–238.

Table 6.10 includes the data about corrosion potentials, which are similar in each case, and the I_r/I_a quotient for each case investigated.

From the research it can be concluded that conditioning the surface of a given previously sensitized austenitic stainless steel after laser remelting has a beneficial effect on intergranular corrosion resistance properties.

REFERENCES

1. Daley, W.W. Laser processing and analysis of materials. *Chapter 1: Lasers and Laser Radiation*; Plenum Press: New York, 1983; 158–162.
2. Koebner, H. Overview. *Chapter 1: Industrial Applications of Lasers*; Koebner, H., Ed.; John Wiley & Sons Ltd.: Chichester, 1984; 1–68.
3. Gregson, V.G. Chapter 4: laser heat treatment. In *Laser Materials Processing*. Bass, M., Ed.; Materials Processing Theory and Practices; North-Holland Publishing Company: Amsterdam, 1983, Vol. 3, 201–234.
4. Sridhar, K.; Khanna, A.S. Laser surface heat treatment. In *Lasers in Surface Engineering*; Dahotre, N.B., Ed.; ASM International: Materials Park, Ohio, 1998; 69–179.
5. Rykalin, N.; Uglov, A.; Kokora, A. Laser melting and welding. *Chapter 3: Heat Treatment and Welding by Laser Radiation*; Mir Publisher: Moscow, 1978; 57–125.
6. Steen, W.M. Laser material processing. *Chapter 6: Laser Surface Treatment*; Springer-Verlag: London, 1996; 172–219.
7. Migliore, L. Heat treatment. In *Laser Materials Processing*. Migliore, L., Ed.; Marcel Dekker: New York, 1996; 209–238.
8. Migliore, L. Considerations for real-world laser beams. In *Laser Materials Processing*. Migliore, L., Ed.; Marcel Dekker: New York, 1996; 49–64.
9. Luxon, J.T. Propagation of laser light. In *Laser Materials Processing*. Migliore, L., Ed.; Marcel Dekker: New York, 1996; 31–48.
10. Bass, M. Chapter 1: lasers for laser materials processing. In *Laser Materials Processing*; Bass, M., Ed.; Materials Processing Theory and Practices; North-Holland Publishing Company: Amsterdam, 1983; Vol. 3, 1–14.
11. Dawes, C. *Laser Welding*; Ablington Publishing and Woodhead Publishing in Association with the Welding Institute: Cambridge, 1992; 1–95.
12. Luxon, J.T. Laser optics/beam characteristic. In *Guide to Laser Materials Processing*. Charschan, S.S., Ed.; CRC Press: Boca Raton, 1993; 57–71.

13. Bolin, S.R. Chapter 8: Nd–YAG laser application survey. In *Laser Materials Processing*; Bass, M., Ed.; Materials Processing Theory and Practices; North-Holland Publishing Company: Amsterdam, 1983; Vol. 3, 407–438.

14. Nonhof, C.J. Materials processing with Nd-lasers. *Chapter 1: Introduction*; Electrochemical Publications Limited: Ayr, Scotland, 1988; 1–40.

15. Steen, W.M. Laser material processing. *Chapter 1: Background and General Applications*; Springer-Verlag: London, 1996; 7–68.

16. Steen, W.M. Laser materials processing. *Chapter 2: Basic Laser Optics*; Springer-Verlag: London, 1996; 40–68.

17. Schuöcker, D. High power lasers in production engineering. *Chapter 3: Beam and Resonators*; Imperial College Press: London and Singapore: World Scientific Publishing Co. Pte. Ltd, 1999; 39–72.

18. Schuöcker, D. High power lasers in production engineering. *Chapter 4: Laser Sources*; Imperial College Press: London and Singapore: World Scientific Publishing Co. Pte. Ltd, 1999; 73–150.

19. Migliore, L. Theory of laser operation. In *Laser Materials Processing*. Migliore, L., Ed.; Marcel Dekker: New York, 1996; 1–30.

20. Steffen, J. Lasers for micromechanical, electronic, and electrical tasks. *Chapter 10*. In *Industrial Applications of Lasers*. Koebner, H., Ed.; John Wiley & Sons: Chichester, 1984; 209–221.

21. Seaman, F.D.; Gnanamuthu, D.S. Using the industrial laser to surface harden and alloy. In *Source Book on Applications of the Laser in Metalworking*; Metzbower, E.A., Ed.; American Society for Metals: Metals Park, Ohio, 1981; 179–184.

22. Charschan, S.S.; Webb, R. Chapter 9: considerations for lasers in manufacturing. In *Laser Materials Processing*; Bass, M., Ed.; Materials Processing Theory and Practices; North-Holland Publishing Company: Amsterdam, 1983; Vol. 3, 439–473.

23. Kawasumi, H. Metal surface hardening CO_2 laser. In *Source Book on Applications of the Laser in Metalworking*; Metzbower, E.A., Ed.; American Society for Metals: Metals Park, Ohio, 1981; 185–194.

24. Amende, W. Transformation hardening of steel and cast iron with high-power lasers. In *Industrial Applications of Lasers*. Koebner, H., Ed.; John Wiley & Sons: Chichester, 1984; 79–99; Chapter 3.

25. Meijer, J.; Kuilboer, R.B.; Kirner, P.K.; Rund, M. Laser beam hardening: transferability of machining parameters. Proceedings of the 26th International CIRP Seminar on Manufacturing Systems—LANE'94. In *Laser Assisted Net Shape Engineering*; Geiger, M., Vollertsen, F., Eds.; Meisenbach-Verlag: Erlangen, Bamberg, 1994; 234–252.

26. Belforte, D.; Levitt, M., Eds.; *The Industrial Laser Handbook*, Section 1, 1992–1993 Ed. Springer-Verlag: New York, 1992; 13–32.

27. Carslaw, H.C.; Jaeger, J.C. Conduction of heat in solids. *Chapter II: Linear* Flow *of Heat: The Infinite edition and Semi-infinite Solid*, 2nd Ed. 50–91 Oxford University Press: 1986; 50–91.

28. Bass, M. Laser heating of Solids. In *Physical Process in Laser–Materials Interactions*; Bertolotti, M., Ed.; Plenum Press: New York and London, 1983; 77–116. Published in Cooperation with NATO Scientific Affairs Division.

29. Ashby, M.F.; Easterling, K.E. The Transformation hardening of steel surface by laser beam: I. hypoeutectoid steels. *Acta Metall.* 1984, *32*, 1933–1948.

30. Li, W.B.; Easterling, K.E.; Ashby, M.F. Laser transformation hardening of steel: II. hypereutectoid steels. *Acta Metall.* 1986, *34*, 1533–1543.

31. Kou, S. Welding glazing, and heat treating—a dimensional analysis of heat flow. *Metall. Trans., A.* 1982, *13A*, 363–371.

32. Kou, S.; Sun, D.K. A fundamental study of laser transformation hardening. *Metall. Trans., A.* 1983, *14A*, 643–653.

33. Shercliff, H.R.; Ashby, M.F. The prediction of case depth in laser transformation hardening. *Metall. Trans., A.* 1991, *22A*, 2459–2466.

34. Festa, R.; Manza, O.; Naso, V. Simplified thermal models in laser and electron beam surface hardening. *Int. J. Heat Mass Transfer.* 1990, *33*, 2511–2518.

35. Mazumder, J. Laser heat treatment: the state of the art. *J. Met.* 1983; 18–26.

36. Gregson, V. Laser heat treatment. Paper no. 12. Proc. 1st USA/Japan Laser Processing Conf., LIA Toledo, Ohio, 1981.
37. Sandven, O.A. Heat flow in cylindrical bodies during laser surface transformation hardening. In *Laser Application in Materials Processing*; Proceedings of the Society of Photo-Optical Instrumentation Engineers, SPIE, Washington; Ready, J.F., Ed.; San Diego, California, 1980; Vol. 198, 138–143.
38. Cline, H.E.; Anthony, T.R. Heat treating and melting material with a scanning laser or electron beam. *J. Appl. Phys.* 1977, *48*, 3895–3900.
39. Grum, J.; Šturm, R. Calculation of temperature cycles heating and quenching rates during laser melt-hardening of cast iron. In *Surface Engineering and Functional Materials*; Proceedings of the 5th European Conference on Advanced Materials and Processes and Applications, Materials, Functionality and Design; Maastricht, NL., Sarton, L.A.J.L., Zeedijk, H.B., Eds.; Published by the Netherlands Society for Materials Science Aj Zwijndrecht; 3/155–3/159.
40. Engel, S.L. Section IV. Surface hardening—basics of laser heat treating. In *Source Book on Applications of the Laser in Metalworking*; Metzbower, E.A., Ed.; American Society for Metals: Metals Park, Ohio, 1981; 149–171.
41. Tizian, A.; Giordano, L.; Ramous, E. Laser surface treatment by rapid solidification. In *Laser in Materials Processing*. Metzbower, E.A., Ed.; Conference Proceedings; American Society for Metals: Metals Park, Ohio, 1983; 108–115.
42. Mordike, S.; Puel, D.R.; Szengel, H. Laser Oberflächenbehandlung—ein Productionsreifes Verfahren für—Vielfältige Anwendungen. In *New Technology for Heat Treating of the Metals, Conference Proceedings*, Liščić, B., Ed.; Croatian Society for Heat Treatment: Zagreb, Croatia, 1990; 1–12.
43. Steen, W.M. Laser cladding, alloying and melting. In *The Industrial Laser Annual Handbook 1986*; Belforte, D., Levitt, M., Eds.; Penn Well Books, Laser Focus: Tulsa Oklahoma, 1986; 158–174.
44. Bramson, M.A. Infrared radiation. *A Handbook for Applications*; Plenum Press: New York, 1968.
45. Nonhof, C.J. Materials Processing with Nd-Lasers. *Chapter 5: Absorption and Reflection of Materials*; Electrochemical Publications Limited: Ayr, Scotland, 1998; 147–163.
46. Rykalin, N.; Uglov, A.; Zuer, I.; Kokora, A. Laser and electron beam material processing handbook. *Chapter 1: Lasers and Laser Radiation*; Mir Publisher: Moscow, 1988; 9–73.
47. Rykalin, N.; Uglov, A.; Kokora, A. Laser melting and welding. *Chapter 1: Basic Physical Effects of Laser Radiation on Opaque Mediums*; Mir Publisher: Moscow, 1978; 9–40.
48. Rykalin, N.; Uglov, A.; Kokora, A. Laser melting and welding. *Chapter 2: Techniques for Studying Laser Radiation Effects on Opaque Materials*; Mir Publisher: Moscow, 1978; 4156.
49. Ready, J.F. Absorption of laser energy. In *Guide to Laser Materials Processing*. Charschan, S.S., Ed.; CRC Press: Boca Raton, 1993; 73–95.
50. Migliore, L. Laser–Material Interactions. In *Laser Materials Processing*, Migliore, L., Ed.; Marcel Dekker: New York, 1996; 65–88.
51. Schuöcker, D. *High Power Lasers in Production Engineering*; Imperial College Press and World Scientific Publishing: London, 1987; 1–448.
52. von Allmen, M.; Blatter, A. *Laser-Beam Interactions with Materials: Physical Principles and Applications*; Springer-Verlag: Berlin, 1999; 6–48.
53. Wissenbach, K.; Gillner, A.; Dausinger, F. *Transformation Hardening by CO_2 Laser Radiation, Laser und Optoelektronic*; AT-Fachferlach, Stuttgart, 1985; Vol. 3, 291–296.
54. Beyer, E.; Wissenbach, K. *Oberflächenbehandlung mit Laserstrahlung*; Springer-Verlag: Allgemaine Grundlagen, Berlin, 1998; 19–83.
55. Guanamuthu, D.S.; Shankar, V. Laser heat treatment of iron-base alloys. In *Laser Surface Treatment of Metals*. Draper, C.V., Mazzoldi, P., Eds.; NATO ASI, Series-No. 115, Martinus Nijhoff Publishers: Dordrecht, 1986; 413–433.
56. Rothe, R.; Chatterjee-Fischer, R.; Sepold, G. Hardening with Laser Beams. Proceedings of the 3rd International Colloquium on Welding and Melting by Electrons and Laser Beams, Organized by Le Commisariat a l'Energie Atomique l'Institut de Soudure; Contre, M., Kuncevic, M., Eds.; Lyon, France, 1983; Vol. 2, 211–218.

57. Kechemair, D.; Gerbet, D. Laser metal hardening: models and measures. In Proc. of the 3rd Int. Conf on Lasers in Manufacturing (LIM-3). Quenzer, A., Ed.; IFS Publications: Bedford; Springer-Verlag, Berlin, 1986; 261–270.

58. Woo, H.G.; Cho, H.S. Estimation of hardened laser dimensions in laser surface hardening processes with variations of coating thickness. *Surf. Coat. Technol.* 1998, *102*, 205–217.

59. Trafford, D.N.H.; Bell, T.; Megaw, J.H.P.C.; Bransden, A.S. heat treatment using a high power laser. *Heat Treatment'79*; The Metal Society: London, 1979; 32–44.

60. Inagaki, M.; Jimbou, R.; Shiono, S. Absorptive surface coatings for CO_2 laser transformation hardening; In *Proceedings of the 3rd International Colloquium on Welding and Melting by Electrons and Laser Beam*, Organized by Le Commisariat a l'Energie Atomique l'Institut de Soudure; Contre, M., Kuncevic, M., Eds.; Lyon, France, 1983; Vol. 1, 183–190.

61. Grum, J.; Božič, S.; Šturm, R. Measuring and analysis of residual stresses after laser remelting of various aluminium alloys. In Proc. of the 7th Int. Seminar of IFHT, Heat Treatment and Surface Engineering of Light Alloys, Budapest, Hungary; Lendvai, J., Reti, T., Eds.; Hungarian Scientific Society of Mechanical Engineering (GTE); 507–516.

62. von Allmen, M. Laser-beam interactions with materials: physical principles and applications. *Chapter 2: Absorption of Laser Light*; Springer-Verlag: Berlin, 1987; 6–48.

63. von Allmen, M. Laser-beam interactions with materials: physical principles and applications. *Chapter 3: Heating by Laser Light*; Springer-Verlag: Berlin, 1987; 49–82.

64. von Allmen, M. Laser-beam interactions with materials: physical principles and applications. *Chapter 4: Melting and Solidification*; Springer-Verlag: Berlin, 1987; 83–145.

65. Daley, W.W. Laser processing and analysis of materials. *Chapter 1: Lasers and Laser Radiation*; Plenum Press: New York, 1983; 1–110.

66. Gay, P. Application of mathematical heat transfer analysis to high-power CO_2 laser material processing: treatment parameter prediction, absorption coefficient measurements. In *Laser Surface Treatment of Metals*. Draper, C.W., Mazzoldi, P., Eds.; Martinus Nijhoff Publishers in cooperation with NATO Scientific Affairs Division: Boston, 1986; 201–212.

67. Rykalin, N.; Uglov, A.; Zuer, I.; Kokora, A. Laser and electron beam material processing handbook. *Chapter 3: Thermal Processes in Interaction Zones*; Mir Publisher: Moscow, 1988; 98–167.

68. Steen, W.M. Laser material processing. *Chapter 5: Heat Flow Theory*; Springer-Verlag: London, 1996; 145–171.

69. Guangjun, Z.; Qidun, Y.; Yungkong, W.; Baorong, S. Laser transformation hardening of precision v-slideway. In *Proceedings of the 3rd International Congress on Heat Treatment of Materials*, Shanghai, 1983. Bell, T., Ed.; The Metals Society London: 1984; 2.9–2.18.

70. Meijer, J.; Seegers, M.; Vroegop, P.H.; Wes, G.J.W. Line hardening by low-power CO_2 lasers. In *Laser Welding, Machining and Materials Processing*. Proceedings of the International Conference on Applications of Lasers and Electro-Optics "ICALEO'85", San Francisco, 1985; Albright, C., Eds.; Springer-Verlag, Laser Institute of America: Berlin, 1986; 229–238.

71. Arata, Y.; Inoue, K.; Maruo, H.; Miyamoto, I. Application of laser for material processing—heat flow in laser hardening. In *Plasma, Electron and Laser Beam Technology, Development and Use in Materials Processing*: Arata, Y., Eds.; American Society for Metals: Metals Park, Ohio, 1986; 550–557.

72. Grum, J.; Žerovnik, P. Laser hardening steels, Part 1. Heat treating. Vol. 25–7, July 16–20, 1993.

73. Gutu, I.; Mihâilescu, I.N.; Comaniciu, N.; Drâgânescu, V.; Denghel, N.; Mehlmann, A. Heat treatment of gears in oil pumping units reductor. In *Proceedings of SPIE—The International Society for Optical Engineering*; Fagan, W.F., Ed.; Washington: Industrial Applications of Laser Technology, Geneva, 1983; Vol. 398, 393–397.

74. Ursu, I.; Nistor, L.C.; Teodorescu, V.S.; Mihâilescu, I.N.; Apostol, I.; Nanu, L.; Prokhorov, A.M.; Chapliev, N.I.; Konov, V.I.; Tokarev, V.N.; Ralchenko, V.G. Continuous wave laser oxidation of copper. In *Industrial Applications of Laser Technology*. Fagan, W.F., Ed.; Proc. of SPIE, The Int. Society for Optical Engineering: Washington, 1983; Vol. 398, 398–402.

75. Engel, S.L. Basics of laser heat treating. In *Source Book on Applications of the Laser in Metalworking. A Comprehensive Collection of Outstanding Articles from the Periodical and Reference Literature*; Metzbower, E.A. Ed.; American Society for Metals: Metals Park, Ohio, 1979; 49–171.
76. Toshiba CO_2 Laser Machining System; Toshiba Corp. Shiyodo-ku: Tokyo, 18 pp.
77. Schachrei, A.; Casbellani, M. Application of high power lasers in manufacturing. Keynote papers. Ann. CIRP 1979, *28*, 457–471.
78. Carroz, J. Laser in high rate industrial production automated systems and laser robotics. In *Laser in Manufacturing*, Proceedings of the 3rd International Conference, Paris; IFS: France, Bedford, 1986, Quenzer, A., Springer-Verlag, Berlin, 1986; 345–354.
79. Mordike, S.; Puel, D.R.; Szengel, H. Laser over-flächenbehandlung-ein Productionsreifes Verfahren für Vielfältige Anwendungen. In *New Technologies in Heat Treating of Metals*; Croatian Society for Heat Treatment: Zagreb, Croatia, Liščić, B., Ed.; 1–12.
80. Marinoin, G.; Maccogno, A.; Robino, E. Technical and economic comparison of laser technology with the conventional technologies for welding. In *Proceedings of the 6th International Conference Lasers in Manufacturing*, Birmingham; Steen, W.M., Eds.; IFS Publication: Bedford Springer-Verlag, Berlin, 1989; 105–120.
81. Pantelis, D.I. Excimer laser surface modification of engineering metallic materials: case studies. In *Lasers in Surface Engineering*. Dahotre, N.B. Ed.; ASM International: Materials Park, Ohio, 1998; 179–204.
82. Steen, W.M. Laser material processing. *Chapter 7: Laser Automation and In-Process Sensing*; Springer-Verlag: London, 1996; 220–243.
83. Sona, A. Lasers for surface engineering: fundamentals and types. In *Lasers in Surface Engineering*. Dahotre, N.B., Ed.; ASM International: Materials Park, Ohio, 1998; 1–33.
84. Shono, S.; Ishide, T.; Mega, M. Uniforming of Laser Beam Distribution and Its Application to Surface Treatment, Takasago Research & Development Center, Mitsubishi Heavy Industries, Ltd, Japan; Institute of Welding; IIW-DOC-IV-450-88, 1988; 1–17.
85. Kreutz, E.V.; Schloms, R.; Wissenbach, K. Absorbtion von laserstrahlung. In *Werkstoffbearbeitung mit Laser- strahlung: Grundlagen - Systeme - Verfahren*; Herziger, G. Loosen, P., Eds.; Carl Hanser Verlag: München, 1993; 78–87.
86. von Allmaen, M. Laser-beam interactions with materials physical principles and applications. *Chapter 2: Absorbtion of Laser Light*; Springer-Verlag: Berlin, 1987; 6–48.
87. Grum, J. Possibilities of kaleidoscope use for low power lasers. In *Conf. Proc. Heat Treating: Equipment and Processes*; Totten, G.E., Wallis, R.A., Eds.; ASM International: Materials Park, Ohio, 1994; 265–274.
88. Field, M.; Kahles, J.F. Review of surface integrity of machined components. *Ann. CIRP* 1970, *20*, 107–108.
88. Field, M.; Kahles, J.F.; Cammet, J.T. Review of measuring method for surface integrity. *Ann. CIRP* 1971, *21*, 219–237.
90. Yang, Y.S.; Na, S.J. A study on the thermal and residual stress by welding and laser surface hardening using a new two-dimensional finite element model. *Proc. Inst. Mech. Eng.* 1990, *204*, 167–173.
91. Li, W.B.; Easterling, K.E. Residual stresses in laser transformation hardened steel. *Surf. Eng.* 1986, *2*, 43–48.
92. Solina, A. Origin and development of residual stresses induced by laser surface hardening treatment. *J. Heat Treat.* 1984, *3*, 193–203.
93. Com-Nougue, J.; Kerrand, E. Laser surface treatment for electromechanical applications. In *Laser Surface Treatment of Metals*. Draper, C.W. Mazzoldi, P., Eds.; Martinus Nijhoff Publishers in Cooperation with NATO Scientific Affairs Division: Boston, 1986; 497–571.
94. Mor, G.P. Residual stresses measurements by means of x-ray diffraction on electron beam welded joints and laser hardened surfaces. In *Proceedings of the 2nd International Conference on Residual Stresses "ICRS2"*. Beck, G., Denis, S., Simon, A., Eds.; 696–702 Elsevier Applied Science: Nancy, London, 1988; 696–702.

95. Ericsson, T.; Chang, Y.S.; Melander, M. Residual Stresses and Microstructures in Laser Hardened Medium and High Carbon Steels. In *Proceedings of the 4th International Congress on Heat Treatment of Materials*, Berlin, Int. Federation for the Heat Treatment of Materials; Vol. 2, 702–733.

96. Cassino, F.S.L.; Moulin, G.; Ji, V. Residual stresses in water-assisted laser transformation hardening of 55C1 steel. In *Proceedings of the 4th European Conference on Residual Stresses "ECRS4"*; Denis, S. Lebrun, J.L., Bourniquel, B., Barral, M., Flavenot, J.F., Eds.; Vol. 2, 839–849.

97. Grevey, D.; Maiffredy, L.; Vannes, A.B. A simple way to estimate the level of the residual stresses after laser heating. *J. Mech. Work. Technol.* 1988, *16*, 65–78.

98. Chabrol, C.; Vannes, A.B. Residual stresses induced by laser surface treatment. In *Laser Surface Treatment of Metals*; Draper, C.W. Mazzoldi, P., Eds.; Martinus Nijhoff Publishers in Cooperation with NATO Scientific Affairs Division: Boston, 1986; 435–450.

99. Ericsson, T.; Lin, R. Influence of laser surface hardening on fatigue properties and residual stress profiles of notched and smooth specimens. In *Proceeding of the Conference "MAT-TEC 91"*, Paris; Vincent, L., Niku-Lari, A., Eds.; Technology Transfer Series, Published by Institute for Industrial Technology Transfer (IITT) Int.; Gowruay-Sur-Marne, France, 1991; 255–260.

100. Bohne, C.; Pyzalla, A.; Reimers, W.; Heitkemper, M.; Fischer, A. *Influence of Rapid Heat Treatment on Micro-Structure and Residual Stresses of Tool Steels*. Eclat—European Conf. on Laser Treatment of Materials, Hanover, 1998; Werkstoff - Informationsgesellschaft GmbH: Frankfurt, 1998; 183–188.

101. Mordike, B.L. Surface treatment of materials using high power lasers, advances in surface treatments, technology—applications—effects. In *Proceedings of the AST World Conf. on Advances in Surface Treatments and Surface Finishing, Paris 1986*. Niku Lari, A. Ed.; Pergamon Press: Oxford, 1996; Vol. 5, 381–408.

102. Grum, J.; Žerovnik, P. *Laser Hardening Steels, Part 2. Heat Treating*; Chilton Publication Company, August, 1993; Vol. 25, No. 8, 32–36.

103. Grum, J; Žerovnik, P. Residual stresses in laser heat treatment of plane surfaces. In *Proceedings of the 1st International Conference on Quenching & Control of Distortion, Chicago, Illinois*; Totten, G., Ed.; ASM International: Materials Park, Ohio, 1992; 333–341.

104. Yang, Y.S.; Na, S.J. A study on residual stresses in laser surface hardening of a medium carbon steel. *Surf. Coat. Technol.* 1989, *38*, 311–324.

105. Lepski, D.; Reitzenstein, W. Estimation and optimization of processing parameters in laser surface hardening. In *Proceedings of the 10th Meeting on Modeling of Laser Material Processing, Igls/Innsbruck*; Kaplan A., Schnöcker D., Eds.; Forschungsinstitut für Hochcleistungsstrahltechnik der TüW Wien, 1995; 18 pp.

106. Meijer, J.; Kuilboer, R.B.; Kirner, P.K.; Rund, M. Laser beam hardening: transferability of machining parameters. *Manuf. Syst.* 1995, *24*, 135–140.

107. Kugler, P.; Gropp, S.; Dierken, R.; Gottschling, S. Temperature controlled surface hardening of industrial tools—experiences with 4kW-diode-laser. In *Proceedings of the 3rd Conference "LANE 2001": Laser Assisted Net Shape Engineering 3, Erlangen*; Geiger, M., Otto, A., Eds.; Meisenbach-Verlag GmbH: Bamberg, 2001; 191–198.

108. Marya, M.; Marya, S.K. Prediction and optimization of laser transformation hardening. In *Proceedings of the 2nd Conference "LANE'97": Laser Assisted Net Shape Engineering 2, Erlangen*; Vollersten, F., Ed.; Meisenbach-Verlag GmbH: Bamberg, 1997; 693–698.

109. Grum, J.; Šturm, R. Residual stress state after the laser surface remelting process. *J. Mater. Eng. Perform.* 2001, *10*, 270–281.

110. Hawkes, I.C.; Steen, W.M.; West, D.R.F. Laser Surface Melt Hardening of S.G. Irons. In *Proceedings of the 1st International Conference on Laser in Manufacturing*, Brighton, UK; Kimmit, M.F., Ed.; Co-published by: IFS (Publications), Bedford, UK, Ltd. and North-Holland Publishing Company: Amsterdam, 1983; 97–108.

111. Roy, A.; Manna, I. Mathematical modeling of localized melting around graphite nodules during laser surface hardening of austempered ductile iron. *Opt. Lasers Eng.* 2000, *34*, 369–383.

112. Grum, J.; Šturm, R. Microstructure analysis of nodular iron 400–12 after laser surface melt hardening. *Mater. Charact.* 1996, *37*, 81–88.

113. Bergmann, H.W. Current status of laser surface melting of cast iron. *Surf. Eng.* 1985, *1*, 137–155.

114. Domes, J.; Müller, D.; Bergmann, H.W. Evaluation of Residual Stresses after Laser Remelting of Cast Iron. In Deutscher Verlag fuer Schweisstechnik (DVS), 272–278.

115. Grum, J.; Šturm, R. Residual stresses on flat specimens of different kinds of grey and nodular irons after laser surface remelting. *Mater. Sci. Technol.* 2001, *17*, 419–424.

116. Grum, J.; Šturm, R. Residual stresses in gray and nodular irons after laser surface melt-hardening. In *Proceedings of the 5th International Conference on Residual Stresses "ICRS-5"*; Ericsson, T., Odén, M., Andersson, A., Eds.; Institute of Technology, Linköpings University: Linköping, 1997; Vol. 1, 256–261.

117. Grum, J.; Šturm, R. Deformation of specimen during laser surface remelting. *J. Mater. Eng. Perform.* 2000, *9*, 138–146.

118. Grum, J.; Šturm, R. Optimization of laser surface remelting process on strain and residual stress criteria. *Mater. Sci. Forum* 2002, *404–407*, 405–412.

119. Grum, J.; Šturm, R.; Žerovnik, P. Optimization of laser surface melt-hardening on gray and nodular iron. In *Surface Treatment: Computer Methods and Experimental Measurements*; Aliabadi, M.H., Brebbia, C.A., Eds.; Computational Mechanics Publications: Boston, 1997; 259–266.

120. Bergmann, H.W. Laser surface melting of iron-base alloys. In *Laser Surface Treatment of Metals*; Draper, C.W., Mazzoldi, P., Eds.; Series E: Applied Science— No. 115, NATO ASI Series; Martinus Nijhoff Publishers: Dordracht, 1986; 351–368.

121. Grum, J.; Šturm, R. Laser surface melt-hardening of gray and nodular iron. In *Proceedings of the International Conference on Laser Material Processing*, Opatija; Croatian Society for Heat Treatment, 1995; 165–172.

122. Hawkes, I.C.; Steen, K.M.; West, D.R.F. Laser surface melt hardening of S.G. irons, In *Proceedings of the 1st International Conference* on Laser in *Manufacturing*, Brighton, UK; Kimmit, M.F., Ed.; Co-published by: IFS (Publications), Bedford, UK, Ltd. and North-Holland Publishing Company: Amsterdam, 1983; 97–108.

123. Ricciardi, G.; Pasquini, P.; Rudilosso, S. Remelting surface hardening of cast iron by CO_2 laser. In *Proceedings of the 1st International Conference on* Laser in *Manufacturing*, Brighton, UK; Kimmit, M.F., Ed.; Co-published by: IFS (Publications), Bedford, UK, Ltd. and North-Holland Publishing Company: Amsterdam, 1983; 87–95.

124. Czichos, H. Basic tribological parameters. ASM Handbook, Volume 18; Friction, Lubrication, and Wear Technology; Volume Chairman PJ Blau; ASM International: Materials Park, Ohio, 1992; 473–479. Printed in the United States of America.

125. Czichos, H. Presentation of friction and wear data. ASM Handbook, Volume 18, Friction, Lubrication, and Wear Technology; Volume Chairman PJ Blau; ASM International: Materials Park, Ohio, 1992; 489–492. Printed in the United States of America.

126. Fukuda, T.; Kikuchi, M.; Yamanishi, A.; Kiguchi, S. Laser hardening of spheroidal graphite cast iron. In *Proceedings of the Third International Congress on Heat Treatment of Materials, Shanghai 1983*; Bell, T., Ed.; The Metals Society: London, 1984; 2.34–2.44.

127. Magnusson, C.F.; Wiklund, G.; Vuorinen, E.; Engström, H.; Pedersen, T.F. Creating tailor-made surfaces with high power CO_2-lasers. In *Proceedings of the 1st ASM Heat Treatment and Surface Engineering Conference Mater. Sci. Forum*; 1992, Vol. 102–104, 443–458.

128. Coquerelle, G.; Fachinetti, J.L. Friction and wear of laser treated aluminium–silicon alloys. Paper Presented at the European Conf. on Laser Treatment of Materials, Bad Nauheim, 1986. In *Laser Treatment of Materials*; Mordike, B.L., Ed.; DGM Informationsgesellschaft Verlag: Oberursel, 1987; 171–178.

129. Antona, P.L.; Appiano, S.; Moschini, R. Laser furface remelting and alloying of aluminum alloys. Paper Presented at the European Conf. on Laser Treatment of Materials, Bad Nauheim, 1986. In *Laser Treatment of Materials*; Mordike, B.L., Ed.; DGM Informationsgesellschaft Verlag: Oberursel, 1987; 133–145.

130. Luft, U.; Bergmann, H.W.; Mordike, B.L. Laser surface melting of aluminium alloys. In *Paper Presented at the European Conference on Laser Treatment of Materials, Bad Nauheim, 1986*. Mordike, B.L., Ed.; Laser Treatment of Materials; DGM Informationsgesellschaft Verlag: Oberursel, 1987; 147–161.

131. Vollmer, H.; Hornbogen, E. Microstructure of laser treated Al–Si-alloys. In *Paper Presented at the European Conference on Laser Treatment of Materials, Bad Nauheim, 1986*. Mordike, B.L., Ed.; Laser Treatment of Materials; DGM Informationsgesellschaft Verlag: Oberursel, 1987; 163–170.

132. Grum, J.; Božiĉ, S.; Šturm, R. Measuring and Analysis of Residual Stresses after Laser Remelting of Various Aluminium Alloys. In *Proceedings of the 7th International seminar of IFHT, Heat Treatment and Surface Engineering of Light Alloys, Budapest, Hungary*; Lendvai, J., Réti, T., Eds.; 1999; 507–516.

133. de Damborena, J.; Vazquez, A.J.; Gonzalez, J.A.; West, D.R.F. Elimination of intergranular corrosion susceptibility of a sensitized 304 steel by subsequent laser surface melting. *Surf. Eng.* 1989, *5*, 235–238.

Section II

Testing

7 Metallurgical Property Testing

Xiwen Xie

CONTENTS

7.1 INTRODUCTION

Various testing methods may be selected to verify the quality of heat-treated parts or specimens. Hardness testing is perhaps the most commonly used testing method for heat-treated parts; however, additional tests (such as impact and tensile tests) are required for some critical parts. In some cases, the microstructure of heat-treated parts should be examined under a metallurgical microscope because any property change during heat treatment is closely related to the microstructural change. Sometimes microstructural measurements are needed in order to obtain a more informative and quantitative result. These can be done manually, however, people often rely on an image analyzer to accomplish the quantitative analysis of microstructures during the last three decades. This chapter provides an overview of the various metallurgical testing methods, how they are carried out, and the physical principles involved in testing. Helpful testing strategies are also provided.

7.2 METALLOGRAPHIC TECHNIQUE FOR STEELS

The true microstructure of a steel specimen can be observed under a microscope only when the necessary preparation procedures are properly performed. In the past, the quality of a prepared metallographic specimen depended on the experience and skillfulness of the operator. This metallographic technique was actually an art, not a science. Besides, traditional methods of specimen preparation involve many superfluous and lengthy steps creating an excessive expenditure of consumables and time.

There has recently been tremendous progress in the understanding of the physical nature of specimen preparation. The principles and techniques of specimen preparation are now scientifically established. Based on extensive work, Nelson [1] developed a new concept of specimen preparation producing superior polished surfaces on all types of materials, with the least number of steps and in the shortest possible time. This is based on the following premises:

1. Each stage of preparation is important. Any mistake at one stage is difficult to correct during subsequent steps, as each stage relies on the quality of the previous stage.
2. It is very important to attain a proper balance between the material removal rate and the depth of the deformed layer that remains at the end of each step.

3. Preparation parameters (surface, abrasive type and size, and lubricant) should be selected and optimized according to the physical properties of the sample material.
4. Other parameters such as pressure, rotation direction and speed, and time should also be carefully optimized and controlled at each step.

Generally, specimen preparation can be divided into several stages, i.e., sampling, sectioning, mounting, planar grinding, integrity, and polishing. The first three stages may be regarded as preliminary stages of specimen preparation. The last three stages are the main stages of specimen preparation through which a damage-free, scratch-free, and smooth surface is obtained. A true microstructure can then be revealed by etching suitable reagents, although in certain cases, the true microstructure of a specimen can be revealed without any etching under the microscope using different illumination methods, i.e., brightfield, darkfield, polarized light, uncrossed polarized light, or differential interference contrast (DIC).

7.2.1 Preliminary Stages of Specimen Preparation

7.2.1.1 Sampling

Although random sampling sounds more reasonable, it is usually impractical. In many cases, the top priority is convenience in sectioning. For some important heat-treated parts, e.g., the turbine axis of a jet engine, test specimen locations are usually specified at critical sites. For the examination of wrought materials, it is important that the orientation of the prepared surface be chosen in accordance with the appropriate specification. Information about metal flow patterns and inclusion deformability can be easily seen on longitudinal sections and is not obtainable on transverse sections. The microstructure often appears more uniform and the grain structure more equiaxed on transverse sections. Sometimes separate test specimens are loaded into the furnace together with the parts to be treated. In this case, the test specimens should be of the same heat and located in the same temperature zone as the parts to be treated. This is because steels of the same grade but different heats differ slightly from each other in chemical composition; although still within allowable limits, their responses to heat treatment may not be the same.

7.2.1.2 Sectioning

The surface condition of a metallographic specimen after sectioning can be considered as the starting point for specimen preparation. The quality of this initial surface condition is often overlooked. If sectioning is done on a handsaw or dry-cutting machine, a very rough surface with a deep deformed layer as large as several hundred micrometers and, in many cases, excessive surface damage due to overheating will be produced. This in turn will lead to extended preparation time, higher consumption of consumables, and incorrect microstructure. Therefore, proper sectioning is critical in specimen preparation. Two basic requirements must be fulfilled: flatness of the resultant surface and minimal depth of the damaged or deformed layer. The sectioning operation must be fast, easy, and inexpensive.

The cutoff wheel usually consists of suitable abrasive particles and bonding material. The common abrasives used for sectioning are aluminum oxide and silicon carbide, the former is used mainly in the 95% pure state, i.e., the so-called brown alumina, a tough abrasive and is used on most ferrous materials. Although silicon carbide can also be used for ferrous materials, its use is limited because of its friability. When the conventional aluminum oxide wheels fail to cut alloy tool steels, cubic boron nitride (CBN) can be chosen as the alternative although CBN is expensive. During sectioning, the worn-out abrasive particles should wear away with the binding material, at the same time, unused, sharp abrasive particles should be showed up for continued sectioning. Therefore, the bond strength of the wheel should be compatible with the abrasion of abrasive particles, which in turn, is material dependent. For example, a rubber–resin-bonded aluminum oxide (Al_2O_3) cutoff wheel is suitable for sectioning tool steels with a hardness of 60 HRC and above, carburized steels, and medium hard steels with a hardness from 35 to 50 HRC. To minimize kerf loss and sectioning deformation, rubber-bonded aluminum oxide cutoff wheels are suitable for sectioning soft or annealed steels with a hardness of 15 to 35 HRC (46 to 90 HRB). For delicate cutting, ultrathin rubber-bonded aluminum oxide cutoff wheel (10-in. diameter) with a thickness of only 1.0 mm is available. During sectioning, effective cooling of the specimen must be maintained.

After selecting the appropriate cutoff wheel for the specific material, efforts should also be made to make use of the abrasive particles in the wheel most efficiently, i.e., the area of contact between the wheel and the sample cut should be kept constant and low. This is the so-called principle of minimum area of contact cutting (MACC).

The simplest way to section a sample is by chop cutting, this is the traditional form of machine operation. As long as the sample size is small, chop cutting will not bring about any problem because the area of contact between the wheel and the sample is kept relatively small. However, as the size of the sample increased, the wheel contact area is increased more and more when the wheel feeds deep into the cut, therefore, causing the wheel to slowdown or even stall. Under such circumstances, the sectioning is of course not efficient; moreover, the induced deformation damage increases significantly.

In order to overcome the shortcomings of chop cutting, measures have been proposed and adopted in some models of commercial production, and the orbital cutting developed recently by Buehler seems to be the most successful one [2]. Figure 7.1 shows the DELTA automatic orbital cutter. During orbital cutting, as the wheel is fed into the sample, the arbor of the wheel is also rotated in a small ellipse planar with the wheel once per second. This allows the wheel to cut a strip of material out of the sample, and the wheel is ready to begin to cut on the original side. Because of the elliptical orbit in conjunction with the controlled wheel feed, MACC is maintained regardless of part size, and the cutter needs only to be told the overall cut depth and desired feed rate if the part size is changed.

Test results showed that, for low hardness (around 20 HRC) materials, the cut times for chop cutting and orbital cutting were nearly identical; however, at a sample hardness of 60 HRC, chop cutting took over 50% longer than orbital cutting. Wheel life can be measured by a parameter called material/wear ratio, which measures the area of material cut relative to the area of wheel wear encountered. Higher material/wear ratio, or M/W ratio, means longer wheel life. As the sample hardness increases from 20 to 60 HRC, the M/W ratio for orbital cutting is 40–600% higher than that for chop cutting, which means that the wheel life for orbital cutting is much longer than that for chop cutting. The deformation created by chop cutting and orbital cutting can be compared by a metallographic examination of the prepared surface perpendicular to the cut surface. For a sample of 0.4%C carbon steel, the surface roughness of the part sectioned by chop cutting was approximately 12 μm, while the surface roughness of the part cut on the DELTA orbital cutter was only 1 to 2 μm. Moreover, the subsurface deformation measured on the chop cut part was 28 μm compared with 12 μm depth of deformation for the orbital cut part, this means a 57% reduction in damage depth.

Another type of cutoff machine is the precision cutoff machine or precision saw. This machine is suitable for sectioning medium-sized (up to 50 mm in diameter) to very small samples. The diameter of the wafering blades and abrasive cutoff wheels varies from 3 to 8 in. (76 to 203 mm). The wafering blades are of the nonconsumable type, and are made of copper-plated steel with diamond or CBN abrasive bonded only to the periphery of the blade. The thickness of the wafering blades varies from 0.006 to 0.030 in. (0.15 to 0.76 mm). The abrasive

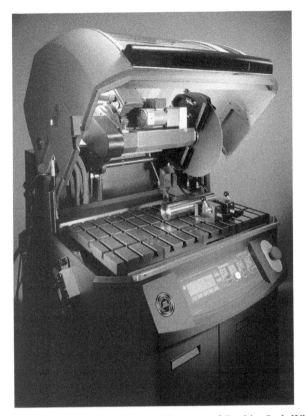

FIGURE 7.1 The DELTA automatic orbital cutter. (Courtesy of Buehler Ltd. With permission.)

cutoff wheels are of the consumable type using rubber-bonded alumina or silicon carbide abrasives. Since blades used in precision saws are much thinner, and the applied load during cutting is much less, the deformed layer of the surface cut is much thinner. For example, the depth of the deformed layer is 25 μm for carbon steel at a low cutting speed of 150 rpm as compared with 120 μm for the same material using a conventional abrasive wheel [3]. The wheel speed may be increased to 5000 rpm. A fivefold increase in the cutting speed reduces cutting time by one fifth without increasing the deformation depth by more than 10%. Figure 7.2 shows the Isomet 5000 linear precision saw.

Table 7.1 shows solutions for problems encountered in the abrasive cutoff sectioning [4].

7.2.1.3 Mounting

The mounting of metallographic specimens is necessary in several cases, i.e.:

1. Small parts or specimens such as thin sheets and coil springs that are difficult or impossible to handle during preparation
2. When edge retention is needed (e.g., case-hardened or decarburized parts), otherwise edge rounding will result
3. Central force type specimen holders of certain semiautomatic and automatic grinding/polishing machines require the use of mounted specimens
4. Mounted specimens are especially suitable for storage

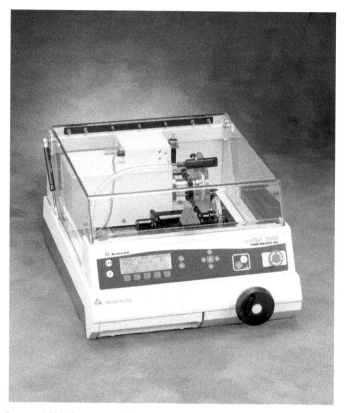

FIGURE 7.2 The Isomet 5000 linear precision saw. (Courtesy of Buehler Ltd. With permission.)

There are two mounting techniques, i.e., hot mounting and cold mounting. Hot mounting or compression mounting requires heat (around 170°C) and pressure (around 290 bar) during molding, whereas cold mounting is conducted at room temperature. Hot mounting is the fastest and most efficient mounting method. It provides the best quality and allows the use of less expensive mounting resins, among which Bakelite or phenolics is most popular and least expensive. There are two types of hot mounting resins, i.e., thermosetting resins that cure at elevated temperatures (e.g., Bakelite and epoxy) and thermoplastic resins that soften at elevated temperatures and harden during cooling (e.g., acrylics). Each resin has its specific characteristics and application, for example, epoxy with higher hardness has better edge retention characteristics (e.g., Buehler's Epomet F and Epomet G). Moreover, conductive resins are necessary for specimens using electrolytic polishing during preparation or for specimens examined under a scanning electron microscope (SEM) (e.g., Buehler's Probemet and Konductomet with copper in the former and graphite in the latter as filler material). During hot mounting, the applied heating time, molding temperature and cooling time, which are different for different resins, can be accurately adjusted. In certain models, the parameters of the whole mounting cycle can be preset and controlled automatically. Figure 7.3 shows an automatic mounting press.

Table 7.2 shows solutions for typical problems of compression mounting materials [5].

Cold mounting or castable mounting is used when the specimen cannot withstand any heat or pressure during mounting, e.g., as-quenched martensite, which would transform to tempered martensite. Materials for cold mounting consist of fluid or powdered resin (e.g., epoxies) and fluid hardener, which are mixed thoroughly at a given proportion before casting.

TABLE 7.1
Solutions for Problems Encountered in Abrasive Sectioning

Fault	Cause	Remedy
Wheel does not cut or cutting ceases after a short time	Incorrect abrasive	Use alternative abrasive or softer grade of wheel
	Wheel has become blunted or glazed	Reduce coolant contact with wheel
Wheel wears rapidly	Wheel is too soft	Use harder grade of wheel
Wheel breaks	Excessive cutting force	Reduce cutting pressure
	Sample moved during cutting	Clamp sample more securely
	Wheel not clamped securely	Tighten wheel flanges more securely but do not overtighten
	Uneven wear on wheel rim causing deflection of cut	Coolant flow not equal on both sides of wheel—check for blockages
	Wheel pinched in cut by sample	a. Sample clamping arrangement causing pinching action on cut when partially complete—reduce pressure on one vice
		b. Internal stresses in sample causing pinching—take step or intermittent cuts
	Wheel vibration	Check wear on cutting shaft bearings
Overheating and burning of cut surface	Abrasive blunted causing excessive friction	Use softer grade of wheel
	Partial or complete failure of coolant supply	Check coolant flow to wheel and sample
	Cutting rate too high	Reduce cutting pressure and feed rate
	Stressed component pinching in on wheel sides	Use softer wheel or use a reciprocating wheel action
Excessive wheel vibration during cutting	Worn cutting shaft bearings	Replace bearings
	Air in hydraulic system (where fitted)	Bleed air from hydraulic system
	Mechanical damper too loose (where fitted)	Tighten mechanical damper
	Wheel too hard	Use softer wheel
Coolant froths	Antifrothing agents in coolant have deteriorated	Clean out coolant system and refill with fresh coolant
	If coolant is new and frothing occurs	Wrong type of coolant used—replace with recommended grade
Cutting motor overheats and lacks power	Excessive cutting pressure causing overloading	Reduce cutting pressure
	One or more of phase voltage low	Check voltages on each phase of three-phase supply to machine
	Failure of one of motor windings	Check motor windings for each phase
	Wheel too hard	Use softer wheel

The specimen is placed in a plastic or rubber mold, and the mixed mounting material is poured over the specimen, an exothermic reaction takes place and cures the resin. After curing, the mount is removed from the mold.

An embedding mold called Monoform from Struers is developed for fast cold mounting. Monoform is a plastic tube with a number of rims on the inside to ensure maximum adhesion between the embedding and the mold (Figure 7.4). Such a mold is prepared together with

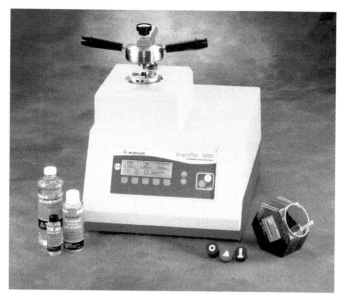

FIGURE 7.3 SIMPLIMET 3000 automatic mounting press. (Courtesy of Buehler Ltd. With permission.)

the specimen. Like hot-mounted specimens, specimens embedded in Monoform can be fit perfectly into specimen holders of a semiautomatic grinding and polishing machine.

It is important that the cold-mounting medium adheres well to the specimen and be free of hardening shrinkage; otherwise cracking will occur [6]. Also, the hardness of the mounting medium should be matched with that of the specimen, as a harder specimen will result in edge

TABLE 7.2
Solutions for Typical Problems of Compression Mounting Materials

Problem	Cause	Solution
Thermosetting resins		
Radial split	Too large a section in the given mold area; sharp corners specimens	Increase mold size; reduce specimen size
Edge shrinkage	Excessive shrinkage of plastic away from sample	Decrease mold temperature; cool mold slightly before ejection
Circumferential splits	Absorbed moisture; entrapped gasses during molding	Preheat powder or premold; momentarily release pressure during fluid state
Burst	Too short a cure period; insufficient pressure	Lengthen cure period; apply sufficient pressure during transition from fluid state to solid state
Unfused	Insufficient molding pressure; insufficient time at cure temperature; increased surface area of powdered materials	Use proper molding pressure; increase cure time; with powders, quickly seal mold closure and apply pressure to eliminate localized curing
Thermoplastic resins		
Cottonball	Powdered media did not reach maximum temperature; insufficient time at maximum temperature	Increase holding time at maximum temperature
Crazing	Inherent stresses relieved upon or after ejection	Allow cooling to a lower temperature before ejection; temper mounts in boiling water

FIGURE 7.4 Embedding molds (Monoform) for fast cold mounting. (Courtesy of Struers, Inc.)

rounding during preparation. A series of sintered aluminum oxide pellets dispersed in the resin will match the hardnesses of a wide range of materials [6].

Table 7.3 shows the solutions for problems of castable mounting materials [5].

If a specimen with quenching cracks or with porous plasma spray coating is to be examined, then vacuum impregnation is a necessary step to remove the air from the cracks or porosity during mounting, thus allowing the epoxy to enter so that complete bonding is effected. Otherwise the crack or porosity may be enlarged and crack or porosity edges may become rounded during later preparation steps [6]. Abrasives, solvent, and etchants may also be entrapped in open cracks, causing staining problems. Resins for vacuum impregnation should be of low viscosity (<250 cP), e.g., Buehler's EPO-THIN, to ease the entering into cracks and porosity. Figure 7.5 shows the vacuum impregnation system from Buehler.

7.2.2 Main Stages of Specimen Preparation

7.2.2.1 Planar Grinding Stage

The aim of planar grinding stage is threefold:

1. When specimen holder of the central loading type is used in semiautomatic grinder and polisher, planar grinding is to make the surface of all the specimens clamped on the holder situated in a common plane with uniform flatness. Planar surface should be achieved in the shortest possible time with minimum residual damage. When multiple individually loaded specimens are prepared, planar grinding is not necessary. When sample has been carefully sectioned this stage can also be omitted for hand preparation
2. When the cut surface is rough.
3. When preset material removal is needed.

In the past, planar grinding was usually accomplished by using 180 or 240 (P180 or P280) grit SiC waterproof abrasive paper for specimens sectioned by an abrasive cutoff wheel. For specimens with rougher cut surfaces, e.g., sectioned by hacksaw, coarser grit size of 120 to 180 grit should be used instead.

TABLE 7.3
Solutions for Problems of Castable Mounting Materials

Problem	Cause	Solution
Acrylics		
Bubbles	Too violent agitation while blending resin and hardener	Blend mixture gently to avoid air entrapment
Polyesters		
Cracking	Insufficient air cure before oven cure; oven cure temperature too high; resin-to-hardener ratio incorrect	Increase air cure time; decrease oven cure temperature; correct resin-to-hardener ratio
Discoloration	Resin-to-hardener ratio incorrect; resin has oxidized	Correct resin-to-hardener ratio; keep containers tightly sealed
Soft mounts	Resin-to-hardener ratio incorrect; incomplete blending of resin-hardener mixture	Correct resin-to-hardener ratio; blend mixture completely
Tacky tops	Resin-to-hardener ratio incorrect; incomplete blending of resin-hardener mixture	Correct resin-to-hardener ratio; blend mixture completely
Epoxies		
Cracking	Insufficient air cure before oven cure; oven cure temperature too high; resin-to-hardener ratio incorrect	Increase air cure time; decrease oven cure temperatures; correct resin-to-hardener ratio
Bubbles	Too violent agitation while blending resin and hardener mixture	Blend mixture gently to avoid air entrapment
Discoloration	Resin-to-hardener ratio incorrect; oxidized hardener	Correct resin-to-hardener ratio; keep containers tightly sealed
Soft mounts	Resin-to-hardener ratio incorrect; incorrect blending of resin-to-hardener mixture	Correct resin-to-hardener ratio; blend mixture completely

FIGURE 7.5 Vacuum impregnation equipment I, excluding the vacuum pump and filtering flask. (Courtesy of Buehler Ltd. With permission.)

TABLE 7.4
USA/European Grit Equivalency Guide

FEPA (Europe)		ANSI/CAMI (USA)	
Grit Number	Size (μm)	Grit Number	Size (μm)
100	148.0	P100	162.0
120	116.0	P120	127.0
180	78.0	P180	78.0
220	66.0	P240	58.5
240	51.8	P280	52.2
		P320	46.2
280	42.3	P360	40.5
320	34.3	P400	35.0
360	27.3	P500	30.2
400	22.1	P600	25.8
		P800	21.8
500	18.2	P1000	18.3
600	14.5	P1200	15.3
800	12.2	P1500	12.6
1000	9.2	P2000	10.3
1200	6.5	P2500	8.4
		P4000[a]	5.0[a]

[a]FEPA grades finer than P2500 are not standardized and are graded at the discretion of the manufacturer. In practice, the above standard values are only guidelines and individual manufactures may work to a different size range and mean value.

The chart shows the midpoints for the size ranges for ANSI/CAMI graded paper according to ANSI standard B74.18-1996 and for FEPA graded paper according to FEPA standard 43-GB-1984 (R-1993). The ANSI/CAMI standard lists SiC particles sizes ranges up to 600 grit paper. For finer grit ANSI/CAMI papers, the particles sizes come from the CAMI booklet, Coated Abrasive (1996).

Source: From *Buehler Consumables Buyer's Guide*, Buehler Ltd., IL, 2000.

The SiC papers manufactured in the United States and Europe have different grit size numbering systems, Table 7.4 shows U.S.–European grit equivalency guide [7].

Grinding is defined as a process in which abrasives are either fixed within a matrix, e.g., abrasive paper, or alternatively charged into, for example, resin of a composite preparation surface. The main factor is abrasive and will be fixed into a substrate during the material removal process. An observation of the ground surface will reveal a series of scratches caused by each individual abrasive as it passes over the specimen. It is important to distinguish between a good scratch pattern and a poor one because the former corresponds to least residual structural damage depth under the ground surface. A good scratch pattern, when observed under a microscope at 100×, will reveal a three-dimensional clear-cut scratches, i.e., one scratch on top of another, irrespective of the direction. However, when grinding abrasives become blunt, scratches appear to be on the same plane and no longer clear; although the surface looks more reflective, the damage depth has increased two to three times than that of specimens with good scratch pattern. When abrasive size becomes finer and finer, the scratch pattern can be barely seen even under a microscope using brightfield illumination, nevertheless, this can be overcome by using darkfield illumination where a good scratch pattern appears very thin but with clear, bright lines in a dark background.

Although SiC paper has been used successfully over many years, its service life is quite short. Especially when used on semiautomatic equipment, it lasts only 1 to 2 min, so even more than one sheet may be needed for specimens clamped on a holder in a single step. This renders the method inconvenient.

During the past two decades, new preparation surfaces have been developed to replace silicon carbide abrasive papers in planar grinding, some of which can also be used in subsequent integrity stage:

1. Alumina grinding stone, several specimens in a specimen holder are ground simultaneously on a high-speed ($\geq$1500 rpm) grinding stone. A special purpose and expensive machine is required, like Struers Abraplan high-capacity, automatic grinding machine for rapid and reproducible planar grinding of all kinds of materials (Figure 7.6). The wheel speed is 1500 rpm, the specimen holder run at 140 rpm, and the pressure on the specimens can be adjusted between 0 and 700 N. Although the material removal rate of such grinding stone is high, the damage depth is also substantial. Besides, in order to maintain the flatness of the grinding stone, its grinding surface must be dressed regularly with a diamond tool.

2. Alumina grinding paper where the abrasive was manufactured by a new process, like Buehler's PLANARMET AL planar grinding disk with 120 grit size, the abrasive is firmly attached to a heavy paper backing to enhance cutting rate and cutting life [8]. Steel specimens with wide range of hardness can be well planar ground on fully automatic and semiautomatic grinding machines. Tests have been conducted on A2 tool steel quenched and tempered to 60 HRC and 4340 alloy steel tempered to 40 and

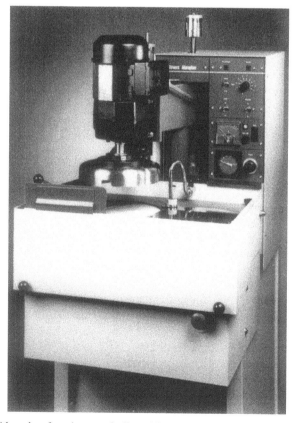

FIGURE 7.6 Struers Abraplan for planar grinding. (Courtesy of Struers, Inc.)

50 HRC, as well as 1045 carbon steel at 20 to 25 HRC and 1018 carbon steel. The results clearly show that PLANARMET AL grinding disk outperforms silicon carbide paper substantially, both in removal rate and life.

3. Metal- or resin-bonded diamond disk with grit sizes from about 70 to 9 μm, the diamond abrasives are spread uniformly over its entire surface. However, cutting efficiency is greatly reduced as swarf is accumulated during grinding.

4. Unlike the metal-bonded diamond disks, some grinding disks have surfaces partially covered by disk surface for lessening the surface tension and improving the removal rate. This new type of disk has small surface pads that contain diamond abrasive of a specified size. An example of such a disk is Buehler's ULTRA-PREP grinding disks, both metal- and resin-bonded disks are available [9]. The metal-bonded disks are available in six diamond sizes from 125 to 6 μm while the resin-bonded disks are available in three diamond sizes from 30 to 3 μm. Planar grinding could be performed using a 45-μm metal-bonded, or a 30-μm resin-bonded ULTRA-PREP disk, depending on the hardness of the specimen.

5. A stainless steel woven mesh cloth on a platen can also be used for planar grinding. The abrasive used is water-based coarse diamond suspension, when sprayed onto the cloth, and as the specimen passes over the cloth, diamond abrasives are impressed into the stainless steel mesh. This kind of cloth can only be used for hard materials, and is an inexpensive alternative to a diamond grinding wheel. An example of such a cloth is Buehler's ULTRA-PLAN cloth.

6. New rigid grinding disk consisting of a thick stainless steel disk upon which a number of approximately 12-mm diameter pads of epoxy are bonded. The epoxy pads contain filler metal and some porosity. The pads cover only a portion of the surface, so that surface tension can be controlled and optimized to achieve optimal removal rate. Since swarf can be easily removed, deformation damage is minimized. Diamond abrasive is added during grinding, rather than embedded, and polycrystalline suspensions are preferred rather than monocrystalline suspensions because of higher removal rates [9]. Examples are BUEHLERHERCULES® H and S rigid grinding disks from Buehler (Figure 7.7). The H disk is suited for the preparation of nearly all types of

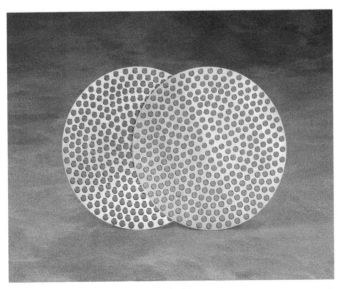

FIGURE 7.7 BUEHLERHERCULES® H and S rigid grinding disk. (Courtesy of Buehler Ltd. With permission.)

steels, while the S disk is designed for soft metals and alloys, it can also be used to prepare harder materials although its wear rate is higher. Struer's Allgro and Largo rigid grinding disks are similar products to BUEHLERHERCULES H and S disks.

7.2.2.2 Integrity Stage

This stage is a progressively structural damage-reducing stage targeting for specimen integrity, i.e., faithful reproduction of the microstructure. Integrity stage is the most important stage of the entire preparation sequence, and any residual structural damage is likely to remain to the end that may cause misinterpretation of the microstructure. Theoretically, integrity is achieved at nil residual structural damage, but in practice this can never be attained, as any material removal process by mechanical means is always accompanied by certain structural damage no matter how small it is. Therefore, during the actual operation, specimen integrity is achieved when the residual structural damage is less than recognizable.

The integrity stage is usually completed by a series of fine grinding steps through which the structural damage of the specimen is minimized to a level adequate for true microstructural analysis. One may ask why not just call the integrity stage the fine grinding stage? The answer is that, traditionally, fine grinding does not mean achieving specimen integrity, at least rough polishing steps should be included in the integrity stage as long as the material removal occurred and scratches produced during these operations.

Due to the introduction of new efficient grinding surfaces using charged diamond abrasives in recent decade, only one to two steps (or at most three steps) are generally required to achieve integrity. This is equivalent to a four-step fine grinding stage plus a one-step rough polishing stage when traditional preparation techniques are employed.

Figure 7.8 schematically shows the behavior of SiC abrasive papers during grinding. As grinding progresses, the initial depth of residual damage (z_0) gradually reduces until a

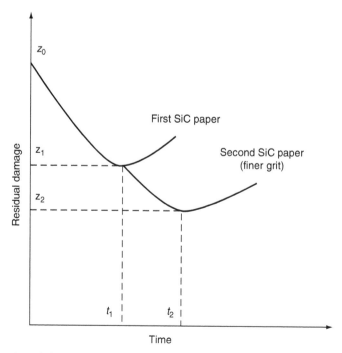

FIGURE 7.8 Behavior of SiC abrasive papers during grinding.

threshold value (z_1) is reached where abrasives break and become blunt. If grinding is continued further, supplementary damage will be induced into the specimen although a brighter shiny smooth surface results. Such abrasives are called degradable abrasives [4].

For low-carbon steels, the threshold positions for 320 and 600 grit SiC abrasive papers are 30 and 20 μm, respectively, while for high-carbon steels, the corresponding values are 25 and 15 μm, respectively. Therefore, the finer the abrasive and the harder the material, the lower the threshold positions. This idea shows that if we want to control the preparation process, grinding should be terminated at the threshold position and not beyond. It is difficult to find the threshold position according to the corresponding time as it will vary with many other factors. However, as mentioned before, the simplest way is to grind the specimen as long as it has a good three-dimensional scratch pattern.

If a second SiC abrasive paper of finer grit is used starting from t_1, then a similar curve will be obtained where a lower new threshold value of z_2 will appear at t_2.

If the first SiC abrasive paper has been used too long beyond t_1, then the time to reach z_2 for the second SiC abrasive paper would increase considerably. Moreover, the downtime needed for changing SiC abrasive papers is not considered yet. Therefore, the drawback of using SiC abrasive papers for grinding is evident, especially when semiautomatic or automatic preparation equipment is used.

Figure 7.9 schematically shows the behavior of diamond-charged preparation surfaces during grinding. It can be seen that when the depth of damage reduces from z_0 to its optimum value of z_1 at t_1, there will be no degrading of the surface after prolonged use. Hence, the advantage of such surface is also evident.

Another term frequently encountered in specimen preparation is lapping, a process in which abrasives can roll across the platen surface as the specimen is passed over, and this kind of surface would, for example, be glass or metal. Because the abrasive is rolling as it removes material, the resultant force exerted on the specimen surface is more to the vertical than is the case in grinding. It is for this reason that this technique must not be used for ductile materials since it would induce impressed abrasives. Lapping therefore is an appropriate technique only

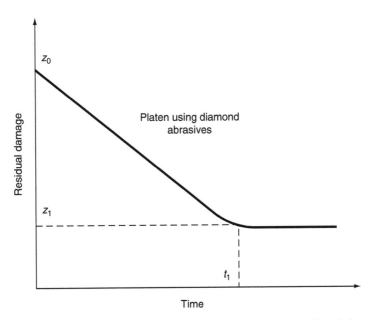

FIGURE 7.9 Behavior of diamond-charged preparation surfaces during grinding (schematic).

for brittle materials. Structural damage caused by lapping is less than an equivalent size abrasive used in the grinding mode. An observation of the lapped surface of a brittle material would give a matte or diffused finish. It is interesting to note that some preparation surfaces bear the name of lapping, but actually the function of these surfaces is to grind and not to lap. An easy way to identify the actual function of such surfaces is to examine the appearance of the specimen surface under a microscope after operation.

Some of the above-mentioned preparation surfaces (also called platens), using embedded or charged diamond abrasives, can also be used in the integrity stage. A napless, acetate woven cloth, e.g., Buehler's Trident Polishing Cloth, is an excellent preparation surface for maintaining edge flatness.

The use of diamonds as a grinding medium has revolutionized metallographic preparation; diamonds are the abrasives most often used for metallographic specimen preparation. There are two different types of synthetic diamonds, i.e., monocrystalline and polycrystalline. Polycrystalline diamonds are far superior because individual polycrystalline diamond grains are produced from a number of very small crystals that are sintered together to form a single grain. Therefore, each grain has many minute cutting edges, resulting in a surface with much finer scratches than can be achieved with monocrystalline diamonds of the same grain size, although the rate of material removal is the same for both types. Figure 7.10 shows a SEM micrograph of 15-μm polycrystalline diamonds. Note the very even grain size.

There are many models of grinder–polisher available, although manual operating grinder–polisher is very easy to operate and inexpensive, consistent results usually cannot be obtained because of the nonreproducibility of the operating parameters. Therefore, semiautomatic grinder–polisher is getting popular nowadays. Figure 7.11 shows (a) the PHOENIX BETA semiautomatic grinder–polisher with VECTOR power head; (b) close-up view of the VECTOR power head with single force specimen holder; and (c) close-up view of the VECTOR power head with central force specimen holder.

The Buehler's PHOENIX BETA grinder–polisher offers the versatility of continuously variable wheel speed from 30 to 600 rpm (Figure 7.11a), and the VECTOR power head enables the grinder–polisher to prepare specimens using either individually or centrally applied pneumatic loading. When a specimen holder with an individual force loading is used (Figure 7.11b), the specimens are held in place loosely. Loading is applied to each

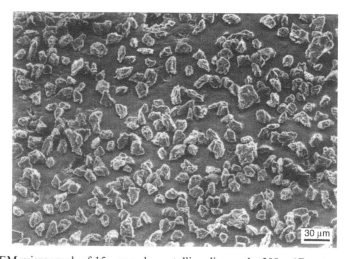

FIGURE 7.10 SEM micrograph of 15-μm polycrystalline diamonds. 300×. (Courtesy of Struers, Inc.)

specimen from above by a piston rod, and the applied force can be adjusted from 0 to 75 N. When the specimen holder with a central force loading is used (Figure 7.11c), all the specimens are held in place rigidly. During operation, the specimen holder with the clamped specimens is pressed downward against the preparation surface, the applied force can be adjusted from 20 to 300 N.

Both specimen-holding systems have advantages and disadvantages. The advantages of loose specimen-holding system are as follows: (a) One or more specimens can be prepared at one time; (b) planar grinding is not necessary; (c) internal stress due to clamping is avoided;

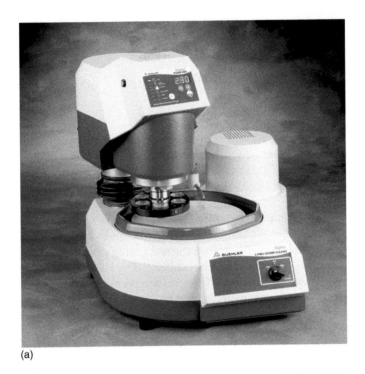

(a)

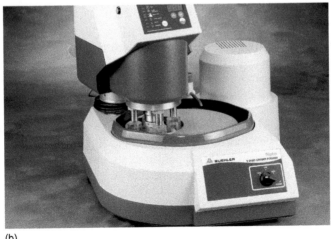

(b)

FIGURE 7.11 (a) The PHOENIX BETA semiautomatic grinder/polisher with VECTOR power head. (b) Close-up view of the specimen holder with individual force loading.

Continued

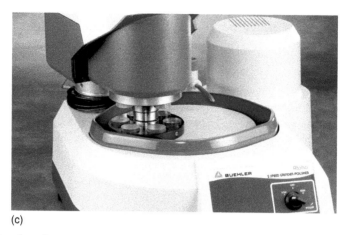

(c)

FIGURE 7.11 (Continued) (c) Close-up view of the specimen holder with central force loading. (Courtesy of Buehler Ltd.)

and (d) no need of using inverted type microscope or special object stage for the examination of specimens held on specimen holder between preparation steps. The disadvantages are as follows: (a) loss of planarity; (b) individual cleaning of specimens after every step is necessary including the specimen holder; and (c) mounted specimens are needed to fit the space of the specimen holder.

The advantages of clamped specimen holding system are as follows: (a) planarity of specimens is obtained after planar grinding; (b) specimens can be cleaned *in situ* with the specimen holder; (c) greater control can be attained during preparation; (d) irregular-shaped specimens and unmounted specimens can be clamped. The disadvantages are as follows: (a) at least three specimens are needed to reach balance; (b) a planar grinding step is required; and (c) if the quality of the prepared specimens after dismounting is found unsatisfactory, the specimens must be placed back in the holder again and the entire preparation procedures repeated.

The PHOENIX series of grinder–polisher is compatible with PRIMET Modular Dispensing System, which delivers different forms of grinding and polishing abrasives at preset time interval and amount, thus specimen consistency increased and waste reduced.

7.2.2.3 Polishing Stage

Polishing is the last stage of specimen preparation, according to the new concept of specimen preparation. Polishing is not used to achieve specimen integrity. At the polishing stage, the integrity has been achieved, therefore, any cosmetic or esthetic requirement beyond the specimen integrity is to be called a polishing stage [4].

During polishing the abrasives charged to a nap type cloth will rise and fall as the specimen passes over, offering a low stress action that takes off the sharp corners of previously left scratches. The result is a scratch free and shiny surface.

The most important parameters during polishing are time, speed, and pressure. Polishing should be completed within the shortest possible time to avoid relief. When diamond is used for polishing, in general 240 rpm for a 300-mm (12 in.) wheel is satisfactory. The pressure used for polishing is generally the half for grinding.

There are many polishing cloths available from various manufacturers. In general, they can be divided into two groups, namely those having a plain smooth surface and those having a plain short nap. The former cloths are suited for hard materials, and the latter with varying degrees of nap are suited for all other materials.

The general usage abrasives for polishing steel specimens are nonagglomerated alumina (0.05 μm) suspensions and colloidal silica (0.06 μm) in basic solution.

Colloidal silica is an extremely useful polishing suspension that is suitable for nearly all materials [10]. Although it has been used for polishing single crystals of silicon for many years, it was not popular with metallographers until recently [11]. With colloidal silica, polishing of almost any material will result in a surface without any scratches, and problems related to edge rounding and poor planarity will not occur due to the short duration, usually from a few seconds to about 3 min of this polishing step.

Colloidal silica contains very fine amorphous, rather than crystalline, particles that remain in suspension for such a long time that shaking or stirring before use is never needed. Figure 7.12 shows a transmission electron micrograph (TEM) of amorphous silica particles, which are nearly spherical in shape, with a mean diameter of 27.2 nm.

Schwuttke and Oster [12] suggested that polishing with colloidal silica involves a combination of chemical and mechanical action due to the high pH (about 9.5) activation. Owing to the development of additives to minimize evaporation, silica crystallization, and freezing problems, colloidal silica is much easier to use. Although the polishing rate is not high, colloidal silica is indispensable for those metallographers who seek the highest quality micrographs [10].

Table 7.5 through Table 7.7 present four- or three-step practices for preparing steels, colloidal silica may be substituted for the Masterprep alumina suspension if preferred [9]. Maraging steels are easily prepared with a four-step method as described in Table 7.5 through

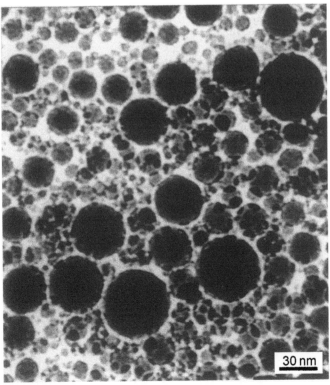

FIGURE 7.12 A TEM micrograph of amorphous silica particles in Syton HT-40. 135000×. (Courtesy of GF Vander Voort, Buehler Ltd. With permission.)

TABLE 7.5
Four-Step Practice for Steels

Stage	Surface	Abrasive/Size	Load, N (lb)	Speed, rpm Direction	Time, min
Planar grinding	SiC paper disk	120 grit (water cooled)	27 (6)	240–300 comp	Until plane
Integrity	BUEHLERHERCULES Rigid grinding disk	Metadi Supreme 9-μm polycrystalline diamond	27 (6)	120–150 comp	5
	Trident cloth	Metadi Supreme 3-μm polycrystalline diamond	27 (6)	120–150 comp	3
Polishing	Microcloth pad	Masterprep 0.05-μm alumina suspension	27 (6)	120–150 contra	2

Note: Comp means specimen holder and platen rotate in same direction. Contra means specimen holder and platen rotate in opposite directions.

Source: From GF Vander Voort, *Adv. Mater. Process.* 158:February (2000).

Table 7.7. The three preparation stages adopted in this book are added to each table by the author.

7.2.3 ETCHING

When a properly prepared as-polished specimen is examined under a metallurgical microscope, it appears essentially featureless. Only features with a 10% or greater difference in reflectivity, such as nonmetallic inclusions, cracks, pores, and scratches, can be detected without etching under brightfield illumination. However, the microstructure of many unetched specimens can be clearly revealed under other illumination conditions like darkfield illumination, crossed polarized illumination, uncrossed polarized illumination, and DIC illumination. All the above-mentioned illumination methods can be reached quite easily by

TABLE 7.6
Three-Step Practice for Steels

Stage	Surface	Abrasive/Size	Load, N (lb)	Speed, rpm Direction	Time, min
Planar grinding	SiC paper disk	120 grit (water cooled)	27 (6)	240–300 comp	Until plane
Integrity	BUEHLERHERCULES Rigid grinding disk	Metadi Supreme 3-μm polycrystalline diamond	27 (6)	120–150 comp	5
Polishing	Microcloth pad	Masterprep 0.05-μm alumina suspension	27 (6)	120–150 contra	5

Source: From GF Vander Voort, *Adv. Mater. Process.* 158:February (2000).

TABLE 7.7
Four-Step Practice for Superalloys and Stainless Steels

Stage	Surface	Abrasive/Size	Load, N (lb)	Speed, rpm Direction	Time, min
Planar grinding	SiC paper disk	120 grit (water cooled)	27 (6)	240–300 comp	Until plane
Integrity	BUEHLERHERCULES Rigid grinding disk	Metadi Supreme 9-μm polycrystalline diamond	27 (6)	120–150 comp	5
	Trident cloth	Metadi Supreme 3-μm polycrystalline diamond	27 (6)	120–150 comp	5
Polishing	Microcloth pad	Masterprep 0.05-μm alumina suspension	27 (6)	120–150 contra	5

Source: From GF Vander Voort, *Adv. Mater. Process.* 158:February (2000).

simply pushing certain accessories into the optical path of the microscope. Therefore, it is wise to rely on optical methods to reveal the microstructure before etching of the specimen.

A vast number of microstructures are visible only after suitable etching, and therefore the purpose of etching is to produce a structure with sufficient contrast to delineate as much detail as possible. Before a specimen is ready for etching, its surface must be flat and free of deformed layers, scratches, pullout, and edge rounding.

7.2.3.1   Chemical Etching

Chemical etching is the oldest and most commonly used etching method in which the specimen is usually immersed in the etchant. Microstructural contrast is produced through selective dissolution among various constituent phases due to differences in the electrochemical potential. When etching occurs, local cells are formed where the more positive constituent phase acts as anode and reacts more strongly with the etchant than the more negative constituent. Potential differences arise chiefly from differences in composition; however, differences in the physical state may also cause potential differences. For example, grain boundaries are etched more easily than bulk grains; also, deformed metal reacts more strongly with the etchant than undeformed metal.

Most etchings is performed at room temperature. As etching proceeds, the surface becomes less reflective or duller, and the appropriateness of etching can be judged by the appearance of the surface that is etched. Owing to the empirical nature of chemical etching, it is difficult to reproduce a distinctive etching contrast.

After etching has been done, the specimen should be rinsed first in distilled water and then in alcohol, which accelerates drying and avoids the formation of water spots. The specimen is then dried in a stream of warm air. For mounted specimens, the use of ultrasonic cleaner is recommended to more effectively clear away the etchant remaining in the pores, cracks, or interfaces between specimen and the mount. Specimens can then be stored in desiccators.

The most widely used general-purpose etchants for steels are nital and picral. The former consists of a 0.5–10% (usually 2–3%) solution of HNO_3 in ethanol or methanol, and the latter consists of a 4% solution of picric acid in ethanol, sometimes with about 0.25–1% zephiran chloride (wetting agent) added to improve structure delineation, etching rate, and uniformity. Nital attacks both ferrite grain boundaries and ferrite–cementite phase boundaries, but picral attacks the latter only. Accordingly, for mild steels, nital should be used, whereas for spheroidized carbide structures, use of picral is preferable because the sensitivity of nital to orientation often produces faint etching at the ferrite–carbide interface within certain grains.

Table 7.8 through Table 7.10 list etchants for microscopic examination of carbon and alloy steels, microetchants for wrought stainless steels, and special-purpose etchants for tool steels [13]. More information can be obtained in Refs. [6,14] and in ASTM Standard E407.

7.2.3.2 Tint Etching

Tint etching produces microstructural contrast through selective coloration rather than selective dissolution as in chemical etching, due to deposition of a thin film of oxide, sulfide, complex molybdate, etc., on the polished surface of the specimen. The deposited film generally has a thickness of 40–500 nm. Coloration is produced by interference, which is visible under bright-field illumination. The color contrast can be enhanced further with the use of polarized light. Film thickness controls film color, which varies with the orientation of the substrate phase and etching time. Tint etching is performed at room temperature by immersing the specimen, never by swabbing. Often the surface is not etched beneath the interference film, therefore, a light pre-etch with nital or picral to reveal phase boundaries is recommended.

Beraha and Shpigler [15] developed a series of tint etchants for steels and many other metals and alloys. The tint etching system can be classified as anodic, cathodic, or complex depending on the nature of film precipitation. Most tint etchants are anodic systems that color the anodic phases only. They are usually acidic solutions using either water or alcohol as the solvent, and sodium metabisulfite ($Na_2S_2O_5$), potassium metabisulfite ($K_2S_2O_5$), and sodium thiosulfate ($Na_2S_2O_3 \cdot 5H_2O$) as common ingredients. Tint etchant based on sodium molybdate ($Na_2MoO_4 \cdot 2H_2O$) generally colors the cathodic phase, such as cementite in steels [16]. Some tint etchants for steels are listed in Table 7.8 and Table 7.9.

When preparing the tint etchants, the formula and the order of mixing should be followed strictly, and caution in handling hazardous reagents should be carefully adhered to. The specimen undergoing a tint etch should be prepared very carefully, since even very light scratches undetectable after polishing will be visible after tint etching. For better results, automatic polishing or vibratory polishing is preferable [17].

7.2.4 ELECTROLYTIC POLISHING

As a specimen preparation method, electrolytic polishing for quality control and research work has been employed in many laboratories. In this method, the specimen surface is leveled by anodic solution in an electrolytic cell consisting of a tank for the electrolyte, a direct current source, an anode (specimen), and a cathode. It is interesting to note that true integrity with zero residual damage can only be achieved by electrolytic polishing.

Although the detailed micromechanism of electropolishing is not clearly understood, the operating conditions of a specific specimen–electrolyte combination can be evaluated by plotting the current density versus the applied voltage curve as shown schematically in Figure 7.13. Etching occurs at the initial segment (A–B) of the curve, polishing occurs at the plateau (C–D) of the curve, and gas evolution and pitting occur at the final part (D–E) of the curve. Optimum polishing effect usually occurs at C.

A desirable electrolyte should be a somewhat viscous solution, a good solvent during electrolysis, and simple to mix, stable, and safe. It should contain one more large ion, e.g., PO_4^{3-}, Cl_4^{4-}, or SO_4^{2-}, or large organic molecules. It should be operable at room temperature and insensitive to temperature changes and should not attack the specimen with the current off.

TABLE 7.8
Etchants for Microscopic Examination of Carbon and Alloy Steels

No.	Etchant	Purpose or Characteristic Revealed
1	Nital: 1 to 5 ml HNO in 99 to 95 ml ethanol (95%) or methanol (95%)	Develops ferrite grain boundaries in low-carbon steels; produces contrast between pearlite and a cementite or ferrite network; develops ferrite boundaries in structures consisting of martensite and ferrite; etches chromium-bearing low-alloy steels resistant to action of picral. Preferred for martensitic structures
2	Picral: 4 g picric acid in 100 ml ethanol (95%) or methanol (95%)	Reveals maximum detail in pearlite, spheroidized carbide structures, and bainite; reveals undissolved carbide particles in martensite; differentiates ferrite, martensite, and massive carbide by coloration; differentiates bainite and fine pearlite; reveals carbide particles in grain boundaries of low-carbon steel. Addition of about 0.5 to 1 ml zephiran chloride wetting agent increases speed of attack
3	Vilella's reagent: 5 ml HCl, 1 g picric acid, 100 ml ethanol (95%) or methanol (95%)	For contrast etching[a]; may reveal prior austenite grains in tempered martensite and in austempered steels; reveals pearlite colonies
4	1 to 1.5 ml HCl (conc); 2 to 4 picric acid 100 ml ethanol (95%)	Reveals pearlite colonies[b]
5	30 g $K_2Cr_2O_7$ in 225 ml hot distilled water; and 30 ml acetic acid (glacial)	Reveals lead inclusions, causing them to appear yellow or gold when specimen is examined under polarized light[c]
6	16 g CrO_3 in 145 ml distilled water; add 80 g NaOH[d]	Reveals intergranular oxidation due to overheating before hot working[e]
7	10 g potassium metabisulfite, 100 ml water	For resolution of hardened structures. Should be preceded by an etch in nital or picral
8	Howarth's reagent: 10 ml H_2SO_4, 10 ml HNO_3, 80 ml water	For detection of overheating and burning, and for examination of steel forgings
9	8 g sodium metabisulfite in 100 ml water	Produces good contrast in as-quenched martensitic structures
10	1 g KCN in 100 ml water mixed with 0.25 g diphenylthiocarbazone in 10 ml chloroform	Reveals lead inclusions by coloring them red; coloration is most visible when specimens are viewed under polarized light
11	Saturated aqueous picric acid plus 1 g/100 ml sodium tridecylbenzene sulfonate	Most successful etch for revealing prior austenite grain boundaries in medium- or high-carbon martensitic steels. Steels should be untempered, or tempered below 540°C (1000°F). Immerse or swab for up to 20 min
12	2 g picric acid, 25 g sodium hydroxide, 100 ml water	Alkaline sodium picrate. Use boiling for 30 s or more to darken cementite. Solution will attack mounts made from Bakelite
13	50 ml cold saturated aqueous sodium thiosulfate, 1 g potassium metabisulfite	Klemm's tint etch; colors ferrite. Immerse for 40 to 100 s until surface is colored. A light pre-etch with nital or picral improves sharpness

[a]Specimen should be tempered for 20 to 30 min at 315°C (600°F).

[b]Immerse specimen for 5 to 10 s in solution at room temperature.

[c]Etch for 10 to 20 s in solution at room temperature, rinse in hot water and dry.

[d]Sodium hydroxide (NaOH) must be added slowly, with constant stirring.

[e]Immerse specimen in boiling solution for 10 to 30 min, rinse in hot water, dry in air blast.

Source: From *Adv. Mater. Process.* 159(12):204–206 (2001).

TABLE 7.9
Microetchants for Wrought Stainless Steels

No.	Composition	Comments
1	1 g picric acid, 5 ml HCl, 100 ml ethanol	Vilella's reagent. Use at room temperature to 1 min. Outlines second-phase particles (carbides, sigma phase, delta ferrite), etches martensite
2	15 g CuCl$_2$ (cupric chloride), 33 ml HCl, 33 ml water[a]	Kalling's No.1 reagent for martensitic stainless steels. Use at room temperature. Martensite dark, ferrite colored, austenite not attacked
3	5 g CuCl$_2$, 100 ml HCl, 100 ml ethanol	Kalling's No.2 reagent. Use at room temperature. Ferrite attacked rapidly, austenite slightly attacked, carbides not attacked
4	5 g CuCl$_2$, 40 ml HCl, 30 ml H$_2$O, 25 ml ethanol	Fry's reagent. For martensitic and precipitation-hardenable grades. Use at room temperature
5	4 g CuSO$_4$, 20 ml HCl, 20 ml H$_2$O	Marble's reagent. Primarily for austenitic grades. Use at room temperature to 10 s. Attacks sigma phase
6	3 parts glycerol, 2–5 parts HCl, 1 part HNO$_3$	Glyceregia. Popular etch for all stainless grades. Higher HCl content reduces pitting tendency. Use fresh, never stored. Discard when reagent is orange colored. Use with care under a hood. Add HNO$_3$ last. Immerse or swab few seconds to a minute. Attacks sigma phase, outlines carbides. Substitution of water for glycerol increases attack rate
7	45 ml HCl, 15 ml HNO$_3$, 20 ml methanol	Methanolic aqua regia. For austenitic grades to reveal grain structure, outline ferrite and sigma phase
8	15 ml HCl, 5 ml HNO$_3$, 20 ml methanol	Dilute aqua regia for austentitic grades. Uniform etching of austenite, outlines carbides, sigma phase, and ferrite (sometimes attacked)
9	4 g KMnO$_4$, 4 g NaOH, 100 ml H$_2$O	Groesbeck's reagent. Use at 60–90°C (140–195°F) for 1 to 10 min. Colors carbides dark, sigma phase gray in duplex alloys. Austenite not attacked
10	30 g KMnO$_4$, 30 g NaOH, 100 ml H$_2$O	Modified Groesbeck's reagent. Use at 90–100°C (195–212°F) for 20 s to 10 min color ferrite dark in duplex alloys. Austenite not affected
11	10 g K$_3$Fe(CN)$_6$, 10 g KOH or 7 g NaOH, 100 ml H$_2$O	Murakami's reagent. Room temperature for 7 to 60 s to reveal carbides; sigma phase faintly revealed by etching to 3 min. Use at 80°C (176°F) to boiling for 2 to 60 min to darken carbides. Sigma may be colored blue, ferrite yellow to yellow-brown, austenite not attacked. Under a hood only
12	30 g KOH, 30 g K$_3$Fe(CN)$_6$, 100 ml H$_2$O	Modified Murakami's reagent. At 95°C (203°F) for 5 s. Colors sigma phase reddish brown, ferrite dark gray, austenite unattacked, carbide black. Use under a hood
13	10 g oxalic acid and 100 ml H$_2$O	Popular electrolytic etch, 6 V dc, 25 mm spacing. 15–30 s reveals carbides; grain boundaries revealed after 45–60 s; sigma phase outlined after 6 s. Lower voltages (1–3 V dc) can be used. Dissolves carbides. Sigma strongly attacked, austenite moderately attacked, ferrite not attacked
14	10 g NaCN (sodium cyanide) and 100 ml H$_2$O	Electrolytic etch at 6 V dc, 25 mm spacing, 5 min, platinum cathode. Sigma darkened, carbides light, ferrite outlined, austenite not attacked. Good for revealing carbides. Use with care under a hood
15	10 ml HCl and 90 ml methanol	Electrolytic etch at 1.5 V dc, 20°C (70°F) to attack sigma phase. Use at 6 V dc for 3–5 s to reveal structure

Continued

TABLE 7.9 (Continued)
Microetchants for Wrought Stainless Steels

No.	Composition	Comments
16	60 ml HNO_3 and 40 ml H_2O	Electrolytic etch to reveal austenite grain boundaries (but not twins) in austenitic grades. With stainless steel cathode, use at 1.1 V dc, 0.075–0.14 A/cm^2 (0.48–0.90 $A/in.^2$), 120 s. With platinum cathode, use at 0.4 V dc, 0.055–0.066 A/cm^2 (0.35–0.43 $A/in.^2$), 45 s. Will reveal prior austenite grain boundaries in solution-treated (but not aged) martensitic precipitation-hardenable alloys
17	50 g NaOH and 100 ml H_2O	Electrolytic etch at 2 V dc, 5–10 s to reveal sigma phase in austenitic grades
18	56 g KOH and 100 ml H_2O	Electrolytic etch at 1.5–3 V dc for 3 s to reveal sigma phase (red-brown) and ferrite (bluish). Chi-phase colored same as sigma
19	20 g NaOH and 100 ml H_2O	Electrolytic etch at 20 V dc for 5–20 s to outline and to color delta-ferrite phase tan
20	NH_4OH (conc)	Electrolytic etch at 1.5–6 V dc for 10–60 s. Very selective. At 1.5 V, carbide completely etched in 40 s; sigma unaffected after 180 s. At 6 V, sigma phase etched after 40 s
21	10 g $(NH_4)_2 S_2O_8$ and 100 ml H_2O	Use at 6 V for 10 s to color carbide dark brown
22	200 ml HCl and 1000 ml H_2O	Baraha's tint etch for austenitic, duplex, and precipitation-hardenable grades. Add 0.5–1.0 g $K_2S_2O_5$ per 100 ml of solution (if etching is too rapid, use a 10% aqueous HCl solution). Immerse at room temperature (never swab) for 30–120 s until surface is reddish. Austenitic phase colored, carbides not colored. Longer immersion colors ferrite lightly. If coloration is inadequate, add 24 g NH_4FxHF (ammonium bifluoride) to stock reagent at left
23	20 g picric acid and 100 ml HCl	Etch by immersion. Develops grain boundaries in austenite and delta ferrite in duplex alloys
24	Saturated aqueous $Ba(OH)_2$ (barium hydroxide)	Attacks carbides well before sigma phase in austenitic grades at 1.5 V dc, but attacks both equally at 3–6 V dc. Has been used to differentiate chi phase and Laves phase (use at 4.3 V dc, platinum cathode, 20 s). Chi is stained mottled-purple, Laves is not colored, ferrite is stained tan
25	50 ml each H_2O, ethanol, methanol, and HCl; plus 1 g $CuCl_2$, 3.5 g $FeCl_3$, 2.5 ml HNO_3	Ralph's reagent. Use by swabbing. Can be stored. General-purpose etchants for most stainless steels. Does not attack sulfides in free-machining grades

[a]When water is specified, use distilled water.

Source: From *Adv. Mater. Process.* 159(12):204–206 (2001).

The advantages of electrolytic polishing are as follows:

1. Fastest and most reproducible specimen preparation method once the polishing parameters have been established and can be precisely controlled.
2. Very suitable for homogeneous materials such as pure metals or single-phase alloys.
3. Eliminates all forms of residual damage or stress or scratches from the observed top surface after completion of the operation.

TABLE 7.10
Special-Purpose Etchants for Tool Steels

Feature	Etchant	Comments
Cementite	2 g picric acid, 25 g NaOH, 100 ml H_2O	Immerse in boiling solution or use electrolytically (6 V dc, 20°C, 30–120 s). Will not darken cementite if it contains substantial amounts of chromium
M_6C, Mo_2C	4 g $KMnO_4$, 4 g NaOH, 100 ml H_2O	Immersion only
MC, M_7C_3	1 g CrO_3, 100 ml H_2O	Electrolytic at 2–3 V dc, 30 s

Source: From *Adv. Mater. Process.* 159(12):204–206 (2001).

4. In some cases, once electropolishing has been done, the cell with the same electrolyte can be immediately switched to the etching condition by simply reducing the applied voltage to about 10% of that required for polishing and continuing electrolysis for a few seconds.

However, one should be aware of the following limitations of electropolishing in order to make full use of its advantages:

1. Only microscratches can be removed. The specimen must undergo grinding with fine grit abrasives before electropolishing.
2. Specimens with multiphase or heterogeneous structures such as nonmetallic inclusions in steels or carbides in stainless steel are not well suited for electropolishing because of the differences in electrochemical potentials among the constituent phases.
3. Some of the electrolytes are toxic, highly corrosive, or otherwise hazardous and can even be explosive. Therefore, precautions for the safe use of the electrolytes should be strictly observed.
4. Edge retention cannot be maintained during electropolishing unless the side of the specimen is protected by a nonconductive epoxy resin.
5. The size of the area that can be polished successfully is limited because of a potential change from the center to the edge of the area to be polished. For example, the polished area is limited to 0.5–2 cm^2 for stainless steel and some high-alloy steels, but up to 5 cm^2 of some carbon steels can be polished.

The mechanical and electrolytic polishing methods each have merits and specific applications. They are complementary and not mutually exclusive. One should select the methods that will give the best result for each specific specimen.

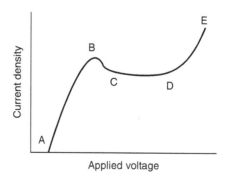

FIGURE 7.13 Typical plot of current density versus applied voltage of an electropolishing solution (schematic).

Table 7.11 provides a listing of electrolytic polishing solutions for steels selected from Ref. [6].

7.2.5 VIBRATORY POLISHING

The vibratory polishing method was first introduced by Krill [18] and Long and Gray [19] in the 1950s. A vibratory polisher is a torsional vibrational system in which the helical motion of the polishing bowl is produced approximately at resonance by electromagnetic force from a

TABLE 7.11
Electrolytic Polishing Solution for Steels

Electrolyte Composition	Current Density (A/cm^2)	Voltage (V, dc)	Temp. ($^\circ$C)	Time (min)	Comments
185 ml H$_3$PO$_4$, 765 ml acetic anhydride	0.04–0.06	50 (ext.)	<30	4–5	Wide applicability. Age solution 24 h before use. Use Fe or Al cathode, 20 times as large as sample. For gamma stainless steel, use 0.1 A/cm^2
54 ml perchloric acid, 146 ml H$_2$O, 800 ml ethanol plus 3% ether	4	110 (ext.)	<35	15 s	Good for many metals. Use Fe or austenitic stainless steel cathode. Pump electrolyte. Cool
420 ml H$_3$PO$_4$, 470 ml glycerol, 150 ml H$_2$O	1.5–2.0		100	8–15	For stainless steel
25 g CrO$_3$, 133 ml acetic acid, 7 ml H$_2$O	0.09–0.22	20	17–19	6	Wide applicability. Dissolve CrO$_3$ in solution using water bath at 60–70°C. Samples mounted in Bakelite can be safely electropolished. Grind to 600 grit. Cool bath during use. Will attack minclusions and cracks. Etch stainless steel samples at 0.025 A/cm^2 for 5–20 min. Store solution in airtight bottle
1000 ml acetic acid, 50 ml perchloric acid, 5–15 ml H$_2$O (optional)	0.01	45 (ext.)	25		Wide applicability. Best polishing without water. Can produce preferential attack in two-phase alloys
650 ml H$_3$PO$_4$, 150 ml H$_2$SO$_4$, 150 ml H$_2$O, 50 g CrO$_3$	66–100		40–60	3–7	For carbon steels up to 1.1% C
62 ml perchloric acid, 700 ml ethanol, 100 ml butyl Cellosolve, 137 ml H$_2$O	1.2			20 s	Wide applicability, including high- and low-carbon steels and high-speed steels. Add perchloric acid carefully to ethanol and water. Add butyl Cellosolve immediately after electropolishing
40% H$_2$SO$_4$, 46% H$_3$PO$_4$, 4% dextrose, 10% H$_2$O	23–70		28–40	5–10	For carbon steels. Etch at 12–16 A/cm^2 for 10 min

Source: From GF Vander Voort, *Metallography—Principles and Practice*, McGraw-Hill, New York, 1984.

solenoid pulsed with a half-wave rectified current. Weighted or unweighted specimens are placed in the polishing bowl. The vibrations cause the specimens to move around the periphery of the bowl and about its own axis, producing a material removal action. The vibrational amplitude of the polishing bowl must be adjusted to ensure smooth movement of the specimen around the bowl without bouncing. The speed of the specimen around the bowl is usually 18–20 rpm, which is much slower than that of a conventional polishing machine.

When the specimen is placed on this vibrating preparation surface, it undergoes a continual series of minute cutting movements by the abrasives, which results in a fairly delicate material removal action and much lower residual damage.

Due to the inherently low polishing rate, vibratory polishing often lasts for many hours. On the other hand, as many as eighteen 40-mm diameter specimens, or specimens of other diameters comprising an equivalent area, can be prepared simultaneously in a single polishing bowl. Once the specimens are loaded, little operator attention is needed, and any specimen can be removed and inspected while others are still prepared. Therefore, the vibratory polisher behaves as a semiautomatic polishing machine.

Figure 7.14 illustrates VIBROMET 2 vibratory polisher. It utilizes a powerful drive mechanism that produces nearly 100% horizontal vibratory motion, the only component of motion causing material removal, with virtually no vertical motion. As compared with its previous model, the vibration frequency is doubled, which results in an exceptional flatness across the entire specimen surface.

The polishing quality obtained on the vibratory polisher is very good when measured in terms of a scratch-free surface, retention of inclusions, specimen flatness, and minimal depth of disturbed layer [20]. Figure 7.15 shows a photomicrograph of an annealed medium-carbon steel specimen, vibratory polished for 20 min after ground with 600 grit SiC abrasive paper, and then etched with 2% nital.

Vander Voort [10] showed that a brief (less than 30 min) final vibratory polish with colloidal silica may be used for carbon and low-alloy steels. The same procedure with longer duration for AISI type 52100 bearing steel (1% C–1.5% Cr) is the only effective way to eliminate minor staining problems around inclusions. Another example is boronated stainless steels that undergo image analysis measurements. A final vibratory polish with colloidal silica

FIGURE 7.14 VIBROMET 2 vibratory polisher. (Courtesy of Buehler Ltd.)

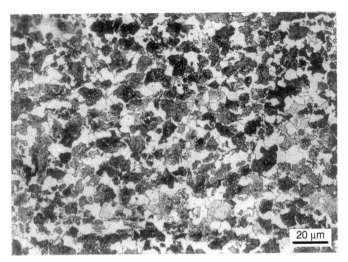

FIGURE 7.15 Photomicrograph of an annealed 0.45% C carbon steel, vibratory polished and etched with 2% nital, 200×. (Reduced 15% in reproduction.)

is also adapted. It seems that vibratory polishing is a useful complement to conventional mechanical polishing.

7.3   THE METALLURGICAL MICROSCOPE

It is well known that the behavior of a steel part depends strongly on the volume fraction, size, and distribution of its constituent phases. During heat treatment, various changes among constituent phases or microstructural constituents may occur. Therefore, the examination of the microstructure of a steel part is a very important and effective measure in both quality inspection of heat-treated parts and related research work. Owing to ease of operation, its larger field of view and relatively lower cost, the optical metallurgical microscope is still one of the most commonly used instruments for routine inspection and research work.

7.3.1   IMAGE FORMATION OF METALLURGICAL MICROSCOPE

The principle of image formation of a metallurgical microscope during visual observation is essentially the same as that of an ordinary microscope, as shown in Figure 7.16. Magnification is accomplished by an objective with very short focus and an ocular or eyepiece with somewhat longer focus. To minimize various aberrations, both objectives and eyepieces are highly corrected compound lenses, with the former far more complicated. For simplicity, both the objective and the eyepiece are treated as simple thin lenses as shown in Figure 7.16. The object (ab) is located just outside the front focal point of the objective (F_{ob}) so that a real magnified image ($a'b'$) can be formed. The intermediate or primary image is located within the focal point of the eyepiece (F_{oc}), which functions as a magnifier and forms a further enlarged but virtual image ($a''b''$). This image becomes the object for the eye itself, which forms the final real image on the retina ($a'''b'''$).

The overall magnification (M) of a microscope is the product of the linear magnification (m_1) of the objective and the angular magnification (m_2) of the eyepiece:

$$M = m_1 \times m_2 \tag{7.1}$$

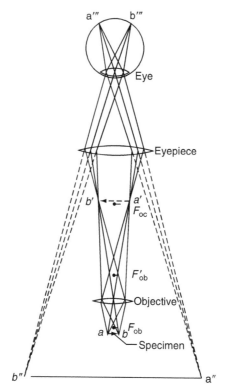

FIGURE 7.16 Image formation of a microscope during visual observation.

From geometrical optics,

$$m_1 = -L/f_1 \quad \text{and} \quad m_2 = 250/f_2 \tag{7.2}$$

where f_1 and f_2 are focal lengths of the objective and eyepiece, respectively; L is the so-called optical tube length of the microscope, which is measured from the back focal point (b.f.p) of the objective to the position where the intermediate image stands; and 250 is the distance of most distinct vision, in mm. f_1, f_2, and L are all expressed in mm. The negative sign of the magnification m_1 signifies an inverted image.

The overall magnification is therefore

$$M = m_1 \times m_2 = -\frac{(L \times 250)}{(f_1 \times f_2)} \tag{7.3}$$

The optical tube length is inconvenient in practical use. Therefore, the mechanical tube length, which is defined as the distance from the screw flange of the objective to the end of the eyepiece tube, is commonly used instead. The mechanical tube length is normally 160 to 250 mm. Because even a slight deviation from the correct tube length will result in a notable reduction of image quality, it is not advisable to use an objective designed for a different tube length from another manufacturer.

Actually, the principle of image formation of a microscope is not as simple as described above. Owing to the fineness of the structure to be resolved by an objective, diffraction and interference must be considered in the theory of image formation first put forward by E. Abbé over a century ago, which shows that the object of the image is the result of interference of direct or undiffracted and diffracted beams [21].

7.3.2 OBJECTIVES

The image quality of a microscope is determined, to a great extent, by the quality of the objective, which forms the primary image of the specimen. Therefore, the objective is the most important part of a microscope.

7.3.2.1 Numerical Aperture

The numerical aperture (NA) of an objective is a measure of its light-collecting ability and can be expressed by the equation

$$NA = n \sin \alpha \tag{7.4}$$

where n is the refractive index of the medium between the specimen and the front lens of the objective and α is half the aperture angle of the most oblique rays entering the front lens of the objective (see Figure 7.17).

Equation 7.4 shows that the larger the angle α, the greater the light-collecting ability of the front lens of the objective. The magnitude of angle α depends on the size of the front lens and the working distance of the objective, i.e., the distance between the specimen and the front surface of the objective when the image is in sharp focus. Since the angle α cannot exceed 90°, the NA of a dry objective cannot exceed 1, and normally it is within 0.95. The NA of oil immersion objectives can be as high as 1.4.

7.3.2.2 Resolution

The resolution or resolving power of an objective is defined as the nearest resolvable distance between two parallel lines or separated points in the object. The smaller the resolvable distance, the higher the resolution. The resolution can be expressed by the equation

$$d = \frac{\lambda}{2NA} \tag{7.5}$$

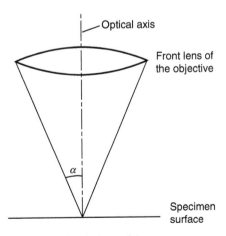

FIGURE 7.17 Aperture angle of an objective (schematic).

where d is the resolution, λ is the wavelength of the light used, and NA is the numerical aperture of the objective as described earlier.

From Equation 7.5 it can be seen that with an incident light of known wavelength, the resolution of an objective is totally determined by its NA. The larger the NA, the higher the resolution. As an example, an objective with the highest attainable NA of 1.4, using yellow-green light of a wavelength of 550 nm, the resolution is about 200 nm or 0.2 μm.

The role of the eyepiece is to magnify the intermediate image formed by the objective. Accordingly, any unresolvable structural details cannot be resolved through the magnification of the eyepiece. Therefore, the resolving power of a microscope is governed chiefly by the resolution of the objective.

To fully utilize the resolution of the objective so that the already resolved structural detail can be viewed clearly by the human eye, hence, suitable magnification of the microscope should be incorporated. According to Abbé's theory, the limit of resolving angle of the human eye under the best illuminating conditions is about 1 min of arc, but more realistic upper and lower limits are 4 and 2 min of arc, which correspond to a separation of about 0.30 and 0.15 mm, respectively, at a distance of 250 mm, i.e., the distance of most distinct vision. Using $\lambda = 550$ nm, the approximate range of magnification M necessary for the microscope can be obtained:

$$550 \text{ NA} < M < 1100 \text{ NA} \tag{7.6}$$

Equation 7.6 is also called the useful range of magnification. Below the lower limit, the human eye cannot clearly see the structural detail already resolved by the objective; above the upper limit, no more structural detail can be observed, and blurring of the image may even occur. This latter kind of magnification, often called empty magnification, should be prevented during observation.

However, Vander Voort [6, p. 282] argued that Equation 7.6 is derived using an ideal contrasting image with an ideal eye and should not be adhered to dogmatically. In fact, resolution depends on both resolving power and image contrast. Besides, illumination, magnification, lens quality, and observation conditions also affect image contrast and therefore resolution. It is suggested that the lowest useful magnification for maximum resolution should be four times greater than under optimum conditions, i.e.,

$$M_{\text{eff}} \approx 2200 \text{ NA} \tag{7.7}$$

and the use of photographs with magnifications up to 4000× or more is fully justified.

Owing to the improved resolution, oil immersion objectives are useful in metallography. However, such lenses should be cleaned with a special solvent after every use, which is inconvenient. Therefore, high-power dry objectives (such as 100/0.95 and 150/0.95) are used more frequently. Recently, dry objectives with even higher magnifications, e.g., 200×, are available from Nikon. With the same NA of 0.95 as the 100× and 150× dry objectives, an overall magnification well above the old criterion (1100 NA) can be easily attained with this 200× objective during visual observation. This further justifies the above viewpoint.

7.3.2.3 Common Types of Objectives

Plano-achromatic and apochromatic objectives are the most commonly used objectives in metallurgical microscopes. The prominent characteristic of a plano-objective is its significantly enlarged flatness of the image field, which is due to an extensive correction of the image field curvature. During visual observation with plano-objectives, fewer fields of view are needed for a given area than during observation with conventional objectives. As a result, eyestrain is notably lessened. Besides, it is self-evident that plano-objectives are particularly useful for photomicrography. However, the accuracy of the sliding mechanism in the object stage and the quality of specimen preparation should be better when plano-objectives are used.

Plano achromatic objectives are corrected spherically for red and blue colors. Therefore, there are still residual chromatic aberrations, but they can be minimized by the use of yellow-green filters.

Plano-apochromatic objectives are corrected chromatically in the range of red, green, and blue colors, i.e., over almost the entire range of visible light. The correction of spherical aberration can reach the range of green and blue colors.

Plano-objectives have more lens elements and are therefore more expensive.

7.3.3 Eyepieces

As stated earlier, the function of an eyepiece is to magnify the intermediate image formed by the objective so that the resolved structural detail of the specimen can be viewed clearly by the human eye. Wide-field compensating eyepieces have been developed to accommodate the plano-objectives. With a significantly larger field of view number (diameter of field diaphragm), such eyepieces can fully utilize the enlarged flat area of the intermediate image. As plano-objectives are undercorrected with respect to lateral chromatic aberration, compensating eyepieces are overcorrected.

Ordinary eyepieces do not suit eyeglass wearers, especially those with astigmatic eyes, because the designed eye clearance (the distance between the eye lens of the eyepiece and the eye) of 10 mm is inadequate for eyeglass viewing. As a result, the image size would be severely limited due to increased eye clearance. If the eyeglass wearer views without eyeglasses, the image quality will be affected by astigmatism. High-eyepoint eyepieces are specially designed for eyeglass wearers. With such eyepieces, the entire field of view can be obtained at a distance of about 20 mm from the eye lens, which is sufficient for viewing with eyeglasses.

7.3.4 Parfocality

Microscopes are often provided with parfocal lens system so that there is no need to refocus after changing the objectives or the eyepieces except for minor adjustment of the fine focusing knob. The optical–mechanical dimensions of both the objectives and the eyepieces should meet the following requirements of parfocality:

1. The distance between the specimen and the intermediate image should be identical for all objectives; for a given mechanical tube length, the distance between the specimen and the screw flange of the objective, also called the parfocal length of the objective, should be held constant. A length of 45 mm is generally taken as the parfocal length.
2. The distance between the intermediate image and the end of the eyepiece tube should be the same for all objectives.
3. The focal plane of all eyepieces must always coincide with the intermediate image plane.

Obviously, parfocality is not an intrinsic property of an objective or an eyepiece, but rather a measure adopted in the design of modern microscopes to ease manipulation and to prevent possible damage to the objective lens or the specimen surface during operation.

7.3.5 Illumination System

7.3.5.1 Principles of Köhler Illumination

Due to the opaqueness of a metal specimen, it is necessary that an incident light illumination system, also called a vertical illuminator, be used in a metallurgical microscope (see the left-hand portion of Figure 7.18). It primarily consists of the following parts: (a) a light source with a collector lens; (b) several auxiliary lenses that provide the desired path of the beam; (c) an aperture diaphragm for controlling the illumination aperture; (d) a field diaphragm for controlling the illuminated area of the object; (e) a plane glass reflector that inclines toward the direction of the beam with an angle of 45° and reflects part of the beam toward the objective; and (f) an objective that acts as a condenser in the illumination system.

Illumination according to Köhler's principles is used in most metallurgical microscopes. To obtain the best image that a given microscope is capable of providing, the requirements of Köhler illumination should be fulfilled. The left-hand portion of Figure 7.18 shows the illuminating beam path, and the right-hand portion shows the part of the imaging beam path that ends at the intermediate image plane. The light source is centered and imaged at the aperture diaphragm, which is then imaged together with the light source image at the b.f.p. of the objective. By adjusting the setting of the aperture diaphragm, the illumination can be matched with the aperture of each objective, bringing the resolution of the objective into full play. However, glare and marginal lens aberrations still exist at this setting. Therefore, one should strive for an ideal compromise between the optimum resolution and the best possible contrast; the aperture diaphragm should be as open as possible to achieve the best resolution and as closed as necessary to achieve the contrast essential for observation of the resolved detail [22].

The aperture diaphragm setting can easily be adjusted by viewing down the eyepiece tube with the eyepiece removed. If there is a built-in Bertrand lens in the microscope, after swinging it into the light path and properly focusing it, a magnified image of the b.f.p. can be seen through the eyepiece, and the aperture diaphragm setting can be adjusted more easily. Although it is usually recommended that the setting of the aperture diaphragm should be such that its image covers about two thirds to three fourths of the rear surface of the objective, this should not be regarded as a hard and fast rule.

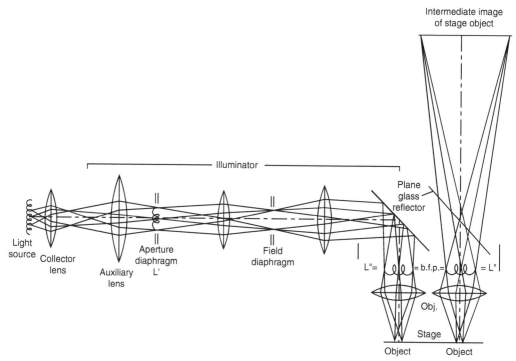

FIGURE 7.18 Ray path in an illumination system according to Köhler's principles. (Courtesy of W. Heffer & Sons, Ltd.)

It should be kept in mind that the aperture diaphragm can never be used to adjust the brightness of the image, which can be done by using neutral density filters. Figure 7.19 shows the effect of aperture diaphragm setting on the image quality of pearlite in steel. In Figure 7.19a, the aperture diaphragm is in its optimum setting, producing the best contrast and good resolution. In Figure 7.19b, the aperture diaphragm is fully open. The contrast is inferior because of excessive glare, and although the resolution is expected to be the best, it is almost the same as in Figure 7.19a. In Figure 7.19c, the aperture diaphragm is closed to its minimum position. Structure with interference fringes can be seen in many places. The impairment of resolution is most clearly illustrated in a pearlite nodule located at the lower right of the figure.

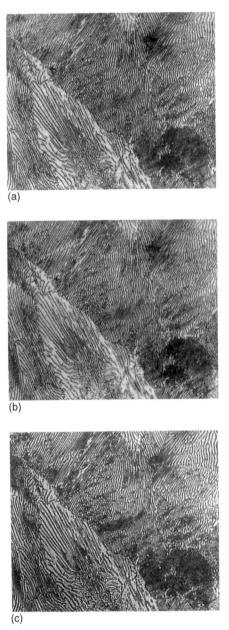

(a)

(b)

(c)

FIGURE 7.19 Effect of aperture diaphragm setting on the image quality of pearlitic structure (600×). (a) Optimum setting; (b) maximum setting; (c) minimum setting. (Reduced 20% in reproduction.)

The field diaphragm is imaged at the object surface, which is then imaged at the intermediate image plane. The field diaphragm should be so adjusted that the image of its edges is located just outside the field of view, thus cutting off unnecessary light and minimizing image glare. With the field diaphragm correctly adjusted, the object field observed is evenly illuminated up to its edge and can be matched exactly with the field of view.

7.3.5.2 Brightfield Illumination

In the above-described method of illumination, the light beam strikes the specimen surface vertically. If the specimen surface has a mirror polish, it will appear bright, because nearly all the reflected light can enter the objective again. After etching, depressions or grooves with sloping surfaces are produced along the grain or phase boundaries, and at the same time, the surface of certain phases becomes somewhat roughened. Part of the reflected light beam from these regions cannot enter the objective. Dark constituents of various gray levels against a bright background are now seen. This kind of illumination is called the brightfield illumination and is the most commonly used illumination method.

7.3.5.3 Darkfield Illumination

Figure 7.20 shows schematically the optical path in darkfield illumination. The objective shown in this figure is of infinite optics, which will be described in Section 7.3.6. With

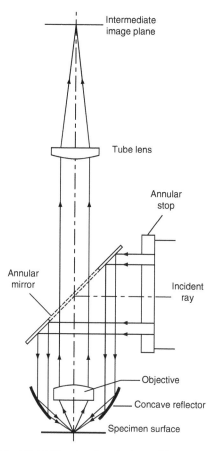

FIGURE 7.20 Optical path of darkfield illumination (schematic).

darkfield illumination, a ring of light, which is produced after passing an annular stop with the aperture fully open, falls on the annular mirror that is located around the plane glass reflector. Thus, the light beam does not pass through the objective but traverses outside the objective and is directed rather obliquely onto the specimen surface by a concave reflector.

If the specimen surface has a mirror polish, it will appear dark because the reflected light does not enter the objective. Structural constituents with sloping surfaces, such as grain and phase boundaries, will appear light because the reflected light may enter the objective, at least partly. The image contrast is completely reversed with darkfield illumination as compared with brightfield illumination. Figure 7.21 shows the prior austenite grain structure of a high-strength low-alloy steel under darkfield illumination.

The advantages of darkfield illumination are as follows:

1. *Higher resolution.* Under darkfield illumination, transparent particles within the specimen that are undetectable under brightfield illumination become self-luminous and are detectable even at 0.006 μm (6 nm).
2. *Higher image contrast.* The objective no longer acts as a condenser in the illumination system, reducing glare.
3. *Transparent inclusions* can be identified accurately according to their inherent colors; e.g., silica (SiO_2) in steels appears dark in brightfield illumination but bright yellow in darkfield illumination.
4. As already shown in Figure 7.21, grain boundaries are more easily delineated in darkfield than in brightfield illumination and can be used for rating grain size when compared with standard charts.

However, there are disadvantages of employing darkfield illumination. Image brightness is much lower, making exposure during photomicrography significantly longer and necessitating high-intensity light sources and automatic exposure devices. A very well prepared surface is necessary for darkfield observation because even the slightest scratches, undetectable under brightfield illumination, will be evident.

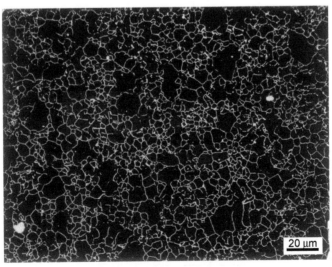

FIGURE 7.21 Prior austenite grain structure of a high-strength low-alloy steel under darkfield illumination (200×). (Reduced 20% in reproduction.)

7.3.6 THE INFINITY-CORRECTED OPTICAL SYSTEM

There is a tendency to use objectives designed for infinite image distance instead of conventional objectives computed for finite image distance, and the color correction of such objectives is also made for infinite image distance. Such optical systems are called infinity-corrected optical systems or simply infinite optics [23]. When the incident light beam is reflected from the specimen surface and passes through an infinite objective, it does not converge and remains a parallel light ray until it passes through the tube lens; then an intermediate image is formed, as shown in Figure 7.22.

The main advantage of infinite optics is that modules of various contrasting modes (such as incident light beam splitter for bright- and darkfield, polarizing beam splitter, Wollaston prisms for interference contrast, analyzers, additional filters, and so on) can be engaged in the extended space of the infinite light path between the flange of the objective and the tube lens. As there is no additional lens system to disturb the part of the light path that is used for both imaging and incident light illumination, the quality of the optical image will not be impaired. Moreover, when infinite optics is used, the tube length coefficient remains at unity, and a fixed lens relay system is not necessary no matter how far the objective is from the eyepiece.

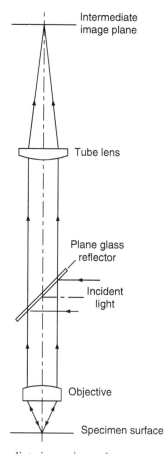

FIGURE 7.22 Formation of intermediate image in a microscope with infinite objective (schematic).

7.3.7   POLARIZED LIGHT ILLUMINATION

Polarized light has been used in metallography to identify inclusions. This method has been replaced by electronic microprobe analysis and energy-dispersive spectroscopy. Polarized light is also very useful for the examination of noncubic anisotropic metals such as beryllium, titanium, zinc, and zirconium. The use of polarized light for the examination of heat-treated specimens is rather limited; therefore, only a brief introduction of polarized light and its characteristics will be given.

Natural lights and most artificial lights are transverse waves vibrating in all directions at right angles to the direction of propagation (Figure 7.23a). Light passing through a polarizer vibrates in only a certain plane along the direction of propagation (Figure 7.23b) and is called plane-polarized or linearly polarized light.

Although both Nicol prisms and polarizing filters can be used as polarizers in microscopes, the latter are most commonly used because of their lower cost. In a polarizing filter, polyvinyl alcohol film is stretched to line up the complex molecules and then impregnated with iodine. This film is usually mounted between two thin plates of optical glass.

Polarized light illumination is attained by placing the polarizer and analyzer, which are made of the same material, into the optical path of the microscope under brightfield illumination. The polarizer is situated ahead of the plane glass reflector and is usually nonrotatable, while the analyzer is placed after the objective and is usually rotatable. When the polarizing direction of the analyzer is at right angles to the polarizer, no light will be observed if a polished optically isotropic metal specimen, e.g., a metal with cubic structure, is used. This condition is known as cross-polarized illumination. However, if a polished optically anisotropic metal specimen is used instead, an image of microstructures such as grain structure, twinning, and the distribution of phases can be delineated.

If an optically isotropic metal specimen is anodized in a standard electropolishing cell, a thin oxide film is deposited on the polished surface of the specimen. The optical anisotropy and surface irregularities of the film may exhibit birefringence, so that the same effect can be produced as with an optically anisotropic metal. Figure 7.24 shows the microstructure of a sand-cast Al–1 wt% Ni aluminum alloy, electropolished and anodized. Figure 7.24a was obtained under brightfield illumination. The specimen shows a typical dendritic structure with an intermetallic phase $NiAl_3$ located at the dendritic arm interface, but the extent of individual grains can hardly be differentiated. Figure 7.24b shows the same field under polarized light illumination. In this case grains with different orientations can easily be recognized according to their gray levels. Figure 7.24c shows grain boundaries of the same field traced out according to Figure 7.24b; the four grains within the figure are denoted by numbers. It can be seen that a dendritic cell of grain 2 is isolated within grain 3, while another dendritic cell of grain 3 is mostly within grain 2 except for a small segment of its boundary neighboring with grain 1.

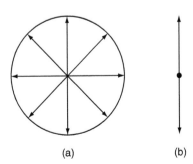

(a)   (b)

FIGURE 7.23 Vibration of (a) ordinary light and (b) plane-polarized wave.

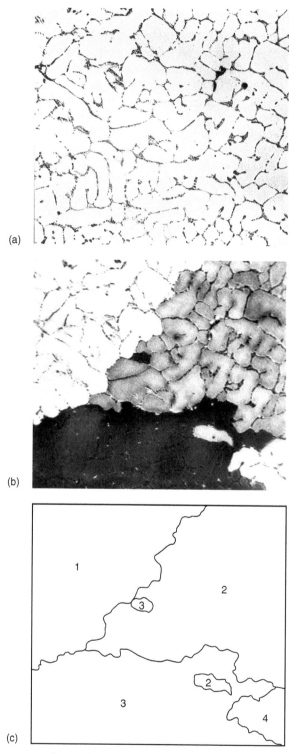

FIGURE 7.24 Microstructure of a sand cast Al–1 wt% Ni aluminum alloy, electropolished and anodized. (a) Brightfield illumination; (b) polarized light illumination; grain boundaries traced out according to (b). (165×) (Reduced 20% in reproduction. Courtesy of Shun-Xian Tang.)

7.3.8 PHASE CONTRAST ILLUMINATION

Although the phase contrast method was developed first and primarily for transmitted light microscopy, it has also found application in incident light microscopy. When both phases in a specimen have similar etching and reflection characteristics, the resulting microstructure shows a lack of intensity contrast. Such a specimen is called a phase object as compared with an amplitude object, in which sufficient contrast is obtained owing to a difference in amplitude or intensity of the reflected light from different phases. For example, in a hyper-eutectoid plain carbon steel containing 1.2% C after spheroidizing, both cementite particles and ferrite matrix appear bright after etching with nital under brightfield illumination, as shown in Figure 7.25a. However, cementite is much harder than ferrite and is less attacked by the etchant; therefore, cementite particles protrude slightly over the surrounding ferritic matrix after polishing even without etching. As a result, there is a path or phase angle difference between the light reflected from cementite and ferrite.

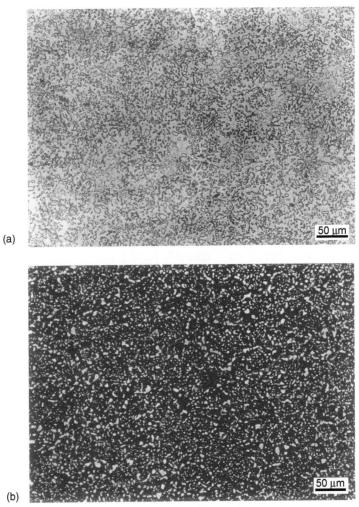

(a)

(b)

FIGURE 7.25 Photomicrograph of a plain carbon steel containing 1.2% C after spheroidizing annealing (500×). (a) Brightfield illumination; (b) phase contrast illumination. (Reduced 30% in reproduction.)

Light with only a slight difference in phase angle cannot be differentiated by the human eye directly. However, under phase contrast illumination, it is possible to transform such a phase angle difference into an amplitude or intensity difference in the final image, so that the image contrast can be enhanced significantly and thus can be detected readily by the human eye. Figure 7.25b shows the photomicrograph of the same specimen and same field of view as in Figure 7.25a but under phase contrast illumination. The bright particles are the protruded cementite phase, while the dark background is the ferritic matrix.

When the height difference between two phases on the specimen surface is around 500 nm, good contrast can be obtained under phase contrast illumination. Usually a light etch of the specimen is needed to attain this condition. Besides, for best results, a higher quality of specimen preparation is required under phase contrast illumination, as even a very light scratch that is undetectable under brightfield illumination will be clearly revealed.

The phase contrast method in metallurgical microscopes has been gradually replaced by the DIC technique, which reveals height differences among phases as a variation in color or gray level over a much wider range. Therefore, we shall not go further into the details of the phase contrast method.

7.3.9 DIFFERENTIAL INTERFERENCE CONTRAST ILLUMINATION

The principles of DIC illumination were set forth by G. Nomarski about 50 years ago. This method uses a modified Wollaston prism located ahead of the objective in conjunction with polarized light at extinction, i.e., the polarizer and the analyzer are in the 90° crossed position as shown in Figure 7.26.

A Wollaston prism [24] is made of a birefringent, uniaxial material such as quartz and consists of two wedge-shaped prisms cemented together (Figure 7.27). The optic axes of the two prisms are at right angles to each other. When a plane-polarized light wave with its vibration plane inclined by 45° to the plane of drawing reaches the Wollaston prism perpendicularly, it will be split into two plane-polarized waves in the lower prism. Each of their vibration planes makes an angle of 45° with the incident wave. At the cemented interface of the Wollaston prism, these two component waves are deflected in two directions at a relatively small angle (less than 0.5 min) and encounter different indices of refraction so that their wave fronts travel at different speeds. It is impossible to locate the interference plane of the Wollaston prism in the image-side focal plane of the objective, at least with

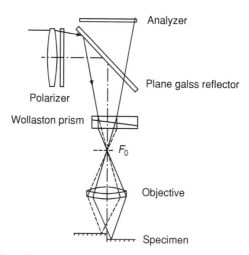

FIGURE 7.26 Beam path in reflected light DIC. (Courtesy of Carl Zeiss.)

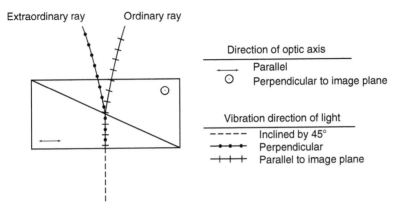

FIGURE 7.27 Beam splitting of plane-polarized light at the interface of the Wollaston prism. (Courtesy of Carl Zeiss.)

high-power objectives. Therefore, a modified Wollaston prism with its interference plane outside the compound prism is used instead and can be located relatively farther from the objective.

The beam from the light source is plane-polarized after passing through the polarizer; then it is deflected to the Wollaston prism by the plane glass reflector of the vertical illuminator as shown in Figure 7.26. After emerging from the objective the two component waves become parallel to each other and with a slight lateral separation. The light rays reflected from the specimen are then recombined by means of the objective and the Wollaston prism, pass through the plane glass reflector, and, with the aid of the analyzer, are brought to travel in one plane again, making them capable of interference. The interference contrast image of the specimen can then be formed in the intermediate plane (not shown) and can be viewed through the eyepiece.

The path difference can be altered by moving the Wollaston prism laterally, i.e., perpendicular to the optical axis. Using white light to illuminate a structureless object, e.g., an optically flat polished surface, the image field (also called the background image) appears dark when the Wollaston prism is at a position with zero path difference. However, once there is a path difference, a colored background image appears; every path difference is associated with a characteristic interference color.

With regard to a phase object, the path difference depends on the geometrical profile of the object. The phase retardation resulting from reflection of the waves from the opaque object, and the relative lateral position of the Wollaston prism.

Figure 7.28 shows schematically the profile of the object, the wave fronts of the DIC image before and after reaching the analyzer, and the resulting intensity distribution. Let phase B be the matrix of the object; then the color of the interference background image depends on the path difference (δ_0). Assuming that the difference in phase retardation resulting from reflection of the waves from the two phases is negligibly small, then the DIC image color of phase A is nearly the same as that of the background (phase B). However, the path differences at phase boundaries vary with δ_0; therefore, by adjusting the path difference δ_0, it is possible to observe images similar to darkfield (when $\delta_0 = 0$), relief-like, and color images. Thus, the best result can be obtained by selecting the type of image suitable for the object to be observed, and the contrast of the object is increased by interference colors.

Figure 7.29 shows a comparison of photomicrographs of a commercial pure iron specimen under (a) brightfield illumination and (b) DIC illumination. The specimen has undergone impact loading at low temperature, and the deformation twins or Neumann bands are more clearly delineated under DIC illumination.

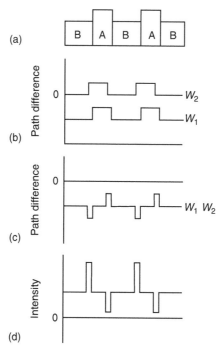

FIGURE 7.28 Schematic representation of the formation of DIC image. (a) Profile of the specimen; (b) and (c) wave fronts of the DIC image before and after reaching the analyzer; (d) resulting intensity distribution.

Under phase contrast illumination, the relief of the specimen surface for producing an image with optimal contrast varies within relatively narrow limits, because the phase shift and absorption of direct light are fixed by the phase contrast accessories and cannot be adjusted. In contrast, by moving the Wollaston prism laterally, DIC images with excellent contrast can be obtained over a far greater range of path differences in the object. As the DIC method is obviously superior to the phase contrast method in incident light microscopy, modern metallurgical microscopes are equipped predominantly with DIC illumination, and phase contrast microscopy can be found only in some older models.

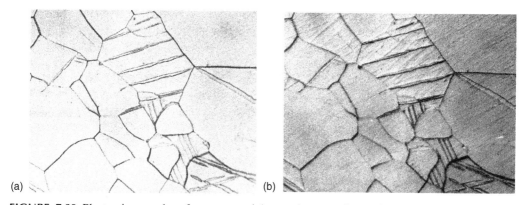

FIGURE 7.29 Photomicrographs of a commercial pure iron specimen that had undergone impact loading at low temperature (600×). (a) Brightfield illumination; (b) DIC illumination. (Reduced 15% in reproduction.) (Courtesy of Shun-Xian Tang.)

7.3.10 PHOTOMICROGRAPHY AND PHOTOMACROGRAPHY

Photomicrography is the reproduction of microstructures observed under a microscope. Photo-macrography is the reproduction of macroetched section or broken parts such as castings, forgings, and heat-treated parts at a magnification from less than 1 to 20 or so with the aid of a camera or stereomicroscope. The aim of photomicrography and photomacrography for heat treaters is to faithfully reproduce what has been observed in heat-treated specimen or parts for reporting, archiving, or publication.

7.3.10.1 Photomicrography

7.3.10.1.1 Equipment for Photomicrography

Every modern metallurgical microscope is equipped with either a camera built into the microscope stand as shown in Figure 7.30 or exits for cameras as shown in Figure 7.31. With even the simplest bench-type microscope, a camera attachment can be fitted onto the binocular tube, forming a rigid link with the microscope as shown in Figure 7.32. The field of view on the film plane of the camera is usually the same as that of the eyepiece, and images at both places are in sharp focus simultaneously so that no special focusing eyepiece is necessary. However, the intermediate image must be located just outside the focal plane of the photo-eyepiece so that a real final image can be formed on the film plane. Two types of cameras are used in photomicrography: large format (4 × 5 in.) and 35 mm.

During recent years, high-resolution digital camera has been introduced to modern metallurgical microscopes. Figure 7.33 shows the GX71 inverted metallurgical microscope from Olympus, the digital camera is attached to the side port below the binocular tube. Besides digital and video cameras, conventional still cameras (large format or 35 mm) can also be attached to the microscope. When both ports are used simultaneously, up to three camera backs can be attached. Another advantage of the GX71 is its erect images, i.e., observed images are not reversed as in old-type microscopes. The specimen's positional characteristics (right–left, up–down) are the same in a photomicrograph as they are in real-time observation of the viewing field. This presents specimen movement in an intuitive and natural way. However, the digital images are still reversed.

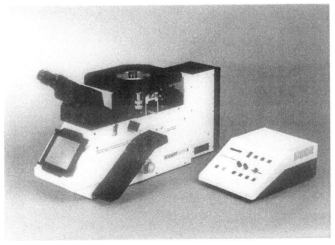

FIGURE 7.30 The Reichert MEF4A wide-field metallograph. (Courtesy of Reichert Division of Leica Aktiengesellschaft.)

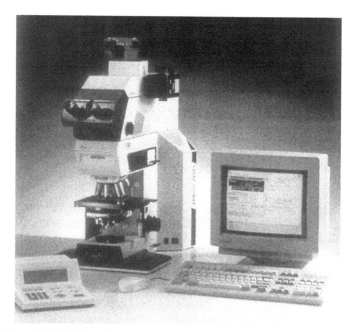

FIGURE 7.31 Leica DM RXE with DM RD and PC. (Courtesy of Leica Mikroskopie und Systeme GmbH.)

7.3.10.1.2 Films Used in Photomicrography

One variety of Polaroid instant film that produces a negative and a positive print simultaneously is perhaps the most extensively used large format film because of its availability, efficiency, and convenience (no darkroom work is necessary), especially in the United States. The negative can be kept for future use when more copies are needed. If only one instant positive print is produced per exposure, the photographic process must be repeated for each copy needed before the specimen is removed from the microscope.

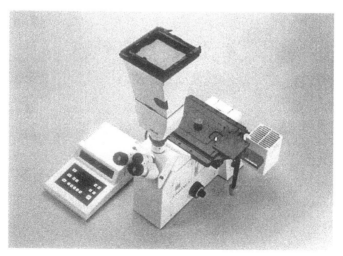

FIGURE 7.32 The MC 80 microscope camera incorporated with Zeiss Axiovert 100A. (Courtesy of Carl Zeiss.)

FIGURE 7.33 The GX71 inverted metallurgical microscope with digital camera, Polaroid large format camera and 35 mm camera attached. (Courtesy of Olympus America, Inc. With permission.)

Increasingly, 35-mm film is used because of lower processing cost and time, but enlargement is necessary for obtaining a print. For best quality of a positive print, large format sheet film is recommended although it is more expensive and requires more darkroom time.

Black-and-white films are more commonly used than color films in photomicrography due to their lower cost. The two major types of black-and-white film used are panchromatic and orthochromatic. Panchromatic films are sensitive to the entire visible spectrum and should be developed in total darkness, whereas orthochromatic films are sensitive to all colors except orange and red and can be developed under dark red safelight. As yellow-green filter is often used with plano-achromatic or achromatic objectives to minimize lens aberrations, orthochromatic films are useful for photomicrography.

Increasingly, color films in photomicrography are used because more information can be obtained in a color photomicrograph. It is known that the human eye can distinguish more than 1000 different colors but can recognize no more than 20 to 30 tonnes of gray. Color photomicrographs are useful for the identification of inclusions and carbides in steels and for the identification of phases in some stainless steels [17].

The speed of a film is an indication of its sensitivity to light and is expressed as a number intended to be used to determine the exposure needed to produce an image of satisfactory quality. The International Organization for Standardization (ISO) adopted both the American Standards Association (ASA), formerly the American National Standards Institute (ANSI), and Deutsche Industrie Norm (DIN) systems for rating the film speed. For example, ISO 100/21° corresponds to an ASA rating of 100 and a DIN rating of 21°.

In the arithmetic expression used by the ASA, the film speed value varies inversely with the exposure required to produce the specified density. For example, films with values of 25, 50, 100, and 200 ASA each require half the exposure of the preceding film; in other words, a speed of 200 ASA is double that of 100 ASA.

In the logarithmic expression used by the DIN, the film speed numbers increase by 3 as the speed is doubled. For example, values of 18°, 21°, and 24° DIN represent materials that each requires half the exposure of the preceding material. In other words, a speed of 24° DIN is double that of 21° DIN.

ASA and DIN numbers can be easily interconnected from the following expressions:

$$DIN = 10 \log_{10} ASA + 1 \tag{7.8}$$

$$ASA = \frac{10^{(DIN-1)}}{10} \tag{7.9}$$

It is desirable that the size of silver grains in the photographic emulsion after development be as small as possible so that it will not mask the details of the image. Higher speed films are always accompanied by coarser grains than lower speed films. Therefore, lower speed films are preferred for photomicrography, where very fine detail is required and slow shutter speeds are possible. There has been tremendous progress in the film emulsion technology so that very fine-grained color print film with a speed of ISO $400/27°$ is available now. Hence, if the negative is not going to be enlarged substantially, good prints with sufficient resolution can be obtained from normal to moderately high-speed film (ISO $100/21°$–ISO $200/24°$), which are more easily available on the market.

There are two types of color films: color slide or reversal films and color print or negative films. Positive color slides are obtained from a color reversal film after processing and are used primarily for illustrating lectures or demonstrations. True color reproduction of the original microstructure can be obtained only at a specific color temperature, which is a scale for rating the color quality of illumination expressed in kelvins (K). Therefore, it is important to choose a type of film to suit the light for which it is balanced. For a xenon light source, daylight film balanced for a color temperature of about 5600 K should be used, and for a halogen light source, artificial light film balanced for a color temperature of about 3200 K should be used. If daylight film is used in halogen light, a suitable filter known as a conversion filter (e.g., Kodak 80B) should be added to alter the color temperature of the incident light and render it suitable for daylight film.

Color print film produces a negative with complementary colors. For example, a blue object is recorded as yellow, but the complementary relationship is partly obscured by the presence of color-masking dyes often used in color negatives. For the same reason as for color reversal film, daylight or artificial light film should be chosen according to the light source used in photomicrography; otherwise, a suitable conversion filter should be used. True color reproduction in the prints of photomicrographs from negatives is not as easy as for ordinary prints because the photolab technicians are not familiar with the subject matter and as a result, proper correction cannot often be made in printing. It would be best if metallographers could do the processing themselves, but this depends much on the volume of work and available funds.

It should be kept in mind that when color film is used in photomicrography, the light source should be operated at its rated voltage so as to attain the correct color temperature. Therefore, lamp voltage must not be used as a means of adjusting the image intensity. In such a case, neutral density filters, which reduce the intensity of incident light without altering its color temperature, with various transmissivities should be used instead.

7.3.10.1.3 Photographic Exposure and the Reciprocity Failure Effect

Correct exposure is an important step for obtaining a negative with all necessary details and good contrast. Photographic exposure (H) is defined as the quantity of light per unit area received by the photographic emulsion of the film and is commonly expressed as the product of illumination (E) and time (t):

$$H = E \times t \tag{7.10}$$

The correct exposure is determined by a light-metering system. The most frequently used system in photomicrography is integrated metering, which measures the light reaching the film, giving an average reading over the entire format area. Sometimes it is desirable to obtain an extremely accurate exposure in a very localized area of the image, which may be darker or brighter than the average reading. In such a case, spot metering can be used in which the angle of acceptance is to narrow that the size of the measured area can be as small as 1% of the total image area. For ordinary photography, however, a center-weighted metering system is used in most single-lens reflex (SLR) cameras. That is, light from all parts of the image field is measured by the silicon photocell, but the influence of the central zone is greatest.

Experiments show that equal exposures do not produce photographic images of equal densities and equal contrast if the time of exposure is extremely long or extremely short (and the corresponding light level is too weak or too strong). This is known as the reciprocity failure effect or Schwartzschild effect. This effect must be taken into account when taking photomicrographs under darkfield, polarized light, differential contrast interference, and phase contrast illumination where a longer or even a much longer exposure is needed. Fortunately, modern automatic photographic systems are capable of correcting for the reciprocity failure effect automatically.

7.3.10.2 Photomacrography

Some metallurgical microscopes are provided with special accessories for photomacrography at magnifications of around 1–10×. In some models, the magnification is continuously adjustable by altering the zoom setting and the object distance, and separate macro illumination is required.

Photomacrography with 35-mm SLR camera at an image ratio of about 1:10 (1/10 life size) or larger is rather easy and convenient. Figure 7.34 shows the cutoff view of a 35-mm SLR camera [25]. Light entering the lens (A) falls on a movable reflex mirror (B), which reflects it upward onto a ground glass screen (C), then enters the pentaprism (D) and reflects twice, so that an upright and correctly oriented image is produced for viewing through the eyepiece (E). The through-the-lens (TTL) light-metering system has a light-sensitive cell (G) that measures the amount of light entering through the lens. When the shutter release button is pressed, the reflex mirror swings upward, allowing the incident light to fall on the film frame (F) and expose the film once the shutter (H) is released.

The focal length of the lens and its maximum aperture, e.g., 50 mm 1:2, are shown on the front of the lens. The focal length of a standard lens of a 35-mm camera is 50 mm, and the angle of view is 46°, roughly the same as that of the human eye. The shorter the focal length, the wider the angle of view, and vice versa.

The brightness of the image on the film is controlled by aperture size expressed by f number, which is defined as the quotient of the lens focal length and the effective aperture. The effective aperture of a lens is the diameter of the light beam entering the lens and just filling the opening in the diaphragm of a camera lens as shown in Figure 7.35.

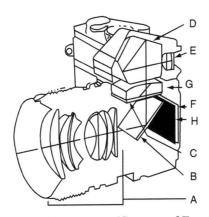

FIGURE 7.34 Cutoff view of a 35-mm SLR camera. (Courtesy of Fountain Press Ltd.)

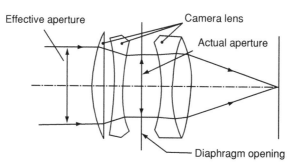

FIGURE 7.35 Effective aperture and actual aperture of a camera (schematic).

The maximum aperture of a lens corresponds to its smallest f number. For example, if the maximum aperture of a 50-mm lens is 25 mm, the corresponding f number is $f2$ or $f/2$, which is usually marked as 1:2.

The f numbers for any lens are engraved on the lens mount in a series as follows: f 1.4, 2, 2.8, 4, 5.6, 8, 11, 16, 22. The series is so arranged that as the f number is increased by one step, the amount of light allowed to enter is halved. In other words, the image illuminance is inversely proportional to the square of the f number, and the required exposure time is directly proportional to the square of the f number.

Lenses of the same f number are said to have the same speed. That is, they produced images of essentially equal luminance.

Aperture size can also be used to control the depth of a field, which is defined as the range of object distances within which objects are imaged with acceptable sharpness [26]. The depth of the field increases as the focal length decreases and as the f number increases.

Several equipment options have been devised for taking close-up photomacrographs with SLR cameras. The basic options are close-up lens, extension tubes or bellows, reversing ring, and macro lens [27,28].

7.3.10.2.1 Close-Up Lens
The close-up lens is by far the cheapest accessory for photomacrography. It works rather like a magnifying glass and can be screwed onto the front rim of the camera lens. It focuses the lens at a closer distance than normal so that a larger image can be formed on the film plane. The close-up lens is inferior optically to the expansive camera lens, so image quality may be slightly compromised. Normally this is hardly noticeable, especially when a large f number, e.g., $f8–f16$, is used.

The strength of a close-up lens is expressed in diopters, which is defined as 1 m divided by the focal length of the lens. Its usual range is 0.4–5 diopters. The larger the number, the closer the lens can focus.

When a close-up lens is not used in photomacrography, something is needed to extend the distance between lens and camera body. For example, in order to get a life-size reproduction (1:1) of a small object with a standard 50-mm lens, an extension of 50 mm between lens and film plane is needed. An extension tube is a tube or set of tubes that can be fitted between lens and camera body as shown in Figure 7.36. The so-called automatic extension tubes are so designed that the action of the fully automatic diaphragm works normally. As extension tubes are not optical devices, no deterioration in lens performance will result.

An extension bellows is basically a continuously variable extension tube mounted on special focusing rail and allowing a full range of extension between 30 and 200 mm as shown in Figure 7.37.

FIGURE 7.36 One or more extension tubes can be fitted between the camera and lens. (Courtesy of AMPHOTO.)

7.3.10.2.2 Reversing Ring

The reversing ring is screwed into the filter ring of the lens and allows the lens to be mounted backward on the camera. As the normal lens is not highly corrected in close-up positions, turning the lens backward changes the optical corrections in the lens and results in better definition of the close-up pictures; hence, the lens behaves as though it were really a

FIGURE 7.37 Camera fitted with extension bellows and focusing rail. (Courtesy of AMPHOTO.)

macro lens (see next paragraph). However, when the lens is mounted backward, the fully automatic diaphragm ceases working, and consequently the aperture must be adjusted manually.

7.3.10.2.3 Macro Lens

A macro lens is so designed and highly corrected as to focus an object closer than normal lenses without special attachments. For example, a 50-mm macro lens can be extended to such a length that an object located about 10 cm from the lens can be focused. Most macro lenses can produce at least an image ratio of 1:2, and some can even produce an image ratio of about 1:1. Macro lenses with focal lengths of 50 and 100 mm are most commonly used. A macro lens usually has maximum aperture of about $f/3.5$ or $f/4$, which is about three to four times slower than an $f/2$ lens. When used with extension tubes or bellows, a macro lens still performs better than normal lenses.

7.4 QUANTITATIVE METALLOGRAPHY

Metallography has long been a means of obtaining qualitative or semiquantitative descriptions of microstructures for quality control in production and research. However, the rapid progress of science and technology over recent decades has placed more stringent demands on the mechanical properties of heat-treated parts and requires a more accurate quantitative description of microstructures. Quantitative metallography deals with the quantitative relationships between measurements made on the two-dimensional plane of polish and the magnitudes of the microstructural features in the three-dimensional metal or alloy [29]. It is an important branch of stereology applied specifically to metals and alloys.

7.4.1 SOME STEREOLOGY BASICS

According to Elias [30], stereology is a system of methods to obtain information about the three-dimensional structure from two-dimensional images, whether sections or projections. Although easily understood and accepted by most practical stereologists and microscopists, this definition is limited because it considers only the procedures from a two-dimensional image to a three-dimensional structure. As indicated by Miles [31] in 1972, all basic stereological equations are special cases of equations in a generalized n-dimensional space. This implies that any n-dimensional structure can be studied by sectioning it with any s-dimensional section $(s < n)$, and the principles involved are always the same. It follows that the increasing attention of mathematicians to the problems in stereology is indispensable for the establishment of a solid foundation for this interdisciplinary field of study.

The basic symbols used in stereology are P, L, A, S, V, and N, which represent point, line, plane surface, curved surface, volume, and number of features, respectively. Compound symbols are used to represent stereological parameters, in which the capital letter refers to the microstructural quantity and the subscript pertains to the test quantity. For example, the symbol P_P refers to point fraction, which is equivalent to P/P_T, with P the number of points (in area features) and P_T the number of test points used during measurement; the symbol S_V refers to surface area per unit test volume. Table 7.12 is a list of basic symbols used in stereology and their definitions quoted from Ref. [32].

Equation 7.11 through Equation 7.13 show statistically exact expressions that relate points, lines, surfaces, and volumes in three-dimensional structures to measurements made on two-dimensional sections [29].

$$V_V = A_A = L_L = P_P \tag{7.11}$$

TABLE 7.12
List of Basic Stereological Symbols and Their Definitions

Symbol	Dimensions[a]	Definition
P		Number of point elements or test points
P_P		Point fraction. Number of points (in real features) per test point
P_L	mm^{-1}	Number of point intersections per unit length of test line
P_A	mm^{-2}	Number of points per unit test area
P_V	mm^{-3}	Number of points per unit test volume
L	mm	Length of lineal elements or test line length
L_L	mm/mm	Length fraction. Length of lineal intercepts per unit length of test line
L_A	mm/mm^2	Length of lineal elements per unit test area
L_V	mm/mm^3	Length of lineal elements per unit test volume
A	mm^2	Planar area of intercepted features, or test area
A_A	mm^2/mm^2	Area fraction. Area of intercepted features per unit test area
S_V	mm^2/mm^3	Surface area per unit test volume
V	mm^3	Volume of three-dimensional features, or test volume
V_V	mm^3/mm^3	Volume fraction. Volume of features per unit test volume
N		Number of features (as opposed to points)
N_L	mm^{-1}	Number of interceptions of features per unit length of test line
N_A	mm^{-2}	Number of interceptions of feature per unit test area
N_V	mm^{-3}	Number of features per unit test volume
$\overline{L}$	mm	Average lineal intercept, L_L/N_L
$\overline{A}$	mm^2	Average areal intercept, A_A/N_A
$\overline{S}$	mm^2	Average surface area, S_V/N_V
$\overline{V}$	mm^3	Average volume, V_V/N_V

[a]Arbitrarily shown in millimeters.

Source: From EE Underwood, *Quantitative Stereology*, Addison-Wesley, Reading, MA, 1970.

$$S_V = (4/\pi)L_A = 2P_L \tag{7.12}$$

$$L_V = 2P_A \tag{7.13}$$

The derivation of the above equations is beyond the scope of this book and can be found in standard textbooks on stereology [32–34]. Stereological parameters with subscripts other than V are measurable and are called measured quantities; stereological parameters with subscript V cannot be measured and are called calculated quantities. However, the measured parameters in Equation 7.11 and Equation 7.12, i.e., A_A, L_L, and L_A, can also be calculated.

The exactness of Equation 7.11 through Equation 7.13 signifies that in the derivation of these equations, no simplified assumptions on the size, shape, distribution, spacing, etc., of microstructural constituents are required. The microstructure under investigation should be representative of the bulk specimen, as microstructures of different fields of view in the same specimen can in no way be identical. Measurements of stereological parameters should be made randomly or with statistical uniformity.

7.4.2 Volume Fraction

Volume fraction is one of the most important and most commonly used stereological measurements. From Equation 7.11 it can be seen that the volume fraction of a phase or microstructural constituent in a microstructure can be obtained by measuring area, lineal, or

FIGURE 7.38 A 16-point (4 × 4) test grid.

point fraction, but for manual operation the use of point counting is most effective and has sufficient accuracy. Usually the intersections of a test grid are used as test points. Figure 7.38 illustrates a 16-point (4 × 4) test grid. The grid can be inserted in the eyepiece of the microscope or superimposed on a photomicrograph. The number of points (P_α) that falls within a certain phase (say α) is counted, the point fraction (P_P) of phase α is P_α divided by the number of grid points P_T. Points that fall on phase boundaries are counted as 1/2 [35]. This process is repeated several times so as to obtain required accuracy (see Section 7.4.3). Figure 7.39 shows an example of point counting, with a 4 × 4 test grid superimposed on a granular microstructure (schematic) with phase α randomly dispersed in a matrix. There are two test points that fall within phase α and five points located on phase boundaries. Therefore, the point fraction of phase α in this field of view is $(2 + 5 \times 1/2)/16 = 0.281$ or 28.1%.

Low-density test grids, such as 3 × 3, 4 × 4, or 5 × 5, are usually employed for point counting. Test grids with more than 100 test points are seldom used because of operator eye fatigue. Generally, higher density test grids should be used for microstructures with lower volume fraction. The magnification of the microstructure should be so chosen that no more than one grid point falls on a given particle of interest [6, p. 427].

Point counting can be used to estimate the carbon content of fully annealed hypoeutectoid carbon steels. For example, in low-carbon steels, pearlite is the minor constituent compared with proeutectoid ferrite. The volume fraction of pearlite can be easily calculated by point counting of pearlite in the microstructure. Let $(P_P)_P$ be the measured point fraction of the pearlite. Then, according to the lever rule, the carbon content x in mass percent of the steel can be obtained from the expression

$$x = 0.77 \, (P_P)_P \qquad\qquad (7.14)$$

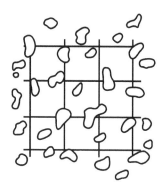

FIGURE 7.39 An example of point counting.

where 0.77 is the mass fraction of carbon in an eutectoid steel. In deriving Equation 7.14, the very small amount of carbon in ferrite is neglected.

Similarly, the mass fraction of carbon in a spheroidized steel can be calculated. Let $(P_P)_{cem}$ be the measured point fraction of cementite particles. Then, according to the lever rule, the mass fraction of carbon in this steel can be obtained from the expression

$$x = 6.69 \, (P_P)_{cem} \tag{7.15}$$

where 6.69 is the mass fraction of carbon in cementite. The mass fraction of carbon in ferrite is also neglected in deriving Equation 7.15.

Volume fraction is such an important stereological measurement that the first standard on quantitative metallography developed by the American Society for Testing and Materials (ASTM) deals with volume fraction. ASTM E562 (Standard Test Method for Determining Volume Fraction by Systematic Manual Point Count) was first adopted in 1976.

7.4.3 STATISTICAL ANALYSIS*

As stated earlier, there is no possibility of having two identical microstructures in a specimen. Therefore, no stereological measurements can be exact, and stereology is deeply statistical in nature [34, Vol. 1, p. 2]. Thus, without proper statistical analysis, stereological measurements and calculations cannot be regarded as complete. We shall not go into details on statistics; interested readers may refer to Ref. [36]. Following the example of the determination of volume fraction by point counting given below, the basic procedures of statistical analysis will be shown.

Suppose the point count for phase α in a specimen were performed on 20 fields of view using a 4×4 test grid like that shown in Figure 7.39. The results of 20 counts of P_α are as follows (the numbers in parentheses are number of occurrences): 3(1), 3.5(3), 3(5), 4.5(5), 5(3), 5.5(3). The mean volume fraction of phase α can be obtained by averaging the 20 counts:

$$V_V = P_P = \frac{\Sigma P_\alpha}{\Sigma P_T} = \frac{87.5}{(20 \times 16)} = 0.273 \text{ or } 27.3\%$$

7.4.3.1 Standard Deviation

The standard deviation s of the data is expressed as

$$s = \left[\frac{\Sigma(x_i - \bar{x})^2}{(n-1)} \right]^{1/2} \tag{7.16}$$

where $\bar{x}$ is the mean value of the individual counts x_i and n is the number of counts. In this example, $\bar{x}$ and x_i correspond to P_α and $(P_\alpha)_i$, respectively, and $n = 20$.

Standard deviation is an absolute measure of data dispersion. It can be easily obtained by using a pocket scientific calculator in the statistical calculation (STAT) mode. In this example, $s = 0.7232$. The standard deviation of point fraction of phase α is $s/P_T = 0.7232/16 = 0.0452$ or 4.52%.

7.4.3.2 Coefficient of Variation

When the mean values of various measurements differ substantially, the comparison of standard deviation among those measurements seems difficult. Therefore, it is more useful

---

*This section is derived from Ref. [6, pp. 428–432].

to use a new parameter, i.e., the coefficient of variation (CV), obtained by normalizing the standard deviation and expressed in percent:

$$CV = \left(\frac{s}{\bar{x}}\right) \tag{7.17}$$

In this example, $CV = (0.0452/0.273) = 0.166$ or 16.6%.

7.4.3.3 95% Confidence Limit

For the nominal 95% confidence limit (95% CL) of the mean,

$$95\%CL = 2s_{\bar{x}} \tag{7.18}$$

where $s_{\bar{x}}$ is called the standard deviation of the mean,

$$s_{\bar{x}} = \frac{s}{(n-1)^{1/2}} \tag{7.19}$$

Therefore,

$$95\% \ CL = \frac{2s}{(n-1)^{1/2}} \tag{7.20}$$

This means that if the test data exhibit a normal distribution about the mean and counting is repeated more times, then 95% of the mean value of the counts will be within 95% CL about the mean. In this example,

$$95\% \ CL = \frac{2 \times 4.52}{\sqrt{19}} = 2.07\%$$

The final volume fraction determination of this example is

$$27.3\% \pm 2.07\% \ (95\% \ CL)$$

7.4.3.4 Percent Relative Accuracy

The percent relative accuracy (% RA) of the test is obtained by using the equation

$$\% \ RA = \left(\frac{95\% \ CL}{\bar{x}}\right) \times 100 \tag{7.21}$$

Percent relative accuracy is also called 95% relative confidence limit or 95% RCL. In this example, we have

$$\% \ RA = \left(\frac{2.07}{27.3}\right) \times 100 = 7.58\%$$

If higher accuracy (i.e., lower % RA) is needed, then the required number of tests can be computed. Substituting n for $n - 1$ in Equation 7.20, we have 95% CL $= 2 \ s/n^{1/2}$;

then by substitution in Equation 7.21, a simple formula proposed by DeHoff [37] can be obtained:

$$n = \left[\left(\frac{200}{\% \, RA}\right) \times \left(\frac{s}{\bar{x}}\right)\right]^2 \tag{7.22}$$

As long as n is not too small, the difference between $(n-1)^{1/2}$ and $n^{1/2}$ is negligible and the above approximation is allowable.

In our example, if a % RA of 5% is desired, the required number of tests can be found from Equation 7.22:

$$n = \left[\left(\frac{200}{5}\right) \times \left(\frac{4.52}{27.3}\right)\right]^2 = 43.86 \approx 44$$

That is, 24 more tests are required to attain a % RA of 5% instead of 7.58% for 20 tests.

7.4.4 SURFACE AREA PER UNIT TEST VOLUME AND LENGTH OF LINEAL TRACES PER UNIT TEST AREA*

Equation 7.12 is the combination of two important equations requiring a P_L measurement.

$$S_V = 2P_L \tag{7.23}$$

$$L_A = \frac{\pi P_L}{2} \tag{7.24}$$

When the surface area per unit volume (S_V) is of interest, Equation 7.23 is used. Figure 7.40 [38] shows a plot of Brinell hardness of a 0.8% C pearlite steel versus S_V. After austenitizing, specimens of the same steel were held isothermally at different temperatures below A_1 in the pearlitic transformation region. As a result, pearlite structures with different lamellar spacing can be obtained. It follows from Figure 7.40 that the Brinell hardness of pearlite is proportional to S_V, the lamellar interface area between ferrite and cementite per unit test volume.

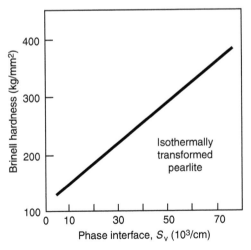

FIGURE 7.40 Brinell hardness of a 0.8% C pearlitic steel as a function of S_V.

---

*This section is based on material in Ref. [29].

The stereological parameter S_V can also be used for analyzing the grain boundary precipitates or transformation products, because grain boundaries are the preferred nucleation sites at a certain stage of precipitation of phase transformation in the solid.

When lineal traces in the plane of polish are of interest, Equation 7.24 is useful. In corrosion studies, for example, grain boundaries are more easily attacked by the corroding medium owing to their higher free energy with respect to their neighboring grains. The magnitude of L_A of the surface exposed to the corroding medium is important.

7.4.5 GRAIN SIZE

During heat treatment of the steels, grain sizes may undergo various changes due to phase transformation or recrystallization. It is well known that the mechanical properties and the behaviors of steels depend strongly on their grain sizes, and in most cases fine grain size is the preferred because of higher strength, ductility, and toughness. Therefore, grain size measurement is indispensable for quality control of heat-treated parts.

7.4.5.1 Grain Size Characterization

Up to now, there has been no statistically exact relationship for the determination of three-dimensional grain size from its two-dimensional sections. Although the first standard chart for rating grain size appeared in 1930, the basic equation used to define the ASTM grain size was not introduced until the adoption of ASTM E91 in 1951 in the form

$$N_{AE} = 2^{G-1} \tag{7.25}$$

where N_{AE} is the number of grains per square inch at $100\times$ magnification. To obtain the number of grains per square millimeter at $1\times$, denoted by N_A, simply multiply N_{AE} by 15.50, i.e.,

$$N_A = 15.50 \times 2^{G-1} \tag{7.26}$$

Equation 7.26 can also be expressed as

$$G = (3.321928 \log_{10} N_A) - 2.954 \tag{7.27}$$

ISO 643 defined grain size with the two equations

$$m = 8 \times 2^G \tag{7.28}$$

or

$$G = -3.0000 + 3.3219 \log_{10} m \tag{7.29}$$

where m is the number of grains per square millimeter at $1\times$ and G is the ISO grain size number. It follows that $G=1$ when $m=16$.

Comparing Equation 7.27 and Equation 7.29, we have

$$G(\text{ASTM}) - G(\text{ISO}) = 0.0458 \tag{7.30}$$

That is, the ISO grain size number is smaller than the ASTM grain size number by 0.0458. This difference is insignificant and hence can be neglected in practice.

To facilitate testing in production, test standards on grain size measurements were first developed by ASTM in the early 1930s. Since then it has been revised more than ten times. The adoption of ASTM E112 (Method for Estimating the Average Grain Size of Metals) in 1961 merged all previous test standards into a single one. It is one of the most extensively quoted ASTM standards [39]. Many other countries have their standards for determining grain size; all are based on ASTM E112 with only minor differences.

Table 7.13 gives grain size relationships computed for uniform, randomly oriented, equiaxed grains from ASTM E112-96. For macrograin size scale, Equation 7.26 is also used, but represents number of grains per square inch at 1×. Macrograin size number is designated as, for example, M-6. For macrograin size numbers higher than M-14, micrograin size number should be used. Micrograin size numbers may be converted to macrograin size numbers by adding 13.288 or $4 \times (1/\log_{10} 2)$.

Table 7.14 gives macroscopic grain size relationships computed for uniform, randomly oriented, equiaxed grains from ASTM E112-96.

7.4.5.2 Comparison Method for Rating Grain Size

The simplest way to estimate grain size is to compare the grain structure under a microscope or a photomicrograph with the standard charts at magnification of 100× provided by ASTM and select the one that tallies most with that of the specimen or interpolate between two standard charts. Estimations should be made on three or more representative areas of the specimen.

When grain size estimations are made by the more convenient comparison method, repeated checks by individuals as well as by interlaboratory tests have shown that unless the appearance of the standard reasonably well approaches that of the sample, errors may occur. To minimize such errors, four categories of standard charts are available from ASTM.

Figure 7.41 shows two standard charts having the same ASTM grain size number ($G = 3$) but from different periods. Chart (a) is from ASTM E19-46; (b) is from ASTM E112-63 and is still in use. The grain size in (a) looks more uniform but too idealized. Grain size distribution in (b) is more realistic and stereologically correct, because when a single-phase grain structure is sectioned randomly the sectioning plane can cut through the grain anywhere between maximum section and extremely small corners but with different probabilities. Therefore, it is natural that a range of grain sizes will be observed even in a specimen composed of extremely uniform grains.

Sometimes mixed grain sizes, i.e., grains with two distinctly different sizes or the so-called bimodal or duplex grain structures, do exist, particularly in hot-worked metals. Averaging from measurements of duplex grain structures is meaningless, because such grain size does not exist in the specimen. In such cases, two ranges of sizes should be reported, e.g., 40% of ASTM Nos. 8–10 and 60% of ASTM Nos. 3–4. Interested readers may refer to ASTM E1181 (Standard Test Methods for Characterizing Duplex Grain Sizes).

There appears to be a general bias in that comparison grain size ratings claim that the grain size is somewhat coarser ($1/2$ to $1 G$ number lower) than it actually is. The repeatability and the reproducibility of comparison chart ratings are generally within ± 1 of the grain size number.

When grain structure of a specimen at magnification other than 100× is compared with the standard charts, the following correction should be made:

$$G = G' + Q \tag{7.31}$$

where G is the actual grain size number after correction, G' is the apparent grain size number obtained by comparing the grain structure as viewed at magnification M with the

TABLE 7.13
Grain Size Relationships Computed for Uniform, Randomly Oriented, Equiaxed Grains from ASTM E112-96

Grain Size No.	N_A Grains/Unit Area		$\bar{A}$ Average Grain Area		$\bar{d}$ Average Diameter		$\bar{l}$ Mean Intercept		$\bar{N}_L$
G	No/in.² at 100×	No/mm² at 1×	mm²	μm²	mm	μm	mm	μm	No/mm
00	0.25	3.88	0.2581	258064	0.5080	508.0	0.4525	452.5	2.21
0	0.50	7.75	0.1290	129032	0.3592	359.2	0.3200	320.0	3.12
0.5	0.71	10.96	0.0912	91239	0.3021	302.1	0.2691	269.1	3.72
1.0	1.00	15.50	0.0645	64516	0.2540	254.0	0.2263	226.3	4.42
1.5	1.41	21.92	0.0456	45620	0.2316	213.6	0.1903	190.3	5.26
2.0	2.00	31.00	0.0323	32258	0.1796	179.6	0.1600	160.0	6.25
2.5	2.83	43.84	0.0228	22810	0.1510	151.0	0.1345	134.5	7.43
3.0	4.00	62.00	0.0161	16129	0.1270	127.0	0.1131	113.1	8.84
3.5	5.66	87.68	0.0114	11405	0.1068	106.8	0.0951	95.1	10.51
4.0	8.00	124.00	0.00806	8065	0.0898	89.8	0.0800	80.0	12.50
4.5	11.31	175.36	0.00570	5703	0.0755	75.5	0.0673	67.3	14.87
5.0	16.00	248.00	0.00403	4032	0.0635	63.5	0.0566	56.6	17.68
5.5	22.63	350.73	0.00285	2851	0.0534	53.4	0.0476	47.6	21.02
6.0	32.00	496.00	0.00202	2016	0.0449	44.9	0.0400	40.0	25.00
6.5	45.25	701.45	0.00143	1426	0.0378	37.8	0.0336	33.6	29.73
7.0	64.00	992.00	0.00101	1008	0.0318	31.8	0.0283	28.3	35.36
7.5	90.51	1402.9	0.00071	713	0.0267	26.7	0.0238	23.8	42.04
8.0	128.00	1984.0	0.00050	504	0.0225	22.5	0.0200	20.0	50.00
8.5	181.02	2805.8	0.00036	356	0.0189	18.9	0.0168	16.8	59.46
9.0	256.00	3968.0	0.00025	252	0.0159	15.9	0.0141	14.1	70.71
9.5	362.04	5611.6	0.00018	178	0.0133	13.3	0.0119	11.9	84.09
10.0	512.00	7936.0	0.00013	126	0.0112	11.2	0.0100	10.0	100.0
10.5	724.08	11223.2	0.000089	89.1	0.0094	9.4	0.0084	8.4	118.9
11.0	1024.00	15872.0	0.000063	63.0	0.0079	7.9	0.0071	7.1	141.4
11.5	1448.15	22446.4	0.000045	44.6	0.0067	6.7	0.0060	5.9	168.2
12.0	2048.00	31744.1	0.000032	31.5	0.0056	5.6	0.0050	5.0	200.0
12.5	2896.31	44892.9	0.000022	22.3	0.0047	4.7	0.0042	4.2	237.8
13.0	4096.00	63488.1	0.000016	15.8	0.0040	4.0	0.0035	3.5	282.8
13.5	5792.62	89785.8	0.000011	11.1	0.0033	3.3	0.0030	3.0	336.4
14.0	8192.00	126976.3	0.000008	7.9	0.0028	2.8	0.0025	2.5	400.0

Note: A calculated G value of -1 corresponds to ASTM $G = 00$.

The average grain area, $\bar{A}$, is the reciprocal of N_A, that is, $1/N_A$, while the mean grain diameter, $\bar{d}$, is the square root of $\bar{A}$. This grain diameter has no physical significance because it represents the side of a square grain of area $\bar{A}$, and there is no grain with square cross section.

TABLE 7.14
Macroscopic Grain Size Relationships Computed for Uniform, Randomly Oriented, Equiaxed Grains from ASTM E112-96

Macro Grain Size No.	$\bar{N}_A$ Grains/Unit Area		$\bar{A}$ Average Grain Area		$\bar{d}$ Average Diameter		$\bar{l}$ Mean Intercept		$\bar{N}_L$	$\bar{N}$
	No/mm²	No/in.²	mm²	in.²	mm	in.	mm	in.	mm⁻¹	100 mm
M-0	0.0008	0.50	1290.3	2.00	35.9	1.41	32.00	1.2	0.031	3.13
M-0.5	0.0011	0.71	912.4	1.41	30.2	1.19	26.91	1.0	0.037	3.72
M-1.0	0.0016	1.00	645.2	1.00	25.4	1.00	22.63	0.89	0.044	4.42
M-1.5	0.0022	1.41	456.2	0.707	21.4	0.841	19.03	0.74	0.053	5.26
M-2.0	0.0031	2.00	322.6	0.500	18.0	0.707	16.00	0.63	0.063	6.25
M-2.5	0.0044	2.83	228.1	0.354	15.1	0.595	13.45	0.53	0.074	7.43
M-3.0	0.0062	4.00	161.3	0.250	12.7	0.500	11.31	0.44	0.088	8.84
M-3.5	0.0088	5.66	114.0	0.177	10.7	0.420	9.51	0.37	0.105	10.51
M-4.0	0.0124	8.00	80.64	0.125	8.98	0.354	8.00	0.31	0.125	12.50
M-4.5	0.0175	11.31	57.02	0.0884	7.55	0.297	6.73	0.26	0.149	14.87
M-5.0	0.0248	16.00	40.32	0.0625	6.35	0.250	5.66	0.22	0.177	17.68
M-5.5	0.0351	22.63	28.51	0.0442	5.34	0.210	4.76	0.18	0.210	21.02
M-6.0	0.0496	32.00	20.16	0.0312	4.49	0.177	4.00	0.15	0.250	25.00
M-6.5	0.0701	45.26	14.26	0.0221	3.78	0.149	3.36	0.13	0.297	29.73
M-7.0	0.099	64.00	10.08	0.0156	3.17	0.125	2.83	0.11	0.354	35.36
M-7.5	0.140	90.51	7.13	0.0110	2.67	0.105	2.38	0.093	0.420	42.05
				×10⁻³	×10⁻³		×10⁻³		×10⁻³	
M-8.0	0.198	128.0	5.04	7.812	2.25	88.4	2.0	78.7	0.500	50.00
M-8.5	0.281	181.0	3.56	5.524	1.89	74.3	1.68	66.2	0.595	59.46
M-9.0	0.397	256.0	2.52	3.906	1.59	62.5	1.41	55.7	0.707	70.71
M-9.5	0.561	362.1	1.78	2.762	1.33	52.6	1.19	46.8	0.841	84.09

Continued

TABLE 7.14 (Continued)
Macroscopic Grain Size Relationships Computed for Uniform, Randomly Oriented, Equiaxed Grains from ASTM E112-96

Macro Grain Size No.	$\overline{N}_A$ Grains/Unit Area		$\overline{A}$ Average Grain Area		$\overline{d}$ Average Diameter		$\overline{l}$ Mean Intercept		$\overline{N}_L$	$\overline{N}$
	No/mm²	No/in.²	mm²	in.²	mm	in.	mm	in.	mm⁻¹	100 mm
M-10.0	0.794	512.0	1.26	1.953	1.12	44.2	1.00	39.4	1.00	100.00
M-10.5	1.122	724.1	0.891	1.381	0.994	37.2	0.841	33.1	1.19	118.9
M-11.0	1.587	1024.1	0.630	0.977	0.794	31.2	0.707	27.8	1.41	141.4
M-11.5	2.245	1448.2	0.0445	0.690	0.667	26.3	0.595	23.4	1.68	168.2
M-12.0	3.175	2048.1	0.315	0.488	0.561	22.1	0.500	19.7	2.00	200.0
M-12.3	3.908	2521.6	0.256	0.397	0.506	19.9	0.451	17.7	2.22	221.9
M-12.5	4.490	2896.5	0.223	0.345	0.472	18.6	0.420	16.6	2.38	237.8
M-13.0	6.349	4096.3	0.157	0.244	0.397	15.6	0.354	13.9	2.83	282.8
M-13.3	7.817	5043.1	0.128	0.198	0.358	14.1	0.319	12.5	3.14	313.8
M-13.5	8.979	5793.0	0.111	0.173	0.334	13.1	0.297	11.7	3.36	336.4
M-13.8	11.055	7132.1	0.091	0.140	0.301	11.8	0.268	10.5	3.73	373.2
M-14.0	12.699	8192.6	0.097	0.122	0.281	11.0	0.250	9.84	4.00	400.0
M-14.3	15.634	10086.3	0.064	0.099	0.253	9.96	0.225	8.87	4.44	443.8

Note: Macroscopically determined grain size numbers M-12.3, M-13.3, M-13.8, and M-14.3 correspond, respectively, to microscopically grain size numbers (*G*) 00, 0, 0.5, and 1.0.

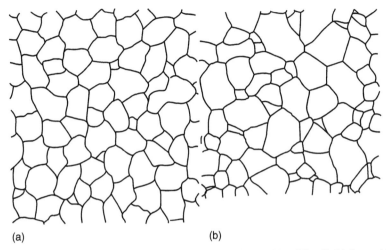

FIGURE 7.41 Standard charts of untwinned grains (flat etch) at $100\times$ ($G=3$) (a) from ASTM E19-46 and (b) from ASTM E112-63. (Reduced 45% in reproduction.)

standard chart at the basic magnification M_b ($100\times$), and Q is the correction factor that should be added to G',

$$Q = 2\log_2\left(\frac{M}{M_b}\right) = 6.6439\log_{10}\left(\frac{M}{M_b}\right) \tag{7.32}$$

In actual practice, the alternative magnifications are usually selected as simple multiples of the basic magnification, and consequently the correction factors are also simple integers. Thus, for magnifications of $400\times$ and $50\times$, the corresponding values of Q are 4 and –2, respectively.

7.4.5.3 Planimetric (or Jefferies') Method

For more accurate rating of grain size, planimetric method was introduced by Jefferies. In this method, the number of grains within a circle or rectangle of known area (usually $5000\,\text{mm}^2$ to simplify the calculations) is counted. The ASTM grain size number can then be computed from Equation 7.27, in which N_A is obtained from the expression given below:

$$N_A = f\left(\frac{N_{\text{inside}} + N_{\text{intercepted}}}{2}\right) \tag{7.33}$$

where f is the Jefferies' multiplier ($f = 0.0002M^2$, M is the magnification used), N_{inside} is the number of grains completely inside the test circle or rectangle, $N_{\text{intercepted}}$ is the number of grains that intercept the test circle or rectangle. Experience suggests that a magnification that produces about 50 grains within the test area is about optimum as to counting accuracy per field. However, when the number of grains within the test area exceeds about 100, counting may become tedious and inaccurate. To ensure a reasonable average, a minimum of three fields should be counted. Fields should be chosen at random without bias. Since it is necessary to mark off the grains to obtain an accurate count, planimetric method is less efficient than intercept method described in the following section.

7.4.5.4 Intercept Method

In the intercept method, the number of grains intercepted by test lines of known length is counted, and the average intercept length of grains can be obtained from the expression:

$$\overline{L}_3 = \frac{1}{N_{\mathrm{L}}} = \frac{1}{P_{\mathrm{L}}} = \frac{L_{\mathrm{T}}}{N \times M} \tag{7.34}$$

where $\overline{L}_3$ is the average intercept length of grains in mm at $1\times$ (this symbol is proposed by Underwood [32], the corresponding symbol used in ASTM E112 is $\overline{l}$); N_{L} is the number of grains intercepted by unit length of test line in mm^{-1}; P_{L} is the number of intersections of grain boundaries cut by unit length of test line in mm^{-1} (for closed test lines, $P_{\mathrm{L}} = N_{\mathrm{L}}$); L_{T} is the total length of test lines in mm; N is the number of intersections of grain boundaries cut by test lines of length L_{T}; and M is the magnification used.

Since the ASTM grain size number was initially computed from N_{A}, and no exact relationship exists between $\overline{L}_3$ and N_{A} (or $\overline{A}$), a formal relation between $\overline{L}_3$ and average grain area $\overline{A}$ was assumed:

$$\overline{L}_3 = \left(\frac{\pi \overline{A}}{4}\right)^{1/2} \tag{7.35}$$

Hence,

$$N_{\mathrm{A}} = \frac{1}{\overline{A}} = \frac{\pi}{4\overline{L}_3} \tag{7.36}$$

Substituting into Equation 7.27,

$$G = -3.3027 - 6.643856 \log_{10} \overline{L}_3 \tag{7.37}$$

Equation 7.37 is precise only for spheres and practically exact only for uniform equiaxed grains. To eliminate the problem of variable conversion factors, the relationship between ASTM grain size number and average intercept length was redefined by ASTM E112 in 1980:

$$G = 2 \log_2 \left(\frac{\overline{L}_0}{\overline{L}_3}\right) \tag{7.38}$$

where $\overline{L}_0$ is the average intercept length for $G = 0$.

For the macrograin size scale, $= \overline{L}_0 = 32$ mm at $1\times$. Substituting into Equation 7.38 we have

$$G = 10.0000 - 6.643856 \log_{10} \overline{L}_3 \quad (\overline{L}_3 \text{ in mm at } 1\times) \tag{7.39}$$

For the micrograin size scale, $\overline{L}_0 = 32$ mm at $100\times$ or 0.32 mm at $1\times$. Then, substituting into Equation 7.38 we have

$$G = 10.0000 - 6.643856 \log_{10} \overline{L}_3 \quad (\overline{L}_3 \text{ in mm at } 100\times) \tag{7.40}$$

or

$$G = -3.2877 - 6.643856 \log_{10} \overline{L}_3 \quad (\overline{L}_3 \text{ in mm at } 1\times) \tag{7.41}$$

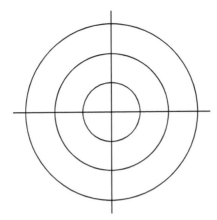

FIGURE 7.42 The three-circle test pattern for intercept count. (Reduced 40% in reproduction.)

Comparing Equation 7.41 with Equation 7.37, it can be seen that for the same value of $\overline{L}_3$, the ASTM grain size number G according to the redefined Equation 7.38 is only 0.015 smaller than that calculated from Equation 7.37. The difference is again insignificant.

Based on an experimentally finding that a total of 500 counts per specimen normally yield acceptable precision, a test pattern most appropriate for manual operation was developed by Abrams [40]. It consists of three concentric and equally spaced circles having a total circumference of 500 mm as shown in Figure 7.42. The magnification needed is so selected that the test pattern will make from 40 to 100 intercepts or intersection counts per placement of the three-circle test grid over the microstructure or photomicrograph.

When the three-circle test pattern is used, the measured value of $\overline{L}_3$ is greater than that using the straight-line test pattern. This is because a straight line connecting two neighboring intercepts is always shorter than a circular arc. For example, when $N = 4$, 6, and 10, the measured values of $\overline{L}_3$ are 10, 4.7, and 1.7% larger, respectively. However, as N increases, the difference decreases significantly. For example, when $N = 18$, $\overline{L}_3$ is only 0.5% larger.

When the test pattern meets a triple point, i.e., the junction of three grains, a score of 1.5 intersections should be recorded. However, when a tally counter is used, score any intersection of the circle with the junction of three grains as two rather than the correct value of 1.5, the error introduced is very small. A tangential intersection with a grain boundary should be scored as 1 intersection.

7.4.5.5 Statistical Analysis of the Grain Size Measurement Data

Usually five randomly selected and widely spaced fields should be measured for every specimen with the three-circle test pattern, and the count of intercepts should be recorded separately for each of the five tests. Then, determine the mean lineal intercept, its standard deviation, 95% confidence limit, and percent relative accuracy. For most work, a relative accuracy of 10% or less represents an acceptable degree of precision. If the calculated relative accuracy is unacceptable for the application, additional fields should be counted until the necessary percent relative accuracy is attained.

Let the measured count of intercepts be denoted by N_1, $N_2 \ldots$, $N_i, \ldots$, N_n. Then the standard deviation (s) and CV of the counts and the 95% confidence limit (95% CL) of the mean $\overline{N}$ can be calculated from Equation 7.16 through Equation 7.20, where x_i, x, and $s_{\overline{N}}$ are replaced by N_i, N, and $s_{\overline{N}}$, respectively. Since grain size is usually characterized by average intercept length $\overline{L}_3$ and ASTM grain size number G, the following procedures of statistical analysis should also be carried out.

7.4.5.5.1 Calculation of the 95% Relative Confidence Limit on $\overline{L}_3$
Since the limiting values of N are $\overline{N} \pm 2\, s_{\overline{N}}$ (95% CL), substituting into Equation 7.34 the limiting values of $\overline{L}_3$ (denoted by $L_{3\pm}$) can be obtained:

$$\overline{L}_{3\pm} = \frac{L_T}{M\overline{N}(1 \mp 2s_{\overline{N}}/\overline{N})} = \frac{\overline{L}_3}{(1 \mp 2s_{\overline{N}}/\overline{N})} \tag{7.42}$$

Averaging the difference on the limiting values of $\overline{L}_3$ and dividing by $\overline{L}_3$, the 95% RCL on $\overline{L}_3$ is obtained:

$$\frac{\Delta\overline{L}_3}{2\overline{L}_3} = \frac{(\overline{L}_{3+} - \overline{L}_{3-})}{2\overline{L}_3} = \frac{2\,s_{\overline{N}}/\overline{N}}{1 - \left(2s_{\overline{N}}/\overline{N}\right)^2} \tag{7.43}$$

From Equation 7.17 and Equation 7.19, we have

$$\frac{s_{\overline{N}}}{\overline{N}} = \frac{CV}{(n-1)^{1/2}} \tag{7.44}$$

Substituting into Equation 7.43, we can get 95% RCL on $\overline{L}_3$ in terms of the CV of the count and number of tests:

$$95\% \text{ RCL on } \overline{L}_3 = \frac{2\,(n-1)\,(CV)}{(n-1) - 4\,(CV)^2} \tag{7.45}$$

7.4.5.5.2 Calculation of the 95% Confidence Limit on G (ΔG)
Substituting the limiting values of (95% CL) into Equation 7.41, the limiting values of G (denoted by $G^\pm$) can be obtained:

$$G^\pm = -3.2877 - 6.6439 \log_{10}\left[\frac{L_T}{M(\overline{N} \pm 2s_{\overline{N}})}\right] = G + 6.6439 \log_{10}\left(\frac{1 \pm 2s_{\overline{N}}}{\overline{N}}\right) \tag{7.46}$$

Hence, the 95% CL on G can be obtained:

$$\begin{aligned}
\Delta G &= (G^+ - G^-)/2 \\
&= 3.3219 \log_{10}\left[(\overline{N} + 2s_{\overline{N}})/(\overline{N} - 2s_{\overline{N}})\right] \\
&= 3.3219 \log_{10}\left[\frac{(n-1)^{1/2} + 2CV}{(n-1)^{1/2} - 2CV}\right]
\end{aligned} \tag{7.47}$$

Equation 7.44 is also used in obtaining the final form of Equation 7.47.

7.4.5.5.3 Calculation of Necessary Number of Tests According to Required 95% CL on G
If the calculated value of 95% CL on G is larger than required, it is possible to calculate the number of tests n from Equation 7.47 at a given value of CV and required 95% CL on G.

From Equation 7.47, the number of tests n can be expressed as a function of G (95% CL) and CV:

$$n = 4 \times \left(\frac{Z+1}{Z-1}\right)^2 \times (\text{CV})^2 + 1 \qquad (7.48)$$

where $Z = 10^{\Delta G/3.3219}$.

For example, if the required accuracy of grain size scale is $0.5G$, then ΔG (95% CL) should be 0.25, and substituting into Equation 7.48 yields

$$n = 23.14 \times (\text{CV})^2 + 1 \qquad (7.49)$$

7.4.5.5.4 Illustrated Example

An example is given to illustrate procedures for the determination of confidence limit from grain size measurements.

The five intercept counts using a three-circle test pattern at a magnification of 100 are $N = 84, 73, 94, 100$, and 79. It follows that

$$\Sigma N_i = 470 \quad \text{and} \quad \bar{N} = \Sigma N_i/n = 94.0$$

1. From Equation 7.34, the average intercept length $\bar{L}_3 = 0.0234\,\text{mm}$ or 23.4 μm, and the corresponding ASTM grain size number can be obtained from Equation 7.41, i.e.,

$$G = 7.5649 \approx 7.6$$

2. Using Equation 7.16 or a pocket calculator, the standard deviation of the counts $s = 10.98$, and from Equation 7.17 the $\text{CV} = 0.1277$ or 12.77%.
3. From Equation 7.20, the 95% confidence limit of the mean is 10.98. Therefore, the limiting values of $\bar{N}$ (95% CL) are 86.0 ± 10.98.
4. From Equation 7.45, the 95% relative confidence limit on $\bar{L}_3$ is found to be 0.1298 or 12.98%. Therefore, the limiting values of $\bar{L}_3$ (95% CL) are: $0.0234 (1 \pm 0.1298) = 0.02340 \pm 0.0030\,\text{mm}$.
5. From Equation 7.47, the 95% confidence limit on G is found to be 0.3705. Hence, the limiting values of G (95% CL) are 7.6 ± 0.4.
6. If the required accuracy of grain size is $0.25G$, then ΔG (95% CL) should be 0.125, and substituting into Equation 7.48 yields $n = (46.20\,\text{CV})^2 + 1$. Using this expression for $\text{CV} = 0.1277$, and the necessary number of tests is found to be $n = 35.81 \approx 36$; i.e., 31 more tests are needed.

7.4.6 Automatic Image Analysis*

As described in earlier sections, quantitative metallography using manual methods has the advantage of convenience at much lower cost because no sophisticated instrument other than a microscope is needed. Besides, the feature recognition ability of the human eye is far superior to that of an image analyzer. However, manual operation is much slower, and it is also tedious, causing operator fatigue.

*This section is based on Refs. [41,42].

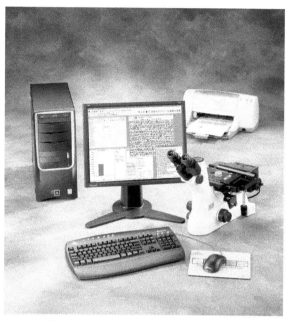

FIGURE 7.43 The OMNIMET® Modular Image Analysis System. (Courtesy of Buehler Ltd. With permission.)

With the advent of the automatic image analyzer and its continuous improvement and perfection over recent decades, this computer-aided alternative to manual image analysis has occupied a firm and increasingly important place in research, quality assurance, and process control. Figure 7.43 shows a modern image processing and analysis system. Automatic image analysis consists of four main steps: image acquisition, image processing, detection of image details, and quantitative measurements.

7.4.6.1 Image Acquisition

The image to be analyzed usually comes from an optical microscope. To obtain the best image contrast from different types of specimens, various methods of illumination, e.g., brightfield and darkfield, polarized light, DIC, and phase contrast illumination, should be provided.

It is of utmost importance that proper specimen preparation to reveal the true structure for image analysis. It would be best that only microstructural features of interest be revealed so that other microstructural constituents would not obscure the analysis and measurement, for example, during evaluation of inclusions in steel or graphite in cast iron, specimen in the as-polished condition should be used. If etching is necessary, then one should keep in mind that over-etching should be prevented. An over-etched specimen makes the surface nonplanar, however, the basic and many equations for stereological measurement are based on the assumption of a planar surface. Although there are equations for stereological measurement of nonplanar surface (e.g., fractured surface) and projected images of microstructural features obtained from thin foils, films, and slices (e.g., specimens for transmitted light microscopy or TEM), they are much more complicate and difficult to employ.

The video or digital camera attached to the optical microscope converts the optical image into a digitized image, which has a discrete number of pixels (picture elements), each containing an intensity value, usually scaled from 0 (black) to 255 (white). Camera resolution is determined by the number of pixels, the more the number of pixels, the higher the resolution.

High-resolution cameras can be used for analyzing large and small features at the same time. It is evident that the resolution of the microscope is also a vital importance in determining the overall resolution of the imaging system. Digitized images can be saved on disk for later processing and analysis or just for archival purposes.

7.4.6.2 Image Processing

Image processing, or image enhancement, is employed to enhance the image details and thus improve the image contrast. To extract maximum information from the image, it is necessary that an image analyzer be equipped with a rich library of image processing capabilities, so that the image can be modified in a controlled manner. Any image processing operation can be shown on the screen and the results displayed in a window. The one operation that is applied to almost every image is the removal of noise and nonuniformity of illumination.

Although the image is changed to some extent as a result of image processing operations, no additional information can be gained that was not present in the original image.

The following is an example of image processing procedures used for grain sizing. Although it is quite easy to determine the ASTM grain size number of an etched single-phase specimen using the manual intercept method as described in earlier sections, automatic grain sizing does encounter difficulties so that some of the following image processing procedures are needed before automatic image analysis [42]:

1. *Grain boundary detection.* Pixels are identified as grain boundary or grain based on local image characteristics. If the grains are lighter or darker than the boundaries as in the case of bright- or darkfield illumination, respectively, thresholding defines the boundary. If contrast etch is used to delineate grains, then edge-enhancement algorithms must be used before thresholding.
2. *Inclusion removal.* Inclusions may be mistaken for incomplete segments of grain boundaries. Therefore, inclusions smaller than a specified size should be eliminated.
3. *Grain boundary completion or reconstruction.* This is the most important procedure in automatic grain sizing. The absence of even a single grain boundary pixel may result in erroneous counting. For example, two neighboring grains may be regarded as one grain if a pixel of their common grain boundary is absent. Figure 7.44 shows the microstructure of a cold-rolled ferritic steel (a) before and (b) after image processing. The measured average grain areas in (a) and (b) are 35.21 and 25.01 μm^2, respectively, which correspond to ASTM grain size number of 11.8 and 12.3, respectively. Using the intercept method, the manually measured ASTM grain size number of the same field is 12.4, which is very close to that of automatic image analysis after image processing.

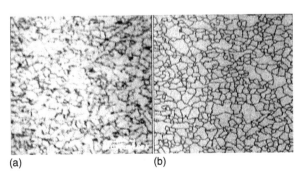

(a)   (b)

FIGURE 7.44 Microstructure of a cold-rolled ferritic steel (a) before and (b) after image processing. (Courtesy of Gamma-Tech, Inc.)

4. *Grain boundary thinning*. This is an operation for minimizing the area fraction of grain boundaries. It is well known that grain boundaries are only a few atomic diameters in width, but the etched grain boundaries observed under the microscope are much thicker. During manual grain sizing the intersections of test lines with grain boundaries are considered points; hence, the thickness of grain boundaries is not a problem. However, in a digital image, the area occupied by grain boundaries cannot be ignored, particularly at small grain sizes. The goal of grain boundary thinning is to minimize the boundaries to a width of 2 pixels. For example, in a grain structure of $G = 6$, the area fraction of the grain boundaries can be reduced from 19 to 8% by a boundary-thinning routine. According to Vander Voort and Friel [43], under these conditions each grain will still be undersized, causing an insignificant increase of about 0.2 in ASTM grain size number.

5. *Detection and removal of twin boundaries*. Annealing twins are often found in certain face-centered cubic metals and alloys, e.g., austenitic stainless steels. During automatic grain sizing, the twin boundaries should not be mistaken for grain boundaries. They should be detected and removed before measurements are made. In austenitic stainless steels most of the twin boundaries can be detected according to the straightness of the boundary and its length, and hence can be removed. However, such straight twin boundaries become curved after cold deformation and cannot be identified. Likewise, lenticular deformation twins such as Neumann bands in ferritic grains (see Figure 7.29) cannot be detected.

7.4.6.3   Detection of Image Details

Image details or features to be measured are selected by applying thresholds to highlight the features of interest. In the case of a multiphase specimen, several thresholds are set, one for each phase. Two phases can be differentiated only if they have different gray levels; therefore, the etching step during specimen preparation is critical.

As stated earlier, the human eye can discriminate only about 20 levels of intensity at the same time, but it can distinguish hundreds of different shades of color. Therefore, pseudocolor can be used in an image to show details that would be lost in the gray scale image. Also, a pseudocolor display of the image gives more visual comfort; even detection on the basis of gray level is satisfactory. However, true color facilities should be used when color is used as the basis for the identification and detection of features. At present image analysis system with true color facilities is available but, of course, at a higher cost.

7.4.6.4   Quantitative Measurements

Usually an image analyzer is capable of providing a generous range of measurement parameters that covers a wide spectrum of applications. In field measurements, each individual feature is measured separately, and the results are summed for each field of view. In feature measurements, multiple features are measured, and a cumulative statistical distribution of results for all features in the measured fields is presented. Results can be printed or filed to disk in the most suitable format.

The following are three ASTM metallographic test standards using automatic image analysis [39]:

1. ASTM E1112-86: Standard Practice for Obtaining JK Inclusion Rating Using Automatic Image Analysis
2. ASTM E1245-00: Standard Practice for Determining the Inclusion or Second-Phase Constituent Content of Metals by Automatic Image Analysis
3. ASTM E1382-91: Standard Test Methods for Determining the Average Grain Size Using Semiautomatic and Automatic Image Analysis

REFERENCES

1. JA Nelson, A new direction for metallography, *Pract. Metallogr.* 26:225 (1989).
2. S Buttitta, N Dougill, *Orbital Technology—the Science of the Cut, Tech-Notes*, Buehler Ltd., Vol. 3, Iss. 4 (2000).
3. P Wellner, Investigation on the effect of the cutting operation on the surface deformation of different materials, *Pract. Metallogr.* 17:525 (1980).
4. B Bousfield, *Surface Preparation and Microscopy of Materials*, John Wiley & Sons, 1992, reprinted with corrections July 1994.
5. Guide to engineering materials (testing/analysis/metallography), *Adv. Mater. Process* 159(12):203 (2001).
6. GF Vander Voort, *Metallography—Principles and Practice*, McGraw-Hill, New York, 1984.
7. *Buehler Consumables Buyers Guide*, Buehler Ltd., IL, 2000.
8. GF Vander Voort, GM Lucas, *Planarmet® AL Paper for Rough Grinding, Tech-Notes*, Buehler Ltd., Vol. 2, Iss. 3 (1998).
9. GF Vander Voort, Trends in specimen preparation, *Adv. Mater. Process.* 158:February (2000).
10. GF Vander Voort, Polishing with colloidal silica, Structure 26, *Struers J. Materialogr.* 2:3 (1992).
11. JA Nelson, New abrasives for metallography, *Microstruct. Sci.* 11:251–261 (1983).
12. GH Schwuttke, A Oster, Damage removal on silicon surfaces: a comparison of polishing techniques. Tech. Rep. 7, Part 1 of Damage Profiles in Silicon and their Impact on Device Reliability, IBM Corp., Ref. No. TR 22, 1989, January 1976. [From Ref. 10.]
13. Guide to engineering materials (testing/analysis/metallography), *Adv. Mater. Process.* 159(12): 204–206 (2001).
14. JH Richardson, *Optical Microscopy for the Materials Sciences*, Marcel Dekker, New York, 1971, pp. 432–449.
15. E Beraha, B Shpigler, *Color Metallography*, American Society for Metals, Materials Park, OH, 1977.
16. E Weck, E Leistner, Molybdic etching reagent almost universal, Structure 11, *Struers Metallogr. News*, August 1985, pp. 11–16.
17. GF Vander Voort, Tint etching, *Metal Prog.* 127(4):31–41 (1985).
18. FM Krill, *Metal Prog.* 70(1):81 (1956).
19. EL Long Jr., RJ Gray, *Metal Prog.* 74(4):145 (1958).
20. P Rothstein, FR Turner, Metallographic specimen preparation by vibratory polishing, in *Symposium on Methods of Metallographic Specimen Preparation*, ASTM STP 285, ASTM, PA, 1960, pp. 90–102.
21. H Modin, S Modin, *Metallurgical Microscopy*, Butterworths, London, 1973, pp. 61–66.
22. O Goldberg, Koehler illumination, *Microscope* 28:15 (1980).
23. P Eutteneuner, A Muller-Rentz, K–H Schade, DELTA—the new system of microscope optics from Leica, *Sci. Tech. Inf.* 10(4):114–122 (1992).
24. W Lang, Nomarski Differential Interference Contrast Microscopy (Reprint), Collection of four articles from Zeiss Information, S 41-210.2-5-e, 1979.
25. E Voogel, P Keyzer, *200 SLR Tips*, Fountain Press, Windsor, 1984, p. 1.
26. L Stroebel, HN Todd, *Dictionary of Contemporary Photography*, Morgan & Morgan, Dobbs Ferry, NY, 1974, p. 49.
27. MJ Langford (Consulting Editor), *The Camera Book*, Ziff-Davis, New York, 1980, pp. 152–173.
28. L Ericksenn, E Sincebough, Adventures of Closeup Photography, AMPHOTO, New York, 1983, p. 10.
29. EE Underwood, Quantitative metallography, in *Metals Handbook*, 9th ed., Vol. 9, *Metallography and Microstructures*, ASM, Materials Park, OH, 1985, pp. 123–134.
30. H Elias, Introduction: problems of stereology, in *Proceedings of the second International Congress on Stereology*, Chicago, April 8–13, 1967, H Elias, Ed., Springer-Verlag, Berlin, 1967, pp. 1–11.
31. RE Miles, Multidimensional perspectives on stereology, *J. Microsc.* 95(2):181–195 (1972).
32. EE Underwood, *Quantitative Stereology*, Addison-Wesley, Reading, MA, 1970.
33. RT DeHoff, FN Rhines (Eds.), *Quantitative Microscopy*, McGraw-Hill, New York, 1968.
34. ER Weibel, *Stereological Methods*, Vol. 1, *Practical Methods for Biological Morphometry*, Vol. 2, *Theoretical Foundations*, Academic Press, New York, 1978, 1980.

35. JE Hilliard, Measurement of volume in volume, in *Quantitative Microscopy*, RT DeHoff, FN Rhines, Eds., McGraw-Hill, New York, 1968, p. 45.
36. RT DeHoff, The statistical background of quantitative metallography, in *Quantitative Microscopy*, RT DeHoff, FN Rhines, Eds., McGraw-Hill, New York, 1968, pp.11–44.
37. RT DeHoff, *Quantitative Metallography, Techniques of Metals Research*, Vol. II, Part 1, Interscience, New York, 1968, pp. 221–253. [Quoted from Ref. 6, p. 503.]
38. FN Rhines, Geometry of microstructures, Part 1, *Metal Prog.* 112(3):60 (1977).
39. GF Vander Voort, Progress in metallographic test standards, *Adv. Mater. Process.* 138:30 (1990).
40. H Abrams, Grain size measurements by the intercept method, *Metallography* 4:59 (1971).
41. KA Leitner, Basics of quantitative image analysis, *Adv. Mater. Process.* 11(5):18–23 (1993).
42. JJ Friel, EB Prestridge, Grain sizing by image analysis, *Adv. Mater. Process.* 139(2):33–37 (1991).
43. GF Vander Voort, JJ Friel, Image analysis measurements of duplex grain structure, *Mater. Charact.* 29:293–312 (1992).

8 Mechanical Property Testing Methods

D. Scott MacKenzie

CONTENTS

8.1   INTRODUCTION

Mechanical testing is necessary to ensure that components will not fail in service. Testing is performed so that the designer can predict the performance of a part or a component in the field. In this chapter, a brief description of design theories and selected testing methods that are available to the designer to predict component serviceability under realistic conditions are provided. Often, mechanical testing is done for quality control. As mechanical testing is a dynamic field, new tests are continually devised, and older tests are revised, and reinterpreted, for cost or performance reasons. Therefore, a review of the literature should be conducted to conserve materials, achieve reliability, and minimize environmental impact.

The prediction of the behavior of a stressed material is dependent on the applied loading and the relative magnitudes of the principal stresses and strains. The material can fail in either a ductile or a brittle fashion. The predictive theory used is dependent on the type of loading and the expected response.

When a component is loaded, it responds either elastically or plastically depending on the material, amount of stress, strain rate, and geometry.

In elastic deformation, there is a temporary change in the distances between atoms or crystallographic planes. In plastic deformation, there is a permanent change in the relative position of the atoms. This is caused by the displacement of atoms within the crystal lattice.

If the loads are sufficiently high to cause failure or fracture, the material behaves in either a brittle or a ductile fashion. Gensamer [1] summarized the terms used to characterize fracture. These terms and the described behavior are shown in Table 8.1.

8.1.1   TYPES OF LOADING

The method of loading affects the manner in which a body reacts under stress. A body can experience four types of loading:

1. *Static load, short time*: The load is gradually applied until failure occurs. The loads on the body are in equilibrium. A tensile test is typical for this type of loading.
2. *Static load, long duration*: The load is applied gradually until it reaches a maximum and is then maintained. Creep testing is done in this manner.

TABLE 8.1
Terms Used to Describe Fracture Surfaces

Crystallographic mode	Shear, cleavage
Appearance	Fibrous, granular
Strain to fracture	Ductile, brittle

3. *Dynamic loading*: The load is applied rapidly, and the loads do not have time to reach equilibrium within the body. The part momentum is considered as the part is accelerated. The stresses are considered static but are applied in a short-time impulse. Charpy impact testing is typical for this type of loading.

4. *Repeated loading*: Loads are applied to the body repetitively, with the load remaining either completely or partially removed after each cycle. There are also two main types of loading: high stress applied for a few cycles or low stress applied for more cycles.

Loadings of various types may be applied at the same time or sequentially. The order is unimportant unless it is necessary to determine the source of failure. The local stress distribution and the relative magnitudes of the stress and strain present are also important.

8.1.2 STRESS AND STRAIN

8.1.2.1 Plane Stress

If a small differential body is stressed (Figure 8.1), there are three main stresses normal to the faces of the cube (σ_x, σ_y, and σ_z) and six shear stresses along each face (τ_{xy}, τ_{xz}, τ_{yz}, τ_{yx}, τ_{zx},

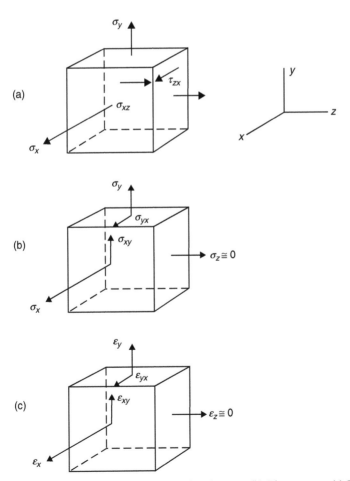

FIGURE 8.1 (a) Principal stresses on a small differential element. (b) Plane stress. (c) Plane strain.

and τ_{zy}). If σ_z is much smaller than σ_x and σ_y, then the stress state is a condition called plane stress, because the normal stresses of significant magnitude lie in only one plane.

Assume that a triaxial stress state occurs at a uniaxially loaded notched plate (Figure 8.2). Because no load is applied to the surface or at the root of the notch, the stress in the z direction is zero. Stress in the z direction rises to a maximum at the center of the plate. If the plate is thin, then the increase in this stress is small relative to stress in the other directions. A plane stress condition exists (Figure 8.1b). This is very common in thin materials such as sheets.

While the stress in the z direction is small compared to that in the x and y directions, there is a large maximum shear stress that lied at a 45°C angle to the load direction. Plastic deformation occurs by shear along this inclined plane, creating a shear lip or slant fracture.

8.1.2.2 Plane Strain

If the strain ε in one direction is much smaller than in the other two directions, then a condition of plane strain exists (Figure 8.1c). If the notched plane is thick, then the strain in the z direction is constrained, resulting in a plane strain condition. Because strain in the z direction is limited, the plastic zone at the notch tip is small. Therefore the axial stresses at the notch tip are large, and the shear stresses are small. As the axial stresses are much larger than the shear stresses, the axial stresses control the fracture. A flat fracture results that is normal to the axial stresses.

8.2 PLASTICITY AND DUCTILE FRACTURE

When a material behaves in a ductile or plastic manner, gross deformation occurs. This may manifest as a failure to obey Hooke's law throughout the load history or as plasticity in an initially elastic region, as is typical in a common tensile test. It occurs by portions of grains sliding over one another. Energy is absorbed by the plastic deformation. Ductile fractures are characterized by yielding along the edges of the fracture. The final fracture is usually at 45° to the original plane of fracture.

8.2.1 DUCTILE FRACTURE APPEARANCE

On a macroscopic scale, a ductile fracture is accompanied by a relatively large amount of plastic deformation before the part fails. After failure, the cross section is reduced or

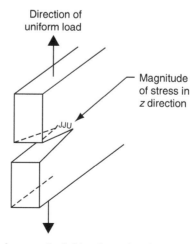

FIGURE 8.2 Stress distribution in a notched thin plate, showing stress distribution at notch.

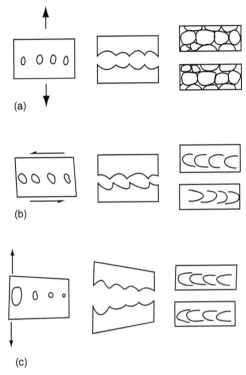

(a)

(b)

(c)

FIGURE 8.3 Schematic representation of the creation of dimples in a loaded member by (a) simple tension, (b) shear loading, and (c) tearing.

distorted. Shear lips are observed at the latter part of the fracture and indicate the final failure of the part. The fracture surface is dull, with a fibrous appearance.

Microscopically, a ductile fracture surface is characterized by dimples. These are voids that join internally under load. The formation of dimples is associated with slip along crystallographic planes, with decohesion at second-phase particles such as inclusions or precipitates. This is shown schematically in Figure 8.3. Dimples can be detected only by using electronic microscopy (see Figure 8.4).

8.2.2 THEORIES OF PLASTICITY AND DUCTILE FRACTURE

Many theories have been put forth in an attempt to explain plasticity and ductile fracture, but only a few have survived rigorous examination. Three of these are discussed below.

8.2.2.1 Maximum Normal Stress Theory

According to the maximum normal stress theory, the failure of a part occurs when the largest principal stress equals the yield strength. Consider a small material element that is loaded as in Figure 8.5. If $\sigma_1 > \sigma_2 > \sigma_3$, then failure occurs when $\sigma_1 = \sigma_y$, where σ_y is the yield strength of the material. This is illustrated schematically in Figure 8.6. This theory is applicable for simple stress states such as tension. It is not applicable to torsion. In torsion, this theory implies that failure occurs when $\tau_{max} = \sigma_y$. This does not fit the experimental data.

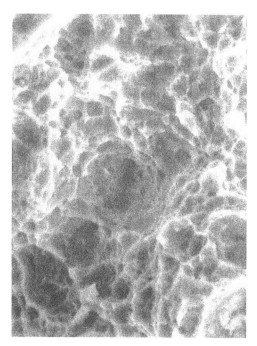

FIGURE 8.4 Scanning electron photomicrograph of dimples in a high-strength steel (5000×).

8.2.2.2 Maximum Shear Stress Theory

The maximum shear stress theory is easy to use and errs on the conservative side. It has been used as the basis for many design codes and adequately predicts yielding. This theory says that yielding occurs whenever the maximum shear stress in a small element is equal to the maximum shear stress in a tensile specimen at the onset of yielding.

If Mohr's circle for a tensile test is considered (Figure 8.7), then yielding occurs when

$$\tau_{\max} = \frac{\sigma_y}{2} \tag{8.1}$$

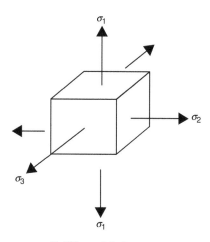

FIGURE 8.5 Principal stresses on a small differential element.

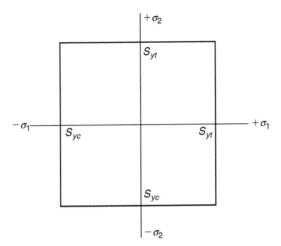

FIGURE 8.6 The maximum normal stress theory of failure based on the yield strength of a material.

For a three-dimensional mechanical element, the shear stresses are

$$\tau_{12} = \frac{\sigma_1 - \sigma_2}{2}, \quad \tau_{23} = \frac{\sigma_2 - \sigma_3}{2}, \quad \text{and} \quad \tau_{13} = \frac{\sigma_1 - \sigma_3}{2} \tag{8.2}$$

The largest of the above stresses is designated as τ_{max}; whenever $\tau_{max} = \sigma_y$, then yielding will occur. This is illustrated in Figure 8.8. The maximum normal stress theory and the maximum shear stress theory are identical when the principal stresses have the same sign.

8.2.2.3 Distortion Energy Theory (von Mices–Hencky Theory)

The von Mices–Hencky theory is the best theory to use for ductile materials because it defines accurately the beginning of yielding. This theory was originally proposed by von Mices [2] in 1914. In essence, the von Mices yield criterion states that all the principals (tensile and shear) must be considered to create the von Mices stress, σ_0:

$$\sigma_0 = \frac{1}{\sqrt{2}} [(\sigma_1 - \sigma_2)^2 + (\sigma_2 - \sigma_3)^2 + (\sigma_3 - \sigma_1)^2 + 6(\tau_{12}^2 + \tau_{23}^2 + \tau_{13}^2)]^{1/2} \tag{8.3}$$

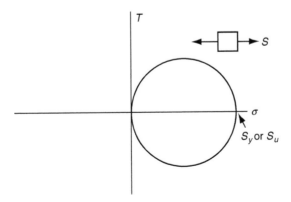

FIGURE 8.7 Mohr's circle for a tensile test.

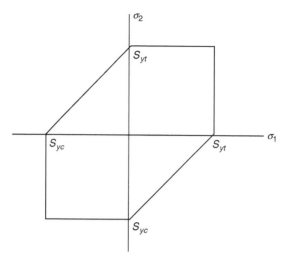

FIGURE 8.8 The maximum shear stress theory of failure.

Yielding is predicted whenever the von Mices stress $\sigma_0 = \sigma_y$. If the stress state is biaxial ($\sigma_3 = 0$) and the member is in torsion ($\sigma_2 = -\sigma_1$ and $\tau = \sigma_1$), then Equation 8.3 shows that

$$\tau_{\max} = 0.577\sigma_y \tag{8.4}$$

This definition of yielding is not dependent on any normal stress or shear stress but depends on all components of the stress. Because the terms are squared, it is independent of the direction of the stress. It is also not necessary to know the largest principal stress to learn if yielding will occur.

8.2.2.4 A Comparison

The three yielding criteria are shown schematically in Figure 8.9. Comparing Equation 8.4 from the distortion energy theory and the maximum shear stress theory (Equation 8.1), the distortion energy theory predicts that significantly higher stresses must be experienced before yielding occurs. The maximum normal stress theory predicts results equivalent to those predicted by the maximum shear stress theory whenever the directions of the principal stresses are the same but fails to accurately predict yielding when the signs of the stresses are opposite. The maximum shear stress theory always gives conservative results. Of the three theories, the distortion energy theory predicts yielding with the greatest accuracy in all four quadrants. The maximum normal stress theory should not be used in the design of a component.

8.3 ELASTICITY AND BRITTLE FRACTURE

Since the early 1940s, there has been tremendous growth in the number of large welded structures. Many of these structures have failed catastrophically in service—most notably the Liberty ships [3] used to transport war materials during World War II. Analysis of the fracture surfaces of the failures [4] indicated that they initiated at a notch and propagated with no plastic deformation. These notches were of three types:

1. *Design features*: Structural members that were rigidly joined at angles less than 90° and then welded.

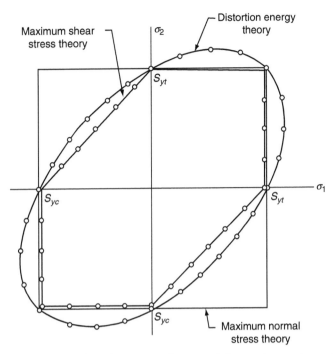

FIGURE 8.9 Comparison of the distortion energy theory and the maximum shear stress theory.

2. *Fabrication details*: Procedures used during the manufacture of the part caused the formation of notches. Welding arc strikes, gouges, and fitting procedures created physical notches. Weld procedures and heat treatment caused metallurgical or micro-structural notches to occur from abrupt changes in microstructure or the production of microstructures that were brittle. Features such as porosity from welding or casting also caused brittle fracture initiation.

3. *Material flaws*: These flaws resulted from melt practice at the mill and appeared as large inclusions, internal oxidation, porosity, or segregation.

In brittle fractures, limited energy is absorbed by the fracture. Energy is absorbed through the regions of small plastic deformation. Individual grains are separated by cleavage along specific crystallographic planes.

8.3.1 Brittle Fracture Appearance

Visually, brittle fractures are characterized by little or no plastic deformation or distortion of the shape of the part. The fracture is usually flat and perpendicular to the stress axis. The fracture surface is shiny, with a grainy appearance. Failure occurs rapidly, often with a loud report. Because the brittle cleavage is crystallographic in nature, the fracture appearance is faceted (Figure 8.10). Often other features are present, such as river patterns [5]. These are shown in Figure 8.11.

8.3.2 Theories of Elasticity and Brittle Fracture

When calculating stresses on a body, it is assumed that the body is elastic and homogeneous and that it conforms to Hooke's law,

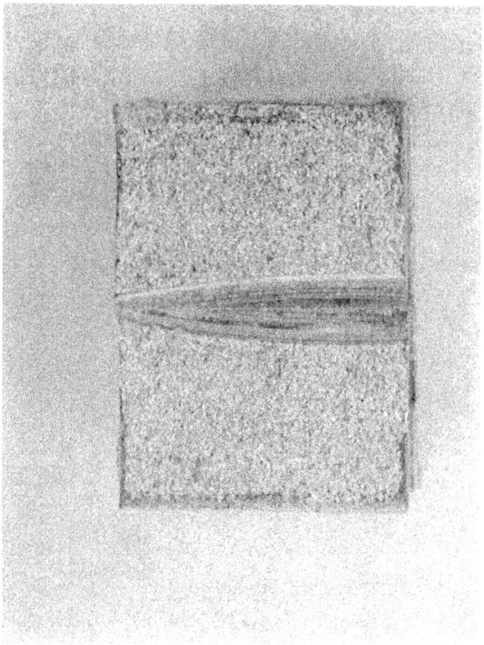

FIGURE 8.10 Fracture surface of AISI 1020 steel broken by impact at −73°C.

$$\varepsilon = \frac{\sigma}{E} \tag{8.5}$$

where ε is the strain (in./in.), σ is the stress (lb/in.^{2}), and E is Young's modulus. However, a material is not completely elastic or completely homogeneous. It is a combination of a collection of fibers or microstructures, either randomly aligned or with an organized orientation. This orientation can provide a directionality to the properties of a material. A material is isotropic (no directionality of properties) if the component is much larger than the constituent parts. If the organization of the constituent parts is nonrandom, the material is

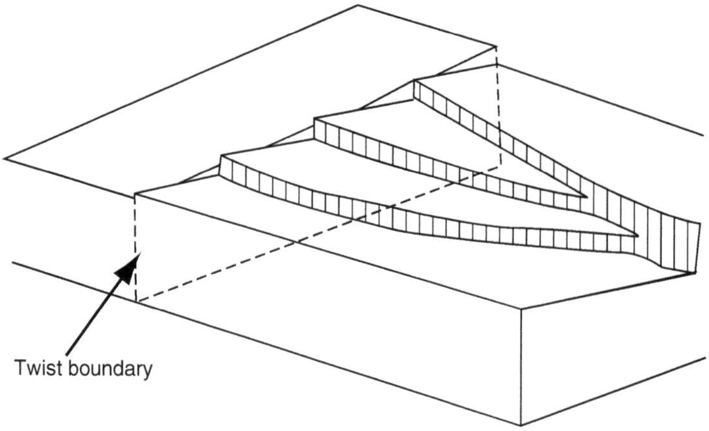
Twist boundary

FIGURE 8.11 Schematic of river patterns formed in brittle materials. (After J. McCall and P. French, *Metallography in Failure Analysis*, Plenum Press, New York, 1978, p. 6.)

anisotropic and will show directionality in properties such as tensile, impact, or electrical conductivity.

If a material is perfectly elastic, there is no permanent deformation, and fracture occurs at the maximum stress, σ_{UTS}, where UTS is the ultimate tensile strength. The characteristics of brittle materials include the following: (1) they do not undergo plastic deformation, i.e., $\sigma_y = \sigma_{UTS}$; (2) they follow Hooke's law until fracture; (3) their compressive strength is much greater than their tensile strength; and (4) their torsional strength is equal to their tensile strength. Most of the theories proposed to explain brittle fracture make the assumption that the material is perfectly elastic. A few of the more important theories explaining brittle fracture are described below.

8.3.2.1 Coulomb–Mohr Theory

The Coulomb–Mohr theory, proposed by Coffin [6], is based on tensile properties and compressive strength. It says that failure will occur for any stress state for an elastic material whenever

$$\frac{\sigma_1}{\sigma_{UTS}} + \frac{\sigma_3}{\sigma_{UCS}} \geq 1 \tag{8.6}$$

This theory was confirmed by Grassi and Cornet [7], who stressed gray cast iron tubes biaxially until failure. Note that the maximum normal stress theory produces similar results when both of the principal stresses, σ_1 and σ_2, are positive. This theory was modified by Burton [8], who produced similar results. The results are not as conservative as the Coulomb–Mohr theory in that fracture was more closely predicted. These theories are shown schematically in Figure 8.12.

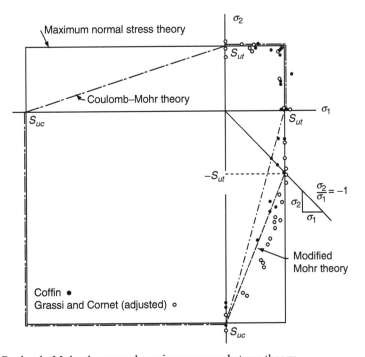

FIGURE 8.12 Coulomb–Mohr theory and maximum normal stress theory.

8.3.2.2 Griffith Microcrack Theory

The theoretical strength of a material is based on the cohesive force between atoms. If the cohesive strength is a sine curve, the following is obtained:

$$\sigma = \sigma_{max} \sin \frac{2\pi x}{\lambda} \tag{8.7}$$

where σ_{max} is the maximum theoretical cohesive strength, x is the atomic displacement due to the applied force, and λ is the wavelength of the lattice spacing. As the changes in the atomic displacement are small, $\sin x \approx x$, resulting in modification of Equation 8.7:

$$\sigma = \sigma_{max} \frac{2\pi x}{\lambda} \tag{8.8}$$

As only brittle materials are considered,

$$\sigma = E\varepsilon = \frac{Ex}{\lambda} \tag{8.9}$$

Combining Equation 8.9 and Equation 8.8 gives

$$\sigma_{max} = \frac{\lambda}{2\pi} \left(\frac{E}{a_0} \right) \tag{8.10}$$

When the part fractures, two surfaces are created. Each of these surfaces has a surface energy γ_s, so that the work done per unit area in creating the fracture surface is the area under the stress–displacement curve:

$$U_0 = \int_0^{1/2} \sigma_{max} \sin \frac{2\pi x}{\lambda} dx = \lambda \frac{\sigma_{max}}{\pi} \tag{8.11}$$

As energy is required to create two fracture surfaces,

$$\lambda = \frac{2\pi \gamma_s}{\sigma_{max}} \tag{8.12}$$

The maximum theoretical stress from atomic cohesive forces is

$$\sigma_{max} = \left(\frac{E\gamma_s}{a_0} \right)^{1/2} \tag{8.13}$$

Using typical values for the variables above and expressing results in terms of the elastic modulus E, estimates of σ_{max} vary between $E/4$ and $E/15$. Engineering steels rarely exceed fracture stresses above 300,000 psi, or $E/100$. Common construction steels typically have a fracture stress of approximately $E/1,000$. The only materials that approach the theoretical values are defect-free metallic whiskers or ceramic fibers. Therefore, this shows that small flaws or cracks are responsible for the tremendous decrease in the theoretical strength. These small flaws or cracks reduce the fracture strength due to stress concentrations [9].

Griffith [10] proposed that the difference between the theoretical strength and the strength realized in practice was due to a population of fine cracks that produce stress concentrations. These stress concentrations cause the theoretical cohesive strength to be achieved in local regions. Griffith crated the criterion that "a crack will propagate when the decrease in strain energy is at least equal to the energy required to create a new crack surface" [11]. This statement is used to establish when a flow of specific size exposed to a tensile stress will propagate in a brittle fashion.

Using the crack model in Figure 8.13, the flaw or crack is assumed to have an elliptical cross section of length $2c$. This shape is typical of many types of flaws. The thickness of the plate is very small when compared to the width and length of the plate (conditions of plane stress predominate). Using the stress concentration of this elliptical crack (determined by Inglis [12]), the maximum stress at the tip of the crack is

$$\sigma_{\max} = \sigma \left[1 + 2 \left(\frac{c}{\rho_t} \right)^{1/2} \right] \approx 2\sigma \left(\frac{c}{\rho_t} \right)^{1/2} \tag{8.14}$$

where ρ_t is the radius of the crack tip. The reduction of the elastic strain energy is equivalent to

$$U_E = \frac{-\pi c^2 \sigma^2}{E} \tag{8.15}$$

where σ the applied tensile stress normal to the crack. As energy is released by propagation of the crack, the right-hand side term is negative. The theoretical stress between the atomic planes needs to be exceeded at only one point, and the applied stress will be significantly lower than the theoretical stress. If the crack has a length of $2c$, and it is elliptical in shape and is relieved of stress in a roughly circular area of radius c, the increase in surface energy is made up for by the decrease in strain energy. The condition when elastic strain energy equals the increase in surface energy due to propagation of the crack is provided by

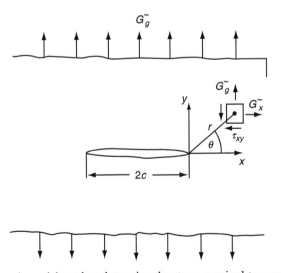

FIGURE 8.13 Griffith crack model used to determine the stress required to propagate a crack.

$$\frac{\mathrm{d}\Delta U}{\mathrm{d}c} = 0 = \frac{\mathrm{d}}{\mathrm{d}c}\left(4c\gamma_s - \frac{\pi c^2 \sigma^2}{E}\right) \tag{8.16a}$$

where

$$4\gamma_s - \frac{\pi c^2 \sigma^2}{E} = 0 \tag{8.16b}$$

This leads to the Griffith formula,

$$\sigma = \left(\frac{2E\gamma_s}{\pi c}\right)^{1/2} \tag{8.17}$$

which is the stress required to propagate a crack of size c in a brittle material. If the material is thick in relation to the crack, and plane strain conditions predominate, then the Griffith equation is

$$\sigma = \left(\frac{2E\gamma_s}{(1-v)^2\pi c}\right)^{1/2} \tag{8.18}$$

where v is Poisson's ratio.

8.4   FRACTURE MECHANICS

Metals that fail in a brittle manner experience plastic deformation before failure [13–15]. There are three types of loading, common in most engineering structures. They are shown in Figure 8.14. Mode I is the most common and is discussed here. Because of plastic deformation before fracture in metals (even when failing in a brittle manner), the Griffith microcrack theory does not apply to metals. One method of making the Griffith criterion of brittle fracture, compatible with the plastic deformation evident in metals, was suggested by Orowan [16]. He suggested the inclusion of a surface energy term due to the plasticity at the crack tip, γ_p. This results in a modification of the Griffith equation:

$$\sigma = \left(\frac{2E(\gamma_s + \gamma_p)}{\pi c}\right)^{1/2} \approx \left(\frac{2E\gamma_p}{c}\right)^{1/2} \tag{8.19}$$

The elastic surface energy term, γ_s, is neglected because $\gamma_p \gg \gamma_s$.

Irwin [17] proposed that stress at the crack tip were a function of the applied stress and the crack size. He developed the relationship:

$$K = \sigma\sqrt{\pi c} \tag{8.20}$$

where K is the stress intensity factor. K is completely defined by the crack geometry, applied stress, and specimen geometry. The value of the stress intensity factor when unstable crack growth occurs is the critical stress intensity factor, K_{Ic} (for mode I), where the value of K_{Ic} is a material property. While the above is for an elliptical flaw, other flaw shapes have also been calculated [18,19].

This assumes that plane strain conditions have been realized. If plane stress conditions are present, then the stress is relaxed by the increased plastic zone at the crack tip. Further, the

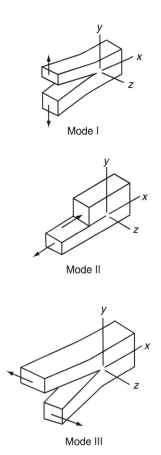

Mode I

Mode II

Mode III

FIGURE 8.14 Types of loading typically experienced by engineering materials.

state of stress is no longer triaxial and is diminished. The cases of plane stress and plane strain for modes I, II, and III are fully described by Hertzberg [20] and Rolfe and Barsom [21].

Toughness is a measure of the energy required to resist fracture in a material. Often this property is more important than the actual tensile properties, particularly if the part is to be used in a dynamic environment. The term impact strength is used to denote the toughness of the material. This term is actually a misnomer; it should really be impact energy. However, the term impact strength is so established that it makes little sense to change it. Toughness is strongly dependent on the rate of loading, temperature, and the presence of stress concentrations. Several standardized tests have been developed since World War II to measure the resistance to brittle fracture, notably the Charpy V notch test [22], the dynamic tear test [23], and the plane strain fracture toughness test [24]. Essentially, these tests are attempts to quantify the behavior of the material in service and how the material is expected to fail in service. The first two tests are discussed elaborately in the later section of this chapter.

Under impact loading, there is a limited time for uniform plastic flow to occur. Locally, the deformation may exceed the fracture stress required at the grain boundaries, geometric irregularities, or other discontinuities. Once the crack has initiated, the crack itself becomes a stress riser, propagating until it is blunted or complete failure has occurred. Fracture toughness of steels is dependent on a variety of variables that affect the mechanical properties such as test, temperature, chemistry and melt practice, strain rate, section size, notch acuity.

As these variables are changed, the transition temperature between ductile and brittle behavior may change. An excellent review of the effect of processing variables on high-strength steels was conducted by Thurston [25]. Over 100 papers were reviewed, and 77 references are provided.

8.4.1 EFFECT OF TEST TEMPERATURE

In body-centered cubic (bcc) metals, during impact loading, a transition from ductile to brittle fracture occurs that is dependent on temperature. The temperature at which it occurs is called the transition temperature. Other variables such as geometry, grain size, and alloying elements affect the ductile to brittle transition temperature, but only within a given alloy. The toughness versus temperature curve (Figure 8.15) has three basic regions: the upper shelf, the lower shelf, and the transition region. The upper shelf is characterized by primarily ductile fracture. High impact energies are associated with this regime. The lower shelf is a region where fracture is brittle; low impact energies are found in this region. The third region is the transition region, which displays a reduction in impact energies required to fracture the specimen. It is in this region that the fracture changes from ductile to brittle. Because this transition often occurs over a wide temperature range, various criteria have been developed to define the transition temperature.

One criterion of the ductile to brittle transition temperature that has common acceptance is the ductility transition temperature [26]. This standard criterion was developed because of spectacular failures of World War II Liberty ships mentioned earlier. This standard criterion established the acceptance criteria of Charpy V notch energy, C_v, greater than 15 ft lb. It was learned that this value would not result in a brittle fracture at a test temperature of $-40°F$. However, it is not applicable for many materials that have lower shelf impact energies greater than 15 ft lb when tested at $-40°F$. The transition temperature is often much lower. For example, AISI 4340 steel quenched and tempered to HRC 35 exhibits a transition temperature of approximately $-150°F$ and at $40°F$ has a C_v of 60 ft lb.

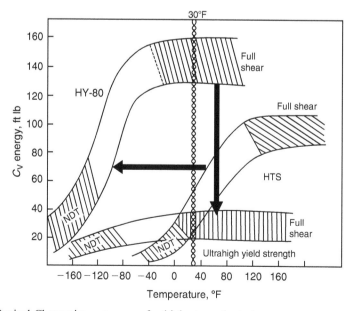

FIGURE 8.15 Typical Charpy impact curves for high-strength steels.

Another definition of the ductile to brittle transition temperature, and the one used in this chapter, is the fracture appearance transition temperature (FATT). For a given test, the ductile to brittle transition temperature is proportional to the plastic zone at the crack tip. As the test temperature is decreased, increasing amounts of cleavage occur. When the temperature is increased, increasing amounts of dimpled rupture occur. At the temperature at which there are equal amounts of cleavage and dimpled rupture, the temperature is known as the 50% FATT, T_{50}. This temperature is determined by plotting the amount of brittle or ductile fracture as a function of temperature. The amount of brittle or ductile fracture is usually decided by looking at the fracture surface either with the naked eye or with a low power (5×) magnifying glass. Ductile or shear fractures appear dull, and brittle fractures usually appear bright and shiny. A schematic example of the fracture surface of an impact test is shown in Figure 8.16.

The percentage of shear fracture in an impact test can be determined by any of several methods. The first method involves measuring the length and width of the cleavage portions of the fracture surface. The amount of ductile fracture is determined from a series of tables (assuming a standardized test geometry), depending on the type of measurement used. The amount of shear can be also measured by magnifying the fracture surface and comparing it to a calibrated overlay, or by photographing the fracture surface and measuring the ductile

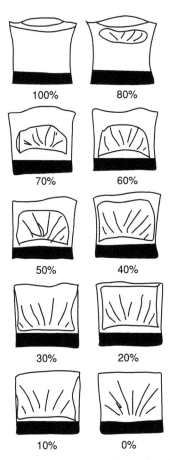

FIGURE 8.16 Schematic of Charpy impact fracture surfaces showing change in percentage of brittle fracture.

fracture by means of a planimeter. Another method compares the fracture surface of the specimen with a comparative fracture appearance feature chart. Because of the subjective nature of the measurements, the transition temperature, defined by the above methods, is not recommended for acceptance testing, material specifications, or procurement documents.

8.4.2 EFFECT OF CHEMISTRY, MELT PRACTICE, AND GRAIN SIZE

Changes in the ductile to brittle transition temperature of up to 100°F can be made by simple changes in the chemistry of the steel [27,28]. The greatest chemical changes are due to the addition of carbon (Figure 8.17) and manganese. It has been found that the transition temperature will increase by 25°F, for each additional 0.1% carbon added to the steel. Cotrell et al. [29] tested a very controlled chemistry 1% Cr–Mo steel and found that the Charpy impact strength sharply decreased as the carbon content was raised from 0.30 to 0.44% C. These findings are similar to tests performed on a 43XX steel [30]. However, for an H-11 steel, the notch strength ratio increased carbon contents up to 0.43% contrary to the results of Cotrell (Espey). Increases in the manganese content will decrease the transition temperature by 10°F for each 0.1% increase of manganese. However, these limits must be kept in mind because of hardenability constraints and the presence of retained austenite in the steel.

Other alloying additions also tend to change the ductile to brittle transition temperature in steels. Nickel decreases the transition temperature, whereas chromium has little effect. Silicon raises the transition temperature, as does phosphorus. Phosphorus in excessive amounts will also tend to form grain boundary precipitates, further increasing the transition temperature. Additions of columbium and vanadium in small amounts to the alloy decrease the transition temperature by forming carbide precipitates. These precipitates play a role in dispersion strengthening and act to pin the grain boundaries, preventing grain growth.

Sulfur greatly affects the fracture toughness of steel. Wei [31] measured the fracture toughness in a 4745 steel under controlled melting conditions at several different sulfur levels and strengths. He found that the fracture toughness decreased smoothly as the sulfur level was increased from 0.008 to 0.049%. This is shown schematically in Figure 8.18. Interestingly, in this study, silicon was found not to increase the fracture toughness.

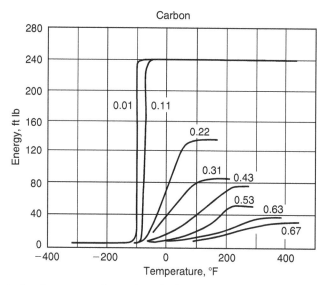

FIGURE 8.17 Effect of carbon content in steel on the impact energy and transition temperature. (After G.E. Dieter, *Mechanical Metallurgy*, McGraw-Hill, New York, 1976, p. 496.)

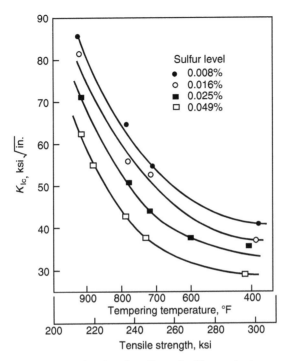

FIGURE 8.18 Charpy impact curves, showing the effect of sulfur content on an AISI 4730 steel. (After G.E. Dieter, *Mechanical Metallurgy*, McGraw-Hill, New York, 1976, p. 496.)

The oxygen content of the steel also plays an important part in determining the ductile to brittle transition temperature. Rimmed steel with a high content of oxygen shows a high transition temperature. Silicon-deoxidized (semikilled) steels have lower transition temperatures, and the oxygen is tied up as inclusions. Fully killed steels with aluminum and silicon have even lower transition temperatures.

Vacuum melting produces alloys with excellent fracture toughness. This is accomplished by reducing the gas content and the number of nonmetallic inclusions. The effect varies with the composition and the heat-treated tensile strength of the alloy. It is not always beneficial. Gilbert and Brown [32] reported an improvement of 100% in the transverse fracture strength, with the longitudinal fracture strength increasing by 50% over that of conventional air-melted 4340 steel heat-treated to 280 ksi UTS. They also found that the property directionality was almost completely removed. Cottrell [33] investigated consumable electrode vacuum melting on 3% Cr–Mo–V steel (yield strength greater than 250 ksi). The fracture strength was increased by almost 20 ksi. Different results were found with HP9-4-30 maraging steel (0.30% C, 9% Ni, and 4% Co). Vacuum remelting reduced directionality, but showed no increase in the notch strength over that of a conventional silicon- and aluminum-deoxidized heat. However, a vacuum carbon deoxidation showed significant improvements in notch strength. Tests on vacuum-remelted 4340 steel showed in excess of 100% improvement over conventional air melting when the steel was heat-treated to the 280 ksi UTS range [34]. The fracture toughness of vacuum-remelted H-11 steel was found to increase by 10% over air-melted H-11 steel heat-treated to 250 ksi ultimate strength. No effect was found when it was heat-treated to 300 ksi ultimate strength.

Grain size of the steel plays an important role in the temperature of the ductile to brittle transition. Decreasing the grain size significantly decreases the transition temperature [35].

Changes in processing that create or promote fine ferrite grain size, such as normalizing, with subsequent tempering will decrease the transition temperature. The rolling temperatures and other processing variables all play important roles in the finished grain size of the product and therefore the grain size. Careful manipulation of the process will yield a low-transition temperature.

Decarburization has a negative influence on fatigue properties and fracture toughness in high-strength steels. Warke and Elsea [36] conducted an extensive review and found decarburization to be advantageous in some materials (due to reducing the yield strength, while significantly reducing the fatigue strength). Banding and segregation effects have been found [37] to have serious detrimental effects, with cracks forming and propagating along segregation bands.

Finally, significant variation in the fracture toughness in connection with chemistry has been found [38] between heat lots and between vendors of the same alloy. Charpy V notch testing was found insensitive to these subtle changes. The K_{Ic} testing was incorporated into the procurement specifications.

8.4.3 MICROSTRUCTURE

The propagation of a crack is impeded by changes in the orientation of the cleavage plane at the grain boundary. This change in orientation is accomplished by Smallmann [39] creating cleavage steps, localized deformation, or tearing near-grain boundaries. These changes in orientation cause additional energy to be expended by the crack tip, slowing the propagation of the crack.

8.4.4 EFFECT OF STRAIN RATE

An increase in the strain rate increases the transition temperature in low-strength steels. This is because the yield point in low-strength steels is dependent on the strain rate. In addition, the transition between ductile and brittle becomes sharper as the strain rate is increased. In high-strength steels (above 200 ksi), the yield strength is not as dependent on the strain rate, primarily due to the presence of additional alloying elements.

There has not been a significant amount of literature regarding the effects of strain rate on the toughness of high-strength steels. A martensitic grade stainless, 422M, with a tensile strength of 250 ksi was tested [40], using center-cracked specimens over a range of temperatures, and it was found that the fracture strength transition temperature was not affected. However, when the material was tested at room temperature, it showed a slight decrease in strength. Marshall [41] found that decreases of up to 43% in a similar steel heat-treated to a higher strength (290 ksi UTS).

8.4.5 EFFECT OF SECTION SIZE

The measurements of transition temperature obtained in different types of tests do not coincide. This is partially because there is no single definition for the transition temperature. For example, for the Charpy V notch test, the ductile to brittle transition can be defined as the temperature at which some specific energy is absorbed or the FATT, where 50% ductile and brittle fracture appears. A ductile to brittle transition temperature occurs because the yield stress σ_y lowers as temperature is increased. The size of the plastic zone surrounding the crack tip also increases as the testing temperature increases. Because of the larger plastic zone surrounding the crack tip, there is a corresponding loss of constraint.

For a given test, the ductile to brittle transition temperature, T_{DB}, is proportional to the thickness B and the size of the plastic zone. In ferritic or martensitic steels, the transition temperature is an increasing function of temperature. To obtain a correlation between the results of the various fracture tests and the size of the plastic zone, Francois [42] defined the parameter β, where

$$B = \beta \left(\frac{K_{Ic}}{\sigma_y} \right)^2 \tag{8.21}$$

For a given thickness B and a chosen value of β, a particular temperature on the K_{Ic} transition curve can be defined as the transition temperature. If β is taken to be equal to 2.5 (the ASTM limit of validity of K_{Ic} specimen), then there will be a certain temperature where plane strain conditions predominate and the test is valid. The transition temperature is then defined in terms of both thickness B and the chosen value of β. The K_{Ic} transition curve can be approximated by various empirical formulas. Ikeda and Kihara [43] developed a good correlation of experimental data,

$$K_{Ic} = A e^{T/T_0} \tag{8.22}$$

Eliminating the fracture toughness between Equation 8.22 and Equation 8.21 leads to a formulation for the transition temperature:

$$T_{DB} = \frac{T_0}{2} \ln \left[\frac{B}{\beta} \left(\frac{\sigma_y}{A} \right)^2 \right] \tag{8.23}$$

This approach has its limitations: it does not make sense that the transition temperature, T_{DB}, disappears when the thickness B becomes smaller or that the transition temperature increases to infinity when B becomes very big. Based on these assertions, the validity of this equation is limited to thicknesses between 5 and 500 mm.

Using the 50% FATT as the basis for the measurement of the ductile to brittle transition temperature, Francois found that the dynamic tear test has a $\beta = 0.6$. For the Charpy V notch test (discussed later), a lower value of $\beta = 0.4$ was determined. This is expected because of the smaller size of the Charpy V notch specimen. Francois showed a reasonable correlation between the observed transition temperatures and those calculated from Equation 8.23. However, the scatter is large, yielding a variation of $\pm 20°C$.

The transition temperatures of different tests are related to the logarithm of the thickness ratio by the expression

$$T_{DB_1} - T_{DB_2} = \frac{T_0}{2} \ln \left(\frac{\beta_1}{\beta_2} \right) \tag{8.24}$$

The transition temperature for a particular test and section thickness can be obtained if the transition temperature has been learned already for two different thicknesses, B_1 and B_2:

$$\frac{T_{DB} - T_{DB_1}}{T_{DB_2} - T_{DB_1}} \frac{\ln(\beta/\beta_1)}{\ln(\beta_2/\beta_1)} = 1 \tag{8.25}$$

For a given steel, the transition temperature of one test can be used to determine the transition temperature of a different test (e.g., Charpy V notch and dynamic tear test) by using

$$T_{DB_1} - T_{DB_2} = \frac{T_0}{2} \ln\left[\frac{B_1\beta_2}{B_2\beta_1}\right] \tag{8.26}$$

If the transition temperature from K_{Ic} testing is unknown, the transition temperature for plane strain fracture could be estimated from knowledge of the transition temperature from other tests, using

$$\frac{T_{K_{Ic}} - T_{t_1}}{T_{t_2} - T_{t_1}} = \frac{\ln(B\beta_1/B\beta)}{\ln(B_2\beta_1/B_1\beta_2)} \tag{8.27}$$

In this case, $\beta_1 = 0.4$ (Charpy) and $\beta_2 = 0.6$ (dynamic tear). This method is very useful in determining K_{Ic} values and critical flaw sizes that, because of size limitations, might preclude K_{Ic} testing (such as a field failure).

Other approaches have been taken to correlate simpler impact tests to K_{Ic}. Hertzberg [20] and Rolfe and Barsom [21] showed that for steel exhibiting 100% brittle fracture, the fracture toughness K_{Ic} can be estimated from the yield strength at the testing temperature:

$$K_{Ic_{est}} = 0.45\sigma_{ys} \tag{8.28}$$

where K_{Ic} is in the customary units of $\sqrt{in.}$ and σ_{ys} is the yield strength at the testing temperature. If the yield strength is not available at the testing temperature, room temperature properties can be used with conservative results. For materials showing 100% ductile failure, the following expression can be used:

$$\left(\frac{K_{Ic}}{\sigma_{ys}}\right)^2 = 5\left(\frac{CVN}{\sigma_{ys}}\right) - 0.05 \tag{8.29}$$

where CVN is the Charpy V notch impact energy (ft lb).

This correlation is useful for learning the effects of different variables such as thickness, microstructure, and transition temperatures from different kinds of tests or for determining the K_{Ic} transition curve from different tests. It should not be used for procurement specifications.

8.5 FATIGUE

Parts are subjected to varying stresses during service. These stresses are often in the form of repeated or cyclic loading. After enough applications of load or stress, the components fail at stresses significantly less than their yield strength. Fatigue is a measure of the decrease in resistance to repeated stresses.

Fatigue failures are brittle appearing, with no gross deformation. The fracture surface is usually normal to the main principal tensile stress. Fatigue failures are recognized by the appearance of a smooth rubbed type of surface, generally in a semicircular pattern. The progress of the fracture (and crack propagation) is generally suggested by beach marks. The initiation site of fatigue failures is generally at some sort of stress concentration site or stress riser. Typical fracture appearance is shown schematically in Figure 8.8.

Three factors are necessary for fatigue to occur. First, the stress must be high enough that a crack is initiated. Second, the variation in the stress application must be large enough that the crack can propagate. Third, the number of stress applications must be sufficiently large that the crack can propagate a significant distance. The fatigue life of a component is affected by a number of variables, including stress concentration, corrosion, temperature, microstructure, residual stresses, and combined stresses.

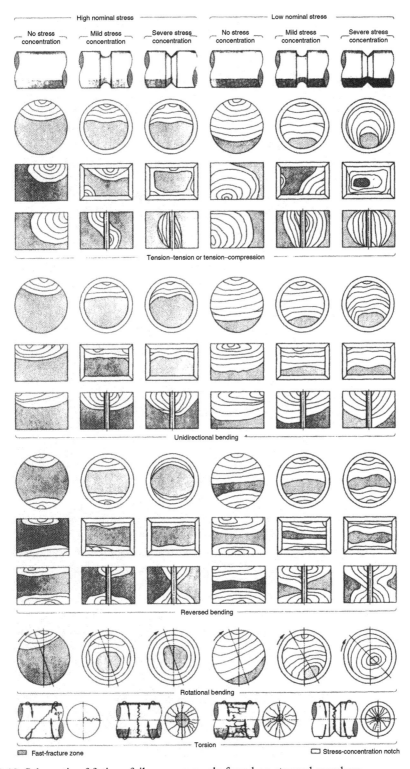

FIGURE 8.19 Schematic of fatigue failures commonly found on stressed members.

8.5.1 FATIGUE MECHANISMS

The structural features of fatigue failures are generally divided into four distinct areas [44]:

1. Crack initiation, the early development of fatigue damage.
2. Slip band crack growth, the early stages of crack propagation. This is often called stage I crack growth.
3. Stable crack growth, which is usually normal to the applied tensile stress. This is called stage II crack growth.
4. Unstable crack growth with final failure from overload. This is called stage III crack growth.

Fatigue usually occurs at a free surface, with the initial features of stage I growth, fatigue cracks, initiated at slip-band extrusions and intrusions [45,46]. Cottrel and Hull [47] proposed a mechanism for the formation of these extrusions and intrusions (shown schematically in Figure 8.20) that depends on the presence of slip, with slip systems at 45° angles to each other, operating sequentially on loading and unloading. Wood [48] suggested that the formation of the intrusions and extrusions was the result of fine slip and buildup of notches (Figure 8.21). The notch created on a microscopic scale would be the initiation site of stable fatigue crack growth.

In stage II, stable fatigue crack growth, striations (Figure 8.22) often show the successive position of the crack front at each cycle of stress. Fatigue striations are usually detected using electron microscopy and are visual evidence that fatigue occurred. The absence of fatigue striations does not preclude the occurrence of fatigue, however.

Striations are formed by a plastic-blunting process [49]. At the end of the stage I crack tip, there exist sharp notches due to the presence of slip. These sharp notches cause stress to be concentrated at the crack tip. The application of a tensile load opens the crack along slip planes by plastic shearing, eventually blunting the crack tip. When the load is released, the slip direction reverses, and the crack tip is compressed and sharpened. This provides a sharp notch at the new crack tip where propagation can occur. This is shown schematically in Figure 8.23.

An alternative hypothesis on striation formation was presented by Forsyth and Ryder [50]. In their model, the triaxial stress state at the crack tip forms a dimple ahead of the crack front. The material between the crack tip and the dimple contracts eventually ruptures, forming a fatigue striation. This is shown schematically in Figure 8.24.

In mild steel, striations are observed, but not as well-defined or as spectacular manner as in aluminum. This was first assumed to be due to the crystal lattice structure, as face-centered cubic austenic steels show well-defined striations, and mild steels (base-structured cubic) do not [51]. Other alloys, such as titanium alloys [52], with a hexagonal close-packed (hcp) crystal structure show very defined striations. However, β-titanium alloys (bcc) [53] show strongly

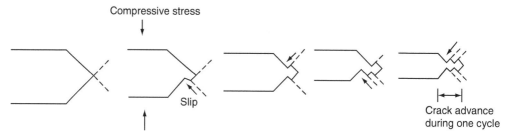

FIGURE 8.20 Schematic representation of the mechanism of fatigue intrusions and extrusions. (After A.H. Cottrel and D. Hull, *Proc. R. Soc. London 242A*:211, 1953.)

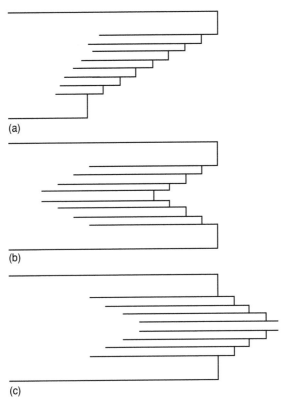

(a)

(b)

(c)

FIGURE 8.21 Mechanism of intrusions and extrusions. (After W.A. Wood, *Bull. Inst. Met. 3*:5, 1955.)

FIGURE 8.22 Typical fatigue striations, these in 7075 aluminum. Striations in steels, particularly high-strength steels, are difficult to observe.

defined striations. Therefore, attributing defined striations to crystal lattice alone was discounted as a viable theory.

Deformation and available slip systems were presumed to be more significant [54]. However, this does not allow, because mild carbon steels are more ductile than austenitic steels. It is now generally accepted that fatigue striations form by the plastic-blunting process.

8.5.2 DESIGN FOR FATIGUE

Fatigue is caused by a series of loading and unloading, in a variety of waveforms. The load application can be sinusoidal, triangular (sawtooth waveform), or spectral (random) loading. The range of stress application, σ_r, is suggested by

$$\sigma_r = \sigma_{max} - \sigma_{min} \tag{8.30}$$

and the alternating stress, σ_a, is represented by

$$\sigma_a = \frac{\sigma_r}{2} \tag{8.31}$$

The mean stress, σ_m, is calculated as

$$\sigma_m = \frac{\sigma_{max} + \sigma_{min}}{2} \tag{8.32}$$

When representing the fatigue data, two quantities, the stress ratio R and the stress amplitude A, completely describe the stress applied. The stress ratio is given by

$$R = \frac{\sigma_{max}}{\sigma_{min}} \tag{8.33}$$

and the amplitude is

$$A = \frac{\sigma_a}{\sigma_m} \tag{8.34}$$

Engineering fatigue data are usually displayed graphically in an $S–N$ curve (Figure 8.25), where the applied stress σ (or S') is plotted against logarithm of the number of cycles. As the stress is increased, the number of cycles until failure decreases. For steels, the $S–N$ curve becomes horizontal at some low-stress level. This is called the fatigue limit. Aluminum alloys do not show a fatigue limit, and failure occurs at some extended number of cycles at low-applied stress.

For steels, the fatigue limit σ_c depends on the UTS [55] in a rotating, reversed fatigue test:

$$\sigma_e = \frac{\sigma_{UTS}}{2} \tag{8.35}$$

In other loading types, the endurance limit or alternating axial loading (completely reversed) is [56]

$$\sigma_e = 0.35\sigma_{UTS} \tag{8.36}$$

Mechanical property testing methods

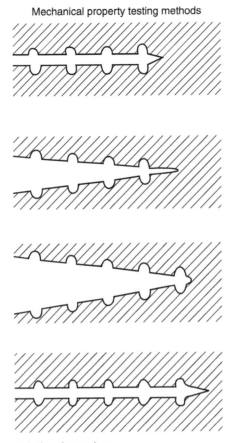

FIGURE 8.23 Mechanism for striation formation.

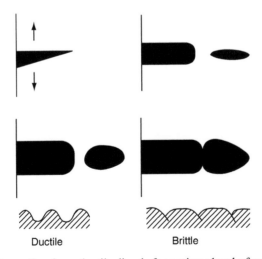

Ductile   Brittle

FIGURE 8.24 Striation formation from ductile dimple formation ahead of crack front.

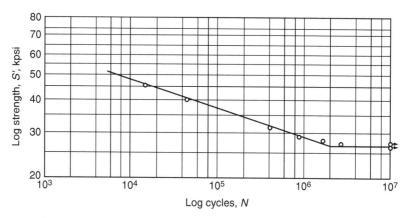

FIGURE 8.25 Typical presentation of fatigue data in an *S–N* diagram. Data shown are for an annealed AISI 1040 steel.

and for reversed torsional testing, the torsional endurance limit is

$$\tau_e = 0.3\sigma_{UTS} \tag{8.37}$$

The fatigue limit and fatigue tests show considerable variation [57–59]. Because of this variability, the fatigue life must be examined in terms of probability. This requires testing of many samples so that the mean and standard deviation can be determined. In one study [60], 200 steel specimens were tested, and it was found that fatigue life followed a Gaussian distribution when it was examined using a log scale. The fatigue endurance limit was examined by Ransom [61]. In this study, ten *S–N* curves were developed using ten specimens of each, for a total of 100 fatigue specimens. The specimens were prepared identically and were from the same bar of material. Ransom found that there was a 20% scatter around the fatigue limit.

When designing for cyclic loading, it is necessary to allow for the factors that influence fatigue. One such method was proposed by Shigley [62]. In this method, correction terms for each of the important deleterious effects are multiplied together to obtain a maximum design stress, σ'_e, for the machine element. This maximum design stress is obtained from data representing a smooth polished rotating beam specimen at the endurance limit, σ_e, as

$$\sigma'_e = k_a k_b k_c k_d k_e K k_n \sigma_e \tag{8.38}$$

The correction terms k_i are related to the stress concentration factor by

$$k_i = \frac{1}{K_i} \tag{8.39}$$

where K_i is the stress concentration.

8.5.2.1 Surface Treatments

As fatigue failures usually begin at the surface, the surface condition is very important. Surface roughness is a key factor influencing fatigue. Highly polished specimens exhibit the longest fatigue life, with increasingly rougher surfaces yielding decreased fatigue life. Figure 8.26 shows

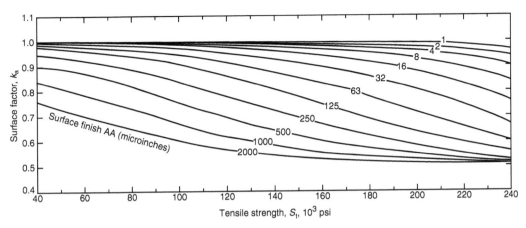

FIGURE 8.26 Fatigue stress concentration due to surface finish and surface roughness. (After J.H. Faupel and F.E. Fisher, *Engineering Design*, Wiley-Interscience, New York, 1981.)

the correction factor for surface roughness as a function of strength. As the measurement of surface roughness is statistical in nature, extremes of machine grooves can occur. Typical depths of surface grooves from machining [63] are shown in Table 8.2. These extremes, as well as the stress concentrations they create, must be considered for fatigue.

The surface treatment of a steel must be accounted for in determining its fatigue life. Electroplating is considered detrimental to fatigue life. Chrome plating has been found [64] to reduce the fatigue strength according to the formula

$$k = 1 + Y \tag{8.40}$$

where Y is the correction factor, which can be given as

$$Y = 0.3667 - 9.193 \times 10^{-3} \sigma_e \tag{8.41}$$

If the part is nickel-plated, the correction factor $Y = -0.01$ for 1008 steel and $Y = -0.33$ for 1063 steel [65]. If shot peening is carried out after nickel or chromium plating, then the fatigue strength is often increased beyond that of the base metal [66]. Cadmium plating is not considered to affect the fatigue life of steels because it is very soft and ductile. The plating conditions can greatly affect the fatigue properties of a steel. A comprehensive review of the effect of plating on fatigue life is given by Hammond and Williams [67].

TABLE 8.2
Machine Grooves Commonly Found on Machined Surfaces

Method	Surface Roughness (rms)	Groove Depth (in. $\times 10^{-3}$)
Polished	8	0.04
Fine grind	10	0.08
Rough grind	70	0.2–0.4
Fine turn	70–90	0.4–0.8
Rough turn	90–500	0.8–2

Source: From J.H. Faupel and F.E. Fisher, *Engineering Design*, Wiley-Interscience, New York, 1981.

TABLE 8.3
Values of Y from Surface Hardening

	Flame and Induction Hardening	Carburizing	Nitriding
Layer thickness	0.125–0.500 (induction) ≈ 0.125 (flame)	0.03–0.1	0.004–0.02
Steel	0.66–0.80	0.62–0.85	
Alloy steel	0.06–0.64	0.02–0.36	0.30–1.00

Source: From J.H. Faupel and F.E. Fisher, *Engineering Design*, Wiley-Interscience, New York, 1981.

Surface treatments of steel are not limited to electroplating. Other surface treatments such as carburizing and nitriding are used to impact surface hardness and improve wear characteristics [68]. Improvements in fatigue life are also obtained. These effects are due to the increased hardness of the surface (with increased endurance limit) and the creation of a favorable residual stress at the surface. Flame hardening and induction hardening have similar effects. Using Equation 8.40 as a basis for determining the surface improvements, values of the correction factor Y for surface treatments have been determined [69]; these are shown in Table 8.3. Because of the compressive residual stress fields at the surface, surface-hardened steels generally do not initiate fatigue at the surface but at the case–core interface.

8.5.2.2 Residual Stresses

As indicated above, residual stresses at the surface can improve or decrease the fatigue life of a steel. Residual stresses occur when a part is plastically deformed nonuniformly over its cross section. If a part is plastically deformed at only the surface (by rolling or peening), a compressive surface stress results. This compressive stress constrains the interior of the part, placing it under a tensile stress. These stresses are superimposed on the applied stress field, reducing the total stress. Unfortunately, a uniform compressive stress field is difficult to produce on a complex geometry and difficult to measure. It is measured using strain gauges on x-ray diffraction or by noting part displacement when material is removed.

Typically, shot peening is used to impart a compressive residual stress at the surface. Again, using Equation 8.40, the correction factor improvement in the fatigue life from shot peening is shown in Table 8.4. Surface rolling creates a thicker compressive residual stress layer (0.040–0.5 in.) and therefore creates a larger improvement in the fatigue life. For steel shafts, $Y \approx 0.2$–0.8, while for polished steel parts, $Y \approx 0.06$–0.5 (Faupel). This is further

TABLE 8.4
Value of Y from Shot Peening for Steels

Surface	Y
Polished	0.04–0.22
Machined	0.25
Rolled	0.25–0.5
Forged	1–2

Source: From J.H. Faupel and F.E. Fisher, *Engineering Design*, Wiley-Interscience, New York, 1981.

discussed by Lipson and Juvinall [70]. It has been shown [71] that grinding produces tensile residual stresses at the surface.

8.5.2.3   Size Effect

In fatigue, the larger the part tested, the lower the fatigue limit. Fatigue starts at the surface, and as part diameter increases the surface area also increases, providing more sites for fatigue initiation. Also because of the larger diameter, the stress gradient across the plastic zone at the crack tip decreases, making it difficult to model larger structures on small specimens. A correction for fatigue in larger parts is provided by Juvinall [72]. For parts with diameters less than or equal to 0.030 in., the correction factor is $k = 1$. If the part diameter is greater than 0.3 in. but less than 2 in., then $k = 0.85$. For parts larger than 2 in., in diameter, the decrement in fatigue strength is [73]:

$$k_b = 1 - \frac{d - 0.03}{15} \tag{8.42}$$

It has also been found [74] that the thicker the test piece, the faster the crack propagation rate. It is likely that the propagation rates for thicker pieces are due to increased plane strain conditions, with a small plastic zone at the crack tip. Because for a small plastic zone there is a greater stress gradient, a faster crack propagation rate might be expected. Also, in thicker panels there is a higher state of triaxial stress, which would also tend to increase crack growth rates.

8.5.2.4   Stress Concentrations

The fatigue life of a component is seriously affected by the presence of stress concentrations. Often, it is found that fatigue initiates at the site of a stress concentration. Stress concentrations are determined from geometry effects and are either calculated from elasticity theory [75] or with numerical methods or determined experimentally using stress analysis techniques. A very comprehensive set of stress concentration factors has been collected by Peterson [76] and Young [77]. Besides the stress concentration, the notch sensitivity of the material is necessary to calculate the effect of the stress concentration. The notch sensitivity of a material is related to the stress concentration by

$$q = \frac{K_f - 1}{K_t - 1} \tag{8.43}$$

where q is the notch sensitivity, K_t is the theoretical stress concentration, and K_f is the stress concentration for fatigue. This is related to the correction factor k for the fatigue endurance limit by

$$K_f = 1 + q(K_t - 1) \tag{8.44}$$

or

$$k = \frac{1}{K_f} \tag{8.45}$$

The notch sensitivity, q, of a material has been determined for many materials [78], and some data are plotted in Figure 8.27. The notch sensitivity varies with the type of notch and notch sensitivity. The notch sensitivity increases with increasing tensile strength.

An alternative approach was taken by Neuber [79], who suggested that the stress concentration for fatigue be expressed as

$$K_f = 1 + \frac{K_t - 1}{1 + (\pi/(\pi - \omega))(a/r)^{1/2}} \qquad (8.46)$$

where r is the root radius of the notch, ω is the angle of the notch, and a is the elementary block size. The elementary block size idea has no real basis but is a convenient empirical tool.

The above is one approach for determining the fatigue life of a component. Fracture mechanics has also been used to study crack propagation [80,81]. The crack propagation rate in a material follows an equation of the form

$$\frac{da}{dN} - C\sigma_a^m a^N \qquad (8.47)$$

where C is a constant based on the geometry of the crack and test, σ_a is the alternating stress, a is the crack length, m is a material constant found during testing, and N is the number of cycles to failure. The growth of a crack can be expressed also as a function of the stress intensity factor K from fracture mechanics,

$$\frac{da}{dN} - C(\Delta K)^n \qquad (8.48)$$

where C and n are material constants found during testing. For martensitic steels, it has been found [82] that

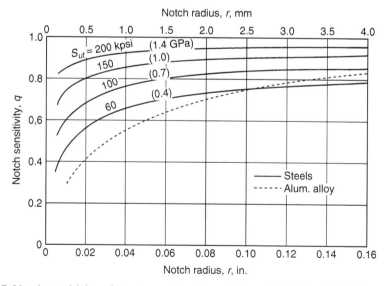

FIGURE 8.27 Notch sensitivity of steels as a function of strength. (After G.E. Dieter, *Mechanical Metallurgy*, McGraw-Hill, New York, 1976, p. 496.)

$$\frac{da}{dN} - 0.66 \times 10^{-8} \Delta K^{2.25} \tag{8.49}$$

and for ferritic–pearlitic steels

$$\frac{da}{dN} = 3.6 \times 10^{-10} \Delta K^{3.0} \tag{8.50}$$

Equation 8.48 is illustrated in Figure 8.28. This has been found to be an extremely useful concept in high-strength steels.

While the above equation provides information regarding how a crack propagates in a solid, it can also be used to predict fatigue life. Assume a solid, with some elliptical flaw. The critical size of the flaw is given by Broek [83] as

$$a_{cr} = \frac{K^2 Q}{1.21 \pi \sigma^2} \tag{8.51}$$

where a_{cr} is the critical flaw depth, K is the critical stress intensity, and σ is the applied stress. The growth rate of a crack is expressed in Equation 8.48 and in its simplest form as

$$\frac{da}{dN} = CK_1^n \tag{8.52}$$

where da/dN is the growth of a flaw per cycle, a is the depth of the flaw, N is the number of cycles sustained, and K_1 is the stress range,

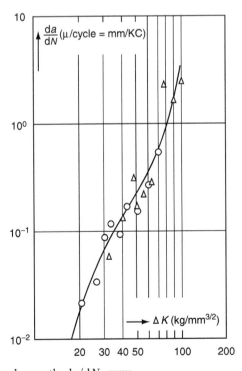

FIGURE 8.28 Schematic crack growth, da/dN, curve.

TABLE 8.5
Comparison of Notch Analysis and Fracture Mechanics in Fatigue Analysis

	Fracture Mechanics	Notch Analysis
Fundamentals	Elastic energy Equivalent stress Concentration	Stress concentration
Applied theory	Linear elastic theory	Linear elastic theory, elastic–plastic behavior
Principles	$\sigma\sqrt{a} = $ constant Correction for plastic zone at crack tip	Size effect

Source: From G. Jacoby, *Application of Microfractography to the Study of Crack Propagation Under Fatigue Stress*, AGARD Rep. 541, NATO, 1966.

$$K_1 = \sigma\left(\frac{1.21\pi a}{Q}\right)^{1/2} \tag{8.53}$$

The stress range is obtained using specially designed specimens related to a specific geometry [84–86]. The material constants, c and n, are obtained through fatigue crack growth testing. The value of the flaw shape factor is obtained from the references cited above as well as others [87–89]. In this instance, the stress range is for an elliptical flaw in the inside diameter of a tube. By combining Equation 8.52 and Equation 8.53 and rearranging, the following equation is obtained:

$$dN = \frac{da}{c\left[\sigma(1.21\pi a/Q)^{1/2}\right]^n} \tag{8.54}$$

The cycle life is obtained by integrating Equation 8.54 between the limits of the initial flaw depth a_i and the critical flaw depth a_{cr}:

$$N = \int_{a_i}^{a_{cr}} da\left\{c\left[\sigma\left(\frac{1.21\pi a}{q}\right)^{1/2}\right]^n\right\}^{-1} \tag{8.55}$$

The initial flaw depth a_i could be either the detectable flaw size from nondestructive testing or an assumed flaw size.

There are very important differences between the two fatigue prediction methods described above. Each has found a home according to the industry served. For the most part, the aerospace industry, because of its use of ultrahigh-strength steels, relies on the fracture mechanism approach, while most other industries tend to use the notch analysis approach. A brief comparison of the two methods is presented in Table 8.5 [90]. A more complete review of the differences is found in Ref. [91].

8.6   CREEP AND STRESS RUPTURE

The effects of elevated temperature on mechanical properties and material behavior are commonplace in everyday living. Examples include pipes bursting in the middle of winter, the expansion of a bridge in the middle of summer, and the sagging of a fireplace grate. Each of these examples is an indication that properties change with temperature. In addition, the discussion above indicated that steels become more brittle as the temperature is decreased. There are many other effects of temperature that have been cited [92]. Even the concept of the elevated temperature is relative [93]. What is considered hot for one material may be considered cold for another; for instance, gallium has a melting point of 30°C and tungsten has a melting point of about 3400°C.

Creep is the continuous deformation of a material as a function of time and temperature. This topic is treated very thoroughly by Finnie and Heller [94]. The creep of a material is shown in Figure 8.29. It can be seen from the figure that creep in a material occurs in three stages:

1. Stage I, where a rapid creep rate is seen at the onset of load application, then gradually decreasing
2. Stage II, where creep remains at a steady-state rate
3. State III, where the creep rate shows an increasing rate until failure occurs

The behavior and the creep rate are sensitive to the temperature to which the material is exposed, the surrounding atmosphere, and the prior strain history. Andrade and Chalmers [95] were pioneers in the study of creep and proposed that creep followed the equation:

$$\varepsilon = \varepsilon_0(1 + \beta t^{1/3})e^{Kt} \tag{8.56}$$

where β and K are material constants that can be evaluated by several different methods [96]. A better fit for the creep of materials was proposed by Garofalo [97]. He indicated that

$$\varepsilon = \varepsilon_0 + \varepsilon_t(1 - e^{-rt}) + \frac{d\varepsilon}{dt}t \tag{8.57}$$

where $d\varepsilon/dt$ is the steady-state creep rate, ε_0 is the strain on loading, r is the ratio of the transient creep rate to the transient creep strain, and ε_t is the transient creep strain.

Very early it was recognized that fractures at elevated temperatures occurred intergranularly [98]. In stage III creep, intergranular wedge cracks and cavities form. Wedge-shaped cracks and creep cavities usually initiate at or near-grain boundary triple points and propagate

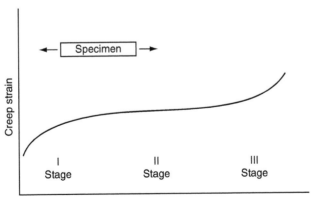

FIGURE 8.29 Creep of a material showing the stages of creep. (After G.E. Dieter, *Mechanical Metallurgy*, McGraw-Hill, New York, 1976, p. 496.)

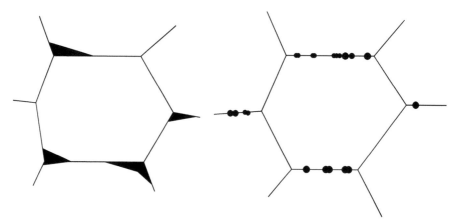

FIGURE 8.30 Creep cavities and wedges forming at grain boundaries.

along grain boundaries normal to the applied tensile stress. Creep cavities form at higher temperatures and lower working stresses. These structural features are shown in Figure 8.30.

Creep testing is usually performed for 1,000–10,000 h with strains of up to 0.5%. Stress rupture testing, or testing to failure, uses much higher loads and temperatures, and the test is usually terminated after 1000 h. In stress rupture testing, the time to failure is measured at a constant stress and constant temperature. This test has gained acceptance for elevated temperature testing of turbine blade materials in jet engines.

Using a tensile machine and high-temperature furnace (Figure 8.31), the strain is measured in creep testing by special extensometers suited for elevated temperatures. In stress rupture testing, simple apparatus such as dial calipers are used, because only the overall strain at constant time and temperature are needed.

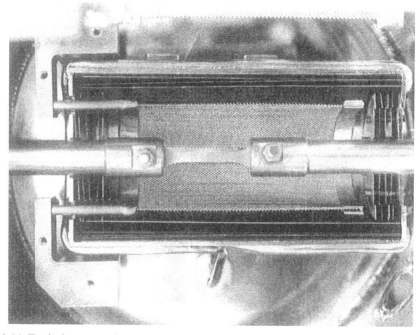

FIGURE 8.31 Typical creep testing apparatus.

8.7 TENSILE TESTING

Tensile testing is the most generally useful of all mechanical tests. Both the strength and a measure of ductility are obtained. In this test, a specimen is loaded axially and the load is increased continuously until failure occurs. The load and elongation are continuously plotted during testing and then converted to engineering stress σ,

$$\sigma = \frac{P}{A_0} \tag{8.58}$$

and engineering strain e,

$$e = \frac{\delta}{L_0} \tag{8.59}$$

where P is the load applied, A_0 is the original cross-sectional area, δ is the measured elongation, and L_0 is the original gauge length. The resulting data are plotted as engineering stress and strain, schematically shown in Figure 8.32. For a rigorous mathematical treatment of the tensile test and the changes that occur, the reader is referred to Nadai [99].

At the beginning of the test, there is a linear region where Hooke's law is followed. The slope of this linear region is the elastic modulus E (for steels, the elastic modulus is about 30×10^6 psi). The yield stress, σ_{ys}, which is the limit of elastic behavior, is defined as the point at which a small amount of permanent deformation occurs. This deformation is defined in the United States as a strain of 0.002 in./in. or 0.2% strain [100]. Loading past this point causes plastic deformation. As the plastic deformation increases, strain hardening occurs, making the material stronger. Eventually, the load reaches a maximum value and failure occurs. The UTS in σ_{UTS} is determined by taking the maximum load experienced and dividing it by the original area:

$$\sigma_{UTS} = \frac{P_{max}}{A_0} \tag{8.60}$$

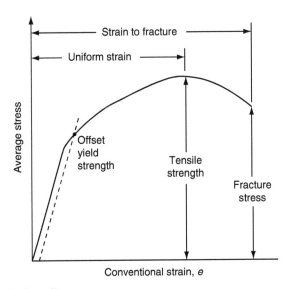

FIGURE 8.32 Schematic of tensile test.

In ductile materials, the strain may increase after the maximum load is reached and the applied load is decreased. At this point is the onset of necking.

The ultimate strength is the most quoted property but is the least useful of all the properties determined from the tensile test. In ductile materials, ultimate strength is a measure of the maximum load that the material can experience in uniaxial loading. However, in brittle materials, the UTS is valid design information. The UTS used to be the basis for many designs and design codes (with a margin of safety), but now design codes rely on the yield strength instead. Because of its reproducibility, the UTS is often used for procurement, specifications, and quality documents.

The yield strength, σ_{ys}, is the stress required to obtain a small (0.2%) permanent strain. In other words, if the specimen were unloaded at the yield stress, it would be 0.2% longer than original strain. The yield strength is now the basis for many design codes and is used extensively in the determination of many other properties such as fracture toughness and fatigue strength.

Ductility in the tensile specimen is measured and reported in two different ways: in terms of elongation,

$$\%E_f = \frac{L_f - L_0}{L_0} \times 100 \tag{8.61}$$

and in terms of the reduction in area,

$$\%RA = \frac{A_0 - A_f}{A_0} \times 100 \tag{8.62}$$

where L_f is the final gauge length after fracture, L_0 is the original gauge length, A_0 is the original cross-sectional area, and A_f is the final cross-sectional area after fracture. As the final strain will be concentrated in the necked region, the elongation is dependent on the gauge length L_0. For that reason, in reporting elongation, the gauge length is always provided. The change in elongation as a function of gauge length is provided in Figure 8.33.

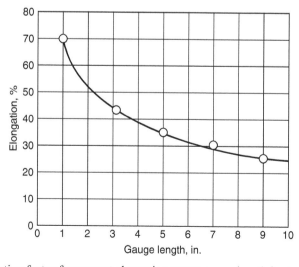

FIGURE 8.33 Correction factor for percent elongation versus gauge length in steels.

FIGURE 8.34 Screw-type tensile machine.

Testing machines used for tensile are generally simple. Either screw-type (Figure 8.34) or hydraulic (Figure 8.35) machines are used. The load is measured by a load cell, composed of strain gauges or a linear velocity displacement transducer (LVDT). This is recorded by a chart recorder as a plot of load versus strain.

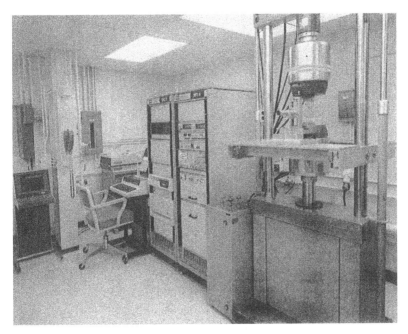

FIGURE 8.35 Hydraulic tensile machine.

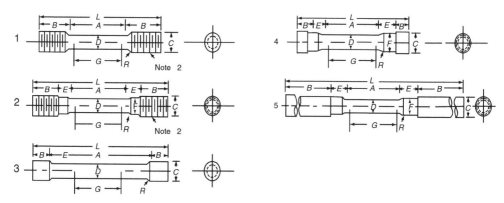

FIGURE 8.36 Typical tensile specimens per ASTM E8.

The strain is measured by extensometers, which measure *in situ* the strain experienced by the tensile specimen. Modern extensometers use strain gauges to provide strain measurements. This is an advantage because it sends an electric signal proportional to the experienced strain to the chart recorder.

The shapes of tensile specimens are standardized by ASTM. Their configurations are shown in Figure 8.36, and their dimensions are given in Table 8.6. Care in fabricating the tensile specimen is necessary to preclude faulty or inaccurate results. Heating and cold working of the specimens must be minimized during fabrication, or inaccurate yield and ultimate stress values may result. The specimens must be straight and flat; otherwise, a distortion of the elastic region and an inaccurate yield strength may result. During machining, the tensile specimens must be symmetrical about the load axis. If not, bending may occur, leading to erroneous results because of the combined stresses of tension and bending.

The tensile specimen is held during testing by either hydraulic or mechanical grips. Both methods grip the specimen by wedges with serrated surfaces. In mechanical grips, the load applied to the specimen forces the wedges tighter against the specimen by inclined surfaces inside the grip. With hydraulic wedges, hydraulic pressure is used to force the wedges against the specimen and hold it in place. For high-strength steels or steels that have hard or highly polished surfaces, hydraulic grips are preferred to prevent the specimen from slipping in the grip.

TABLE 8.6
Dimensions of Full-Size Tensile Specimens per ASTM E8

	Specimen Dimensions (mm)				
	1	2	3	4	5
G, gauge length	50	50	50	50	50
D, diameter	12.5	12.5	12.5	12.5	12.5
R, fillet radius	10	10	2	10	10
A, length of reduced section	60	60	100	60	60
L, overall length	125	140	140	120	240
B, length of end section	35	25	20	13	75
C, diameter of end section	20	20	18	22	20
E, length of shoulder	—	16	—	20	16
F, diameter of shoulder	—	16	—	16	15

The tensile properties of annealed and normalized steels are dependent on the flow characteristics of ferrite. The strength of ferrite is a function of the alloying elements in solid solution in it and the ferrite grain size [101]. The percent carbon in the steel has a strong effect because it influences the amount of cementite (Fe_3C) present as pearlite or spherodite. As a general rule, the strength increases and the ductility decreases as the percent carbon is increased.

Normalized steels tend to show higher strengths than annealed steels, as a finer pearlite spacing results from the faster cooling rate in normalizing. Expressions for the UTS, yield strength, and reduction in area have been published [102] and are reproduced here:

$$\sigma_{UTS} \text{ [ksi]} = 42.8 + 4(\%Mn) + 12(\%Si) + 5.64 V_p + 0.224 d^{-1/2} \qquad (8.63)$$

$$\sigma_{ys} \text{ [ksi]} = 15 + 4.73(\%Mn) + 12.2(\%Si) + 0.55^{-1/2} \qquad (8.64)$$

$$\%RA - 78.5 + 5.39(\%Mn) - 0.53 V_p - 8.399 d \qquad (8.65)$$

where alloying elements are expressed as weight percent, V_p is the volume percent pearlite, and d is the grain intercept in inches.

Quenched and tempered steels, heat-treated to tempered martensite, offer the best balance of strength and ductility. For a given composition, the properties are altered by changing the tempering temperature. Composition plays a minor role, with the function of alloying elements to increase the hardenability of the steel or to increase the depth to which a fully hardened structure can be obtained. Relationships for the strength of steel as a function of quench rate, tempering temperature, and composition are published elsewhere [103,104].

8.8   HARDNESS TESTING

Hardness testing is probably the most common type of mechanical test performed in the United States, perhaps the most common worldwide. The term hardness is poorly defined and is relative to the measuring device. There are three basic types of hardness tests: the scratch test, the indenter, and the dynamic rebound test. The scratch test is familiar to mineralogists, who use the Mohs scale.

The Mohs scale is the relative hardness of ten minerals arranged in order. Talc is the softest (Mohs 1), and diamond is the hardest (Mohs 10). Most hard metals fall in the range of Mohs 4–8. There is inadequate differentiation along the scale to be of much use to a metallurgist. In dynamic tests, an indenter is dropped on the material, and hardness is defined as the energy of impact. The Mohs scale is commonly used for rubbers and polymers. One exception is the Shore sceleroscope, which is used for metals. The indenter type of hardness test is the most widely accepted for metals.

In the indenter type of hardness test, an indenter is pressed into the material and released, and either the diameter or the depth of the impression is measured. The load and the impression measurement determine the hardness. As a hardness impression is made, there is a plastic zone around the hardness indentation that is surrounded by undisturbed elastic material. This elastic zone hinders plastic flow. As the plastic region is constrained by the elastic region, the compressive strength of the material in the area of the hardness impression is higher than the value of simple compression. This is a classic problem in plasticity and should be able to be explained by slip line theory. The load required to indent a specific distance δ by a punch is given by Hill [105] as

$$P = \frac{4a\sigma_{ys}}{\sqrt{3}} \left[1 + \frac{\pi}{2} - \left(\frac{\delta}{2R} \right)^{1/2} \right]$$ (8.66)

where R is the radius of the punch, P is the applied load, and a is the resulting radius after punching. Nadai [99] determined the pressure to indent, using slip line theory, and found

$$p = \frac{2\sigma_{ys}}{\sqrt{3}} \left[\frac{1}{2} + \frac{\pi}{4} + \alpha + \frac{\cot \alpha}{2} \right]$$ (8.67)

where α is the included angle of the conical indenter.

In hardness testing, the impression is asymmetric, so that the slip line theory is not applicable [106,107], but an elastic–plastic boundary problem, best explained from the Hertz theory of contact stresses [108]. This model accounts for the material displaced by the indenter by the decrease (by compression) in volume of the elastic underlying material. No upward flow around the indenter is predicted, which agrees with observation. This explanation is the basis for all indention hardness tests used for metallic materials.

8.8.1 Brinnell Test

The Brinnell test was first proposed in 1900 and has since become widely accepted throughout the world. It is accomplished by indenting the surface with a 10-mm steel ball with a 3000 kg load. For soft metals a 500 kg load is usually used because otherwise the impression is too deep. For hard metals, a tungsten-carbide ball is used to prevent distortion of the indenter. The diameter of the round indentation is measured with a low power microscope after loading. At least two measurements of the diameter are made, and the results are averaged. To ensure that accurate measurements are made, the surface must be free from dirt and scale. The hardness expressed as a Brinnell hardness number (BHN) is determined from the equation

$$\text{BHN } [\text{kg/mm}^2] = \frac{P}{\pi D/2} [D - (D^2 - d^2)^{1/2}]$$ (8.68)

where P is the applied load (kg), D is the diameter of the indenter (10 mm), and d is the measured diameter of the impression. The BHN could also be calculated by measuring the depth of the impression, t:

$$\text{BHN} = \frac{P}{\pi D t}$$ (8.69)

or

$$\text{BHN} = \frac{P}{(\pi/2)D^2(1 - \cos \phi)}$$ (8.70)

as $d = D \sin \phi$, where ϕ is the included angle of the chord of the impression.

Because of the large size of the impression made in the Brinnell test, it averages out any local inhomogeneities. It also precludes the testing of small objects or objects in which the Brinnell impression can be a site of crack initiation. Therefore, it is commonly used for castings, forgings, or raw stock. The size of the piece tested for Brinnell hardness should be at least ten times the depth of the impression. Because of the plastic zone surrounding the impression and the elastic constraint, additional Brinnell hardnesses should not be measured any closer to the impression than 4 times the impression diameter. The distance from an edge when taking a hardness reading should also be at least 4 times the impression diameter.

8.8.2 VICKERS HARDNESS OR DIAMOND PYRAMID HARDNESS

In the Vickers hardness test, a square-based pyramid-shaped diamond penetrator is used. The included angle of the pyramid is 136° between opposite faces. The test is taken by indenting, under load, with the penetrator, and after release of the load, measuring the width of the diagonals. The Vickers hardness number or diamond pyramid hardness (DPH) is calculated as

$$\text{DPH} = \frac{2P \sin(\theta/2)}{L^2} \tag{8.71}$$

where P is the load (kg), L is the average length of the diagonals, and θ is the angle between opposite faces (136°).

The Vickers hardness test has gained wide acceptance in the world for research because the loads can be varied between 1 and 120 kg. It is a continuous scale of hardness that is internally consistent; i.e., hardness numbers determined from one test with one load can be compared to those determined in another test using a different load (except for very light loads) because the impressions are geometrically similar. Other tests require changing loads and indenters, and the hardnesses obtained cannot be strictly compared with others.

This test has not been widely accepted in the United States because it is slow and requires careful surface preparation. In addition, the opportunity exists for operator error when measuring the diagonals.

Sometimes, impressions are obtained that are not perfectly square. These fall into two categories: pincushion-shaped impressions and barrel-shaped impressions (Figure 8.37). The pincushion-shaped impression is caused by the tested metal sinking in around the flat faces of the pyramid. This often occurs when testing very soft or annealed metals, with the result that inaccurate low hardness numbers are obtained by overestimating the diagonal lengths. Barrell-shaped impressions are usually found when testing highly cold worked materials and are caused by the piling up of material around the indenter. This produces low diagonal values, with the hardnesses erring on the high side.

8.8.3 ROCKWELL HARDNESS TEST

The Rockwell hardness test is very widely used in the United States because of its speed and its freedom from errors by operating technicians. The impression is small, making it possible to test a wide variety of parts.

In this test, a minor load of 10 kg is applied to the part to seat the indenter and part; then the major load is applied. After 30 s the depth of the impression is measured and exhibited on

FIGURE 8.37 Vickers indenter defects. (After G.L. Kehl, *The Principles of Metallographic Laboratory Practice*, McGraw-Hill, New York, 1949.)

TABLE 8.7
Description of Standard Rockwell Hardness Scales

	Rockwell A	Rockwell B	Rockwell C
Indenter	Brale	1/16 in.-diameter ball	Brale
Major load (kg)	60	100	150
Application	Very hard materials; tungsten carbide, hard thin materials, etc.	Rolled steel sheet, brass, aluminum, annealed steels, etc.	Fully hardened steels, quenched and tempered steels, etc.

a rotary dial or, in newer machines, on a digital readout. Some machines are equipped with printer ports or RS-232C interfaces for communicating with a computer.

The scale used in Rockwell testers is based on 100 divisions, with each division equal to a depth of 0.00008 in. The scale is reversed, so that a small impression results in a high hardness number. The number from the Rockwell test is an arbitrary number that is only consistent within the same scale. This is unlike the Brinnell and Vickers hardness tests, which provide numbers that are based on mass per unit area (kg/mm^2).

More than one indenter can be used with the Rockwell test. The most common is the Brale indenter, which is a 120° diamond cone and is used for the Rockwell C test. Other indenters used are 1/16- and 1/8 in.-diameter steel balls. Major loads can be varied from 60 to 150 kg. Table 8.7 lists the application, indenter, and major load used for the three types of Rockwell hardness tests.

If a carburized surface is measured, the hardness reading may be lower than expected because of a greater depth of impression. The material supporting the test piece may also influence the hardness reading. As a rule of thumb, it is wise to test only test pieces that are at least 10 times as thick as the depth of the impression [109]. If the impression of the hardness indenter shows on the other side of the tested piece, the test is obviously invalid.

Because the scale on a Rockwell hardness machine is arbitrary, it is necessary to ensure that one machine records the same hardness as another machine. This is accomplished by the use of standardized test blocks calibrated by the manufacturer. Generally, three test blocks of values throughout the range are adequate to maintain the hardness machine in calibration.

8.8.4 ROCKWELL SUPERFICIAL HARDNESS TEST

The superficial tester operates in an identical fashion to the Rockwell hardness tester. In fact, they look very similar. Two types of indenters are used: a 1/16 in.-diameter steel ball (used for surface hardnesses of brasses, aluminum sheet, etc.) and the Brale penetrator. This Brale penetrator is similar to the one used in the Rockwell hardness tests except that its spherical end is shaped to higher tolerances. This penetrator is designated the N-Brale indenter.

The minor load used to seat the indenter and specimen is 3 kg. The major loads used are 15, 30, and 45 kg. The major load is applied for 30 s, and the depth of the penetration is measured. The scale of the superficial Rockwell test is arbitrary like that of the Rockwell test, except that each division represents 0.001 in. in depth. The scale is reversed, so deeper impressions mean lower hardness numbers. As the loads used on the superficial Rockwell tester are different from those used for the Rockwell tests, different scales were established defined by the load and indenter as shown in Table 8.8.

As the impression left by the superficial test is not very deep, it is important that the test surface be smoother than is necessary for the standard Rockwell test. The superficial hardness test is more accurate when the part has a good surface finish. The test is sensitive to dirt and

TABLE 8.8
Indenters and Major Loads for the Rockwell Superficial Hardness Test Scales

Scale	Penetrator	Major Load (kg)
15-N	N Brale	15
30-N	N Brale	30
45-N	N Brale	45
15-T	1/16 in.-diameter steel ball	15
30-T	1/16 in.-diameter steel ball	30
45-T	1/16 in.-diameter steel ball	45

hard particles under the test piece or the indenter, giving unusually high readings. Soft particles underlying the test piece would provide lower readings.

8.8.5 TUKON MICROHARDNESS TEST

The Tukon or Knoop microhardness test was developed in 1939 [110] by the National Bureau of Standards as a method of measuring the hardness of very small constituent phases, segregation effects, and hard and brittle materials. With proper selection of the load, the depth of the impression will not exceed 1 μm.

The Knoop penetrator is a pyramidal diamond, cut to have an included transverse angle of 130° and an included longitudinal transverse angle of 170° 30′. The resulting impression is rhombic, with the long diagonal approximately 7 times the length of the transverse diagonal, as shown in Figure 8.38.

The test is conducted by placing the prepared metallographic specimen on a microscope stage, and the desired location for the impression is determined with the aid of a metallographic microscope. Once located, the indenter stage is located over the desired area, and the tester is actuated. The specimen is moved upward automatically by means of an elevating screw until it makes contact with the indenter. The preload is applied for approximately 20 s, then the selected major load is applied gradually, reaching maximum load in approximately 20 s. The load is removed, and the specimen is lowered. The operator moves the microscope to view the impression, and the impression's long diagonal is measured.

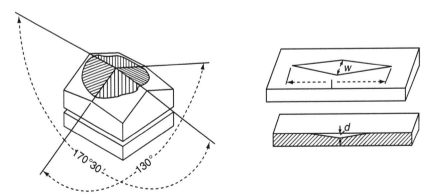

FIGURE 8.38 Schematic of Knoop indenter. (After G.L. Kehl, *The Principles of Metallographic Laboratory Practice*, McGraw-Hill, New York, 1949.)

The longitudinal diagonal of the Knoop impression is measured with a filar eyepiece, which is part of the microhardness tester. This eyepiece has a fixed micrometer scale with a movable vertical hairline. The micrometer screw is divided into 100 divisions, with each division corresponding to a lateral movement of 0.01 mm. Movement is maintained in one direction to eliminate any errors due to backlash in the micrometer gear train.

Using the filar eyepiece, the long diagonal is measured, and the Knoop hardness number (KHN) is calculated:

$$KHN = \frac{P}{0.7028l^2} \qquad (8.72)$$

where P is the applied load and l is the length of the long diagonal (mm). As the impression left by the microhardness tester is very small, it is necessary that the surface to be measured be prepared metallographically and be free of any surface scratches or other detrimental defects. The surface may be either etched or unetched. It is very important that the specimen be perpendicular to the penetrator. If the impression is lopsided, the measurements of the long diagonal will be inaccurate because of asymmetric elastic recovery. The specimen must be removed and releveled and oriented properly so that accurate measurements can be taken.

8.9   TOUGHNESS TESTING

8.9.1   Charpy V Notch Test

The Charpy V notch test is a test for measuring impact strength in which a small notched bar is loaded dynamically in three-point bending. The specimen has a square cross section of 10 mm and a length of 55 mm. The bar contains a sharp notch with an included angle of 45° and a depth of 2 mm. The notch radius is small, 0.25 mm. The use of subsize specimens is permitted provided that they conform to ASTM E23-88.

European specifications for Charpy-type impact testing include the ISO-U and ISO-V specifications. The ISO-U specification calls for a U-shaped notch, which shows a gradual degradation in the impact energy absorbed as the testing temperature is lowered. The ISO-V test is identical to the U.S. standard ASTM E23-88. There have been studies [111] that indicate that in the upper shelf area the absorbed energy is directly proportional to the width of the specimen. In other words, a specimen 5 mm across with a 10-mm depth will show 50% of the impact strength exhibited by the full-size specimen [112]. This is applicable to both transverse and longitudinal loading. There is very little difference in the lower shelf region.

Details of the testing procedure are covered in ASTM E23-88 [113]. In this test, the test specimen is removed from its cooling (or heating) bath and placed on the specimen fixture (Figure 8.39). The pendulum is released, and the specimen is broken within 5 s after removal from the bath. The calibrated dial of the impact machine is read, and the broken specimen is retrieved. If high-strength, low-energy specimens are tested at low temperatures, the specimens have a tendency to leave the machine perpendicular to the swing of the pendulum. This may cause errors in reading (as well as pose hazards to the operator) from the specimens hitting the pendulum. Because of conservation of energy, the specimens may leave the machine at speeds in excess of 50 ft/s. If the specimen hits the pendulum with sufficient energy, the pendulum will slow down and the machine will record a higher impact energy absorbed than truly occurred. This has been cited in ASTM E23-88 as the cause of much of the scatter in Charpy V notch testing in the 10–25 ft lb range.

Being hit by the pendulum forces the specimen to bend and fracture. The strain rate of loading is high, approximately 10^3 s^{-1}. Because of the high-strain rate, a considerable plastic constraint

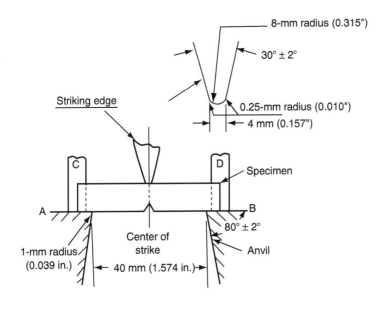

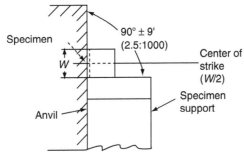

FIGURE 8.39 Fixture used for holding Charpy impact specimens.

exists at the notch. This plastic constraint yields a triaxial stress state at the notch tip. Dieter [26] indicates that the maximum plastic stress concentration at the notch tip is given by

$$K_\sigma = 1 + \frac{\pi}{2} + \frac{\omega}{2} \tag{8.73}$$

where ω is the included angle of the notch. Because of the high-strain rate, section thickness, and notch radius, a large amount of plane strain loading and triaxial stresses exist. There have been numerous correlations of the Charpy V notch and plane strain fracture toughness K_{Ic} testing [114,115].

8.9.2 IZOD TEST

The Izod notched bar test was developed at the beginning of this century for measuring notch toughness. It was assumed originally that the impact energy of a material was proportional to its notch toughness and that there was advantage to testing at temperatures other than room temperature. Because the Izod specimen is held in a large vise (see schematic, Figure 8.40), there is a large heat sink present, making testing at temperatures other than room temperature problematic. Now the test is used primarily for the plastics industry. It is rarely used for the testing of metallic materials.

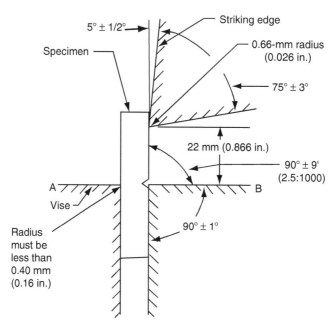

FIGURE 8.40 Izod device.

8.9.3 DYNAMIC TEAR TEST

The dynamic tear test was developed to measure the transition temperature of the steels used in the pipe industry. It is an ASTM standard [116] and is essentially a large-scale Charpy V notch test. ASTM E604 has established the width of the specimen to 0.625 in., but thicknesses as great as 12 in. have been tested. Typical specimen dimensions are shown in Figure 8.41.

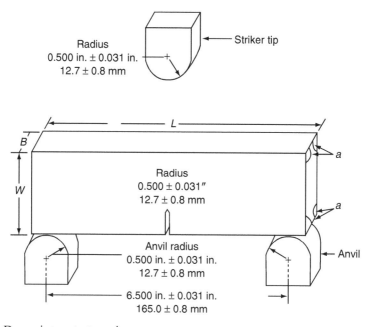

FIGURE 8.41 Dynamic tear test specimens.

The test involves the dynamic tear specimen being impact loaded in three-point bending, and the energy absorbed is measured. Other measurements performed include the percentage of brittle or ductile failure to determine the FATT temperature. The testing machine can either be as pendulum-type like a Charpy V notch machine or a drop-weight type with the capacity to break the specimen in a single blow. For most steels, with a 0.625-in. (16 mm) thick specimen, the capacity necessary to conduct the dynamic tear test is approximately 2700 J (2000 ft lb$_f$).

The notch on the specimen is prepared by machining a notch 0.475 in. deep with a radius of 0.0625 in. The angular root section of the notch is usually made by a precisely ground saw or an electric discharge machine. The notch if further sharpened by pressing a hardened high-speed tool steel (HRC 60 min) knife blade into the machined notch to a depth of 0.130 in. Notches have also been made with an electron beam welder. The high-energy density of the beam creates a very sharp, brittle, and well-defined heat-affected zone. The notch is also embrittled by the use of alloying elements.

The dynamic tear test is conducted at a variety of temperatures, and the percentage of brittle failure is plotted as a function of temperature. This determines the transition temperature (FATT) of the material being tested. The energy absorbed during fracture is also plotted, yielding similar information.

Dimensions and Tolerance for Specimen Blank

Parameter	Units	Dimension	Tolerance
Length L	in.	7.125	0.125
	mm	181.0	3.2
Width W	in.	1.60	0.10
	mm	38.0	2.5
Thickness B	in.	0.625	0.033
	mm	15.8	0.8
Angularity α	deg.	90	2

The strain rate of the dynamic tear test is similar to that of the Charpy V notch test at approximately 10^{-3} s^{-1}. Because of the greater section thickness, a higher state of triaxial stress exists in the specimen during testing than in the Charpy V notch test. A plane strain condition will be reached earlier than in the Charpy test as the testing temperature is decreased.

8.9.4 Fracture Toughness (K_{Ic}) Testing

As indicated above, the use of fracture mechanics is important in determining the maximum flaw size that a material can withstand before failing catastrophically. As has been noted, cracking in a thick plate is worse than in a thin plate. This is because of plane strain conditions. At the crack tip, the plastic zone is small, with a high-stress gradient across the plastic zone. A schematic of the plastic zone is shown in Figure 8.42. In addition, very high triaxial stresses are present. Because of this, the fracture appearance changes with specimen thickness. This is shown in Figure 8.43.

Mechanical property testing method

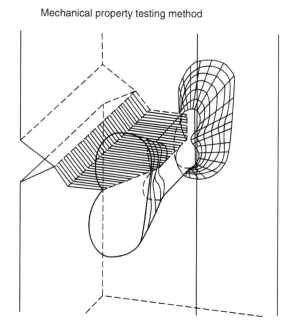

FIGURE 8.42 Plastic zone at crack tip.

In thin plates, the fracture is characterized by a mixed mode ductile and brittle fracture, with the presence of shear lips. Under plane strain conditions, when the plate is thick enough the fracture is flat, and the fracture stress is a constant with increasing thickness. The minimum thickness for plane strain conditions to occur is given by

$$B = 2.5 \left(\frac{K_{Ic}}{\sigma_{ys}} \right)^2 \tag{8.74}$$

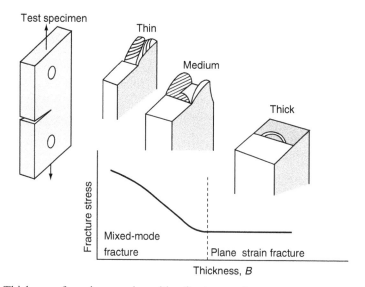

FIGURE 8.43 Thickness of specimen and resulting fracture surface.

Different configurations are used to determine the plane strain fracture toughness, K_{Ic}. These are shown in Figure 8.44. Other specimens include the center-cracked plate [117]. The center-cracked plate is also used in determining the fatigue crack growth rate.

The notch is machined in the specimen and made sharper by fatiguing at low cycle, high strain until the crack is about the width of the test specimen. The initial crack length is measured by including the length of the fatigue crack and the notch.

Testing of the specimen is accomplished by loading the specimen in tension, with the load and crack opening displacement continuously recorded until failure. In general, there are three types of load responses to the testing, shown in Figure 8.45. Type I loading is characteristic of ductile materials, with no onset of unstable brittle fracture. For this type of

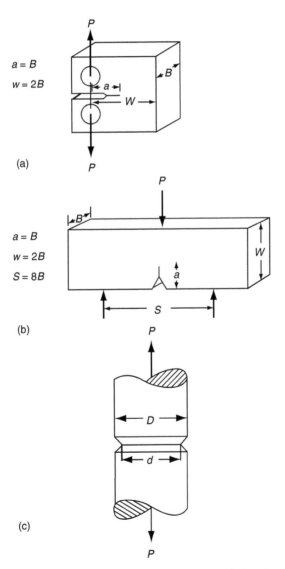

FIGURE 8.44 Typical K_{Ic} specimens: (a) compact tension specimen, (b) bend specimen, and (c) notched round specimen. (After G.E. Dieter, *Mechanical Metallurgy*, McGraw-Hill, New York, 1976, p. 496.)

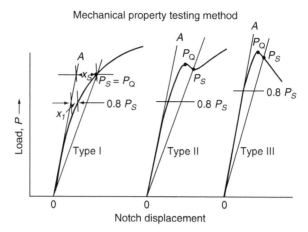

FIGURE 8.45 Typical K_{Ic} load responses to determine fracture toughness.

response, the K_{Ic} value is difficult to obtain, and very thick specimens must be used before plane strain conditions occur. Type II response shows a sharp drop in the load with some load recovery. In this mode, the crack becomes unstable and propagates partially through the material until the crack front is blunted by plastic tearing. For this response, P_Q is considered the maximum load for K_{Ic} determination. In type III loading, a maximum load is reached, with brittle crack propagation occurring rapidly. In this response, plane strain conditions exist.

A conditional value of fracture toughness, K_Q is calculated using (for the compact specimen)

$$K_Q = \frac{P_Q S}{Bb}\left[2.9\left(\frac{a}{b}\right)^{1/2} - 4.6\left(\frac{a}{b}\right)^{3/2} + 21.8\left(\frac{a}{b}\right)^{5/2} - 37.6\left(\frac{a}{b}\right)^{7/2} + 38.7\left(\frac{a}{b}\right)^{9/2}\right] \quad (8.75)$$

For the three-point bending specimen,

$$K_Q = \frac{P_Q}{Bb^{1/2}}\left[2.9.6\left(\frac{a}{b}\right)^{1/2} - 185.5\left(\frac{a}{b}\right)^{3/2} + 655.7\left(\frac{a}{b}\right)^{5/2} - 1017\left(\frac{a}{b}\right)^{7/2} + 638.9\left(\frac{a}{b}\right)^{9/2}\right] \quad (8.76)$$

TABLE 8.9
Typical Fracture Toughness K_{Ic} of Some Steels

Steel	Yield Strength σ_{ys} (ksi)	Fracture Toughness K_{Ic} (ksi$\sqrt{\text{in.}}$)
SAE 4340	230	60
Cr–Mo–V	101	100
Ni–Mo–V	89	80
304 Stainless	30	100
HP9-4-20 (9%Ni–4%Co–20%C)	180	170

For the notched round specimen,

$$K_Q = \frac{P}{D^{3/2}} \left(1.72 \frac{D}{d} - 1.27 \right) \tag{8.77}$$

To determine the fracture toughness, K_{Ic}, the crack length is measured and B is calculated:

$$B = 2.5(K_Q/\sigma_{ys})^{1/2} \tag{8.78}$$

If both B and a are less than the width b of the specimen, then $K_Q = K_{Ic}$. If not, then a thicker specimen is required, and K_Q is used to determine the new thickness using Equation 8.78. Typical K_{Ic} values for steels are shown in Table 8.9.

REFERENCES

1. M. Gensamer, *Fatigue and Fracture in Metals*, Wiley, New York, 1952.
2. R. von Mices, Der kritischer aussendruck Zylinderscher Rohre, *Z. Ver. Deut. Ing. 58*:750 (1914).
3. M.L. Williams, Analysis of brittle behavior in ship plates, in *Symposium on the Effect of Temperature on the Brittle Behavior of Metals with Particular Reference to Low Temperatures*, ASTM STP 158, 1954, pp. 11–44.
4. M.E. Shank, A critical survey of brittle fracture in carbon steel structures other than ships, in *Symposium on the Effect of Temperature on the Brittle Behavior of Metals with Particular Reference to Low Temperatures*, ASTM STP 158, 1954, pp. 45–110.
5. J. McCall and P. French, *Metallography in Failure Analysis*, Plenum Press, New York, 1978, p. 6.
6. L.F. Coffin, Flow and fracture of a brittle material, *J. Appl. Mech. 17*:233 (1950).
7. R.C. Grassi and I. Cornet, Fracture of gray cast iron tubes under biaxial stresses, *J. Appl. Mech. 16*:178 (1949).
8. P. Burton, A modification of the Coulomb–Mohr theory of fracture, *J. Appl. Mech., Ser. E 28*:259 (1961).
9. E. Orowan, *Welding J. 34*:157 (1955).
10. A. Griffith, *Phil. Trans. R. Soc. London 221A*:163 (1920).
11. G. Dieter, *Mechanical Metallurgy*, McGraw-Hill, New York, 1976, p. 253.
12. C.E. Inglis, *Trans. Inst. Nav. Archit. 55*:219 (1913).
13. E. Klier, *Trans. Am. Soc. Met. 43*:935 (1951).
14. L.C. Chang, *J. Mech. Phys. Solids 3*:212 (1955).
15. D.K. Felbeck and E. Orowan, *Welding J. 34*:570s (1955).
16. E. Orowan, Ed., *Fatigue and Fracture of Metals*, Symposium at MIT, Wiley, New York, 1952.
17. G.R. Irwin, Fracture, in *Encyclopedia of Physics*, Vol. VI, Springer, Heidelberg, 1958, p. 561.
18. P.C. Paris and G.C. Sih, *Fracture Toughness Testing*, ASTM STP 381 (J.E. Shawley and W.F. Brown, Eds.), American Society for Testing and Materials, PA, 1965.
8. J.W. Faupel and F.E. Fisher, *Engineering Design*, Wiley, New York, 1981.
20. R.W. Hertzberg, *Deformation and Fracture Mechanics of Engineering Materials*, New York, 1976.
21. S.T. Rolfe and J.M. Barsom, *Fracture and Fatigue Control in Structures*, Prentice-Hall, Englewood Cliffs, NJ, 1977.
22. Anon., Standard Test Methods for Notched Bar Impact Testing of Metallic Methods, E23-88, American Society for Testing and Materials, Philadelphia, PA.
23. Anon., Standard Test Method for Dynamic Tear Testing of Metallic Materials, E604-83, American Society for Testing and Materials, PA.
24. Anon., Standard Test Method for Plane Strain Fracture Toughness, K_{Ic}, of Metallic Materials, ASTM E399-74, American Society for Testing and Materials, PA.

25. R.C.A. Thurston, The notch toughness of ultrahigh-strength steels in relation to design considerations, *Problems in the Load-Carrying Applying of High Strength Steels*, DMIC Report 210, October 26, 1964.
26. G.E. Dieter, *Mechanical Metallurgy*, McGraw-Hill, New York, 1976, p. 496.
27. J.A. Rinebolt and W.J. Harris Jr., *Trans. ASM 43*:1175 (1951).
28. A.S. Tetelman and A.J. McEvily, *Fracture of Structural Materials*, Wiley, New York, 1967.
29. C.L. Cotrell, P.F. Langstone, and J.H. Rendall, *Iron Steel Inst. 20*(1):1032–1037 (1963).
30. E.P. Klier, ASTM STP 287, 1961, p. 196.
31. R.P. Wei, ASTM Preprint, ASTM Annual Meeting, June 1964.
32. L.L. Gilbert and J.A. Brown, *High Strength Steels for the Missile Industry*, ASM 3–39, 1961.
33. C.L. Cottrell, *High Strength Steels*, Iron Steel Inst. Spec. Rep. 76, 1962, pp. 1–6.
34. J.L. Shawin, G.B. Espey, L.J. Repko, and W.F. Brown, *Proc. ASTM 60*:761–777 (1960).
35. W.S. Owen, P.H. Whitman, M. Cohen, and R.L. Averbach, *Welding J. 36*:503s (1957).
36. W.R. Warke and A.R. Elsea, DMIC 154, Battelle Memorial Institute, Columbus, OH, June 18, 1962.
37. G.E. Pellissier, Third Annual Maraging Steel Project Review, RTD-TDR-63-4048, November 1963, p. 407.
38. W.F. Payne, ASTM Annual Meeting, 1964.
39. R.E. Smallmann, *Modern Physical Metallurgy*, Butterworth-Heinemann, Oxford, U.K., 1985, p. 483.
40. J.E. Shawley and C.D. Beacham, NRL Rep. 5127, April 9, 1958.
41. C.W. Marshall, DMIC 147, Battelle Memorial Institute, Columbus, OH, 6 February 1961.
42. D. Francois, Relation between various fracture transition temperatures and the K_{Ic} fracture toughness transition curve, *Eng. Frac. Mech. 23*(2):455–465 (1986).
43. K. Ikeda and H. Kihara, *Proceedings of the Second International Conference on Fracture* (P.L. Pratt et al., Eds.), Chapman and Hall, London, 1969, pp. 851–867.
44. W.J. Plumbridge and D.A. Ryder, *Metall. Rev. 14*:136 (1969).
45. P.J. Frsyth and C.A. Stubbington, *J. Inst. Met. 83*:395 (1955).
46. W.A. Wood, *Some Basic Studies of Fatigue in Metals*, Wiley, New York, 1959.
47. A.H. Cottrel and D. Hull, *Proc. R. Soc. London 242A*:211 (1953).
48. W.A. Wood, *Bull. Inst. Met. 3*:5 (1955).
49. C. Laird, *Fatigue Crack Propagation*, in ASTM STP 415, American Society for Testing and Materials, PA, 1967, p. 136.
50. P.J. Forsyth and D.A. Ryder, Some results of the examination of aluminum specimen fracture surfaces, *Metallurgica 63*:117 (1961).
51. G. Jacoby, Observations of crack propagation on the fracture surface, in *Current Aeronautical Fatigue Problems* (J. Schijve, Ed.), Pergamon Press, New York, 1965, p. 78.
52. W.R. Warke and J.M. McCall, *Fractography Using the Electron Microscope*, ASM Tech. Rep. W3-2-65, American Society of Metals, Metals Park, OH, 1965.
53. G. Jacoby, Fractographic methods, *Exp. Mech. 5*:65 (1965).
54. P.J. Forsyth, A two stage process of fatigue crack growth, *Symposium Crack Propagation*, Vol. II, Cranfield, U.K., 1961, p. 76.
55. H.J. Grover, S.A. Gordon, and L.R. Jackson, *Fatigue of Metals and Structures*, NAVWEPS Rep. 00-25-534, Bureau of Naval Weapons, Department of the Navy, Washington, D.C., 1960.
56. K.E. Thelning, *Steel and Its Heat Treatment*, Butterworths, London, U.K., 1984.
57. J.T. Ransom and R.F. Mehl, *Trans. AIME 185*:364 (1949).
58. P.H. Armitage, *Metall. Rev. 6*:353 (1964).
59. R.E. Little and E.H. Jebe, *Statistical Design of Fatigue Experiments*, Wiley, New York, 1975.
60. H. Muller-Stock, *Mitt. Kohle Eisenforsch GmbH 8*:83 (1938).
61. J.T. Ransom, ASTM STP 121, 1952, p. 59.
62. J.E. Shigley, *Mechanical Engineering Design*, McGraw-Hill, New York, 1977.
63. P.G. Forest, *Fatigue in Metals*, Addison-Wesley, Reading, PA, 1962.
64. N.E. Frost, K.J. Marsh, and L.P. Pook, *Metal Fatigue*, Oxford University Press, London, U.K., 1974.
65. L. Sors, *Fatigue Design of Machine Components*, Pergamon Press, Oxford, U.K., 1971.
66. V.M. Fairies, *Design of Machine Elements*, Macmillan, New York, 1965.

67. R.A. Hammond and C. Williams, *Metall. Rev. 5*:165 (1960).
68. Anon., *Fatigue Durability of Carburized Steel*, American Society for Metals, Metals Park, OH, 1957.
69. J.H. Faupel and F.E. Fisher, *Engineering Design*, Wiley-Interscience, New York, 1981.
70. C. Lipson and R.C. Juvinall, *Stress and Strength*, Macmillan, New York, 1963.
71. L.P. Tarasov, W.S. Hyler, and H.R. Letner, *ASTM Proc. 57*:601 (1957).
72. R.C. Juvinall, *Stress, Strain and Strength*, McGraw-Hill, New York, 1967.
73. G. Castleberry, *Mach. Des. 50*:108 (1978).
74. D. Broek and J. Schijve, *The Influence of Sheet Thickness in the Fatigue Crack Propagation in 2024-T3 Alcad Sheet Material*, NLR Tech. Rep. M2129, Amsterdam, 1963.
75. I.S. Sokolnikoff, *Mathematical Theory of Elasticity*, McGraw-Hill, New York, 1956.
76. R.E. Peterson, *Stress Concentration Design Factors*, Wiley, New York, 1974.
77. W.C. Young, *Roark's Formulas for Stress and Strains*, McGraw-Hill, New York, 1989.
78. R.E. Peterson, in *Metal Fatigue* (G. Sines and J.L. Waisman, Eds.), McGraw-Hill, New York, 1959, p. 138.
79. H. Neuber, *Theory of Notch Stresses*, JW Edwards, Ann Arbor, MI, 1946.
80. D. Walton and E.G. Ellison, *Inst. Metal Rev. 17*:100 (1972).
81. T.J. Crooker and E.A. Lange, *Inst. Metal Rev. 17*:94 (1972).
82. R.I. Stephens, Linear elastic fracture mechanics and its application to fatigue, SAE Paper 740220, Automotive Engineering Congress, Detroit, MI, 1974.
83. D. Broek, *Elementary Engineering Fracture Mechanics*, Noordhoff, Leyden, The Netherlands, 1974.
84. W.F. Brown and J.E. Shawley, *Plane Strain Fracture Toughness Testing of High Strength Metallic Materials*, ASTM STP 410, ASTM, PA, 1966.
85. A.S. Kobayashi, Ed., *Experimental Techniques in Fracture Mechanics*, Vols. I and II, Society for Experimental Stress Analysis, Westport, CT, 1975.
86. J.E. Campbell, W.E. Berry, and C.E. Feddergen, Eds., *Damage Tolerance Handbook—A Compilation of Fracture and Crack Growth Data for High Strength Alloys*, MCIC-HB-01, MCIC, Battelle Memorial Institute, Columbus, OH, 1972.
87. T.P. Rich and D.J. Cartwright, Eds., *Case Studies in Fracture Mechanics*, AMMRC MS 77–5, Army Materials and Mechanics Research Center, Watertown, MA, 1977.
88. H. Tada, P.C. Paris, and G.R. Irwin, *The Stress Analysis of Cracks Handbook*, Del Research Corporation, Hellertown, PA, 1973.
89. H. Liebowitz, Ed., *Fracture Mechanics of Aircraft Structures*, AGARD-AG-176, NTIS, Springfield, VA, 1974.
90. G. Jacoby, *Application of Microfractography to the Study of Crack Propagation Under Fatigue Stress*, AGARD Rep. 541, NATO, 1966.
91. P. Kuhn, A comparison of fracture mechanics and notch analysis, presented to the ASTM Special Committee on Fracture Testing, January 1965.
92. J.E. Dorn, Ed., *Mechanical Behavior of Materials at Elevated Temperatures*, McGraw-Hill, New York, 1961.
93. R.W. Guard, *Prod. Eng. 27*(10):160–174 (1956).
94. I. Finnie and W.R. Heller, *Creep of Engineering Materials*, McGraw-Hill, New York, 1959.
95. E.N. da C. Andrade and B. Chalmers, *Proc. R. Soc London 138A*:348 (1932).
96. J.B. Conway, *Trans. Metall. Soc. AIME 223*:2018 (1965).
97. F. Garofalo, *Properties of Crystalline Solids*, ASTM STP 283, ASTM, PA, 1965.
98. W. Rosenhahn and D. Ewen, *J. Inst. Met. 10*:119 (1913).
99. A. Nadai, *Theory of Flow and Fracture of Solids*, Vol. I, McGraw-Hill, New York, 1950.
100. Anon., *Standard Methods of Tension Testing Metallic Materials*, ASTM E8, American Society for Testing and Materials, PA, 1979.
101. C.E. Lacy and M. Gensamer, *Trans. ASM 32*:88 (1944).
102. E.C. Bain and H.W. Paxton, *Alloying Elements in Steel*, ASM, Metals Park, OH, 1966.
103. P. Maynier, B. Jungmann, and J. Dollet, Creusot-Loire system for the prediction of the mechanical properties of low alloy steel products, in *Hardenability Concepts with Applications to Steel* (D.V. Doane and J.S. Kirkaldy, Eds.), Met. Soc. AIME, 1978, p. 518.

104. D. Venugopalan and J.S. Kirkardy, New relations for predicting the mechanical properties of quenched and tempered low alloy steels, in *Hardenability Concepts with Applications to Steel* (D.V. Doane and J.S. Kirkaldy, Eds.), Met. Soc. AIME, 1978, p. 249.
105. R. Hill, *The Mathematical Theory of Plasticity*, Clarendon Press, Oxford, UK, 1950.
106. M.C. Shaw and G.J. DeSalvo, *J. Eng. Ind. 92*:469 (1970).
107. M.C. Shaw and G.J. DeSalvo, *Met. Eng. Q. 121*(5):1 (1972).
108. S. Tmoshenko and J.N. Goodier, *Theory of Elasticity*, 2nd ed., McGraw-Hill, New York, 1972, p. 372.
109. G.L. Kehl, *The Principles of Metallographic Laboratory Practice*, McGraw-Hill, New York, 1949.
110. F.C. Knoop, C.G. Peters, and W.B. Emerson, A sensitive pyramidal-diamond tool for indentation measurements, *J. Res. Natl. Bur. Stand. 23*:49 (1959).
111. G. Robiller, Influence on the width of the specimen on the results of the notched bar impact bending test, *Stahl Eisen 100*(19):1132–1138 (1955).
112. M. Lai and W. Ferguson, Effect of specimen thickness on fracture toughness, *Eng. Fracture Mech. 23*(4):649 (1986).
113. Anon., Standard Test for Notched Bar Impact Testing of Metallic Materials, ASTM E23-88, American Society for Testing and Materials, PA, 1988.
114. R.H. Sailors and H.T. Corten, Relationship between material facture toughness using fracture mechanics and transition temperature test, in *Fracture Toughness*, Proceedings of the 1971 National Symposium Fracture Mechanics, Part II, ASTM STP 514, American Society for Testing and Materials, 1972, pp. 164–191.
115. D. Francois and A. Krasowsky, Relation between various fracture transition temperatures and the K_{Ic} fracture toughness transition curve, *Eng. Fracture Mech. 23*(2):455 (1986).
116. Anon., Standard Test Method for Dynamic Tear Testing of Metallic Materials, E604-83, American Society for Testing and Materials, PA, 1983.
117. K.M. Kraft, *Techniques of Metal Research*, Vol. V, Wiley, New York, 1971, Chap. 7.

Index

Related Titles

Handbook of Induction Heating
Valery Rudnev, Don Loveless, Raymond Cook, and Micah Black
ISBN: 0-8427-0848-2

Phase Transformations of Elements under High Pressure
E. Yu Tonkov and E.G. Ponyatovsky
ISBN: 0-8493-3367-9

Handbook of Aluminum, Volume 1: Physical Metallurgy and Processes
George Totten and Scott D. MacKenzie
ISBN: 0-8427-0494-0

Handbook of Aluminum, Volume 2: Alloy Production and Materials Manufacturing
George Totten and Scott D. MacKenzie
ISBN: 0-8427-0896-2

Coming Soon

Heat Treatment of Aluminum Alloys
Leonid B. Ber, Nikolai Kolobnev, and Evgeniy Kablov
ISBN: 0-8493-7610-6

Self-Organizing Layers and Nanotribology in Advanced Tooling Materials and Systems
George Totten and German Fox-Rabinvich
ISBN: 1-57444-719X

Milton Keynes UK
Ingram Content Group UK Ltd.
UKHW052031071024
449327UK00027B/2512

9 780367 390280